D0622338

THE CAMBRIDGE DICTIONARY OF STATISTICS
FOURTH EDITION

If you work with data and need easy access to clear, reliable definitions and explanations of modern statistical and statistics-related concepts, then look no further than this dictionary. Nearly 4000 terms are defined, covering medical, survey, theoretical and applied statistics, including computational and graphical aspects. Entries are provided for standard and specialized statistical software. In addition, short biographies of over 100 important statisticians are given. Definitions provide enough mathematical detail to clarify concepts and give standard formulae when these are helpful. The majority of definitions then give a reference to a book or article where the user can seek further or more specialized information, and many are accompanied by graphical material to aid understanding.

B. S. EVERITT is Professor Emeritus of King's College London. He is the author of almost 60 books on statistics and computing, including *Medical Statistics from A to Z*, also from Cambridge University Press.

A. SKRONDAL is Senior Statistician in the Division of Epidemiology, Norwegian Institute of Public Health and Professor of Biostatistics in the Department of Mathematics, University of Oslo. Previous positions include Professor of Statistics and Director of The Methodology Institute at the London School of Economics.

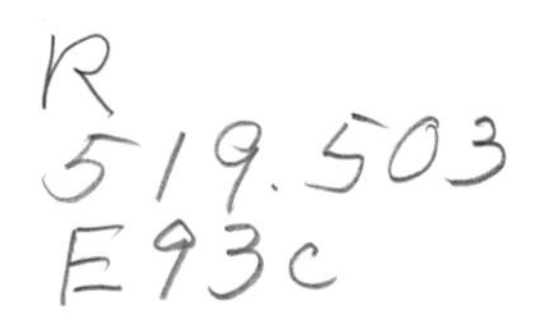

THE CAMBRIDGE DICTIONARY OF Statistics

Fourth Edition

B. S. EVERITT
Institute of Psychiatry, King's College London

A. SKRONDAL
Norwegian Institute of Public Health
Department of Mathematics, University of Oslo

CAMBRIDGE UNIVERSITY PRESS
Cambridge, New York, Melbourne, Madrid, Cape Town, Singapore,
São Paulo, Delhi, Dubai, Tokyo, Mexico City

Cambridge University Press
The Edinburgh Building, Cambridge CB2 8RU, UK

Published in the United States of America by Cambridge University Press, New York

www.cambridge.org
Information on this title: www.cambridge.org/9780521766999

First published 1998
Reprinted 1999
Second edition published 2002
Reprinted with corrections 2003, 2005
Third edition published 2006
Fourth edition published 2010

Printed in the United Kingdom at the University Press, Cambridge

A catalogue record for this publication is available from the British Library

ISBN 978-0-521-76699-9 Hardback

To the memory of my dear sister Iris

B. S. E.

To my children Astrid and Inge

A. S.

Preface to fourth edition

In the fourth edition of this dictionary many new entries have been added reflecting, in particular, the expanding interest in Bayesian statistics, causality and machine learning. There has also been a comprehensive review and, where thought necessary, subsequent revision of existing entries. The number of biographies of important statisticians has been increased by including many from outside the UK and the USA and by the inclusion of entries for those who have died since the publication of the third edition. But perhaps the most significant addition to this edition is that of a co-author, namely Professor Anders Skrondal.

Preface to third edition

In this third edition of the Cambridge Dictionary of Statistics I have added many new entries and taken the opportunity to correct and clarify a number of the previous entries. I have also added biographies of important statisticians whom I overlooked in the first and second editions and, sadly, I have had to include a number of new biographies of statisticians who have died since the publication of the second edition in 2002.

B. S. Everitt, 2005

Preface to first edition

The Cambridge Dictionary of Statistics aims to provide students of statistics, working statisticians and researchers in many disciplines who are users of statistics with relatively concise definitions of statistical terms. All areas of statistics are covered, theoretical, applied, medical, etc., although, as in any dictionary, the choice of which terms to include and which to exclude is likely to reflect some aspects of the compiler's main areas of interest, and I have no illusions that this dictionary is any different. My hope is that the dictionary will provide a useful source of reference for both specialists and non-specialists alike. Many definitions necessarily contain some mathematical formulae and/or nomenclature, others contain none. But the difference in mathematical content and level among the definitions will, with luck, largely reflect the type of reader likely to turn to a particular definition. The non-specialist looking up, for example, **Student's t-tests** will hopefully find the simple formulae and associated written material more than adequate to satisfy their curiosity, while the specialist seeking a quick reminder about **spline functions** will find the more extensive technical material just what they need.

The dictionary contains approximately 3000 headwords and short biographies of more than 100 important statisticians (fellow statisticians who regard themselves as 'important' but who are *not* included here should note the single common characteristic of those who are). Several forms of cross-referencing are used. Terms in *slanted roman* in an entry appear as separate headwords, although headwords defining relatively commonly occurring terms such as **random variable**, **probability**, **distribution**, **population**, **sample**, etc., are *not* referred to in this way. Some entries simply refer readers to another entry. This may indicate that the terms are synonyms or, alternatively, that the term is more conveniently discussed under another entry. In the latter case the term is printed in *italics* in the main entry.

Entries are in alphabetical order using the letter-by-letter rather than the word-by-word convention. In terms containing numbers or Greek letters, the numbers or corresponding English word are spelt out and alphabetized accordingly. So, for example, 2×2 table is found under **two-by-two table**, and α-trimmed mean, under **alpha-trimmed mean**. Only headings corresponding to names are inverted, so the entry for William Gosset is found under **Gosset, William** but there is an entry under **Box–Müller** transformation *not* under **Transformation, Box–Müller**.

For those readers seeking more detailed information about a topic, many entries contain either a reference to one or other of the texts listed later, or a more specific reference to a relevant book or journal article. (Entries for software contain the appropriate address.) Additional material is also available in many cases in either the *Encyclopedia of Statistical Sciences*, edited by Kotz and Johnson, or the *Encyclopedia of Biostatistics*, edited by Armitage and Colton, both published by Wiley. Extended biographies of many of the people included in this dictionary can also be found in these two encyclopedias and also in *Leading Personalities in Statistical Sciences* by Johnson and Kotz published in 1997 again by Wiley.

Lastly and paraphrasing Oscar Wilde 'writing one dictionary is suspect, writing two borders on the pathological'. But before readers jump to an obvious conclusion I would like to make it very clear that an anorak has never featured in my wardrobe.

B. S. Everitt, 1998

Acknowledgements

Firstly I would like to thank the many authors who have, unwittingly, provided the basis of a large number of the definitions included in this dictionary through their books and papers. Next thanks are due to many members of the 'allstat' mailing list who helped with references to particular terms. I am also extremely grateful to my colleagues, Dr Sophia Rabe-Hesketh and Dr Sabine Landau, for their careful reading of the text and their numerous helpful suggestions. Lastly I have to thank my secretary, Mrs Harriet Meteyard, for maintaining and typing the many files that contained the material for the dictionary and for her constant reassurance that nothing was lost!

Notation

The transpose of a matrix $\mathbf{A}$ is denoted by $\mathbf{A}'$.

Sources

Altman, D. G. (1991) *Practical Statistics for Medical Research*, Chapman and Hall, London. (SMR)

Chatfield, C. (2003) *The Analysis of Time Series: An Introduction*, 6th edition, Chapman and Hall, London. (TMS)

Evans, M., Hastings, N. and Peacock, B. (2000) *Statistical Distributions*, 3rd edition, Wiley, New York. (STD)

Krzanowski, W. J. and Marriot, F. H. C. (1994) *Multivariate Analysis, Part 1*, Edward Arnold, London. (MV1)

Krzanowski, W. J. and Marriot, F. H. C. (1995) *Multivariate Analysis, Part 2*, Edward Arnold, London. (MV2)

McCullagh, P. M. and Nelder, J. A. (1989) *Generalized Linear Models*, 2nd edition, Chapman and Hall, London. (GLM)

Rawlings, J. O., Sastry, G. P. and Dickey, D. A. (2001) *Applied Regression Analysis: A Research Tool*, Springer-Verlag, New York. (ARA)

Stuart, A. and Ord, K. (1994) *Kendall's Advanced Theory of Statistics, Volume 1, 6th edition*, Edward Arnold, London. (KA1)

Stuart, A. and Ord, K. (1991) *Kendall's Advanced Theory of Statistics, Volume 2, 5th edition*, Edward Arnold, London. (KA2)

Aalen–Johansen estimator: An estimator of the *survival function* for a set of *survival times*, when there are competing causes of death. Related to the *Nelson–Aalen estimator*. [*Scandinavian Journal of Statistics*, 1978, **5**, 141–50.]

Aalen's linear regression model: A model for the *hazard function* of a set of *survival times* given by

$$\alpha(t; \mathbf{z}(t)) = \alpha_0(t) + \alpha_1(t)z_1(t) + \cdots + \alpha_p(t)z_p(t)$$

where $\alpha(t)$ is the hazard function at time t for an individual with covariates $\mathbf{z}(t)' = [z_1(t), \ldots, z_p(t)]$. The 'parameters' in the model are functions of time with $\alpha_0(t)$ the baseline hazard corresponding to $\mathbf{z}(t) = \mathbf{0}$ for all t, and $\alpha_q(t)$, the excess rate at time t per unit increase in $z_q(t)$. See also **Cox's proportional hazards model**. [*Statistics in Medicine*, 1989, **8**, 907–25.]

Abbot's formula: A formula for the proportion of animals (usually insects) dying in a toxicity trial that recognizes that some insects may die during the experiment even when they have not been exposed to the toxin, and among those who have been so exposed, some may dic of natural causes. Explicitly the formula is

$$p_i^* = \pi + (1 - \pi)p_i$$

where p_i^* is the observable response proportion, p_i is the expected proportion dying at a given dose and π is the proportion of insects who respond naturally. [*Modelling Binary Data*, 2nd edition, 2003, D. Collett, Chapman and Hall/CRC Press, London.]

ABC method: Abbreviation for **approximate bootstrap confidence method**.

Ability parameter: See **Rasch model**.

Absolute deviation: Synonym for **average deviation**.

Absolute risk: Synonym for **incidence**.

Absorbing barrier: See **random walk**.

Absorbing Markov chains: A state of a *Markov chain* is absorbing if it is impossible to leave it, i.e. the probability of leaving the state is zero, and a Markov chain is labelled 'absorbing' if it has at least one absorbing state. [*International Journal of Mathematical Education in Science and Technology*, 1996, **27**, 197–205.]

Absorption distributions: Probability distributions that represent the number of 'individuals' (e.g. particles) that fail to cross a specified region containing hazards of various kinds. For example, the region may simply be a straight line containing a number of 'absorption' points. When a particle travelling along the line meets such a point, there is a probability p that it will be absorbed. If it is absorbed it fails to make any further progress, but also the point is incapable of absorbing any more particles. When there are M active absorption points, the probability of a particle being absorbed is $[1 - (1 - p)^M]$. [*Naval Research Logistics Quarterly*, 1966, **13**, 35–48.]

Abundance matrices: Matrices that occur in ecological applications. They are essentially two-dimensional tables in which the classifications correspond to site and species. The value in the *ij*th cell gives the number of species *j* found at site *i*. [*Ecography*, 2006, **29**, 525–530.]

Accelerated failure time model: A general model for data consisting of *survival times*, in which explanatory variables measured on an individual are assumed to act multiplicatively on the time-scale, and so affect the rate at which an individual proceeds along the time axis. Consequently the model can be interpreted in terms of the speed of progression of a disease. In the simplest case of comparing two groups of patients, for example, those receiving treatment A and those receiving treatment B, this model assumes that the survival time of an individual on one treatment is a multiple of the survival time on the other treatment; as a result the probability that an individual on treatment A survives beyond time t is the probability that an individual on treatment B survives beyond time ϕt, where ϕ is an unknown positive constant. When the end-point of interest is the death of a patient, values of ϕ less than one correspond to an acceleration in the time of death of an individual assigned to treatment A, and values of ϕ greater than one indicate the reverse. The parameter ϕ is known as the *acceleration factor*. [*Modelling Survival Data in Medical Research*, 2nd edition, 2003, D. Collett, Chapman and Hall/CRC Press, London.]

Accelerated life testing: A set of methods intended to ensure product reliability during design and manufacture in which stress is applied to promote failure. The applied stresses might be temperature, vibration, shock etc. In order to make a valid inference about the normal lifetime of the system from the accelerated data (accelerated in the sense that a shortened time to failure is implied), it is necessary to know the relationship between time to failure and the applied stress. Often parametric statistical models of the time to failure and of the manner in which stress accelerates aging are used. [*Accelerated Testing*, 2004, W. Nelson, Wiley, New York.]

Acceleration factor: See **accelerated failure time model**.

Acceptable quality level: See **quality control procedures**.

Acceptable risk: The risk for which the benefits of a particular medical procedure are considered to outweigh the potential hazards. [*Acceptable Risk*, 1984, B. Fischoff, Cambridge University Press, Cambridge.]

Acceptance region: A term associated with statistical significance tests, that gives the set of values of a *test statistic* for which the null hypothesis is not rejected. Suppose, for example, a *z-test* is being used to test the null hypothesis that the mean blood pressure of men and women is equal against the alternative hypothesis that the two means are not equal. If the chosen significance level of the test is 0.05 then the acceptance region consists of values of the test statistic z between −1.96 and 1.96. [*Encyclopedia of Statistical Sciences*, 2006, eds. S. Kotz, C. B. Read, N. Balakrishnan and B. Vidakovic, Wiley, New York.]

Acceptance–rejection algorithm: An algorithm for generating random numbers from some probability distribution, $f(x)$, by first generating a random number from some other distribution, $g(x)$, where f and g are related by

$$f(x) \leq kg(x) \text{ for all } x$$

with k a constant. The algorithm works as follows:

- let r be a random number from $g(x)$;
- let s be a random number from a *uniform distribution* on the interval (0,1);
- calculate $c = ksg(r)$;
- if $c > f(r)$ reject r and return to the first step; if $c \leq f(r)$ accept r as a random number from f. [*Statistics in Civil Engineering*, 1997, A. V. Metcalfe, Edward Arnold, London.]

Acceptance sampling: A type of *quality control procedure* in which a sample is taken from a collection or batch of items, and the decision to accept the batch as satisfactory, or reject them as unsatisfactory, is based on the proportion of defective items in the sample. [*Quality Control and Industrial Statistics*, 4th edition, 1974, A. J. Duncan, R. D. Irwin, Homewood, Illinois.]

Accident proneness: A personal psychological factor that affects an individual's probability of suffering an accident. The concept has been studied statistically under a number of different assumptions for accidents:

- pure chance, leading to the *Poisson distribution*;
- true contagion, i.e. the hypothesis that all individuals initially have the same probability of having an accident, but that this probability changes each time an accident happens;
- apparent contagion, i.e. the hypothesis that individuals have constant but unequal probabilities of having an accident.

The study of accident proneness has been valuable in the development of particular statistical methodologies, although in the past two decades the concept has, in general, been out of favour; attention now appears to have moved more towards risk evaluation and analysis. [*Accident Proneness*, 1971, L. Shaw and H. S. Sichel, Pergamon Press, Oxford.]

Accidentally empty cells: Synonym for **sampling zeros**.

Accrual rate: The rate at which eligible patients are entered into a *clinical trial*, measured as persons per unit of time. Often disappointingly low for reasons that may be both physician and patient related. [*Journal of Clinical Oncology*, 2001, **19**, 3554–61.]

Accuracy: The degree of conformity to some recognized standard value. See also **bias**.

ACE: Abbreviation for **alternating conditional expectation**.

ACE model: A biometrical genetic model that postulates additive genetic factors, common environmental factors, and specific environmental factors in a phenotype. The model is used to quantify the contributions of genetic and environmental influences to variation. [*Encyclopedia of Behavioral Statistics, Volume 1*, 2005, eds. B. S. Everitt and D. C. Howell, Wiley, Chichester.]

ACES: Abbreviation for **active control equivalence studies**.

ACF: Abbreviation for **autocorrelation function**.

ACORN: An acronym for 'A Classification of Residential Neighbourhoods'. It is a system for classifying households according to the demographic, employment and housing characteristics of their immediate neighbourhood. Derived by applying *cluster analysis* to 40 variables describing each neighbourhood including age, class, tenure, dwelling type and car ownership. [*Statistics in Society*, 1999, eds. D. Dorling and S. Simpson, Arnold, London.]

Acquiescence bias: The bias produced by respondents in a survey who have the tendency to give positive responses, such as 'true', 'like', 'often' or 'yes' to a question. At its most extreme, the person responds in this way irrespective of the content of the item. Thus a person may respond 'true' to two items like 'I always take my medication on time' and 'I often forget to take my pills'. See also **end-aversion bias**. [*Journal of Intellectual Disability Research*, 1995, **39**, 331–40.]

Action lines: See **quality control procedures**.

Active control equivalence studies (ACES): *Clinical trials* in which the object is simply to show that the new treatment is at least as good as the existing treatment. Such studies are becoming more widespread due to current therapies that reflect previous successes in the development of new treatments. The studies rely on an implicit historical control assumption, since to conclude that a new drug is efficacious on the basis of this type of study requires a fundamental assumption that the active control drug would have performed better than a placebo, had a placebo been used in the trial. [*Statistical Issues in Drug Development*, 2nd edition, 2008, S. Senn, Wiley-Blackwell, Chichester.]

Active control trials: *Clinical trials* in which the trial drug is compared with some other active compound rather than a placebo. [*Annals of Internal Medicine*, 2000, **135**, 62–4.]

Active life expectancy (ALE): Defined for a given age as the expected remaining years free of disability. A useful index of public health and quality of life for populations. A question of great interest is whether recent trends towards longer *life expectancy* have been accompanied by a comparable increase in ALE. [*New England Journal of Medicine*, 1983, **309**, 1218–24.]

Actuarial estimator: An estimator of the *survival function, S(t)*, often used when the data are in grouped form. Given explicitly by

$$S(t) = \prod_{\substack{j \geq 0 \\ t_{(j+1)} \leq t}} \left[1 - \frac{d_j}{N_j - \frac{1}{2} w_j} \right]$$

where the ordered *survival times* are $0 < t_{(1)} < \cdots < t_{(n)}$, N_i is the number of people at risk at the start of the interval $t_{(i)}, t_{(i+1)}$, d_i is the observed number of deaths in the interval and w_i the number of censored observations in the interval. [*Survival Models and Data Analysis*, 1999, R. G. Elandt–Johnson and N. L. Johnson, Wiley, New York.]

Actuarial statistics: The statistics used by actuaries to evaluate risks, calculate liabilities and plan the financial course of insurance, pensions, etc. An example is *life expectancy* for people of various ages, occupations, etc. See also **life table**. [*Financial and Actuarial Statistics: An Introduction*, 2003, D. S. Borowiak and A. F. Shapiro, CRC Press, Boca Raton.]

Adaptive cluster sampling: A procedure in which an initial set of subjects is selected by some sampling procedure and, whenever the variable of interest of a selected subject satisfies a given criterion, additional subjects in the neighbourhood of that subject are added to the sample. [*Biometrika*, 1996, **84**, 209–19.]

Adaptive designs: *Clinical trials* that are modified in some way as the data are collected within the trial. For example, the allocation of treatment may be altered as a function of the response to protect patients from ineffective or toxic doses. [*Controlled Clinical Trials*, 1999, **20**, 172–86.]

Adaptive estimator: See **adaptive methods**.

Adaptive lasso: See **lasso**.

Adaptive methods: Procedures that use various aspects of the sample data to select the most appropriate type of statistical method for analysis. An *adaptive estimator*, T, for the centre of a distribution, for example, might be

$$\begin{aligned} T &= \text{mid-range when } k \leq 2 \\ &= \text{arithmetic mean when } 2<k<5 \\ &= \text{median when } k \geq 5 \end{aligned}$$

where k is the sample *kurtosis*. So if the sample looks as if it arises from a short-tailed distribution, the average of the largest and smallest observations is used; if it looks like a long-tailed situation the median is used, otherwise the mean of the sample is calculated. [*Journal of the American Statistical Association*, 1967, **62**, 1179–86.]

Adaptive methods of treatment assignment: Any method of treatment allocation in a *clinical trial* that uses accumulating outcome data to affect the treatment selection, for example, the *O'Brien-Fleming method*. [*Biometrika*, 1977, **64**, 191–199.]

Adaptive sampling design: A *sampling design* in which the procedure for selecting *sampling units* on which to make observations may depend on observed values of the variable of interest. In a survey for estimating the abundance of a natural resource, for example, additional sites (the sampling units in this case) in the vicinity of high observed abundance may be added to the sample during the survey. The main aim in such a design is to achieve gains in precision or efficiency compared to conventional designs of equivalent sample size by taking advantage of observed characteristics of the population. For this type of sampling design the probability of a given sample of units is conditioned on the set of values of the variable of interest in the population. [*Adaptive Sampling*, 1996, S. K. Thompson and G. A. F. Seber, Wiley, New York.]

Added variable plot: A graphical procedure used in all types of regression analysis for identifying whether or not a particular explanatory variable should be included in a model, in the presence of other explanatory variables. The variable that is the candidate for inclusion in the model may be new or it may simply be a higher power of one currently included. If the candidate variable is denoted x_i, then the *residuals* from the regression of the response variable on all the explanatory variables, save x_i, are plotted against the residuals from the regression of x_i on the remaining explanatory variables. A strong linear relationship in the plot indicates the need for x_i in the regression equation (Fig. 1). [*Regression Analysis*, Volume 2, 1993, edited by M. S. Lewis-Beck, Sage Publications, London.]

Addition rule for probabilities: For two events, A and B that are *mutually exclusive*, the probability of either event occurring is the sum of the individual probabilities, i.e.

$$\Pr(A \text{ or } B) = \Pr(A) + \Pr(B)$$

where $\Pr(A)$ denotes the probability of event A etc. For k mutually exclusive events A_1, $A_2, \ldots, A_k$, the more general rule is

$$\Pr(A_1 \text{ or } A_2 \ldots \text{ or } A_k) = \Pr(A_1) + \Pr(A_2) + \cdots + \Pr(A_k)$$

See also **multiplication rule for probabilities** and **Boole's inequality**. [KA1 Chapter 8.]

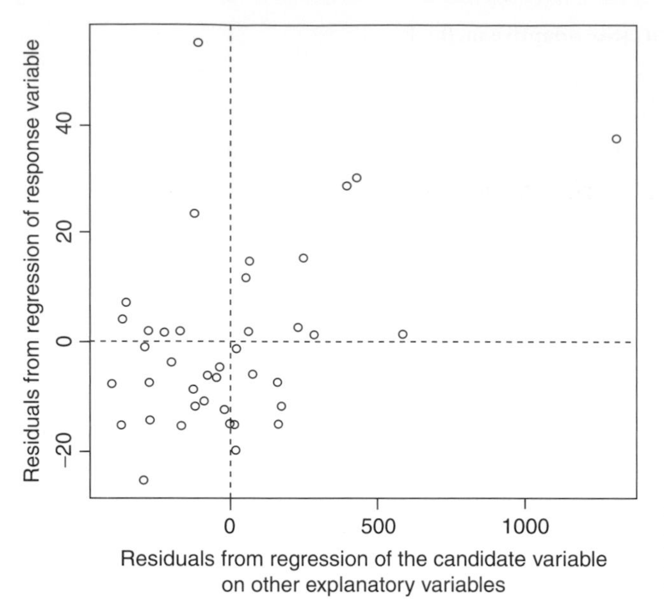

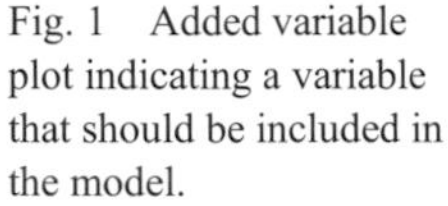
Fig. 1 Added variable plot indicating a variable that should be included in the model.

Additive clustering model: A model for *cluster analysis* which attempts to find the structure of a *similarity matrix* with elements s_{ij} by fitting a model of the form

$$s_{ij} = \sum_{k=1}^{K} w_k p_{ik} p_{jk} + \epsilon_{ij}$$

where K is the number of clusters and w_k is a weight representing the salience of the property corresponding to cluster k. If object i has the property of cluster k, then $p_{ik} = 1$, otherwise it is zero. [*Psychological Review*, 1979, **86**, 87–123.]

Additive effect: A term used when the effect of administering two treatments together is the sum of their separate effects. See also **additive model**. [*Journal of Bone Mineral Research*, 1995, **10**, 1303–11.]

Additive genetic variance: The variance of a trait due to the main effects of *genes*. Usually obtained by a factorial *analysis of variance* of trait values on the genes present at one or more loci. [*Statistics in Human Genetics*, 1998, P. Sham, Arnold, London.]

Additive model: A model in which the explanatory variables have an *additive effect* on the response variable. So, for example, if variable A has an effect of size a on some response measure and variable B one of size b on the same response, then in an assumed additive model for A and B their combined effect would be $a+b$.

Additive outlier: A term applied to an observation in a *time series* which is affected by a non-repetitive intervention such as a strike, a war, etc. Only the level of the particular observation is considered affected. In contrast an *innovational outlier* is one which corresponds to an extraordinary shock at some time point T which also influences subsequent observations in the series. [*Journal of the American Statistical Association*, 1996, **91**, 123–31.]

Additive tree: A connected, *undirected graph* where every pair of nodes is connected by a unique path and where the distances between the nodes are such that

a. Dissimiliaritles

	A	B	C	D	E
Worker A	--				
Worker B	15	--			
Worker C	20	25	--		
Worker D	18	23	6	--	
Worker E	20	25	20	18	--

b. Additive Tree

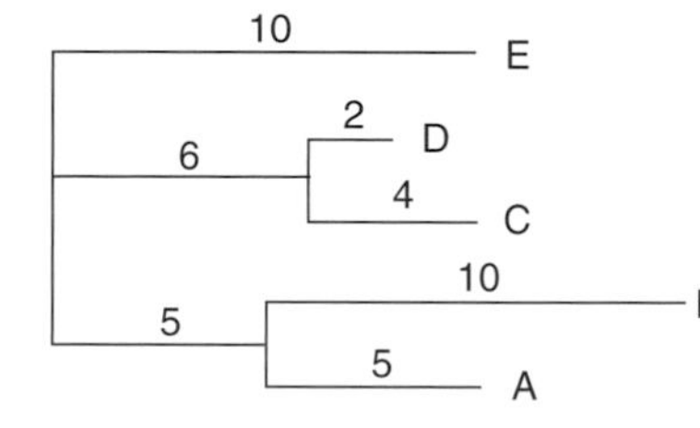

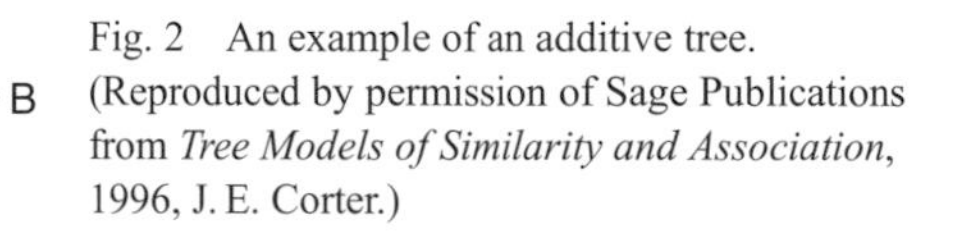

Fig. 2 An example of an additive tree. (Reproduced by permission of Sage Publications from *Tree Models of Similarity and Association*, 1996, J. E. Corter.)

$$d_{xy} + d_{uv} \leq \max[d_{xu} + d_{yv}, d_{xv} + d_{yu}] \text{ for all } x, y, u, \text{ and } v$$

An example of such a tree is shown in Fig. 2. See also **ultrametric tree**. [*Tree Models of Similarity and Association*, 1996, J. E. Corter, Sage University Papers 112, Sage Publications, Thousand Oaks.]

Adelstein, Abe (1916–1993): Born in South Africa, Adelstein studied medicine at the University of the Witwatersrand. In the 1960s he emigrated to Manchester where he worked in the Department of Social Medicine. Later he was appointed Chief Medical Statistician for England and Wales. Adelstein made significant contributions to the classification of mental illness and to the epidemiology of suicide and alcoholism.

Adequate subset: A term used in regression analysis for a subset of the explanatory variables that is thought to contain as much information about the response variable as the complete set. See also **selection methods in regression**.

Adjacency matrix: A matrix with elements, x_{ij}, used to indicate the connections in a *directed graph*. If node i relates to node j, $x_{ij} = 1$, otherwise $x_{ij} = 0$. For a simple graph with no self-loops, the adjacency matrix must have zeros on the diagonal. For an undirected graph the adjacency matrix is symmetric. [*Introductory Graph Theory*, 1985, G. Chartrand, Dover, New York.]

Adjusted correlation matrix: A *correlation matrix* in which the diagonal elements are replaced by *communalities*. The basis of *principal factor analysis*.

Adjusted treatment means: Usually used for estimates of the treatment means in an *analysis of covariance*, after adjusting all treatments to the same mean level for the covariate(s), using the estimated relationship between the covariate(s) and the response variable. [*Biostatistics: A Methodology for the Health Sciences*, 2nd edn, 2004, G. Van Belle, L. D. Fisher, P. J. Heagerty and T. S. Lumley, Wiley, New York.]

Adjusting for baseline: The process of allowing for the effect of *baseline characteristics* on the response variable usually in the context of a *longitudinal study*. See also **Lord's paradox**

and **baseline balance**. [*Statistical Issues in Drug Development*, 2nd edition, 2008, S. Senn, Wiley-Blackwell, Chichester.]

Administrative databases: *Databases* storing information routinely collected for purposes of managing a health-care system. Used by hospitals and insurers to examine admissions, procedures and lengths of stay. [*Healthcare Management Forum*, 1995, **8**, 5–13.]

Admissibility: A very general concept that is applicable to any procedure of statistical inference. The underlying notion is that a procedure is admissible if and only if there does not exist within that class of procedures another one which performs uniformly at least as well as the procedure in question and performs better than it in at least one case. Here 'uniformly' means for all values of the parameters that determine the probability distribution of the random variables under investigation. [KA2 Chapter 31.]

Admixture in human populations: The inter-breeding between two or more populations that were previously isolated from each other for geographical or cultural reasons. Population admixture can be a source of spurious associations between diseases and *alleles* that are both more common in one ancestral population than the others. However, populations that have been admixed for several generations may be useful for mapping disease *genes*, because spurious associations tend to be dissipated more rapidly than true associations in successive generations of random mating. [*Statistics in Human Genetics*, 1998, P. Sham, Arnold, London.]

Adoption studies: Studies of the rearing of a nonbiological child in a family. Such studies have played an important role in the assessment of genetic variation in human and animal traits. [*Foundations of Behavior Genetics*, 1978, J. L. Fulker and W. R. Thompson, Mosby, St. Louis.]

Adverse selection: A term used in insurance when the insurer cannot distinguish between members of good- and poor-risk categories for a certain hazard and the poor-risks are the only purchasers of coverage with the consequence that the insurer expects to lose money on each policy sold. [*Quarterly Journal of Economics*, 1976, **90**, 629–650.]

Aetiological fraction: Synonym for **attributable risk**.

Affine invariance: A term applied to statistical procedures which give identical results after the data has been subjected to an *affine transformation*. An example is *Hotelling's T^2 test*. [*Canadian Journal of Statistics*, 2003, **31**, 437–55.]

Affine transformation: The transformation, $\mathbf{Y} = \mathbf{AX} + \mathbf{b}$ where $\mathbf{A}$ is a nonsingular matrix and $\mathbf{b}$ is any vector of real numbers. Important in many areas of statistics particularly *multivariate analysis*.

Age-dependent birth and death process: A *birth and death process* where the birth rate and death rate are not constant over time, but change in a manner which is dependent on the age of the individual. [*Stochastic Modelling of Scientific Data*, 1995, P. Guttorp, Chapman and Hall/CRC Press, London.]

Age heaping: A term applied to the collection of data on ages when these are accurate only to the nearest year, half year or month. Occurs because many people (particularly older people) tend not to give their exact age in a survey. Instead they round their age up or down to the nearest number that ends in 0 or 5. See also **coarse data** and **Whipple index**. [*Population Studies*, 1991, **45**, 497–518.]

Age–period–cohort model: A model important in many *observational studies* when it is reasonable to suppose that age, number of years exposed to risk factor, and age when first exposed to risk factor, all contribute to disease risk. Unfortunately all three factors cannot be entered simultaneously into a model since this would result in *collinearity*, because 'age first exposed to risk factor'+'years exposed to risk factor' is equal to 'age'. Various methods have been suggested for disentangling the dependence of the factors, although most commonly one of the factors is simply not included in the modelling process. See also **Lexis diagram**. [*Statistics in Medicine*, 1984, **3**, 113–30.]

Age-related reference ranges: Ranges of values of a measurement that give the upper and lower limits of normality in a population according to a subject's age. [*Archives of Disease in Childhood*, 2005, **90**, 1117–1121.]

Age-specific death rates: Death rates calculated within a number of relatively narrow age bands. For example, for 20–30 year olds,

$$DR_{20,30} = \frac{\text{number of deaths among } 20 - 30 \text{ year olds in a year}}{\text{average population size in } 20 - 30 \text{ year olds in the year}}$$

Calculating death rates in this way is usually necessary since such rates almost invariably differ widely with age, a variation not reflected in the *crude death rate*. See also **cause-specific death rates** and **standardized mortality ratio**. [*Biostatistics*, 2nd edition, 2004, G. Van Belle, L. D. Fisher, P. J. Heagerty and T. S. Lumley, Wiley, New York.]

Age-specific failure rate: A synonym for *hazard function* when the time scale is age. [*Statistical Methods for Survival Data Analysis*, 3rd edn, E. T. Lee and J. W. Wang, Wiley, New York.]

Age-specific incidence rate: *Incidence rates* calculated within a number of relatively narrow age bands. See also **age-specific death rates**. [*Cancer Epidemiology Biomarkers and Prevention*, 2004, **13**, 1128–1135.]

Agglomerative hierarchical clustering methods: Methods of *cluster analysis* that begin with each individual in a separate cluster and then, in a series of steps, combine individuals and later, clusters, into new, larger clusters until a final stage is reached where all individuals are members of a single group. At each stage the individuals or clusters that are 'closest', according to some particular definition of distance are joined. The whole process can be summarized by a *dendrogram*. Solutions corresponding to particular numbers of clusters are found by 'cutting' the dendrogram at the appropriate level. See also **average linkage, complete linkage, single linkage, Ward's method, Mojena's test, K-means cluster analysis** and **divisive methods**. [MV2 Chapter 10.]

Agreement: The extent to which different observers, raters or diagnostic tests agree on a binary classification. Measures of agreement such as the *kappa coefficient* quantify the relative frequency of the diagonal elements in a two-by-two contingency table, taking agreement due to chance into account. It is important to note that strong agreement requires strong *association* whereas strong association does not require strong agreement. [*Statistical Methods for Rates and Proportions*, 2nd edn, 2001, J. L.Fleiss, Wiley, New York.]

Agresti's α: A generalization of the *odds ratio* for *2×2 contingency tables* to larger *contingency tables* arising from data where there are different degrees of severity of a disease and differing amounts of exposure. [*Analysis of Ordinal Categorical Data*, 1984, A. Agresti, Wiley, New York.]

Agronomy trials: A general term for a variety of different types of agricultural field experiments including fertilizer studies, time, rate and density of planting, tillage studies, and pest and weed control studies. Because the response to changes in the level of one factor is often conditioned by the levels of other factors it is almost essential that the treatments in such trials include combinations of multiple levels of two or more production factors. [*An Introduction to Statistical Science in Agriculture*, 4th edition, 1972, D. J. Finney, Blackwell, Oxford.]

AI: Abbreviation for **artificial intelligence**.

AIC: Abbreviation for **Akaike's information criterion**.

Aickin's measure of agreement: A chance-corrected measure of *agreement* which is similar to the *kappa coefficient* but based on a different definition of agreement by chance. [*Biometrics*, 1990, **46**, 293–302.]

AID: Abbreviation for **automatic interaction detector**.

Aitchison distributions: A broad class of distributions that includes the *Dirichlet distribution* and *logistic normal distributions* as special cases. [*Journal of the Royal Statistical Society, Series B*, 1985, **47**, 136–46.]

Aitken, Alexander Craig (1895–1967): Born in Dunedin, New Zealand, Aitken first studied classical languages at Otago University, but after service during the First World War he was given a scholarship to study mathematics in Edinburgh. After being awarded a D.Sc., Aitken became a member of the Mathematics Department in Edinburgh and in 1946 was given the Chair of Mathematics which he held until his retirement in 1965. The author of many papers on least squares and the fitting of polynomials, Aitken had a legendary ability at arithmetic and was reputed to be able to dictate rapidly the first 707 digits of π. He was a Fellow of the Royal Society and of the Royal Society of Literature. Aitken died on 3 November 1967 in Edinburgh.

Ajne's test: A *distribution free method* for testing the uniformity of a *circular distribution*. The test statistic A_n is defined as

$$A_n = \int_0^{2\pi} [N(\theta) - n/2]^2 \mathrm{d}\theta$$

where $(N\theta)$ is the number of sample observations that lie in the semicircle, θ to $\theta + \pi$. Values close to zero lead to acceptance of the hypothesis of uniformity. [*Annals of Mathematical Statistics*, 1972, **43**, 468–479.]

Akaike's information criterion (AIC): An index used in a number of areas as an aid to choosing between competing models. It is defined as

$$-2L_m + 2m$$

where L_m is the maximized *log-likelihood* and m is the number of parameters in the model. The index takes into account both the statistical goodness of fit and the number of parameters that have to be estimated to achieve this particular degree of fit, by imposing a penalty for increasing the number of parameters. Lower values of the index indicate the preferred model, that is, the one with the fewest parameters that still provides an adequate fit to the data. See also **parsimony principle** and **Schwarz's criterion**. [MV2 Chapter 11.]

ALE: Abbreviation for **active life expectancy**.

Algorithm: A well-defined set of rules which, when routinely applied, lead to a solution of a particular class of mathematical or computational problem. [*Introduction to Algorithms*, 1989, T. H. Cormen, C. E. Leiserson, and R. L. Rivest, McGraw-Hill, New York.]

Aliasing: Occurs when the estimate of a parameter is wholly confounded with other parameters because sufficient information is not available. *Extrinsic aliasing* is due to lack of adequate data, such as *missing values* and *collinearity*. *Intrinsic aliasing* is due to lack of identification of the specified statistical model, for example a regression model where a categorical explanatory variable is represented by as many dummy variables as there are categories.

Allele: The DNA sequence that exists at a genetic location that shows sequence variation in a population. Sequence variation may take the form of insertion, deletion, substitution, or variable repeat length of a regular motif, for example, CACACA. [*Statistics in Human Genetics*, 1998, P. Sham, Arnold, London.]

Allocation ratio: Synonym for **treatment allocation ratio**.

Allocation rule: See **discriminant analysis**.

Allometry: The study of changes in shape as an organism grows. [MV1 Chapter 4.]

All possible comparisons (APC): A procedure for analysing small unreplicated factorial experiments which used *likelihood ratio* tests to compare competing models. See also **Lenth's method**. [*Technometrics*, 2005, **47**, 51–63.]

All subsets regression: A form of regression analysis in which all possible models are considered and the 'best' selected by comparing the values of some appropriate criterion, for example, *Mallow's C_p statistic*, calculated on each. If there are q explanatory variables, there are a total of $2^p - 1$ models to be examined. The *leaps-and-bounds algorithm* is generally used so that only a small fraction of the possible models have to be examined. See also **selection methods in regression**. [ARA Chapter 7]

Almon lag technique: A method for estimating the coefficients, $\beta_0, \beta_1, \ldots, \beta_r$, in a model of the form

$$y_t = \beta_0 x_t + \cdots + \beta_r x_{t-r} + \epsilon_t$$

where y_t is the value of the dependent variable at time t, $x_t, \ldots, x_{t-r}$ are the values of the explanatory variable at times $t, t-1, \ldots, t-r$ and ϵ_t is a disturbance term at time t. If r is finite and less than the number of observations, the regression coefficients can be found by *least squares estimation*. However, because of the possible problem of a high degree of *multicollinearity* in the variables $x_t, \ldots, x_{t-r}$ the approach is to estimate the coefficients subject to the restriction that they lie on a polynomial of degree p, i.e. it is assumed that there exist parameters $\lambda_0, \lambda_1, \ldots, \lambda_p$ such that

$$\beta_i = \lambda_0 + \lambda_1 i + \cdots + \lambda_p i^p, \; i = 0, 1, \ldots, r, \; p \leq r$$

This reduces the number of parameters from $r+1$ to $p+1$. When $r = p$ the technique is equivalent to least squares. In practice several different values of r and/or p need to be investigated. [*A Guide to Econometrics*, 1986, P. Kennedy, MIT Press.]

Almost sure convergence: A type of convergence that is similar to pointwise convergence of a sequence of functions, except that the convergence need not occur on a set with probability zero. A formal definition is the following: The sequence $\{X_t\}$ converges almost sure to μ, if there exists a set M such that P(M)=1 and for every $\omega \in N$ we have $X_t(\omega) \rightarrow \mu$. [*Parametric Statistical Inference*, 1999, J. K. Lindsey, Oxford University Press, Oxford.]

Alpha(α): The probability of a type I error. See also **significance level**.

Alpha factoring: A method of *factor analysis* in which the variables are considered samples from a population of variables. [*Psychometrika*, 1965, **30**, 1–14.]

Alpha spending function: An approach to *interim analysis* in a *clinical trial* that allows the control of the type I error rate while giving flexibility in how many interim analyses are to be conducted and at what time. [*Statistics in Medicine*, 1996, **15**, 1739–46.]

Alpha(α)-trimmed mean: A method of estimating the mean of a population that is less affected by the presence of *outliers* than the usual estimator, namely the sample average. Calculating the statistic involves dropping a proportion α (approximately) of the observations from both ends of the sample before calculating the mean of the remainder. If $x_{(1)}, x_{(2)}, \ldots, x_{(n)}$ represent the ordered sample values then the measure is given by

$$\alpha_{\text{trimmed mean}} = \frac{1}{n-2k} \sum_{i=k+1}^{n-k} x_{(i)}$$

where k is the smallest integer greater than or equal to $\alpha\, n$. See also **M-estimators**. [*Biostatistics*, 2nd edition, 2004, G. Van Belle, L. D. Fisher, P. J. Heagerty and T. S. Lumley, Wiley, New York.]

Alpha(α)-Winsorized mean: A method of estimating the mean of a population that is less affected by the presence of *outliers* than the usual estimator, namely the sample average. Essentially the k smallest and k largest observations, where k is the smallest integer greater than or equal to αn, are respectively increased or reduced in size to the next remaining observation and counted as though they had these values. Specifically given by

$$\alpha_{\text{Winsorized mean}} = \frac{1}{n}\left[(k+1)\left(x_{(k+1)} + x_{(n-k)}\right) + \sum_{i=k+2}^{n-k-1} x_{(i)}\right]$$

where $x_{(1)}, x_{(2)}, \ldots, x_{(n)}$ are the ordered sample values. See also **M-estimators**. [*Biostatistics: A Methodology for the Health Sciences*, 2nd edn, 2004, G. Van Belle, L. D. Fisher, P. J. Heagerty and T. S. Lumley, Wiley, New York.]

Alshuler's estimator: An estimator of the *survival function* given by

$$\prod_{j=1}^{k} \exp(-d_j/n_j)$$

where d_j is the number of deaths at time $t_{(j)}$, n_j the number of individuals alive just before $t_{(j)}$ and $t_{(1)} \leq t_{(2)} \leq \ldots \leq t_{(k)}$ are the ordered *survival times*. See also **product limit estimator**. [*Modelling Survival Data in Medical Research*, 2nd edition, 2003, D. Collett, Chapman and Hall/CRC Press, London.]

Alternate allocations: A method of allocating patients to treatments in a *clinical trial* in which alternate patients are allocated to treatment A and treatment B. Not to be recommended since it is open to abuse. [SMR Chapter 15.]

Alternating conditional expectation (ACE): A procedure for estimating optimal transformations for regression analysis and correlation. Given explanatory variables $x_1, \ldots, x_q$ and response variable y, the method finds the transformations $g(y)$ and $s_1(x_1), \ldots, s_q(x_q)$ that maximize the correlation between y and its predicted value. The technique allows for arbitrary, smooth transformations of both response and explanatory variables. [*Biometrika*, 1995, **82**, 369–83.]

Alternating least squares: A method most often used in some methods of *multidimensional scaling*, where a goodness-of-fit measure for some configuration of points is minimized in a series of steps, each involving the application of *least squares*. [MV1 Chapter 8.]

Alternating logistic regression: A method of *logistic regression* used in the analysis of *longitudinal data* when the response variable is binary. Based on *generalized estimating equations*. [*Analysis of Longitudinal Data*, 2nd edition, 2002, P. J. Diggle, P. J. Heagerty, K.-Y. Liang and S. L. Zeger, Oxford Science Publications, Oxford.]

Alternative hypothesis: The hypothesis against which the null hypothesis is tested.

Aly's statistic: A statistic used in a *permutation test* for comparing variances, and given by

$$\delta = \sum_{i=1}^{m-1} i(m-i)(X_{(i+1)} - X_{(i)})$$

where $X_{(1)} < X_{(2)} < \ldots < X_{(m)}$ are the *order statistics* of the first sample. [*Statistics and Probability Letters*, 1990, **9**, 323–5.]

Amersham model: A model used for *dose–response curves* in immunoassay and given by

$$y = 100(2(1-\beta_1)\beta_2)/(\beta_3+\beta_2+\beta_4+x+[(\beta_3-\beta_2+\beta_4+x)^2+4\beta_3\beta_2]^{\frac{1}{2}})+\beta_1$$

where y is percentage binding and x is the analyte concentration. Estimates of the four parameters, $\beta_1, \beta_2, \beta_3, \beta_4$, may be obtained in a variety of ways. [*Medical Physics*, 2004 **31**, 2501–8.]

AML: Abbreviation for **asymmetric maximum likelihood**.

Amplitude: A term used in relation to *time series*, for the value of the series at its peak or trough taken from some mean value or trend line.

Amplitude gain: See **linear filters**.

Analysis as-randomized: Synonym for **intention-to-treat analysis**.

Analysis of covariance (ANCOVA): Originally used for an extension of the *analysis of variance* that allows for the possible effects of continuous concomitant variables (covariates) on the response variable, in addition to the effects of the factor or treatment variables. Usually assumed that covariates are unaffected by treatments and that their relationship to the response is linear. If such a relationship exists then inclusion of covariates in this way decreases the *error mean square* and hence increases the sensitivity of the *F-tests* used in assessing treatment differences. The term now appears to also be more generally used for almost any analysis seeking to assess the relationship between a response variable and a number of explanatory variables. See also **parallelism in ANCOVA**, **generalized linear model** and **Johnson–Neyman technique**. [KA2 Chapter 29.]

Analysis of dispersion: Synonym for **multivariate analysis of variance**.

Analysis of variance (ANOVA): The separation of variance attributable to one variable from the variance attributable to others. By partitioning the total variance of a set of observations into parts due to particular factors, for example, sex, treatment group etc., and comparing variances (mean squares) by way of *F-tests*, differences between means can be assessed. The simplest analysis of this type involves a *one-way design*, in which N subjects are allocated, usually at random, to the k different levels of a single factor. The total variation in the observations is then divided into a part due to differences between level means (the *between groups sum of squares*) and a part due to the differences between subjects in the same group (the *within groups sum of squares*, also known as the *residual sum of squares*). These terms are usually arranged as an *analysis of variance table.*

Source	**df**	**SS**	**MS**	**MSR**
Bet. grps.	$k-1$	SSB	SSB/$(k-1)$	$\frac{\text{SSB}/(k-1)}{\text{SSW}/(N-k)}$
With. grps.	$N-k$	SSW	SSW/$(N-k)$	
Total	$N-1$			

SS = sum of squares; MS = mean square; MSR = mean square ratio.

If the means of the populations represented by the factor levels are the same, then within the limits of random variation, the *between groups mean square* and *within groups mean square*, should be the same. Whether this is so can, if certain assumptions are met, be assessed by a suitable F-test on the mean square ratio. The necessary assumptions for the validity of the F-test are that the response variable is normally distributed in each population and that the populations have the same variance. Essentially an example of the *generalized linear model* with an identity *link function* and normally distributed error terms. See also **analysis of covariance, parallel groups design** and **factorial designs**. [SMR Chapter 9.]

Analysis of variance table: See **analysis of variance**.

Analytic epidemiology: A term for epidemiological studies, such as *case-control studies*, that obtain individual-level information on the association between disease status and exposures of interest. [*Journal of the National Cancer Institute*, 1996, **88**, 1738–47.]

Ancillary statistic: A term applied to the statistic C in situations where the *minimal sufficient statistic*, S, for a parameter θ, can be written as $S=(T, C)$ and C has a *marginal distribution* not depending on θ. For example, let N be a random variable with a known distribution $p_n = \Pr(N = n)(n = 1, 2, \ldots)$, and let $Y_1, Y_2, \ldots, Y_N$ be independently and identically distributed random variables from the *exponential family distribution* with parameter, θ. The *likelihood* of the data $(n, y_1, y_2, \ldots, y_n)$ is

$$p_n \exp\left\{a(\theta)\sum_{j=1}^{n} b(y_j) + nc(\theta) + \sum_{j=1}^{n} d(y_j)\right\}$$

so that $S = [\sum_{j=1}^{N} b(Y_j), N]$ is sufficient for θ and N is an ancillary statistic. Important in the application of *conditional likelihood* for estimation. [KA2 Chapter 31.]

ANCOVA: Acronym for **analysis of covariance**.

Andersen, Erling Bernhard (1934–2004): Andersen graduated in 1963, the first Danish graduate with a formal degree in mathematical statistics. In 1965 he received a gold medal from the University of Copenhagen for his work on the *Rasch model.* Andersen received his doctorate in 1973, the topic being conditional inference. He became professor of statistics in the Department of Statistics of the University of Copenhagen in 1974. His most important contributions to statistics were his work in the area of *item-response theory* and Rasch models. Andersen died on the 18th September, 2004.

Andersen–Gill model: A model for analysing *multiple time response data* in which each subject is treated as a multi-event *counting process* with essentially independent increments. [*Annals of Statistics*, 1982, **10**, 1100–20.]

Anderson–Darling test: A test that a given sample of observations arises from some specified theoretical probability distribution. For testing the normality of the data, for example, the test statistic is

$$A_n^2 = -\frac{1}{n}\left[\sum_{i=1}^{n}(2i-1)\{\log z_i + \log(1 - z_{n+1-i})\}\right] - n$$

where $x_{(1)} \leq x_{(2)} \leq \cdots \leq x_{(n)}$ are the ordered observations, s^2 is the sample variance, and

$$z_i = \Phi\left(\frac{x_{(i)} - \bar{x}}{s}\right)$$

where

$$\Phi(x) = \int_{-\infty}^{x} \frac{1}{\sqrt{2\pi}} e^{-\frac{1}{2}u^2} du$$

The null hypothesis of normality is rejected for 'large' values of A_n^2. Critical values of the test statistic are available. See also **Shapiro–Wilk test**. [*Journal of the American Statistical Society*, 1954, **49**, 765–9.]

Anderson-Hsiao estimator: An *instrumental variables estimator* for *dynamic panel data models* with subject-specific intercepts. [*Analysis of Panel Data*, 2nd edn, 2003, C. Hsiao, Cambridge University Press, Cambridge]

Anderson, John Anthony (1939–1983): Anderson studied mathematics at Oxford, obtaining a first degree in 1963, and in 1968 he was awarded a D.Phil. for work on statistical methods in medical diagnosis. After working in the Department of Biomathematics in Oxford for some years, Anderson eventually moved to Newcastle University, becoming professor in 1982. Contributed to *multivariate analysis*, particularly *discriminant analysis* based on *logistic regression*. He died on 7 February 1983, in Newcastle.

Anderson, Oskar Nikolayevick (1887–1960): Born in Minsk, Byelorussia, Anderson studied mathematics at the University of Kazan. Later he took a law degree in St Petersburg and travelled to Turkestan to make a survey of agricultural production under irrigation in the Syr Darya River area. Anderson trained in statistics at the Commercial Institute in Kiev and from the mid-1920s he was a member of the Supreme Statistical Council of the Bulgarian government during which time he successfully advocated the use of sampling techniques. In 1942 Anderson accepted an appointment at the University of Kiel, Germany and from 1947 until his death he was Professor of Statistics in the Economics Department at the University of Munich. Anderson was a pioneer of applied sample-survey techniques.

Andrews' plots: A graphical display of *multivariate data* in which an observation, $\mathbf{x}' = [x_1, x_2, \ldots, x_q]$ is represented by a function of the form

$$f_{\mathbf{x}}(t) = x_1/\sqrt{2} + x_2 \sin(t) + x_3 \cos(t) + x_4 \sin(2t) + x_5 \cos(2t) + \cdots$$

plotted over the range of values $-\pi \leq t \leq \pi$. A set of multivariate observations is displayed as a collection of these plots and it can be shown that those functions that remain close together for all values of t correspond to observations that are close to one another in terms of their *Euclidean distance*. This property means that such plots can often be used to both detect groups of similar observations and identify *outliers* in multivariate data. The example shown at Fig. 3 consists of plots for a sample of 30 observations each having five variable values. The plot indicates the presence of three groups in the data. Such plots can cope only with a

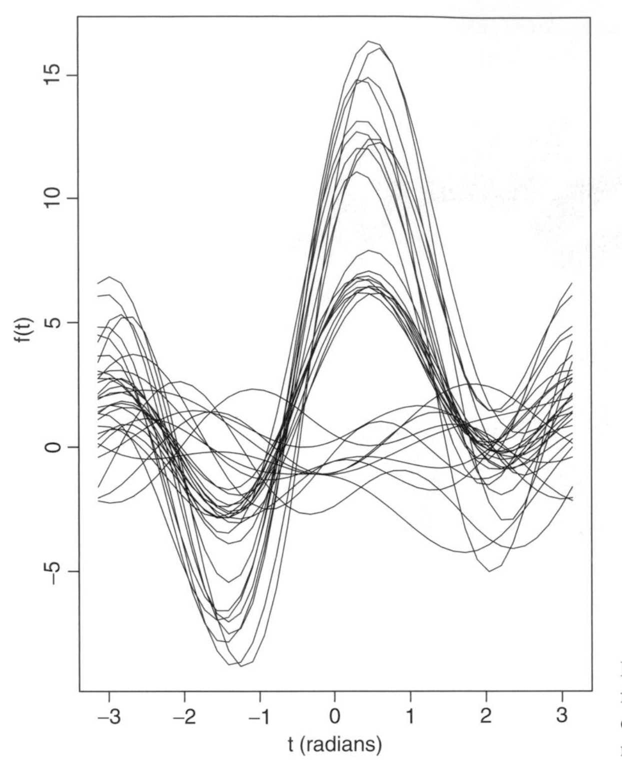

Fig. 3 Andrews' plot for 30, five-dimensional observations constructed to contain three relatively distinct groups.

moderate number of observations before becoming very difficult to unravel. See also **Chernoff faces** and **glyphs** [MV1 Chapter 3.]

Angle count method: A method for estimating the proportion of the area of a forest that is actually covered by the bases of trees. An observer goes to each of a number of points in the forest, chosen either randomly or systematically, and counts the number of trees that subtend, at that point, an angle greater than or equal to some predetermined fixed angle 2α. [*Spatial Data Analysis by Example*, Volume 1, 1985, G. Upton and B. Fingleton, Wiley, New York.]

Angler survey: A survey used by sport fishery managers to estimate the total catch, fishing effort and catch rate for a given body of water. For example, the total effort might be estimated in angler-hours and the catch rate in fish per angler-hour. The total catch is then estimated as the product of the estimates of total effort and average catch rate. [*Fisheries Techniques*, 1983, L. A. Nielson and D. C. Johnson, eds., American Fisheries Society, Bethesda, Maryland.]

Angular histogram: A method for displaying *circular data*, which involves wrapping the usual histogram around a circle. Each bar in the histogram is centred at the midpoint of the group interval with the length of the bar proportional to the frequency in the group. Figure 4 shows such a display for arrival times on a 24 hour clock of 254 patients at an intensive care unit, over a period of 12 months. See also **rose diagram**. [*Statistical Analysis of Circular Data*, 1993, N. I. Fisher, Cambridge University Press, Cambridge.]

Angular transformation: Synonym for **arc sine transformation**.

Angular uniform distribution: A probability distribution for a *circular random variable*, θ, given by

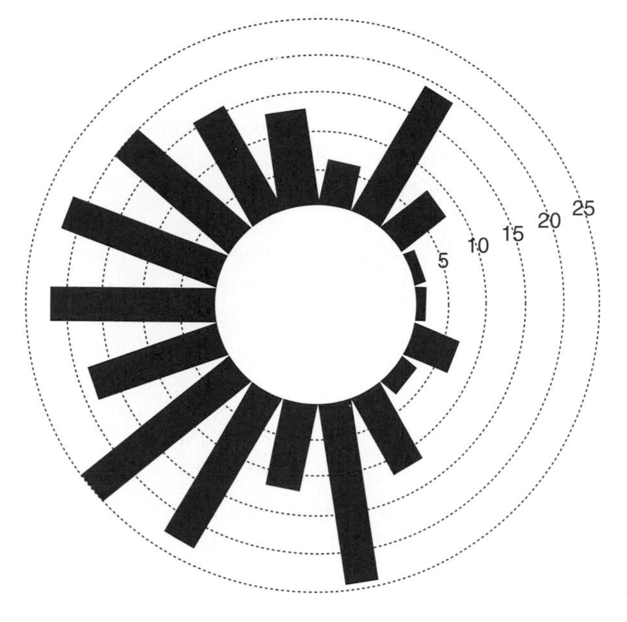

Fig. 4 Angular histogram for arrival times at an intensive care unit. (Reproduced by permission of Cambridge University Press from *Statistical Analysis of Circular Data*, 1993, N. I. Fisher.)

$$f(\theta) = \frac{1}{2\pi}, \quad 0 \leq \theta \leq 2\pi$$

[*Statistical Analysis of Circular Data*, 1993, N. I. Fisher, Cambridge University Press, Cambridge.]

Annealing algorithm: Synonym for **simulated annealing**.

ANOVA: Acronym for **analysis of variance**.

Ansari–Bradley test: A test for the equality of variances of two populations having the same median. The test has rather poor power relative to the *F-test* when the populations are normal. See also **Conover test** and **Klotz test**. [*Annals of Mathematical Statistics*, 1960, **31**, 1174–89.]

Anscombe residual: An alternative to the usual residual for regression models where the random error terms are not normally distributed, for example, logistic regression. The aim is to produce 'residuals' that have near-normal distributions. The form of such a residual depends on the error distribution assumed; in the case of a Poisson distribution, for example, it takes the form $3(y^{2/3}-\hat{y}^{2/3})/2\hat{y}^{1/6}$ where y and $\hat{y}$ are, respectively the observed and fitted values of the response. [*Modelling Binary Data*, 2nd edition, 2003, D. Collett, Chapman and Hall/CRC Press, London.]

Antagonism: See **synergism**.

Antidependence models: A family of structures for the *variance-covariance matrix* of a set of *longitudinal data*, with the model of order r requiring that the sequence of random variables, $Y_1, Y_2, \ldots, Y_T$ is such that for every $t > r$

$$Y_t | Y_{t-1}, Y_{t-2}, \ldots, Y_{t-r}$$

is conditionally independent of $Y_{t-r-1}, \ldots, Y_1$. In other words once account has been taken of the r observations preceding Y_t, the remaining preceding observations carry no additional information about Y_t. The model imposes no constraints on the constancy of variance or covariance with respect to time so that in terms of second-order moments, it is not *stationary*. This is a very useful property in practice since the data from many longitudinal studies often have increasing variance with time. [MV2 Chapter 13.]

Anthropometry: A term used for studies involving measuring the human body. Direct measures such as height and weight or indirect measures such as surface area may be of interest. See also **body mass index**. [*CLEP Human Growth and Development*, 8th edn, 2008, P. Heindel, Research and Education Association.]

Anti-ranks: For a random sample $X_1, \ldots, X_n$, the random variables $D_1, \ldots, D_n$ such that

$$Z_1 = |X_{D_1}| \leq \cdots \leq Z_n = |X_{D_n}|$$

If, for example, $D_1 = 2$ then X_2 is the smallest absolute value and Z_1 has rank 1. [*Robust Nonparametric Statistical Methods*, 1998, T. P. Hettmansperger and J. W. McKean, Arnold, London.]

Antithetic variable: A term that arises in some approaches to *simulation* in which successive simulation runs are undertaken to obtain identically distributed unbiased run estimators that rather than being independent are negatively correlated. The value of this approach is that it results in an unbiased estimator (the average of the estimates from all runs) that has a smaller variance than would the average of identically distributed run estimates that are independent. For example, if r is a random variable between 0 and 1 then so is $s = 1 - r$. Here the two simulation runs would involve $r_1, r_2, \ldots, r_m$ and $1 - r_1, 1 - r_2, \ldots, 1 - r_m$, which are clearly not independent. [*Proceedings of the Cambridge Philosophical Society*, 1956, **52**, 449–75.]

A-optimal design: See **criteria of optimality**.

APC: Abbreviation for **all possible comparisons**.

A posteriori comparisons: Synonym for **post-hoc comparisons**.

Apparent error rate: Synonym for **resubstitution error rate**.

Approximate bootstrap confidence (ABC) method: A method for approximating confidence intervals obtained by using the *bootstrap* approach, that do not use any Monte Carlo replications. [*An Introduction to the Bootstrap*, 1993, B. Efron and R. J. Tibshirani, Chapman and Hall/CRC Press.]

Approximation: A result that is not exact but is sufficiently close for required purposes to be of practical use.

A priori comparisons: Synonym for **planned comparisons**.

Aranda–Ordaz transformations: A family of transformations for a proportion, p, given by

$$y = \ln\left[\frac{(1-p)^{-\alpha} - 1}{\alpha}\right]$$

When $\alpha = 1$, the formula reduces to the *logistic transformation* of p. As $\alpha \rightarrow 0$ the result is the *complementary log-log transformation*. [*Modelling Binary Data*, 2nd edition, 2003, D. Collett, Chapman and Hall/CRC Press, London.]

Arbuthnot, John (1667–1735): Born in Inverbervie, Grampian, Arbuthnot was physician to Queen Anne from 1709 until her death in 1714. A friend of Jonathan Swift who is best known to posterity as the author of satirical pamphlets against the Duke of Marlborough and creator of the prototypical Englishman, John Bull. His statistical claim to fame is based on a short note published in the *Philosophical Transactions of the Royal Society* in 1710, entitled 'An argument for Divine Providence, taken from the constant regularity observ'd in the births of both sexes.' In this note he claimed to demonstrate that divine providence, not chance governed the sex ratio at birth, and presented data on christenings in London for the eighty-two-year period 1629–1710 to support his claim. Part of his reasoning is recognizable as what would now be known as a *sign test.* Arbuthnot was elected a Fellow of the Royal Society in 1704. He died on 27 February 1735 in London.

Archetypal analysis: An approach to the analysis of *multivariate data* which seeks to represent each individual in the data as a mixture of individuals of pure type or archetypes. The archetypes themselves are restricted to being mixtures of individuals in the data set. Explicitly the problem is to find a set of $q \times 1$ vectors $\mathbf{z}_1, \ldots, \mathbf{z}_p$ that characterize the archetypal patterns in the multivariate data, $\mathbf{X}$. For fixed $\mathbf{z}_1, \ldots, \mathbf{z}_p$ where

$$\mathbf{z}_k = \sum_{j=1}^{n} \beta_{kj}\mathbf{x}_j \qquad k = 1, \ldots, p$$

and $\beta_{ki} \geq 0, \ \sum_i \beta_{ki} = 1$, define $\{\alpha_{ik}\}, \ k = 1, \ldots, p$ as the minimizers of

$$\left\| \mathbf{x}_i - \sum_{k=1}^{p} \alpha_{ik}\mathbf{z}_k \right\|^2$$

under the constraints, $\alpha_{ik} \geq 0, \ \sum \alpha_{ik} = 1$. Then define the archetypal patterns or archetypes as the mixtures $\mathbf{z}_1, \ldots, \mathbf{z}_p$ that minimize

$$\sum_i \left\| \mathbf{x}_i - \sum_{k=1}^{p} \alpha_{ik}\mathbf{z}_k \right\|^2$$

For $p > 1$ the archetypes fall on the *convex hull* of the data; they are extreme data values such that all the data can be represented as convex mixtures of the archetypes. However, the archetypes themselves are not wholly mythological because each is constrained to be a mixture of points in the data. [*Technometrics*, 1994, **36**, 338–47.]

ARCH models: Abbreviation for **autoregressive conditional heteroscedastic models**.

Arc sine distribution: A *beta distribution* with $\alpha = \beta = 0.5$.

Arc sine law: An approximation applicable to a simple *random walk* taking values 1 and −1 with probabilities $\frac{1}{2}$ which allows easy computation of the probability of the fraction of time that the accumulated score is either positive or negative. The approximation can be stated thus; for fixed α $(0 < \alpha < 1)$ and $n \rightarrow \infty$ the probability that the fraction k/n of time that the accumulated score is positive is less than α tends to

$$2\pi^{-1}\arcsin(\alpha^{\frac{1}{2}})$$

For example, if an unbiased coin is tossed once per second for a total of 365 days, there is a probability of 0.05 that the more fortunate player will be in the lead for more than 364 days

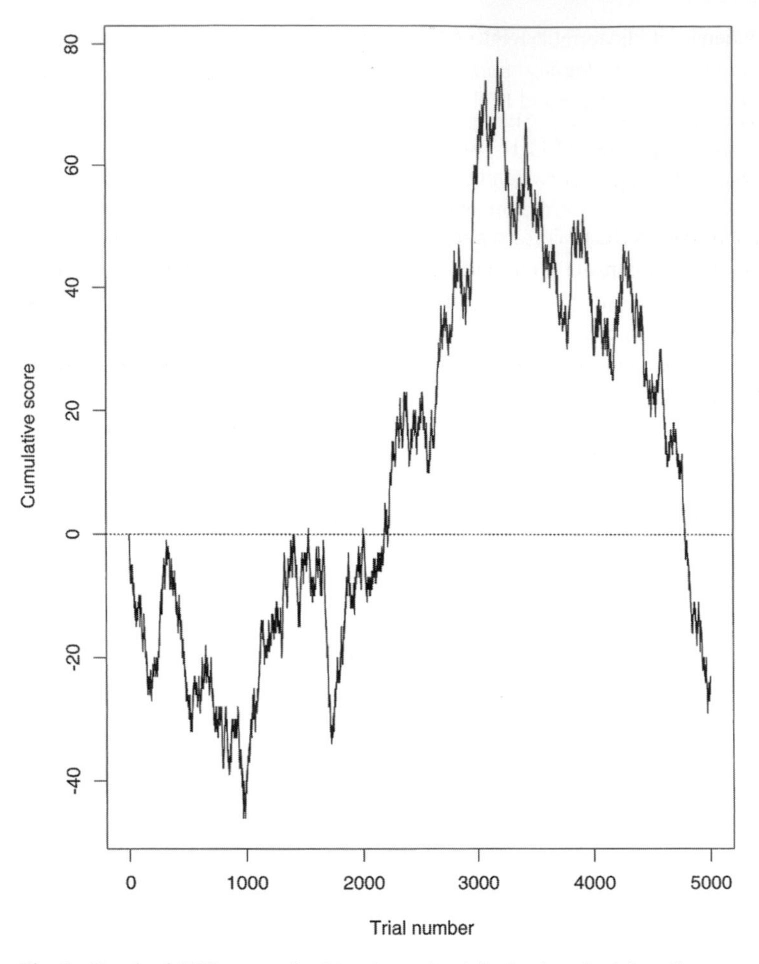

Fig. 5 Result of 5000 tosses of a fair coin scoring 1 for heads and −1 for tails.

and 10 hours. Few people will believe that a perfect coin will produce sequences in which no change of lead occurs for millions of trials in succession and yet this is what such a coin will do rather regularly. Intuitively most people feel that values of k/n close to $\frac{1}{2}$ are most likely. The opposite is in fact true. The possible values close to $\frac{1}{2}$ are least probable and the extreme values $k/n=1$ and $k/n=0$ are most probable. Figure 5 shows the results of an experiment simulating 5000 tosses of a fair coin (Pr(Heads)=Pr(Tails)=$\frac{1}{2}$) in which a head is given a score of 1 and a tail −1. Note the length of the waves between successive crossings of $y=0$, i.e., successive changes of lead. [*An Introduction to Probability Theory and its Applications*, Volume 1, 3rd edition, 1968, W. Feller, Wiley, New York.]

Arc sine transformation: A transformation for a proportion, p, designed to stabilize its variance and produce values more suitable for techniques such as *analysis of variance* and regression analysis. The transformation is given by

$$y = \sin^{-1}\sqrt{p}$$

[*Modelling Binary Data*, 2nd edition, 2003, D. Collett, Chapman and Hall/CRC Press, London.]

ARE: Abbreviation for **asymptotic relative efficiency**.

Area sampling: A method of sampling where a geographical region is subdivided into smaller areas (counties, villages, city blocks, etc.), some of which are selected at random, and the chosen areas are then subsampled or completely surveyed. See also **cluster sampling**. [*Handbook of Area Sampling*, 1959, J. Monroe and A. L. Fisher, Chilton, New York.]

Area under curve (AUC): Often a useful way of summarizing the information from a series of measurements made on an individual over time, for example, those collected in a *longitudinal study* or for a *dose–response curve*. Usually calculated by adding the areas under the curve between each pair of consecutive observations, using, for example, the *trapezium rule*. Often a predictor of biological effects such as toxicity or efficacy. See also **C_{max}**, **response feature analysis** and **T_{max}**. [SMR Chapter 14.]

Arfwedson distribution: The probability distribution of the number of zero values (M_0) among k random variables having a *multinomial distribution* with $p_1 = p_2 = \cdots = p_k$. If the sum of the k random variables is n then the distribution is given by

$$\Pr(M_0 = m) = \binom{k}{m} \sum_{i=0}^{m} (-1)^i \binom{m}{i} \left(\frac{m-i}{k}\right)^n \qquad m = 0, 1, \ldots, k-1$$

[*Combinatorial Methods in Discrete Distribution*, 2005, C. A. Charalambides, Wiley, New York.]

ARIMA: Abbreviation for **autoregressive integrated moving-average model**.

Arithmetic mean: See **mean**.

Arjas plot: A procedure for checking the fit of *Cox's proportional hazards model* by comparing the observed and expected number of events, as a function of time, for various subgroups of covariate values. [*Journal of the American Statistical Association*, 1988, **83**, 204–12.]

ARMA: Abbreviation for **autoregressive moving-average model**.

Armitage–Doll model: A model of carcinogenesis in which the central idea is that the important variable determining the change in risk is not age, but time. The model proposes that cancer of a particular tissue develops according to the following process:

- a normal cell develops into a cancer cell by means of a small number of transitions through a series of intermediate steps;
- initially, the number of normal cells at risk is very large, and for each cell a transition is a rare event;
- the transitions are independent of one another.

[*Statistics in Medicine*, 2006, **9**, 677–679.]

Armitage–Hill test: A test for *carry-over effect* in a *two-by-two crossover design* where the response is a *binary variable*. [*British Journal of Clinical Pharmacology*, 2004, **58**, 718–719.]

Arrelano–Bond estimator: A *generalized method of moments* estimator for *dynamic panel data models* with subject-specific intercepts. More efficient than the *Anderson-Hsiao estimator* since it uses more instruments. [*Analysis of Panel Data*, 2nd edition, 2003, C. Hsiao, Cambridge University Press, Cambridge]

Arrhenius lifetime model: A model commonly to assess the relationship between product lifetime and temperature; an example of an *accelerated life testing* model. [*Accelerated Testing*, 2004, W. Nelson, Wiley, New York.]

Artificial intelligence: A discipline that attempts to understand intelligent behaviour in the broadest sense, by getting computers to reproduce it, and to produce machines that behave intelligently, no matter what their underlying mechanism. (Intelligent behaviour is taken to include reasoning, thinking and learning.) Closely related to *pattern recognition, machine learning* and *artificial neural networks*. [*Introducing Artificial Intelligence*, 2008, H. Brighton, Totem Books.]

Artificial neural network: A mathematical structure modelled on the human neural network and designed to attack many statistical problems, particularly in the areas of *pattern recognition, multivariate analysis*, learning and memory. The essential feature of such a structure is a network of simple processing elements (*artificial neurons*) coupled together (either in the hardware or software), so that they can cooperate. From a set of 'inputs' and an associated set of parameters, the artificial neurons produce an 'output' that provides a possible solution to the problem under investigation. In many neural networks the relationship between the input received by a neuron and its output is determined by a *generalized linear model*. The most common form is the *feed-forward network* which is essentially an extension of the idea of the *perceptron*. In such a network the vertices can be numbered so that all connections go from a vertex to one with a higher number; the vertices are arranged in layers, with connections only to higher layers. This is illustrated in Fig. 6. Each neuron sums its inputs to form a total input x_j and applies a function f_j to x_j to give output y_j. The links have weights w_{ij} which multiply the signals travelling along them by that factor. Many ideas and activities familiar to statisticians can be expressed in a neural-network notation, including regression analysis, *generalized additive models*, and *discriminant analysis*. In any practical problem the statistical equivalent of specifying the architecture of a suitable network is specifying a suitable model, and training the network to perform well with reference to a training set is equivalent to estimating the parameters of the model given a set of data. [*Pattern Recognition and Neural Networks*, 1996, B. D. Ripley, Cambridge University Press, Cambridge.]

Artificial neuron: See **artificial neural network**.

Artificial pairing: See **paired samples**.

Ascertainment bias: A possible form of bias, particularly in *retrospective studies*, that arises from a relationship between the exposure to a risk factor and the probability of detecting an event

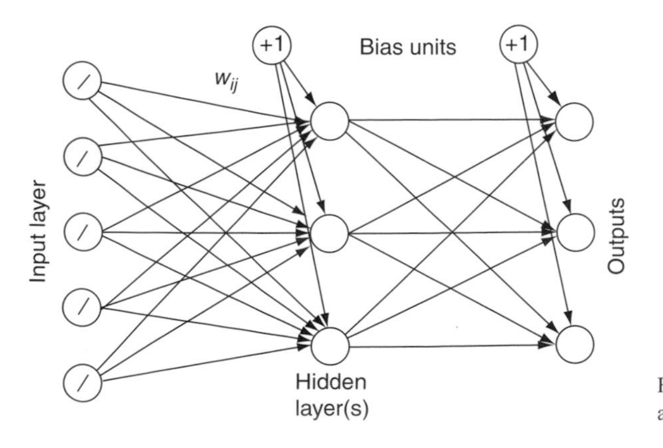

Fig. 6 A diagram illustrating a feed-forward network.

of interest. In a study comparing women with cervical cancer and a control group, for example, an excess of oral contraceptive use among the cases might possibly be due to more frequent screening for the disease among women known to be taking the pill. [SMR Chapter 5.]

ASN: Abbreviation for **average sample number**.

As-randomized analysis: Synonym for **intention-to-treat analysis**.

Assignment method: Synonym for **discriminant analysis**.

Association: A general term used to describe the relationship between two variables. Essentially synonymous with correlation. Most often applied in the context of binary variables forming a *two-by-two contingency table*. See also **phi-coefficient** and **Goodman–Kruskal measures of association**. [SMR Chapter 11.]

Assortative mating: A form of non-random mating where the probability of mating between two individuals is influenced by their *phenotypes* (*phenotypic assortment*), *genotypes* (*genotypic assortment*) or environments (*cultural assortment*). [*Statistics in Human Genetics*, 1998, P. Sham, Arnold, London.]

Assumptions: The conditions under which statistical techniques give valid results. For example, *analysis of variance* generally assumes normality, homogeneity of variance and independence of the observations.

Asymmetrical distribution: A probability distribution or frequency distribution which is not symmetrical about some central value. Examples include the *exponential distribution* and *J-shaped distribution*. [KA1 Chapter 1.]

Asymmetric maximum likelihood (AML): A variant of *maximum likelihood estimation* that is useful for estimating and describing *overdispersion* in a *generalized linear model*. [*IEEE Proceedings Part F – Communications, Radar and Signal Processing*, 1982, **129**, 331–40.]

Asymmetric proximity matrices: *Proximity matrices* in which the off-diagonal elements, in the ith row and jth column and the jth row and ith column, are not necessarily equal. Examples are provided by the number of marriages between men of one nationality and women of another, immigration/emigration statistics and the number of citations of one journal by another. *Multidimensional scaling* methods for such matrices generally rely on their canonical decomposition into the sum of a symmetric matrix and a skew symmetric matrix. [MV1 Chapter 5.]

Asymptotically unbiased estimator: An estimator of a parameter which tends to being unbiased as the sample size, n, increases. For example,

$$s^2 = \frac{1}{n}\sum_{i=1}^{n}(x_i - \bar{x})^2$$

is not an unbiased estimator of the population variance σ^2 since its expected value is

$$\frac{n-1}{n}\sigma^2$$

but it is asymptotically unbiased. [*Normal Approximation and Asymptotic Expansions*, 1976, R. N. Bhattacharya and R. Rao, Wiley, New York.]

Asymptotic distribution: The limiting distribution a sequence of n random variables converges to as $n \to \infty$. For example, the mean of n random variables from a *uniform distribution* has a normal distribution for large n. [KA2 Chapter 25.]

Asymptotic efficiency: A term applied when the estimate of a parameter has a variance achieving the *Cramér–Rao lower bound.* See also **superefficient**. [KA2 Chapter 25.]

Asymptotic methods: Synonym for **large sample methods**.

Asymptotic relative efficiency: The *relative efficiency* of two estimators of a parameter in the limit as the sample size increases. [KA2 Chapter 25.]

Atlas mapping: A biogeographical method used to investigate species-specific distributional status, in which observations are recorded in a grid of cells. Such maps are examples of *geographical information systems.* [*Biometrics*, 1995, **51**, 393–404.]

Atomistic fallacy: A fallacy that arises because the association between two variables at the individual level may differ from the association between the same two variables measured at the group level and using the individual level results for inferences about the aggregated results may be misleading. The term is the opposite of **ecological fallacy**. [*American Journal of Public Health*, 1998, **88**, 216–222.]

Attack rate: A term often used for the *incidence* of a disease or condition in a particular group, or during a limited period of time, or under special circumstances such as an epidemic. A specific example would be one involving outbreaks of food poisoning, where the attack rates would be calculated for those people who have eaten a particular item and for those who have not. [*Epidemiology Principles and Methods*, 1970, B. MacMahon and T. F. Pugh, Little, Brown and Company, Boston.]

Attenuation: A term applied to the correlation between two variables when both are subject to measurement error, to indicate that the value of the correlation between the 'true values' is likely to be underestimated. See also **regression dilution**. [*Biostatistics: A Methodology* for the Health Sciences, 2nd edn, 2004, G. Van Belle, L. D. Fisher, P. J. Heagerty and T. S. Lumley, Wiley, New York.]

Attitude scaling: The process of estimating the positions of individuals on scales purporting to measure attitudes, for example a *liberal–conservative scale*, or a *risk-willingness scale*. Scaling is achieved by developing or selecting a number of stimuli, or items which measure varying levels of the attitude being studied. See also **Likert scale** and **multidimensional scaling**. [*Sociological Methodology*, 1999, **29**, 113–46.]

Attributable response function: A function $N(x, x_0)$ which can be used to summarize the effect of a numerical covariate x on a binary response probability. Assuming that in a finite population there are $m(x)$ individuals with covariate level x who respond with probability $\pi(x)$, then $N(x, x_0)$ is defined as

$$N(x, x_0) = m(x)\{\pi(x) - \pi(x_0)\}$$

The function represents the response attributable to the covariate having value x rather than x_0. When plotted against $x \geq x_0$ this function summarizes the importance of different covariate values in the total response. [*Biometrika*, 1996, **83**, 563–73.]

Attributable risk: A measure of the association between exposure to a particular factor and the risk of a particular outcome, calculated as

$$\frac{\text{incidence rate among exposed} - \text{incidence rate among nonexposed}}{\text{incidence rate among exposed}}$$

Measures the amount of the *incidence* that can be attributed to one particular factor. See also **relative risk** and **prevented fraction**. [*An Introduction to Epidemiology*, 1983, M. Alderson, Macmillan, London.]

Attrition: A term used to describe the loss of subjects over the period of a *longitudinal study*. May occur for a variety of reasons, for example, subjects moving out of the area, subjects dropping out because they feel the treatment is producing adverse side effects, etc. Such a phenomenon may cause problems in the analysis of data from such studies. See also **missing values** and **Diggle–Kenward model for dropouts**.

AUC: Abbreviation for **area under curve**.

Audit in clinical trials: The process of ensuring that data collected in complex *clinical trials* are of high quality. [*Controlled Clinical Trials*, 1995, **16**, 104–36.]

Audit trail: A computer program that keeps a record of changes made to a *database*.

Autocorrelation: The internal correlation of the observations in a *time series*, usually expressed as a function of the time lag between observations. Also used for the correlations between points different distances apart in a set of *spatial data* (*spatial autocorrelation*). The autocorrelation at lag k, $\gamma(k)$, is defined mathematically as

$$\gamma(k) = \frac{E(X_t - \mu)(X_{t+k} - \mu)}{E(X_t - \mu)^2}$$

where $X_t, t = 0, \pm 1, \pm 2, \ldots$ represent the values of the series and μ is the mean of the series. E denotes expected value. The corresponding sample statistic is calculated as

$$\hat{\gamma}(k) = \frac{\sum_{i=1}^{n-k}(x_t - \bar{x})(x_{t+k} - \bar{x})}{\sum_{i=1}^{n}(x_t - \bar{x})^2}$$

where $\bar{x}$ is the mean of the series of observed values, $x_1, x_2, \ldots, x_n$. A plot of the sample values of the autocorrelation against the lag is known as the *autocorrelation function* or *correlogram* and is a basic tool in the analysis of time series particularly for indicating possibly suitable models for the series. An example is shown in Fig. 7. The term in the

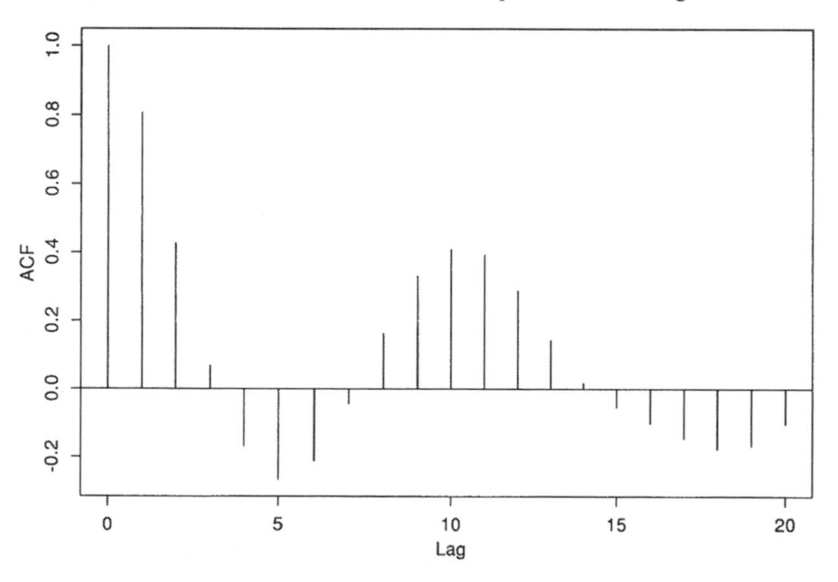

Fig. 7 An example of an autocorrelation function.

numerator of $\gamma(k)$ is the *autocovariance*. A plot of the autocovariance against lag is called the *autocovariance function*. [TMS Chapter 2.]

Autocorrelation function: See **autocorrelation**.

Autocovariance: See **autocorrelation**.

Autocovariance function: See **autocorrelation**.

Auto-logistic regression: A type of regression suitable for modelling georeferenced binary data exhibiting spatial dependence. [*Journal of the Royal Statistical Society, Series B*, 1972, **34**, 75–83.]

Automatic interaction detector (AID): A method that uses a set of categorical explanatory variables to divide data into groups that are relatively homogeneous with respect to the value of some continuous response variable of interest. At each stage, the division of a group into two parts is defined by one of the explanatory variables, a subset of its categories defining one of the parts and the remaining categories the other part. Of the possible splits, the one chosen is that which maximizes the between groups sum of squares of the response variable. The groups eventually formed may often be useful in predicting the value of the response variable for some future observation. See also **classification and regression tree technique** and **chi-squared automated interaction detector**. [*Encyclopedia of Statistical Sciences*, 2006, eds. S. Kotz, C. B. Read, N. Balakrishnan and B. Vidakovic, Wiley, New York.]

Autoregressive conditional heteroscedastic (ARCH) models: Models that attempt to capture the varying (conditional) variance or volatility of *time series*. The models are defined as

$$X_t = \sigma_t \varepsilon_t \text{ where } \sigma_t^2 = \beta_0 + \beta_1 X_{t-1}^2 + \ldots \beta_q X_{t-q}^2$$

Such models are needed because in economic and finance time series it is often found that larger values in the series also have larger variance. The models are important in the analysis of *nonlinear time series*. [*Nonlinear Time Series*, 2003, J.Fan and Q.Yao, Springer, New York.]

Autoregressive integrated moving-average models: See **autoregressive moving-average model**.

Autoregressive model: A model used primarily in the analysis of *time series* in which the observation, x_t, at time t, is postulated to be a linear function of previous values of the series. So, for example, a *first-order autoregressive model* is of the form

$$x_t = \phi x_{t-1} + a_t$$

where a_t is a random disturbance and ϕ is a parameter of the model. The corresponding model of order p is

$$x_t = \phi_1 x_{t-1} + \phi_2 x_{t-2} + \cdots + \phi_p x_{t-p} + a_t$$

which includes the p parameters, $\phi_1, \phi_2, \ldots, \phi_p$. [TMS Chapter 4.]

Autoregressive moving-average model: A model for a *time series* that combines both an *autoregressive model* and a *moving-average model*. The general model of order p,q (usually denoted ARMA(p, q)) is

$$x_t = \phi_1 x_{t-1} + \phi_2 x_{t-2} + \cdots + \phi_p x_{t-p} + a_t - \theta_1 a_{t-1} - \cdots - \theta_q a_{t-q}$$

where $\phi_1, \phi_2, \ldots, \phi_p$ and $\theta_1, \theta_2, \ldots, \theta_q$ are the parameters of the model and $a_t, a_{t-1}, \ldots$ are a *white noise sequence.* In some cases such models are applied to the time series observations after *differencing* to achieve *stationarity,* in which case they are known as *autoregressive integrated moving-average models*. [TMS Chapter 4.]

Auxiliary variable techniques: Techniques for improving the performance of *Gibbs sampling* in the context of *Bayesian inference* for *hierarchical models.* [*Journal of the Royal Statistical Society, Series B*, 1993, **55**, 25–37.]

Available case analysis: An approach to handling *missing values* in a set of multivariate data, in which means, variances, covariances, etc., are calculated from all available subjects with non-missing values for the variable or pair of variables involved. Although this approach makes use of as much of the data as possible it has disadvantages. One is that summary statistics will be based on different numbers of observations. More problematic however is that this method can lead to *variance–covariance matrices* and *correlation matrices* with properties that make them unsuitable for many methods of multivariate analysis such as *principal components analysis* and *factor analysis.* [*Analysis of Incomplete Multivariate Data*, 1997, J. L. Schafer, Chapman and Hall/CRC Press, London.]

Average: Most often used for the arithmetic mean of a sample of observations, but can also be used for other measures of location such as the median.

Average age at death: A flawed statistic summarizing *life expectancy* and other aspects of mortality. For example, a study comparing average age at death for male symphony orchestra conductors and for the entire US male population showed that, on average, the conductors lived about four years longer. The difference is, however, illusory, because as age at entry was birth, those in the US male population who died in infancy and childhood were included in the calculation of the average life span, whereas only men who survived to become conductors could enter the conductor cohort. The apparent difference in longevity disappeared after accounting for infant and perinatal mortality. [*Methodological Errors in Medical Research*, 1990, B. Andersen, Blackwell Scientific, Oxford.]

Average deviation: A little-used measure of the spread of a sample of observations. It is defined as

$$\text{Average deviation} = \frac{\sum_{i=1}^{n} |x_i - \bar{x}|}{n}$$

where $x_1, x_2, \ldots, x_n$ represent the sample values, and $\bar{x}$ their mean.

Average linkage: An *agglomerative hierarchical clustering method* that uses the average distance from members of one cluster to members of another cluster as the measure of inter-group distance. This distance is illustrated in Fig. 8. [MV2 Chapter 10.]

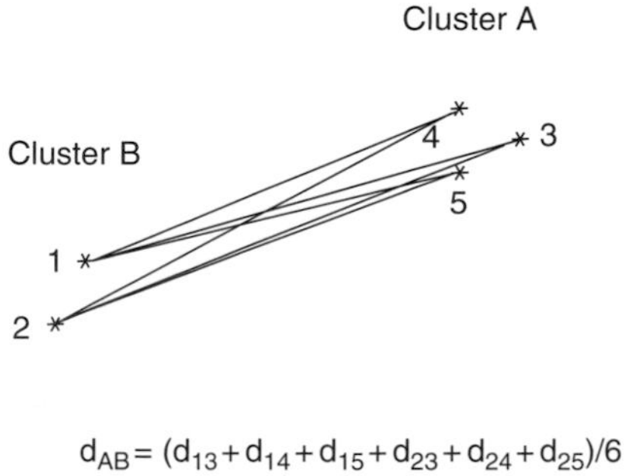

Fig. 8 Average linkage distance for two clusters.

Average man: See **Quetelet, Adolphe (1796–1874)**.

Average sample number (ASN): A quantity used to describe the performance of a *sequential analysis* given by the expected value of the sample size required to reach a decision to accept the null hypothesis or the alternative hypothesis and therefore to discontinue sampling. [KA2 Chapter 24.]

b632 method: A procedure of *error rate estimation* in *discriminant analysis* based on the *bootstrap*, which consists of the following steps:

(1) Randomly sample (with replacement) a *bootstrap sample* from the original data.
(2) Classify the observations omitted from the bootstrap sample using the classification rule calculated from the bootstrap sample.
(3) Repeat (1) and (2) many times and calculate the mean bootstrap *classification matrix*, C_b.
(4) Calculate the resubstitution classification matrix, C_r, based on the original data.
(5) The b632 estimator of the classification matrix is $0.368C_r + 0.632C_b$, from which the required error rate estimate can be obtained. [*Technometrics*, 1996, **38**, 289–99.]

B_k method: A form of *cluster analysis* which produces overlapping clusters. A maximum of $k - 1$ objects may belong to the overlap between any pair of clusters. When $k = 1$ the procedure becomes *single linkage clustering.* [*Classification*, 2nd edition, 1999, A. D. Gordon, Chapman and Hall/CRC Press, London.]

Babbage, Charles (1792–1871): Born near Teignmouth in Devon, Babbage read mathematics at Trinity College, Cambridge, graduating in 1814. His early work was in the theory of functions and modern algebra. Babbage was elected a Fellow of the Royal Society in 1816. Between 1828 and 1839 he held the Lucasian Chair of Mathematics at Trinity College. In the 1820s Babbage developed a 'Difference Engine' to form and print mathematical tables for navigation and spent much time and money developing and perfecting his calculating machines. His ideas were too ambitious to be realized by the mechanical devices available at the time, but can now be seen to contain the essential germ of today's electronic computer. Babbage is rightly seen as the pioneer of modern computers.

Back-calculation: Synonym for **back-projection**.

Backcasting: A technique sometimes used in the analysis of *time series* in which a model is used to make 'forecasts' of the past values of a series and these are then used to improve the estimation of future values of the series. [*Time Series and its Applications: With R Examples*, 2nd edition, 2006, R. H. Shumway and D. S. Stoffer, Springer, New York.]

Backfitting: An algorithm for fitting *generalized additive models.* [*Generalized Additive Models*, 1999, T. Hastie and R. Tibshirani, Chapman and Hall, London.]

Back-projection: A term most often applied to a procedure for reconstructing plausible HIV incidence curves from AIDS incidence data. The method assumes that the probability distribution of the *incubation period* of AIDS has been estimated precisely from separate *cohort studies* and uses this distribution to project the AIDS incidence data backwards to reconstruct an HIV *epidemic curve* that could plausibly have led to the observed AIDS incidence data. [*Statistics in Medicine*, 1994, **13**, 1865–80.]

Before	Stem	After
	12	9
	13	1 6
8 6	14	5 7 7
4	15	1 7
9 7 0	16	5 6 8 8
6 4 3	17	9
7 5	18	0
8	19	
0	20	1

Fig. 9 Back-to-back stem-and-leaf plot of systolic blood pressure of fifteen subjects before and two hours after taking the drug captoril.

Back-to-back stem-and-leaf plots: A method for comparing two distributions by 'hanging' the two sets of leaves in the *stem-and-leaf plots* of the two sets of data, off either side of the same stem. An example appears in Fig. 9.

Backward elimination procedure: See **selection methods in regression**.

Backward-looking study: An alternative term for *retrospective study.*

Backward shift operator: A mathematical operator denoted by B, met in the analysis of *time series.* When applied to such a series the operator moves the observations back one time unit, so that if x_t represents the values of the series then, for example,

$$Bx_t = x_{t-1}$$
$$B(Bx_t) = B(x_{t-1}) = x_{t-2}$$

Baddeley's metric: A way of measuring the 'error' in an image processing technique. The metric is derived using a fundamental theory from stochastic geometry and has the right topological properties for practical use but is robust against 'noise'. In addition to image processing the metric has proved useful in the analysis of *spatial data* and weather prediction. [*Monthly Weather Review*, 2008, **136**, 1747–1757.]

Bagging: A term used for producing replicates of the *training set* in a classification problem and producing an *allocation rule* on each replicate. The basis of *bagging predictions* which involve multiple versions of a predictor that are used to get an aggregated predictor. [*Statistical Pattern Recognition*, 1999, A. Webb, Arnold, London.]

Bagging predictions: See **bagging**.

Bagplot: An approach to detecting *outliers* in *bivariate data.* The plot visualizes location, spread, correlation, *skewness* and the tails of the data without making assumptions about the data being symmetrically distributed. [*American Statistician*, 1999, **53**, 382–7.]

Bahadur, Raghu Raj (1924–1997): Born in Dehli, India, Bahadur graduated from St. Stephen's College, University of Dehli in 1943 with a first class honours degree in mathematics. He continued his studies at Dehli and received his MA in 1945 again in mathematics. In 1946 Bahadur was awarded a scholarship for graduate studies and after spending one year at the Indian Statistical Institute in Calcutta he left India for Chapel Hill, North Carolina, to study mathematical statistics. By 1948 Bahadur had completed his Ph.D on decision theoretic problems for k populations, a problem suggested by *Hotelling.* In 1950 Bahadur joined a new group of statisticians being formed at the University of Chicago. Over the next decade he left Chicago twice to continue his career in Dehli, but eventually returning to stay in Chicago in 1961. Bahadur's research in the 1950s and 1960s played a fundamental part in the development of mathematical statistics over that period with work on the conditions under which *maximum likelihood estimators* will be consistent and the asymptotic theory of *quantiles*. Bahadur was President of the IMS in 1974–1975. He died on June 7th 1997.

Balaam's design: A design for testing differences between two treatments A and B in which patients are randomly allocated to one of four sequences, AA, AB, BA, or BB. See also **crossover design**. [*Statistics in Medicine*, 1988, **7**, 471–82.]

Balanced design: A term usually applied to any experimental design in which the same number of observations is taken for each combination of the experimental factors.

Balanced incomplete block design: A design in which not all treatments are used in all *blocks*. Such designs have the following properties:

- each block contains the same number of units;
- each treatment occurs the same number of times in all blocks;
- each pair of treatment combinations occurs together in a block the same number of times as any other pair of treatments.

In medicine this type of design might be employed to avoid asking subjects to attend for treatment an unrealistic number of times, and thus possibly preventing problems with missing values. For example, in a study with five treatments, it might be thought that subjects could realistically only be asked to make three visits. A possible balanced incomplete design in this case would be the following:

Patient	Visit 1	Visit 2	Visit 3
1	T_4	T_5	T_1
2	T_4	T_2	T_5
3	T_2	T_4	T_1
4	T_5	T_3	T_1
5	T_3	T_4	T_5
6	T_2	T_3	T_1
7	T_3	T_1	T_4
8	T_3	T_5	T_2
9	T_2	T_3	T_4
10	T_5	T_1	T_2

[*Experimental Designs*, 2nd edition, 1992, W. Cochran and G. Cox, Wiley, New York.]

Balanced incomplete repeated measures design (BIRMD): An arrangement of N randomly selected experimental units and k treatments in which every unit receives k_1 treatments $1 \leq k_1 < k$, each treatment is administered to r experimental units and each pair of treatments occurs together the same number of times. See also **balanced incomplete blocks**. [*Journal of the American Statistical Association*, 1996, **91**, 1619–1625.]

Balanced longitudinal data: See **longitudinal data**.

Balanced repeated replication (BRR): A popular method for variance estimation in surveys which works by creating a set of 'balanced' pseudoreplicated datasets from the original dataset. For an estimator, $\hat{\theta}$, of a parameter, θ, the estimated variance is obtained as the average of the squared deviations, $\hat{\theta}^{(r)} - \hat{\theta}$, where $\hat{\theta}^{(r)}$ is the estimate based on the rth replicated data set. See also **jackknife**. [*Biometrika*, 1999, **86**, 403–415.]

Balancing score: Synonymous with **propensity score**.

Ballot theorem: Let $X_1, X_2, \ldots, X_n$ be independent random variables each with a *Bernoulli distribution* with $\Pr(X_i = 1) = \Pr(X_i = -1) = \frac{1}{2}$. Define S_k as the sum of the first k of the observed values of these variables, i.e. $S_k = X_1 + X_2 + \cdots + X_k$ and let a and b be non-negative integers such that $a - b > 0$ and $a + b = n$, then

$$\Pr(S_1 > 0, S_2 > 0, \ldots, S_n > 0 | S_n = a - b) = \frac{a - b}{a + b}$$

If $+1$ is interpreted as a vote for candidate A and -1 as a vote for candidate B, then S_k is the difference in numbers of votes cast for A and B at the time when k votes have been recorded; the probability given is that A is always ahead of B given that A receives a votes in all and B receives b votes. [*An Introduction to Probability Theory and its Applications*, Volume 1, 3rd edition, 1968, W. Feller, Wiley, New York.]

BAN: Abbreviation for **best asymptotically normal estimator**.

Banach's match-box problem: A person carries two boxes of matches, one in their left and one in their right pocket. Initially they contain N matches each. When the person wants a match, a pocket is selected at random, the successive choices thus constituting *Bernoulli trials* with $p = \frac{1}{2}$. On the first occasion that the person finds that a box is empty the other box may contain $0, 1, 2, \ldots, N$ matches. The probability distribution of the number of matches, R, left in the other box is given by:

$$\Pr(R = r) = \binom{2N - r}{N} \frac{1}{2}^{(2N-r)}$$

So, for example, for $N = 50$ the probability of there being not more than 10 matches in the second box is 0.754. [*An Introduction to Probability Theory and its Applications*, Volume 1, 3rd edition, 1968, W. Feller, Wiley, New York.]

Bancroft, Theodore Alfonso (1907–1986): Born in Columbus, Mississippi, Bancroft received a first degree in mathematics from the University of Florida. In 1943 he completed his doctorate in mathematical statistics with a dissertation entitled 'Tests of Significance Considered as an Aid in Statistical Methodology'. In 1950 he became Head of the Department of Statistics of the Iowa Agriculture and Home Economics Experiment Station. His principal area of research was incompletely specified models. Bancroft served as President of the American Statistical Association in 1970. He died on 26 July 1986 in Ames, Iowa.

Band matrix: A matrix that has its nonzero elements arranged uniformly near the diagonal, such that $a_{ij} = 0$ if $(i - j) > ml$ or $(j - i) > mu$ where a_{ij} are the elements of the matrix and ml and mu are the upper and lower band widths respectively and ml+mu+1 is the total band width. An example of such a matrix is the following square matrix, $\mathbf{A}$, with n rows and columns and band widths ml=q−1 and mu=p−1.

$$\mathbf{A} = \begin{bmatrix} a_{11} \; a_{12} \; a_{13} \ldots a_{1p} \; 0 \ldots 0 \\ a_{21} \; a_{22} \; a_{23} \ldots\ldots\ldots 0 \ldots \\ a_{31} \; a_{32} \; a_{33} \ldots\ldots\ldots\ldots 0. \\ \vdots \qquad \ddots \\ \qquad\qquad \ddots \\ a_{q1} \\ 0 \\ \vdots \;\; 0 \\ \vdots \quad\; 0 \\ 0 \ldots\ldots 0 \ldots\ldots\ldots\ldots \; a_{nn} \end{bmatrix}$$

Such matrices often arise in numerical analysis. Diagonal and upper and lower triangular matrices are special cases of band matrices. [*Integral Equations and Operator Theory*, 1987, **10**, 83–95.]

Bandwidth: See **kernel estimation**.

Barahona-Poon test: A test used to differentiate nonlinear deterministic dynamics or *chaos* from random noise in a *time series*. The method compares linear and nonlinear models fitted to the data and rejects the null hypothesis that the time series is stochastic with linear dynamics if at least one nonlinear model is significantly more predictive than all the linear models considered. [*Nature*, 1996, **381**, 215–217.]

Bar chart: A form of graphical representation for displaying data classified into a number of (usually unordered) categories. Equal-width rectangular bars are constructed over each category with height equal to the observed frequency of the category as shown in Fig. 10. See also **histogram** and **component bar chart**.

Barnard, George Alfred (1915–2002): Born in Walthamstow in the east end of London, Barnard gained a scholarship to St. John's College, Cambridge, where he graduated in mathematics in 1936. For the next three years he studied mathematical logic at Princeton, New Jersey, and then in 1940 joined the engineering firm, Plessey. After three years acting as a mathematical consultant for engineers, Barnard joined the Ministry of Supply and it was here that his interest in statistics developed. In 1945 he went to Imperial College London, and then in 1966 he moved to a chair in the newly created University of Essex, where he stayed until his retirement in 1975. Barnard made major and important contributions to several fundamental areas of inference, including likelihood and 2 × 2 tables. He was made President of the Royal Statistical Society in 1971–2 and also received the Society's Guy medal in gold. He died in Brightlingsea, Essex, on 30 July 2002.

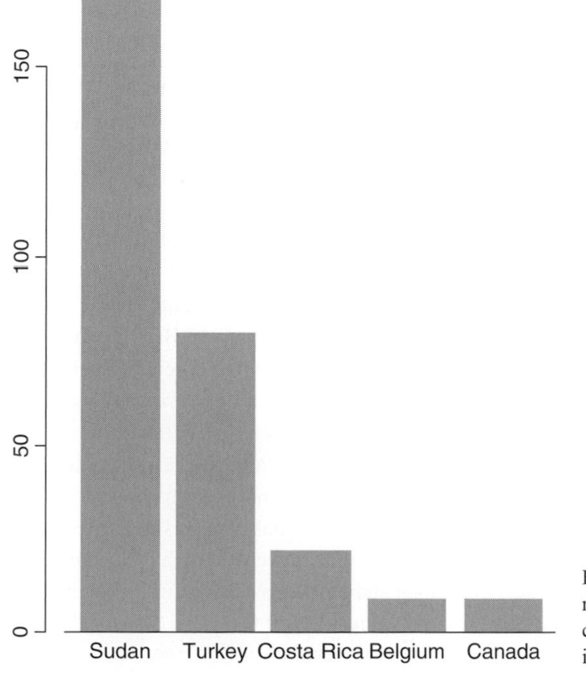

Fig. 10 Bar chart of mortality rates per 1000 live births for children under five years of age in five different countries.

Barrett and Marshall model for conception: A biologically plausible model for the probability of conception in a particular menstrual cycle, which assumes that batches of sperm introduced on different days behave independently. The model is

$$P(\text{conception in cycle } k|\{X_{ik}\}) = 1 - \prod_i (1 - p_i)^{X_{ik}}$$

where the X_{ik} are 0,1 variables corresponding to whether there was intercourse or not on a particular day relative to the estimated day of ovulation (day 0). The parameter p_i is interpreted as the probability that conception would occur following intercourse on day i only. See also **EU model**. [*Biometrics*, 2001, **57**, 1067–73.]

Bartholomew's likelihood function: The joint probability of obtaining the observed known-complete *survival times* as well as the so-far survived measurements of individuals who are still alive at the date of completion of the study or other endpoint of the period of observation. [*Journal of the American Statistical Association*, 1957, **52**, 350–5.]

Bartlett decomposition: The expression for a random matrix **A** that has a *Wishart distribution* as a product of a triangular matrix and its transpose. Letting each of $\mathbf{x}_1, \ldots, \mathbf{x}_n$ be independently distributed as q-dimensional multivariate normal variables with zero means and an identity *variance-covariance matrix* we can write

$$\mathbf{A} = \sum_{i=1}^{n} \mathbf{x}_i \mathbf{x}_i' = \mathbf{TT}'$$

where $t_{ii} = 0, i < j$, and $t_{ii} > 0, i = 1, \ldots, q$. Here $t_{11}, t_{21}, \ldots, t_{qq}$ are independently distributed, the t_{ij} have standard normal distributions for $i > j$ and the t_{ii}^2 have *chi-squared distributions* with $n - i + 1$ degrees of freedom. The decomposition is useful for simulating data from a Wishart distribution. [*An Introduction to Multivariate Statistical Analysis*, 3rd edition, 2003, T. W. Anderson, Wiley, New York.]

Bartlett, Maurice Stevenson (1910–2002): Born in Chiswick, London, Bartlett won a scholarship to Latymer Upper School, where his interest in probability was awakened by a chapter on the topic in Hall and Knight's *Algebra*. In 1929 he went to Queen's College, Cambridge to read mathematics, and in his final undergraduate year in 1932 published his first paper (jointly with *John Wishart*), on second-order moments in a normal system. On leaving Cambridge in 1933 Bartlett became Assistant Lecturer in the new Statistics Department at University College London, where his colleagues included *Egon Pearson*, *Fisher* and *Neyman*. In 1934 he joined Imperial Chemical Industries (ICI) as a statistician. During four very creative years Bartlett published some two-dozen papers on topics as varied as the theory of inbreeding and the effect of non-normality on the t-distribution. From ICI he moved to a lectureship at the University of Cambridge, and then during World War II he was placed in the Ministry of Supply. After the war he returned to Cambridge and began his studies of *time series* and diffusion processes. In 1947 Bartlett was given the Chair of Mathematical Statistics at the University of Manchester where he spent the next 13 years, publishing two important books, *An Introduction to Stochastic Processes* (in 1955) and *Stochastic Population Models in Ecology and Epidemiology* (in 1960) as well as a stream of papers on *stochastic processes*, etc. It was in 1960 that Bartlett returned to University College taking the Chair in Statistics, his work now taking in stochastic path integrals, spatial patterns and *multivariate analysis*. His final post was at Oxford where he held the Chair of Biomathematics from 1967 until his retirement eight years later. Bartlett received many honours and awards in his long and productive career, including being made a Fellow of the Royal Society in 1961 and being President of the Royal Statistical Society for 1966–7. He died on 8 January 2002, in Exmouth, Devon.

Bartlett method: See **factor scores**.

Bartlett's adjustment factor: A correction term for the *likelihood ratio* that makes the *chi-squared distribution* a more accurate approximation to its probability distribution. [*Multivariate Analysis*, 1979, K. V. Mardia, J. T. Kent, and J. M. Bibby, Academic Press, London.]

Bartlett's identity: A matrix identity useful in several areas of *multivariate analysis* and given by

$$(\mathbf{A}+c\mathbf{b}\mathbf{b}')^{-1}=\mathbf{A}^{-1}-\frac{c}{1+c\mathbf{b}'\mathbf{A}^{-1}\mathbf{b}}\mathbf{A}^{-1}\mathbf{b}\mathbf{b}'\mathbf{A}^{-1}$$

where $\mathbf{A}$ is $q\times q$ and nonsingular, $\mathbf{b}$ is a $q\times 1$ vector and c is a scalar.

Bartlett's test for eigenvalues: A large-sample test for the null hypothesis that the last $(q-k)$ eigenvalues, $\lambda_{k+1},\ldots,\lambda_q$, of a *variance–covariance matrix* are zero. The test statistic is

$$X^2=-\nu\sum_{j=k+1}^{q}\ln(\lambda_j)+\nu(q-k)\ln\left[\frac{\sum_{j=k+1}^{q}\lambda_j}{q-k}\right]$$

Under the null hypothesis, X^2 has a *chi-squared distribution* with $(1/2)(q-k-1)(q-k+2)$ degrees of freedom, where ν is the degrees of freedom associated with the covariance matrix. Used mainly in *principal components analysis.* [MV1 Chapter 4.]

Bartlett's test for variances: A test for the equality of the variances of a number (k) of populations. The *test statistic* is given by

$$B=\left[\nu\ln s^2+\sum_{i=1}^{k}\nu_i\ln s_i^2\right]/C$$

where s_i^2 is an estimate of the variance of population i based on ν_i degrees of freedom, and ν and s^2 are given by

$$\nu=\sum_{i=1}^{k}\nu_i$$

$$s^2=\frac{\sum_{i=1}^{k}\nu_i s_i^2}{\nu}$$

and

$$C=1+\frac{1}{3(k-1)}\left[\sum_{i=1}^{k}\frac{1}{\nu_i}-\frac{1}{\nu}\right]$$

Under the hypothesis that the populations all have the same variance, B has a *chi-squared distribution* with $k-1$ degrees of freedom. Sometimes used prior to applying *analysis of variance* techniques to assess the assumption of homogeneity of variance. Of limited practical value because of its known sensitivity to non-normality, so that a significant result might be due to departures from normality rather than to different variances. See also **Box's test** and **Hartley's test**. [SMR Chapter 9.]

Baseline balance: A term used to describe, in some sense, the equality of the observed *baseline characteristics* among the groups in, say, a *clinical trial.* Conventional practice dictates that before proceeding to assess the treatment effects from the clinical outcomes, the groups must be shown to be comparable in terms of these baseline measurements and observations, usually by carrying out appropriate significant tests. Such tests are frequently criticized by statisticians who usually prefer important prognostic variables to be identified prior to the trial and then used in an *analysis of covariance*. [SMR Chapter 15.]

Baseline characteristics: Observations and measurements collected on subjects or patients at the time of entry into a study before undergoing any treatment. The term can be applied to demographic characteristics of the subject such as sex, measurements taken prior to treatment of the same variable which is to be used as a measure of outcome, and measurements taken prior to treatment on variables thought likely to be correlated with the response variable. At first sight, these three types of baseline seem to be quite different, but from the point-of-view of many powerful approaches to analysing data, for example, *analysis of covariance*, there is no essential distinction between them. [SMR Chapter 1.]

Baseline hazard function: See **Cox's proportional hazards model**.

BASIC: Acronym for Beginners All-Purpose Symbolic Instruction Code, a programming language once widely used for writing microcomputer programs.

Basic reproduction number: A term used in the theory of infectious diseases for the number of secondary cases which one case would produce in a completely susceptible population. The number depends on the duration of the *infectious period*, the probability of infecting a susceptible individual during one contact, and the number of new susceptible individuals contacted per unit time, with the consequence that it may vary considerably for different infectious diseases and also for the same disease in different populations. [*Applied Statistics*, 2001, **50**, 251–92.]

Basu's theorem: This theorem states that if T is a complete *sufficient statistic* for a family of probability measures and V is an *ancillary statistic*, then T and V are independent. The theorem shows the connection between sufficiency, ancillarity and independence, and has led to a deeper understanding of the interrelationship between the three concepts. [*Journal of Statistical Planning and Inference*, 2007, **137**, 945–952.]

Bathtub curve: The shape taken by the *hazard function* for the event of death in human beings; it is relatively high during the first year of life, decreases fairly soon to a minimum and begins to climb again sometime around 45–50. See Fig. 11. [*Technometrics*, 1980, **22**, 195–9.]

Battery reduction: A general term for reducing the number of variables of interest in a study for the purposes of analysis and perhaps later data collection. For example, an overly long questionnaire may not yield accurate answers to all questions, and its size may need to be reduced. Techniques such as *factor analysis* and *principal component analysis* are generally

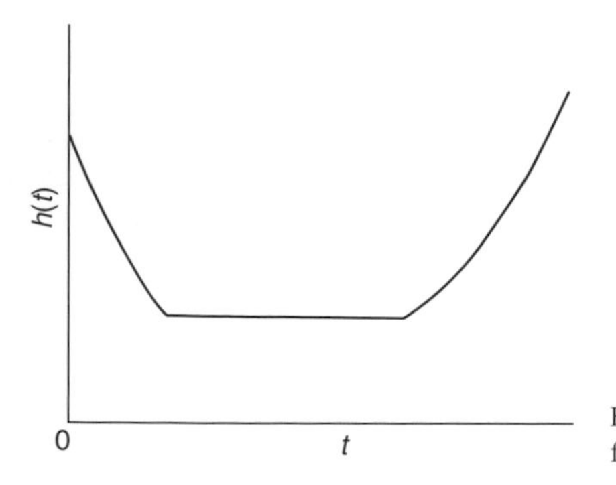

Fig. 11 Bathtub curve shown by hazard function for death in human beings.

used to achieve the required reduction. [*Encyclopedia of Statistics in Behavioral Science*, 2005, eds. B. S. Everitt and D. C. Howell, Wiley, Chichester.]

Bayes factor: A summary of the evidence for a model M_1 against another model M_0 provided by a set of data D, which can be used in model selection. Given by the ratio of posterior to prior odds,

$$B_{10} = \frac{\Pr(D|M_1)}{\Pr(D|M_0)}$$

Twice the logarithm of B_{10} is on the same scale as the *deviance* and the *likelihood ratio* test statistic. The following scale is often useful for interpreting values of B_{10};

$2 \ln B_{10}$	Evidence for M_1
< 0	Negative (supports M_0)
0–2.2	Not worth more than a bare mention
2.2–6	Positive
6–10	Strong
> 10	Very strong

Very sensitive to the assumed *prior distribution* of the models. [*Markov Chain Monte Carlo in Practice*, 1996, eds. W. R. Gilks, S. Richardson and D. Spiegelhalter, Chapman and Hall/CRC Press, London.]

Bayesian confidence interval: An interval of a *posterior distribution* which is such that the density at any point inside the interval is greater than the density at any point outside and that the area under the curve for that interval is equal to a prespecified probability level. For any probability level there is generally only one such interval, which is also often known as the *highest posterior density region*. Unlike the usual *confidence interval* associated with *frequentist inference*, here the intervals specify the range within which parameters lie with a certain probability. [KA2 Chapter 20.]

Bayesian inference: An approach to inference based largely on *Bayes' Theorem* and consisting of the following principal steps:

(1) Obtain the *likelihood*, $f(\mathbf{x}|\boldsymbol{\theta})$ describing the process giving rise to the data $\mathbf{x}$ in terms of the unknown parameters $\boldsymbol{\theta}$.
(2) Obtain the *prior distribution*, $f(\boldsymbol{\theta})$ expressing what is known about $\boldsymbol{\theta}$, prior to observing the data.
(3) Apply Bayes' theorem to derive the *posterior distribution* $f(\boldsymbol{\theta}|\mathbf{x})$ expressing what is known about $\boldsymbol{\theta}$ after observing the data.
(4) Derive appropriate inference statements from the posterior distribution. These may include specific inferences such as point estimates, interval estimates or probabilities of hypotheses. If interest centres on particular components of $\boldsymbol{\theta}$ their posterior distribution is formed by integrating out the other parameters.

This form of inference differs from the classical form of *frequentist inference* in several respects, particularly the use of the prior distribution which is absent from classical inference. It represents the investigator's knowledge about the parameters before seeing the data. Classical statistics uses only the likelihood. Consequently to a Bayesian every problem is unique and is characterized by the investigator's beliefs about the parameters expressed in the prior distribution for the specific investigation. [KA2 Chapter 31.]

Bayesian information criterion (BIC): An index used as an aid to choose between competing statistical models that is similar to *Akaike's information criterion* (AIC) but penalizes models of higher dimensionality more than the AIC. Can be derived as an approximation to the *Bayes factor* under a particular prior distribution. Essentially the BIC is equivalent to *Schwarz's criterion.* [*Journal of the American Statistical Association*, 1996, **64**, 103–37.]

Bayesian model averaging: Incorporating model uncertainty into conclusions about parameters and prediction by averaging over competing models. Requires specification of the *prior probability* that each model is true (given that one of them is true). [*Statistical Science*, 1999, **14**, 382–417.]

Bayesian network: Essentially an *expert system* in which uncertainty is dealt with using conditional probabilities and *Bayes' Theorem.* Formally such a network consists of the following:

- A set of variables and a set of directed edges between variables.
- Each variable has a finite set of mutually exclusive states.
- The variables together with the directed edges form a *conditional independence graph.*
- To each variable A with parents $B_1, \ldots, B_n$ there is attached a conditional probability table $\Pr(A|B_1, B_2, \ldots, B_n)$.

An example is shown in Fig. 12 in which cancer is independent of age group given exposure to toxics and smoking.

Bayesian persuasion probabilities: A term for particular *posterior probabilities* used to judge whether a new therapy is superior to the standard, derived from the priors of two hypothetical experts, one who believes that the new therapy is highly effective and another who believes that it is no more effective than other treatments. The *persuade the pessimist probability* is the posterior probability that the new therapy is an improvement on the standard assuming the sceptical experts prior, and the *persuade the optimist probability*; is the posterior probability that the new therapy gives no advantage over the standard assuming the enthusiasts prior. Large values of these probabilities should persuade the *a priori* most opinionated parties to change their views. [*Statistics in Medicine*, 1997, **16**, 1792–802.]

Bayes, Reverend Thomas (1702–1761): Born in London, Bayes was one of the first six Nonconformist ministers to be publicly ordained in England. Reputed to be a skilful mathematician although oddly there is no sign of him having published any scientific work before his election to the Royal Society in 1741. Principally remembered for his

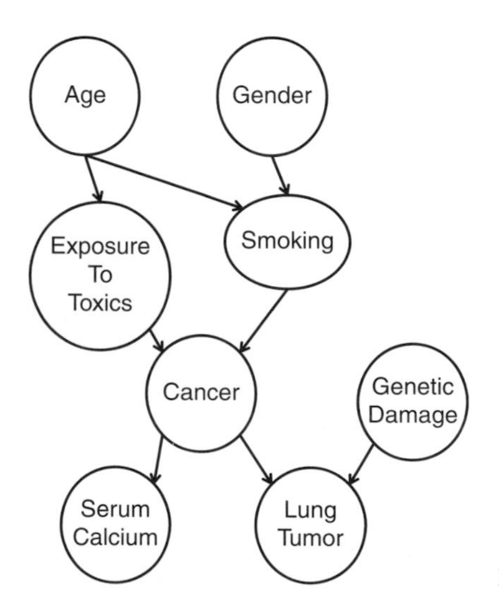

Fig. 12 A Bayesian network for cancer.

posthumously published *Essay Towards Solving a Problem in the Doctrine of Chance* which appeared in 1763 and, heavily disguised, contained a version of what is today known as *Bayes' Theorem.* Bayes died on 7 April 1761 in Tunbridge Wells, England.

Bayes' Theorem: A procedure for revising and updating the probability of some event in the light of new evidence. The theorem originates in an essay by the Reverend Thomas Bayes. In its simplest form the theorem may be written in terms of *conditional probabilities* as,

$$\Pr(B_j|A) = \frac{\Pr(A|B_j)\Pr(B_j)}{\sum_{j=1}^{k} \Pr(A|B_j)\Pr(B_j)}$$

where $\Pr(A|B_j)$ denotes the conditional probability of event A conditional on event B_j and $B_1, B_2, \ldots, B_k$ are *mutually exclusive* and *exhaustive events.* The theorem gives the probabilities of the B_j when A is known to have occurred. The quantity $\Pr(B_j)$ is termed the *prior probability* and $\Pr(B_j|A)$ the *posterior probability.* $\Pr(A|B_j)$ is equivalent to the (normalized) *likelihood,* so that the theorem may be restated as

$$\text{posterior} \propto (\text{prior}) \times (\text{likelihood})$$

See also **Bayesian inference**. [KA1 Chapter 8.]

BC$_a$: Abbreviation for **bias-corrected percentile interval**.

Beattie's procedure: A continous process-monitoring procedure that does not require 100% inspection. Based on a *cusum* procedure, a constant sampling rate is used to chart the number of percent of nonconforming product against a target reference value. [*Applied Statistics*, 1962, **11**, 137–47.]

Behrens–Fisher problem: The problem of testing for the equality of the means of two normal distributions that do not have the same variance. Various *test statistics* have been proposed, although none are completely satisfactory. The one that is most commonly used however is given by

$$t = \frac{\bar{x}_1 - \bar{x}_2}{\sqrt{\frac{s_1^2}{n_1} + \frac{s_2^2}{n_2}}}$$

where $\bar{x}_1, \bar{x}_2, s_1^2, s_2^2, n_1$ and n_2 are the means, variances and sizes of samples of observations from each population. Under the hypothesis that the population means are equal, t has a *Student's t-distribution* with ν degrees of freedom where

$$\nu = \left[\frac{c^2}{n_1 - 1} + \frac{(1-c)^2}{n_2 - 1}\right]^{-1}$$

and

$$c = \frac{s_1^2/n_1}{s_1^2/n_1 + s_2^2/n_2}$$

See also **Student's *t*-test** and **Welch's statistic**. [MV1 Chapter 6.]

Belief functions: A non-Bayesian way of using mathematical probability to quantify subjective judgements. The theory of belief functions is based on two ideas: the idea of obtaining degrees of belief for one question of interest from subjective probabilities for related questions and then combining such degrees of belief when they are based on independent items of evidence. See also **imprecise probabilities**. [*A Mathematical Theory of Evidence*, 1976, G. Shafer, Princeton University Press.]

Believe the negative rule: See **believe the positive rule**.

Believe the positive rule: A rule for combining two *diagnostic tests*, A and B, in which 'disease present' is the diagnosis given if either A or B or both are positive. An alternative *believe the negative rule* assigns a patient to the disease class only if both A and B are positive. These rules do not necessarily have better predictive values than a single test; whether they do depends on the association between test outcomes.

Bell–Doksum test: A test obtained from the *Kruskal–Wallis test* by replacing ranks with values of ordered unit normal random variables. The test can be used to test the hypothesis that random samples of observations from *k* groups have the same population distribution function. [*Australian and New Zealand Journal of Statistics*, 2008, **25**, 105–108.]

Bellman–Harris process: An age-dependent *branching process* in which individuals have independent, identically distributed lifespans, and at death split into independent identically distributed numbers of offspring. [*Branching Processes with Biological Applications*, 1975, P. Jagers, Wiley, Chichester.]

Bell-shaped distribution: A probability distribution having the overall shape of a vertical cross-section of a bell. The normal distribution is the most well known example, but *Student's t-distribution* is also this shape.

Benchmarking: A procedure for adjusting a less reliable series of observations to make it consistent with more reliable measurements or *benchmarks*. For example, data on hospital bed occupation collected monthly will not necessarily agree with figures collected annually and the monthly figures (which are likely to be less reliable) may be adjusted at some point to agree with the more reliable annual figures. See also **Denton method**. [*International Statistical Review*, 1994, **62**, 365–77.]

Benchmarks: See **benchmarking**.

Benchmark dose: A term used in risk assessment studies where human, animal or ecological data are used to set safe low dose levels of a toxic agent, for the dose that is associated with a particular level of risk. [*Applied Statistics*, 2005, **54**, 245–58.]

Benford distribution: The distribution of the initial digit in data sets that describe populations, lengths and durations and that have no built-in maximum or minimum values. Counter-intuitively the distribution is uneven with the probabilities of the first digit being as follows:

Initial Digit	Probability
1	0.301
2	0.176
3	0.125
4	0.097
5	0.079
6	0.067
7	0.058
8	0.052
9	0.046

The non-randomness of the digit frequencies has been used in auditing and fraud detection. [*Proceedings of the American Philosophical Society*, 1938, **78**, 551–572.]

Benini, Rodolpho (1862–1956): Born in Cremona, Italy, Rodolpho was appointed to the Chair of History of Economics at Bari at the early age of 27. From 1928 to his death in 1956 he was Professor of Statistics at Rome University. One of the founders of *demography* as a separate science.

Benjamin, Bernard (1910–2002): Benjamin was educated at Colfe's Grammar School in Lewisham, South London, and later at Sir John Cass College, London, where he studied physics. He began his working life as an actuarial assistant to the London County Council pension fund and in 1941 qualified as a Fellow of the Institute of Actuaries. After World War II he became Chief Statistician at the General Register Office and was later appointed as Director of Statistics at the Ministry of Health. In the 1970s Benjamin joined City University, London as the Foundation Professor of Actuarial Science. He published many papers and books in his career primarily in the areas of *actuarial statistics* and *demography*. Benjamin was made President of the Royal Statistical Society from 1970–2 and received the society's highest honour, the Guy medal in gold, in 1986.

Benjamini and Hochberg step-up methods: Methods used in *bioinformatics* to control the *false discovery rate* when calculating *p*-values from *g* tests under *g* individual null hypotheses, one for each gene. [*Journal of the Royal Statistical Society, Series B*, 1995, **57**, 289–300.]

Bennett's bivariate sign test: A test of the null hypothesis that a *bivariate distribution* has specified marginal medians. [*Metrika*, 1967, **12**, 22–28.]

Bentler–Bonnett index: A goodness of fit measure used in *structural equation modelling*. [*Modelling Covariances and Latent Variables using EQS*, 1993, G. Dunn, B. Everitt and A. Pickles, CRC/Chapman and Hall, London.]

Berkson, Joseph (1899–1982): Born in New York City, Berkson studied physics at Columbia University, before receiving a Ph.D. in medicine from Johns Hopkins University in 1927, and a D.Sc. in statistics from the same university in 1928. In 1933 he became Head of Biometry and Medical Statistics at the Mayo Clinic, a post he held until his retirement in 1964. His research interests covered all aspects of medical statistics and from 1928 to 1980 he published 118 scientific papers. Involved in a number of controversies particularly that involving the role of cigarette smoking in lung cancer, Berkson enjoyed a long and colourful career. He died on 12 September 1982 in Rochester, Minnesota.

Berkson's fallacy: The existence of artifactual correlations between diseases or between a disease and a risk factor arising from the interplay of differential admission rates from an underlying population to a select study group, such as a series of hospital admissions. In any study that purports to establish an association and where it appears likely that differential rates of admission apply, then at least some portion of the observed association should be suspect as attributable to this phenomenon. See also **Simpson's paradox** and **spurious correlation**. [SMR Chapter 5.]

Berman's diagnostic: An ad hoc method for model checking when point process models are fitted to spatial point pattern data. [*Journal of the Royal Statistical Society, Series B*, 2005, **67**, 617–666.]

Berman-Turner device: A method for maximizing the *likelihoods* of inhomogeneous spatial *Poisson processes*, using standard software for *generalized linear models*. [*Journal of the Royal Statistical Society, Series C*, 1992, **41**, 31–38.]

Bernoulli distribution: The probability distribution of a binary random variable, X, where $\Pr(X = 1) = p$ and $\Pr(X = 0) = 1 - p$. Named after Jacques Bernoulli (1654–1705). All moments of X about zero take the value p and the variance of X is $p(1 - p)$. The distribution is negatively skewed when $p > 0.5$ and is positively skewed when $p < 0.5$. [STD Chapter 4.]

Bernoulli, Jacques (1654–1705) (also known as James or Jakob): Born in Basel, Switzerland, the brother of Jean Bernoulli and the uncle of Daniel Bernoulli the most important members of a family of Swiss mathematicians and physicists. Destined by his father to become a theologian, Bernoulli studied mathematics in secret and became Professor of Mathematics at Basel in 1687. His book *Ars Conjectandi* published in 1713, eight years after his death, was an important contribution to probability theory. Responsible for the early theory of permutations and combinations and for the famous *Bernoulli numbers*.

Bernoulli–Laplace model: A probabilistic model for the flow of two liquids between two containers. The model begins by imagining r black balls and r white balls distributed between two boxes. At each stage one ball is chosen at random from each box and the two are interchanged. The state of the system can be specified by the number of white balls in the first box, which can take values from zero to r. The probabilities of the number of white balls in the first box decreasing by one ($p_{i,i-1}$), increasing by one ($p_{i,i+1}$) or staying the same ($p_{i,i}$) at the stage when the box contains i white balls, can be shown to be

$$p_{i,i-1} = \left(\frac{i}{r}\right)^2$$

$$p_{i,i+1} = \left(\frac{r-i}{r}\right)^2$$

$$p_{i,i} = 2\frac{i(r-i)}{r^2}$$

[*Probability and Measure*, 1995, P. Billingsley, Wiley, New York.]

Bernoulli numbers: The numerical coefficients of $t^r/r!$ in the expansion of $t/(e^t - 1)$ as a power series in t. Explicitly, $B_0 = 1, B_1 = -\frac{1}{2}, B_2 = \frac{1}{6}, B_3 = 0, B_4 = -\frac{1}{30}$, etc.

Bernoulli trials: A set of n independent binary variables in which the jth observation is either a 'success' or a 'failure', with the probability of success, p, being the same for all trials

Bernstein polynomial prior: A nonparametric *prior probability distribution* for probability densities on the unit interval. [*Scandinavian Journal of Statistics*, 1999, **26**, 373–93.]

Bernstein-von Mises theorem: Asserts that if there are many independent and identically distributed observations governed by a smooth and finite-dimensional statistical model, the Bayes estimate and the *maximum likelihood estimate* will be close. Furthermore, the *posterior distribution* of the parameter vector around the posterior mean will be close to the distribution of the maximum likelihood estimate around the true parameter values. The theorem provides a detailed asymptotic relation between frequentist and *Bayesian statistics* and validates the interchange of Bayesian credible sets and frequentist confidence regions. It does not hold for infinite dimensional models. [*Annals of Statistics,* 1999, **27**, 1119–1140.]

Berry–Esseen theorem: A theorem relating to how rapidly the distribution of the mean approaches normality. See also **central limit theorem**. [KA1 Chapter 8.]

Berstein, Sergei Natanovich (1880–1968): Born in Odessa, Ukraine, Berstein received his mathematical education in Paris. In 1924 he obtained a doctorate from the Sorbonne and in 1913 a second one from Kharkov. From 1908 until 1933 Berstein taught at Kharkov

University and then moved to the Mathematical Institute of the USSR Academy of Sciences. Berstein made major contributions to theoretical probability. He died on October 26th in Moscow.

Bessel function distributions: A family of probability distributions obtained as the distributions of linear functions of independent random variables, X_1 and X_2, each having a *chi-squared distribution* with common degrees of freedom ν. For example the distribution of $Y = a_1X_1 + a_2X_2$ with $a_1 > 0$ and $a_2 > 0$ is $f(y)$ given by

$$f(y) = \frac{(c^2-1)^{m+1/2}}{\pi^{\frac{1}{2}}2^m b^{m+1}\Gamma(m+\frac{1}{2})} y^m e^{-cy/b} I_m(y/b), \ y>0$$

where $b = 4a_1a_2(a_1 - a_2)^{-1}$, $c = (a_1 + a_2)/(a_1 - a_2)$, $m = 2\nu + 1$ and

$$I_m(x) = (\tfrac{1}{2}x)^m \sum_{j=0}^{\infty} \frac{(x/2)^{2j}}{j!\Gamma(m+j+1)}$$

[*Handbook of Mathematical Functions*, 1964, M. Abramowitz and I. A. Stegun, National Bureau of Standards, Washington.]

Best asymptotically normal estimator (BAN): A *CAN estimator* with minimal asymptotic *variance-covariance matrix*. The notion of minimal in this context is based on the following order relationship among symmetric matrices: $\mathbf{A} \leq \mathbf{B}$ if $\mathbf{B} - \mathbf{A}$ is nonnegative definite. [*Annals of Statistics*, 1983, **11**, 183–96.]

Best linear unbiased estimator (BLUE): A linear estimator of a parameter that has lower variance than any similar estimator of the parameter. See also **best linear unbiased predictor**.

Best linear unbiased predictor (BLUP): A prediction of a *random effect* in a *linear mixed model* that has lower *mean squared error* than any other linear unbiased prediction of the effect. See also **best linear unbiased estimator**. [*Statistical Science*, 1991, **6**, 15–32.]

Beta-binomial distribution: The probability distribution, $f(x)$, found by averaging the parameter, p, of a *binomial distribution* over a *beta distribution*, and given by

$$f(x) = \binom{n}{x} \frac{B(\alpha + x, n + \beta - x)}{B(\alpha, \beta)}$$

where B is the *beta function*. The mean and variance of the distribution are as follows:

$$\text{mean} = n\alpha/(\alpha + \beta)$$
$$\text{variance} = n\alpha\beta(n + \alpha + \beta)/[(\alpha + \beta)^2(1 + \alpha + \beta)]$$

Also known as the *Polyá distribution*. For integer α and β, corresponds to the *negative hypergeometric distribution*. For $\alpha = \beta = 1$ corresponds to the *discrete rectangular distribution*. [STD Chapter 5.]

Beta coefficient: A regression coefficient that is standardized in an attempt to allow for a direct comparison between explanatory variables as to their relative explanatory power for the response variable. Calculated from the raw regression coefficients by multiplying them by the standard deviation of the corresponding explanatory variable. [*Regression Analysis*, Volume 2, 1993, eds. M. S. Lewis-Beck, Sage Publications, London.]

Beta distribution: The probability distribution

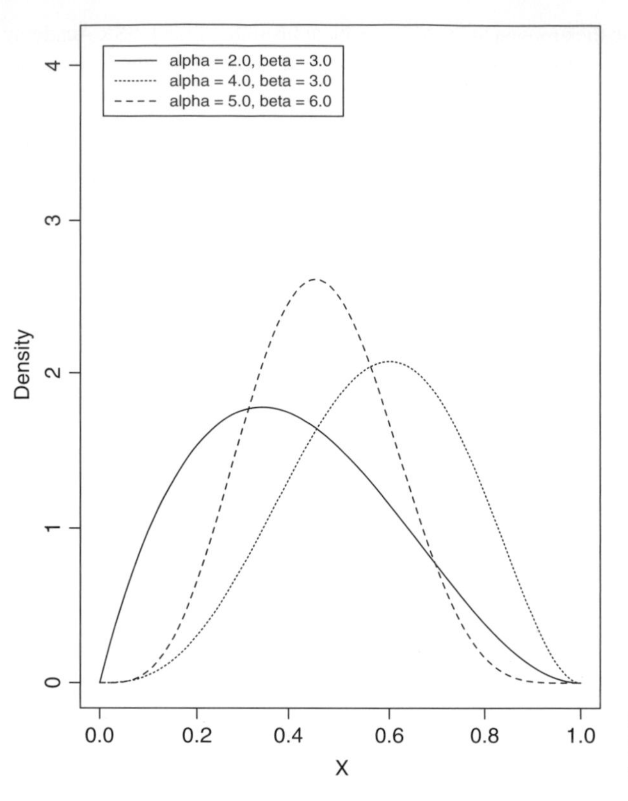

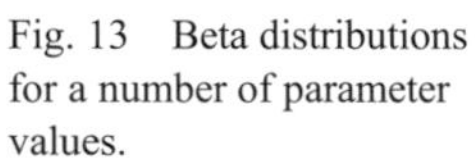
Fig. 13 Beta distributions for a number of parameter values.

$$f(x) = \frac{x^{\alpha-1}(1-x)^{\beta-1}}{B(\alpha,\beta)} \qquad 0 \le x \le 1, \alpha > 0, \beta > 0$$

where B is the *beta function*. Examples of the distribution are shown in Figure 13. The mean, variance, *skewness* and *kurtosis* of the distribution are as follows:

$$\text{mean} = \frac{\alpha}{(\alpha+\beta)}$$

$$\text{variance} = \frac{\alpha\beta}{[(\alpha+\beta)^2(\alpha+\beta+1)]}$$

$$\text{skewness} = \frac{2(\beta-\alpha)(\alpha+\beta+1)^{\frac{1}{2}}}{[(\alpha+\beta+2)(\alpha\beta)^{\frac{1}{2}}]}$$

$$\text{kurtosis} = 3 + \frac{6(\alpha-\beta)^2(\alpha+\beta+1)}{\alpha\beta(\alpha+\beta+2)(\alpha+\beta+3)} - \frac{6}{\alpha+\beta+3}$$

A *U-shaped distribution* if $(\alpha-1)(\beta-1) < 0$. [STD Chapter 5.]

Beta(β)-error: Synonym for **type II error**.

Beta function: The function $B(\alpha,\beta), \alpha > 0, \beta > 0$ given by

$$B(\alpha,\beta) = \int_0^1 u^{\alpha-1}(1-u)^{\beta-1}\mathrm{d}u$$

Can be expressed in terms of the *gamma function* Γ as

$$B(\alpha, \beta) = \frac{\Gamma(\alpha)\Gamma(\beta)}{\Gamma(\alpha + \beta)}$$

The integral

$$\int_0^T u^{\alpha-1}(1-u)^{\beta-1} du$$

is known as the *incomplete beta function*.

Beta-geometric distribution: A probability distribution arising from assuming that the parameter, p, of a *geometric distribution* has itself a *beta distribution*. The distribution has been used to model the number of menstrual cycles required to achieve pregnancy. [*Statistics in Medicine*, 1993, **12**, 867–80.]

Between groups matrix of sums of squares and cross-products: See **multivariate analysis of variance**.

Between groups mean square: See **mean squares**.

Between groups sum of squares: See **analysis of variance**.

BGW: Abbreviation for **Bienaymé–Galton–Watson process**.

Bhattacharyya bound: A better (i.e. greater) lower bound for the variance of an estimator than the more well-known *Cramér-Rao lower bound*. [*Sankhya*, 1946, **8**, 1–10.]

Bhattacharyya's distance: A measure of the distance between two populations with probability distributions $f(x)$ and $g(x)$ respectively. Given by

$$\cos^{-1} \int_{-\infty}^{\infty} [f(x)g(x)]^{\frac{1}{2}} dx$$

See also **Hellinger distance**. [MV2 Chapter 14.]

Bias: In general terms, deviation of results or inferences from the truth, or processes leading to such deviation. More specifically, the extent to which the statistical method used in a study does not estimate the quantity thought to be estimated, or does not test the hypothesis to be tested. In estimation usually measured by the difference between the expected value of an estimator and the true value of the parameter. An estimator for which $E(\hat{\theta}) = \theta$ is said to be *unbiased*. See also **ascertainment bias, recall bias, selection bias** and **biased estimator**.

Bias-corrected percentile interval (BC_a): An improved method of calculating *confidence intervals* when using the *bootstrap*. [*An Introduction to the Bootstrap*, 1993, B. Efron and R. J. Tibshirani, Chapman and Hall, London.]

Bias/variance tradeoff: A term that summarizes the fact that if you want less *bias* in the estimate of a model parameter, it usually costs you more variance.

Biased coin method: A method of random allocation sometimes used in a *clinical trial* in an attempt to avoid major inequalities in treatment numbers. At each point in the trial, the treatment with the fewest number of patients thus far is assigned a probability greater than a half of being allocated the next patient. If the two treatments have equal numbers of patients then simple randomization is used for the next patient. [*Statistics in Medicine*, 1986, **5**, 211–30.]

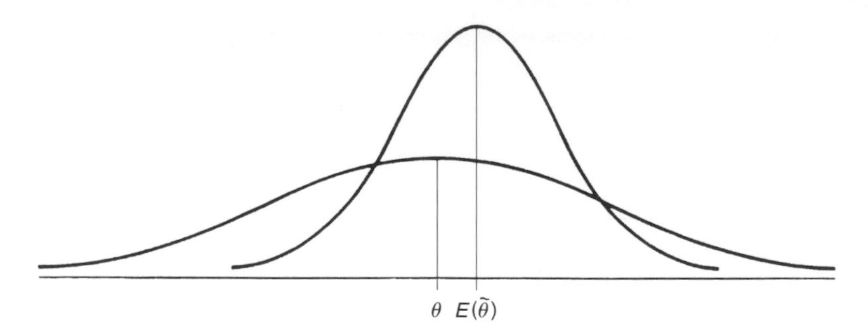

Fig. 14 Biased estimator diagram showing advantages of such an estimator.

Biased estimator: Formally an estimator $\hat{\theta}$ of a parameter, θ, such that

$$E(\hat{\theta}) \neq \theta$$

The motivation behind using such estimators rather than those that are unbiased, rests in the potential for obtaining values that are closer, on average, to the parameter being estimated than would be obtained from the latter. This is so because it is possible for the variance of such an estimator to be sufficiently smaller than the variance of one that is unbiased to more than compensate for the bias introduced. This possible advantage is illustrated in Fig. 14. The normal curve centred at θ in the diagram represents the probability distribution of an unbiased estimator of θ with its expectation equal to θ. The spread of this curve reflects the variance of the estimator. The normal curve centred at $E(\tilde{\theta})$ represents the probability distribution of a biased estimator with the bias being the difference between θ and $E(\tilde{\theta})$. The smaller spread of this distribution reflects its smaller variance. See also **ridge regression**. [ARA Chapter 5.]

BIC: Abbreviation for Bayesian information criterion.

Bienaymé–Galton–Watson process (BGW): A simple *branching process* defined by

$$Z_k = \sum_{i=1}^{Z_{k-1}} X_{ki}$$

where for each k the X_{ki} are independent, identically distributed random variables with the same distribution, $p_r = \Pr(X_{ij} = r)$, called the *offspring distribution*. [*Branching Processes with Biological Applications*, 1975, P. Jagers, Wiley, New York.]

Bienaymé, Jules (1796–1878): Born in Paris, Bienaymé studied at the École Polytechnique and became lecturer in mathematics at Saint Cyr, the French equivalent of West Point, in 1818. Later he joined the civil service as a general inspector of finance and began his studies of actuarial science, statistics and probability. Made contributions to the theory of *runs* and discovered a number of important inequalities.

Big Mac index: An index that attempts to measure different aspects of the economy by comparing the cost of hamburgers between countries. [*Measurement Theory and Practice*, 2004, D. J. Hand, Arnold, London.]

Bilinear loss function: See **decision theory**.

Bimodal distribution: A probability distribution, or a frequency distribution, with two modes. Figure 15 shows an example of each. [KA1 Chapter 1.]

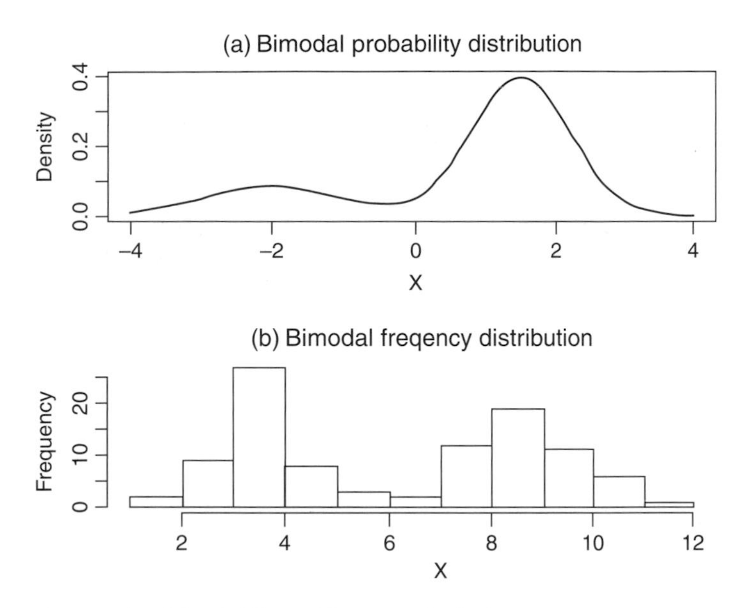

Fig. 15 Bimodal probability and frequency distributions.

Binary variable: Observations which occur in one of two possible states, these often being labelled 0 and 1. Such data is frequently encountered in medical investigations; commonly occurring examples include 'dead/alive', 'improved/not improved' and 'depressed/not depressed.' Data involving this type of variable often require specialized techniques for their analysis such as *logistic regression.* See also **Bernoulli distribution**. [SMR Chapter 2.]

Binning: A term most frequently used in imaging studies to denote that several pixels are grouped together to reduce the impact of read noise on the signal to noise ratio.

Binomial coefficient: The number of ways that k items can be selected from n items irrespective of their order. Usually denoted $C(n,k)$ or $\binom{n}{k}$ and given by

$$C(n,k) = \frac{n!}{k!(n-k)!}$$

See also **Pascal's triangle** and **multinomial coefficient**.

Binomial confidence interval: See **Clopper–Pearson interval**.

Binomial confidence interval when no events are observed: See **rule of three**.

Binomial distribution: The distribution of the number of 'successes', X, in a series of n-independent *Bernoulli trials* where the probability of success at each trial is p and the probability of failure is $q = 1 - p$. Specifically the distribution is given by

$$\Pr(X = x) = \frac{n!}{x!(n-x)!} p^x q^{n-x}, \quad x = 0, 1, 2, \ldots, n$$

The mean, variance, *skewness* and *kurtosis* of the distribution are as follows:

$$\text{mean} = np$$
$$\text{variance} = npq$$
$$\text{skewness} = (q - p)/(npq)^{\frac{1}{2}}$$
$$\text{kurtosis} = 3 - \frac{6}{n} + \frac{1}{npq}$$

See also **beta binomial distribution** and **positive binomial distribution**. [STD Chapter 6.]

Binomial failure rate model: A model used to describe the underlying failure process in a system whose components can be subjected to external events ('shocks') that can cause the failure of two or more system components. For a system of m indentical components, the model is based on $m + 1$ independent *Poisson processes*, $\{N_S(t),\ t \geq 0\}$, $\{N_i(t),\ t \geq 0\}$, $i = 1, \ldots, m$. The quantities N_i determine individual component failures and have equal intensity rate λ. The variables N_S are system shocks with rate μ that represent the external events affecting all components within the system. Given the occurrence of a system shock, each component fails with probability p, independently of the other components, hence the number of failures due to the system shock has a *binomial distribution* with parameters (m, p). [*Nuclear Systems Reliability Engineering and Risk Assessment*, 1977, J. B. Fussell and G. R. Burdick, eds., Society for Industrial and Applied Mathematics, Philadelphia.]

Binomial index of dispersion: An index used to test whether k samples come from populations having *binomial distributions* with the same parameter p. Specifically the index is calculated as

$$\sum_{i=1}^{k} n_i(P_i - P)^2/[P(1 - P)]$$

where $n_1, n_2, \ldots, n_k$ are the respective sample sizes, $P_1, P_2, \ldots, P_k$ are the separate sample estimates of the probability of a 'success', and P is the mean proportion of successes taken over all samples. If the samples are all from a binomial distribution with parameter p, the index has a *chi-squared distribution* with $k - 1$ degrees of freedom. See also **index of dispersion**. [*Biometrika*, 1966, **53**, 167–82.]

Binomial test: A procedure for making inferences about the probability of a success p, in a series of independent repeated *Bernoulli trials*. The test statistic is the observed number of successes, B. For small sample sizes critical values for B can be obtained from appropriate tables. A large-sample approximation is also available based on the asymptotic normality of B. The test statistic in this case is

$$z = \frac{B - np_0}{[np_0(1 - p_0)]^{\frac{1}{2}}}$$

where p_0 is the hypothesised value of p. When the hypothesis is true z has an asymptotic standard normal distribution. See also **Clopper-Pearson interval**. [NSM Chapter 2.]

Bioassay: An abbreviation of biological assay, which in its classical form involves an experiment conducted on biological material to determine the relative potency of test and standard preparations. Recently however, the term has been used in a more general sense, to denote any experiment in which responses to various doses of externally applied agents are observed in animals or some other biological system. See also **calibration** and **probit analysis**.

Bioavailability: The rate and extent to which an active ingredient of a drug product is absorbed and becomes available at the site of drug action. [*Design and Analysis of Bioavailability Studies*, 1992, S. C. Chow and J. P. Liu, Marcel Dekker, New York.]

Bioavailability study: A study designed to assess the pharmacological characteristics of a new drug product during Phase I clinical development or to compare the *bioavailability* of different formulations of the same drug (for example, tablets versus capsules). [*Design and*

Availability of Bioequivalence Studies, 1992, S. C. Chow and J. P. Liu, Marcel Dekker, New York.]

Bioequivalence: The degree to which clinically important outcomes of treatment by a new preparation resemble those of a previously established preparation. [*Design and Analysis of Bioavailability and Bioequivalence Studies*, 1992, S. C. Chow and J. P. Liu, Marcel Dekker, New York.]

Bioequivalence trials: *Clinical trials* carried out to compare two or more formulations of a drug containing the same active ingredient, in order to determine whether the different formulations give rise to comparable blood levels. [*Statistics in Medicine*, 1995, **14**, 853–62.]

Bioinformatics: Essentially the application of *information theory* to biology to deal with the deluge of information resulting from advances in molecular biology. The main tasks in this field are the creation and maintenance of databases of biological information, particularly nucleic acid sequences, finding the genes in the DNA sequences of various organisms and clustering protein sequences into families of related sequences. [*Understanding Bioinformatics*, 2007, M. Zvelebil and J. Baum, Garland Science.]

(Bio)sequence analysis: The statistical analysis of nucleotide sequence data of deoxyribonucleic acid (DNA) or amino acid sequence data of polypeptides, with the purpose of predicting structure or function, or identifying similar sequences that may represent homology (i.e. share comon evolutionary origin). [*Biological Sequence Analysis*, 1998, R. Durbin, S. R. Eddy, A. Kragh, G. Mitchison, Cambridge University Press, Cambridge.]

Biostatistics: A narrow definition is 'the branch of science which applies statistical methods to biological problems'. More commonly, however, it also includes statistics applied to medicine and health sciences. [*Fundamentals of Biostatistics* 6th edn, 2005, B. Rosner, Duxbury Press.]

Biplots: The multivariate analogue of scatterplots, which approximate the multivariate distribution of a sample in a few dimensions, typically two, and superimpose on this display representations of the variables on which the samples are measured. The relationships between the individual sample points can be easily seen and they can also be related to the values of the measurements, thus making such plots useful for providing a graphical description of the data, for detecting patterns, and for displaying results found by more formal methods of analysis. Figure 16 shows the two-dimensional biplot of a set of *multivariate data* with 34 observations and 10 variables. See also **principal components analysis** and **correspondence analysis**. [MV1 Chapter 4.]

Bipolar factor: A factor resulting from the application of *factor analysis* which has a mixture of positive and negative loadings. Such factors can be difficult to interpret and attempts are often made to simplify them by the process of *factor rotation*.

BIRMD: Abbreviation for **balanced incomplete repeated measures design**.

Birnbaum, Allan (1923–1976): Born in San Francisco, Birnbaum studied mathematics at the University of California at Berkeley, obtaining a Ph.D. degree in mathematical statistics in 1954. His initial research was on *classification techniques* and *discriminant analysis* but later he turned to more theoretical topics such as the *likelihood* principle where he made a number of extremely important contributions. Birnbaum died on 1 July 1976 in London.

Birnbaum–Saunders distribution: The probability distribution of

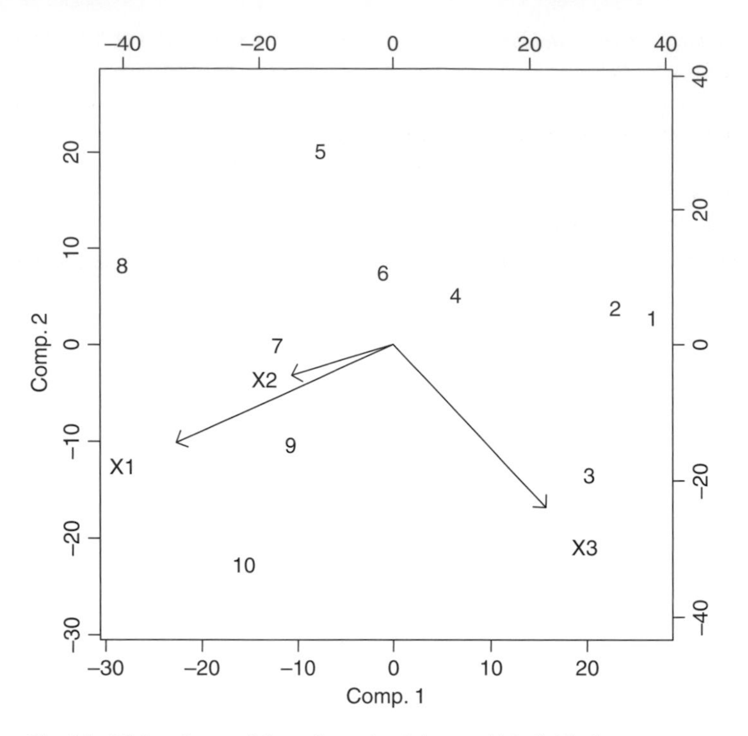

Fig. 16 Biplot of a set of three-dimensional data on 10 individuals.

$$T = \beta \left[\tfrac{1}{2} Z\alpha + \sqrt{\left(\tfrac{1}{2} Z\alpha\right)^2 + 1} \right]^2$$

where α and β are positive parameters and Z is a random variable with a standard normal distribution. Used in models representing time to failure of material subjected to a cyclically repeated stress. The distribution is given explicitly by

$$f(x) = \frac{e^{1/\alpha^2}}{2\alpha\sqrt{2\pi\beta}} x^{-\frac{3}{2}}(x+\beta)\exp\left\{-\frac{1}{2\alpha^2}\left(\frac{x}{\beta}+\frac{\beta}{x}\right)\right\} \quad x>0,\ \alpha>0,\ \beta>0.$$

[*Technometrics*, 1991, **33**, 51–60.]

Birth-cohort study: A *prospective study* of people born in a defined period. They vary from large studies that aim to be nationally representative to those that are area based with populations as small as a 1000 subjects. An example is the Avon longitudinal study of births occurring in one English county used to collect data on risk exposure during pregnancy. [*Paediataric and Perinatal Epidemiology*, 1989, **3**, 460–9.]

Birthday problems: The birthdays of r people form a sample of size r from the population of all days in the year. As a first approximation these may be considered as a random selection of birthdays from the year consisting of 365 days. Then the probability, p, that all r birthdays are different is

$$p = \left(1 - \frac{1}{365}\right)\left(1 - \frac{2}{365}\right)\cdots\left(1 - \frac{r-1}{365}\right)$$

For example, when $r = 23$, $p < 0.5$, so that for 23 people the probability that no two people have the same birthday is less than a half, so the probability that at least two of the twenty-three people share a birthday is greater than a half; most people when asked to make a guess of how many people are needed to achieve a greater than 50% chance of at least two of them sharing a birthday, put the figure higher than 23. Matching triplets of birthdays are much harder to observe and it can be shown that in a group of 23 people the probability of at least one triplet of matching birthdays is only 0.013; now there has to be a group of 88 people before the probability becomes larger than 0.5. [*Chance Rules*, 2nd edn. 2008, B. S. Everitt, Springer, New York.]

Birth–death process: A *continuous time Markov chain* with infinite *state space* in which changes of state are always from n to $n+1$ or n to $n-1$. The state of such a system can be viewed as the size of a population that can increase by one (a 'birth') or decrease by one (a 'death'). [*Stochastic Modelling of Scientific Data*, 1995, P. Guttorp, Chapman and Hall/CRC Press, London.]

Birth–death ratio: The ratio of number of births to number of deaths within a given time in a population. In 2007 the birth–death ratio in Afghanistan was 2.3 and in the United Kingdom it was a little over one.

Birth defect registries: Organized *databases* containing information on individuals born with specified congenital disorders. Important in providing information that may help to prevent birth defects. [*International Journal of Epidemiology*, 1981, **10**, 247–52.]

Birth rate: The number of births occurring in a region in a given time period, divided by the size of the population of the region at the middle of the time period, usually expressed per 1000 population. For example, the birth rates for a number of countries in 2005 were as follows:

Country	Birth rate/1000 total Population
Pakistan	27.1
South Africa	22.3
Switzerland	9.2
Japan	8.3

Biserial correlation: A measure of the strength of the relationship between two variables, one continuous (y) and the other recorded as a binary variable (x), but having underlying continuity and normality. Estimated from the sample values as

$$r_b = \frac{\bar{y}_1 - \bar{y}_0}{s_y}\frac{pq}{u}$$

where $\bar{y}_1$ is the sample mean of the y variable for those individuals for whom $x = 1$, $\bar{y}_0$ is the sample mean of the y variable for those individuals for whom $x = 0$, s_y is the standard deviation of the y values, p is the proportion of individuals with $x = 1$ and $q = 1 - p$ is the proportion of individuals with $x = 0$. Finally u is the ordinate (height) of the standard normal distribution at the point of division between the p and q proportions of the curve. See also **point-biserial correlation**. [KA2 Chapter 26.]

Bisquare regression estimation: *Robust estimation* of the parameters $\beta_1, \beta_2, \ldots, \beta_q$ in the *multiple regression* model

$$y_i = \beta_1 x_{i1} + \cdots + \beta_q x_{iq} + \epsilon_i$$

The estimators are given explicitly by the solutions of the q equations

$$\sum_i x_{ij}\psi[r_i/(kS)] = 0, \; j = 1, \ldots, q$$

where S is the median absolute residual, k is a given positive parameter and

$$\psi(u) = u(1 - u^2)^2, \qquad |u| < 1$$
$$= 0, \quad |u| \geq 1$$

The estimation equations are solved by an iterative process. See also **M-estimators**. [*Least Absolute Deviations, Theory, Applications and Algorithms*, 1983, P. Bloomfield and W. L. Steiger, Birkhauser, Boston.]

Bit: A unit of information, consisting of one binary digit.

Bivar criterion: A weighted sum of squared *bias* and variance in which the relative weights reflect the importance attached to these two quantities. Sometimes used for selecting explanatory variables in regression analysis. [*Technometrics*, 1982, **24**, 181–9.]

Bivariate beta distribution: A *bivariate distribution*, $f(x,y)$, given by

$$f(x,y) = \frac{\Gamma(\alpha+\beta+\gamma)}{\Gamma(\alpha)\Gamma(\beta)\Gamma(\gamma)} x^{\alpha-1} y^{\beta-1} (1-x-y)^{\gamma-1}$$

where $x \geq 0, y \geq 0, x+y \leq 1$ and $\alpha, \beta, \gamma > 0$. *Perspective plots* of a number of such distributions are shown in Fig. 17. [*Communications in Statistics – Theory and Methods*, 1996, **25**, 1207–22.]

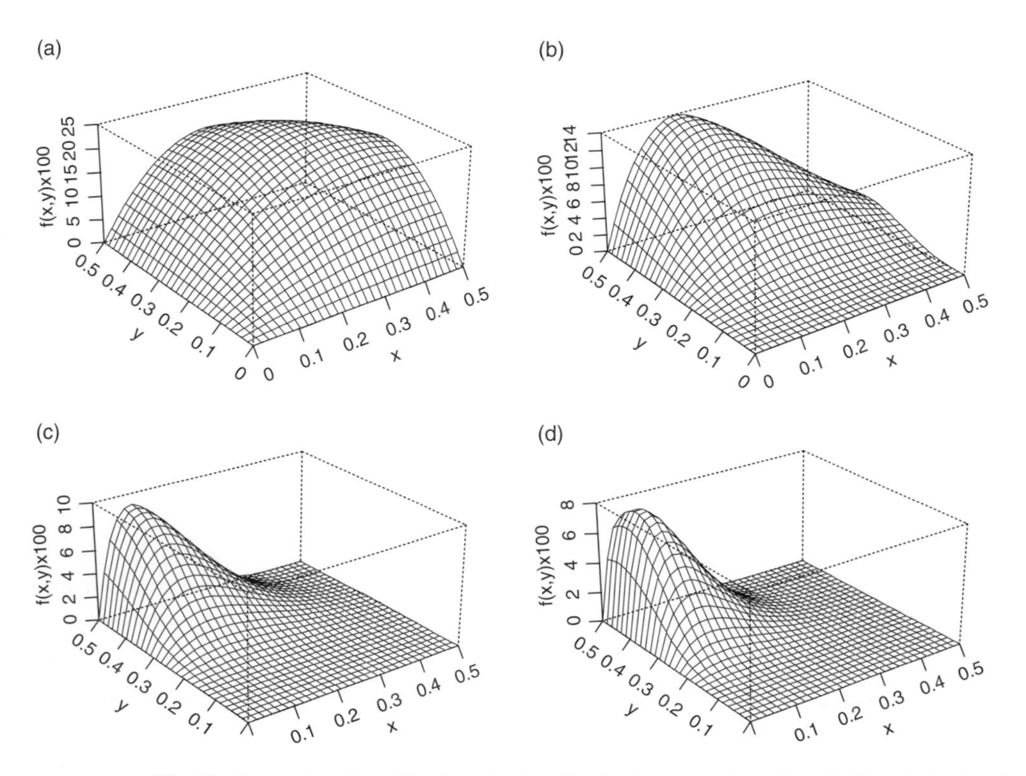

Fig. 17 Perspective plots of four bivariate beta distributions: (a) $\alpha=2, \beta=2, \gamma=2$; (b) $\alpha=2, \beta=4, \gamma=3$; (c) $\alpha=2, \beta=7, \gamma=6$; (d) $\alpha=2, \beta=8, \gamma=9$.

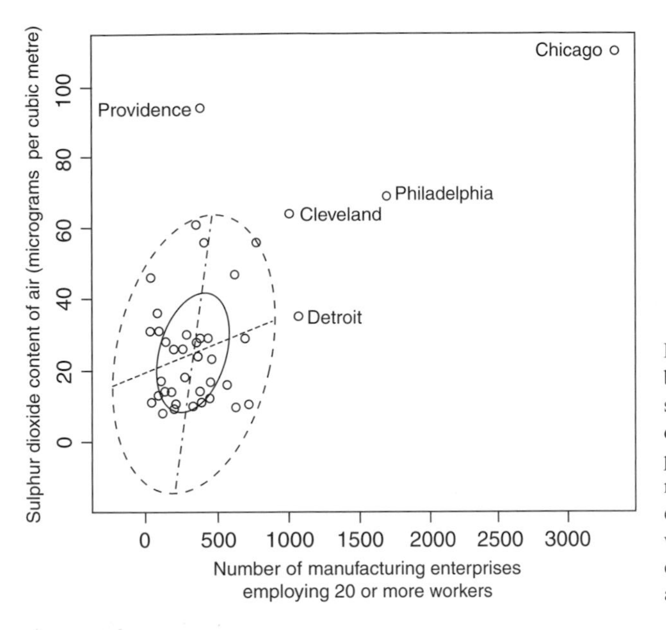

Fig. 18 Bivariate boxplot shown on the scatterplot of sulphur dioxide concentration plotted against number of manufacturing enterprises employing 20 or more workers for a number of cities in the USA. Outliers are labelled.

Bivariate boxplot: A bivariate analogue of the *boxplot* in which an inner region contains 50% of the data, and a 'fence' helps identify potential outliers. Robust methods are generally used to find estimates of the location, scale and correlation parameters required to construct the plot, so as to avoid the possible distortions that outlying obervations may cause. An example of such a plot for number of manufacturing enterprises and sulphur dioxide concentration recorded for a number of cities in the USA with cities deemed outliers labelled with city name. [*Technometrics*, 1992, **34**, 307–20.]

Bivariate Cauchy distribution: A *bivariate distribution*, $f(x,y)$, given by

$$f(x,y) = \frac{\alpha}{2\pi(\alpha^2 + x^2 + y^2)^{3/2}}, \qquad -\infty < x, y < \infty, \ \alpha > 0$$

Perspective plots of a number of such distributions are shown in Fig. 19. [KA1 Chapter 7.]

Bivariate data: Data in which the units each have measurements on two variables.

Bivariate distribution: The *joint distribution* of two random variables, x and y. A well known example is the *bivariate normal distribution* which has the form

$$f(x,y) = \frac{1}{2\pi\sigma_1\sigma_2\sqrt{1-\rho^2}} \times \exp\left\{-\frac{1}{1-\rho^2}\left[\frac{(x-\mu_1)^2}{\sigma_1^2} - 2\rho\frac{(x-\mu_1)(y-\mu_2)}{\sigma_1\sigma_2} + \frac{(y-\mu_2)^2}{\sigma_2^2}\right]\right\}$$

where $\mu_1, \mu_2, \sigma_1, \sigma_2, \rho$ are, respectively, the means, standard deviations and correlation of the two variables. Perspective plots of a number of such distributions are shown in Fig. 20. See also **bivariate beta distribution, bivariate Cauchy distribution, bivariate Gumbel distribution** and **bivariate exponential distribution**. [KA1 Chapter 7.]

Bivariate exponential distribution: Any *bivariate distribution*, $f(x,y)$, that has an *exponential distribution* for each of its *marginal distributions*. An example is

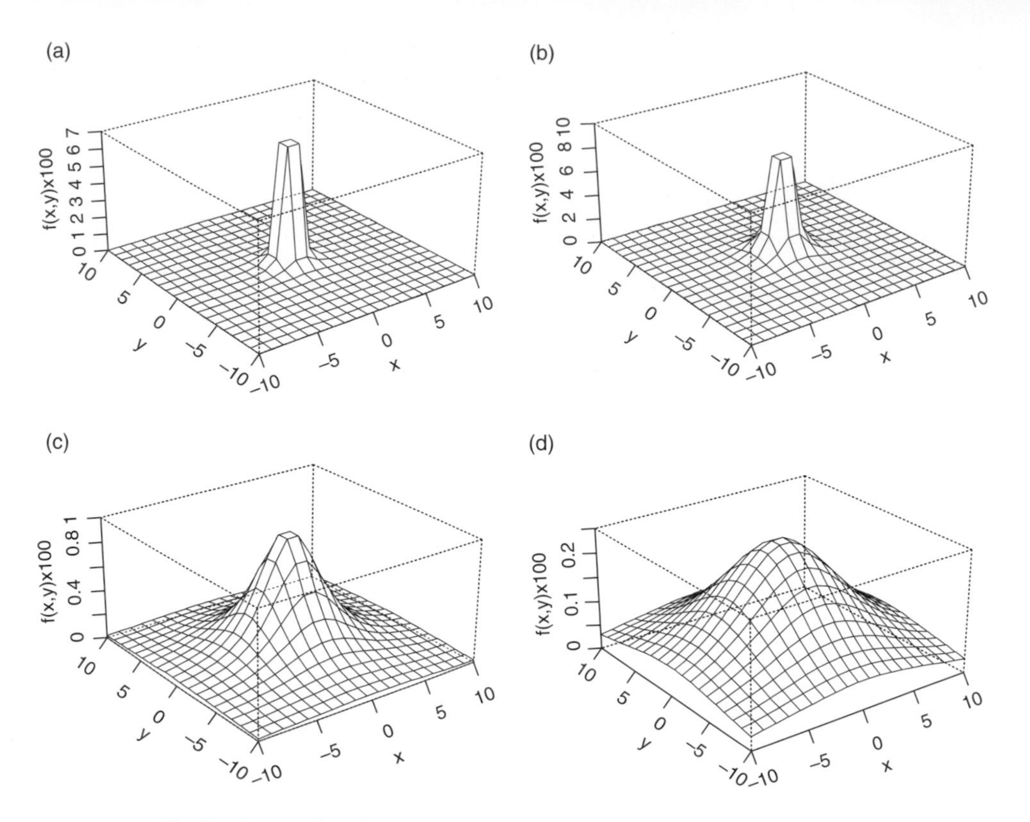

Fig. 19 Perspective plots of four bivariate Cauchy distributions (a) $\alpha = 0.2$; (b) $\alpha = 1.0$; (c) $\alpha = 4.0$; (d) $\alpha = 8.0$.

$$f(x, y) = [(1 + \alpha x)(1 + \alpha y) - \alpha]e^{-x-y-\alpha xy}, \qquad x \geq 0, y \geq 0, \alpha \geq 0$$

Perspective plots of a number of such distributions are shown in Fig. 21. [KA1 Chapter 7.]

Bivariate Gumbel distribution: A *bivariate distribution*, $f(x,y)$, having the form

$$f(x, y) = e^{-(x+y)}[1 - \theta(e^{2x} + e^{2y})(e^x + e^y)^{-2} + 2\theta e^{2(x+y)}(e^x + e^y)^{-3} \\ \times \theta^2 e^{2(x+y)}(e^x + e^y)^{-4}]\exp[-e^{-x} - e^{-y} + \theta(e^x + e^y)^{-1}]$$

where θ is a parameter which can vary from 0 to 1 giving an increasing positive correlation between the two variables. The distribution is often used to model combinations of extreme environmental variables. A perspective plot of the distribution is shown in Fig. 22. [*Statistics in Civil Engineering*, 1997, A. V. Metcalfe, Arnold, London.]

Bivariate normal distribution: See **bivariate distribution**.

Bivariate Oja median: An alternative to the more common *spatial median* as a measure of location for bivariate data. Defined as the value of θ that minimizes the objective function, $T(\theta)$, given by

$$T(\theta) = \sum_{i<j} A(\mathbf{x}_i, \mathbf{x}_j, \theta)$$

where $A(a,\ b,\ c)$ is the area of the triangle with vertices a, b, c and $\mathbf{x}_1, \mathbf{x}_2, \ldots, \mathbf{x}_n$ are n bivariate observations. [*Canadian Journal of Statistics*, 1993, **21**, 397–408.]

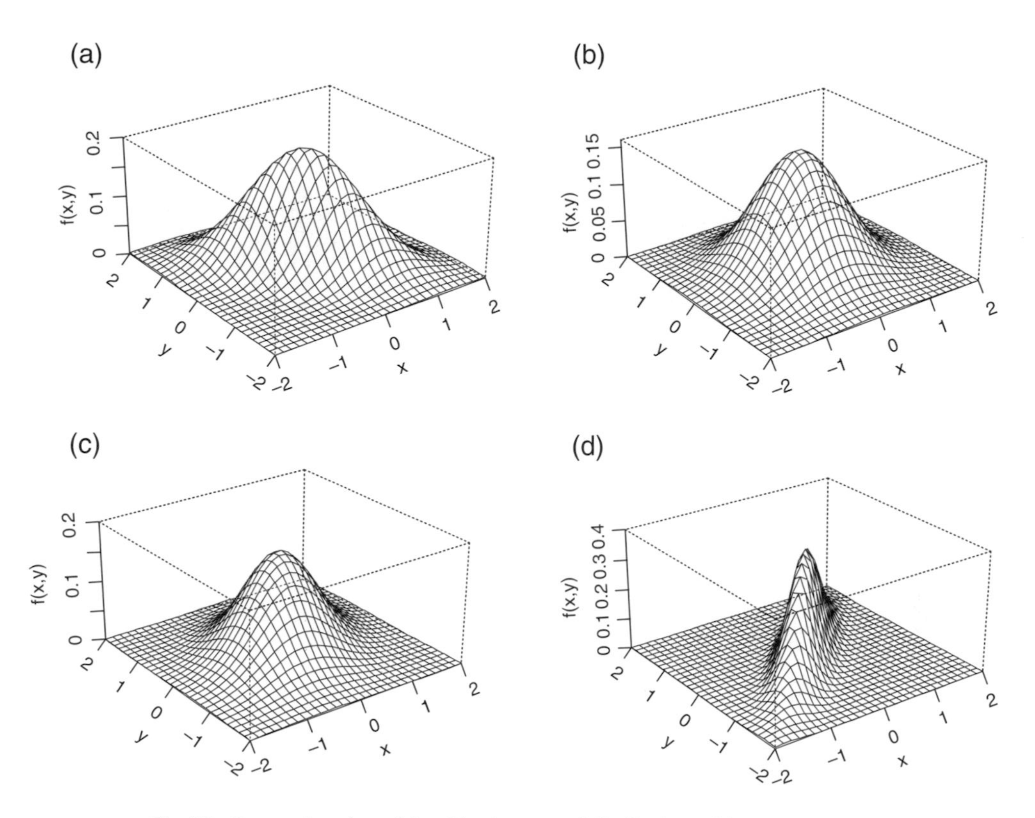

Fig. 20 Perspective plots of four bivariate normal distributions with zero means and unit standard deviations. (a) $\rho = 0.6$; (b) $\rho = 0.0$; (c) $\rho = 0.3$; (d) $\rho = 0.9$.

Bivariate Pareto distribution: A probability distribution useful in the modeling of life times of two-component systems working in a changing environment. The distribution is specified by

$$\Pr(X_1 > x_1, X_2 > x_2) = \left[\frac{x_1}{\beta}\right]^{-\lambda_1}\left[\frac{x_2}{\beta}\right]^{\lambda_2} \max\left[\frac{x_1}{\beta}, \frac{x_2}{\beta}\right]^{-\lambda_0}$$

where $\beta \leq \min(x_1, x_2) < \infty$. [*Communications in Statistics-Theory and Methods*, 2004, **33**, 3033–3042.]

Bivariate Poisson distribution: A bivariate probability distribution based on the joint distribution of the variables X_1 and X_2 defined as

$$X_1 = Y_1 + Y_{12} \qquad X_2 = Y_2 + Y_{12}$$

where Y_1, Y_2 and Y_{12} are mutually independent random variables each having a *Poisson distribution* with means λ_1, λ_2 and λ_3 respectively. The joint distribution of X_1 and X_2 is

$$\Pr(X_1 = x_1, X_2 = x_2) = e^{-(\lambda_1+\lambda_2+\lambda_3)} \sum_{i=0}^{\min(x_1,x_2)} \frac{\lambda_1^{x_1-i}\lambda_2^{x_2-i}\lambda_3^{i}}{(x_1-i)!(x_2-i)!i!}$$

[*Biometrika*, 1964, **51**, 241–5.]

Bivariate survival data: Data in which two related *survival times* are of interest. For example, in familial studies of disease incidence, data may be available on the ages and causes of death of fathers and their sons. [*Statistics in Medicine*, 1993, **12**, 241–8.]

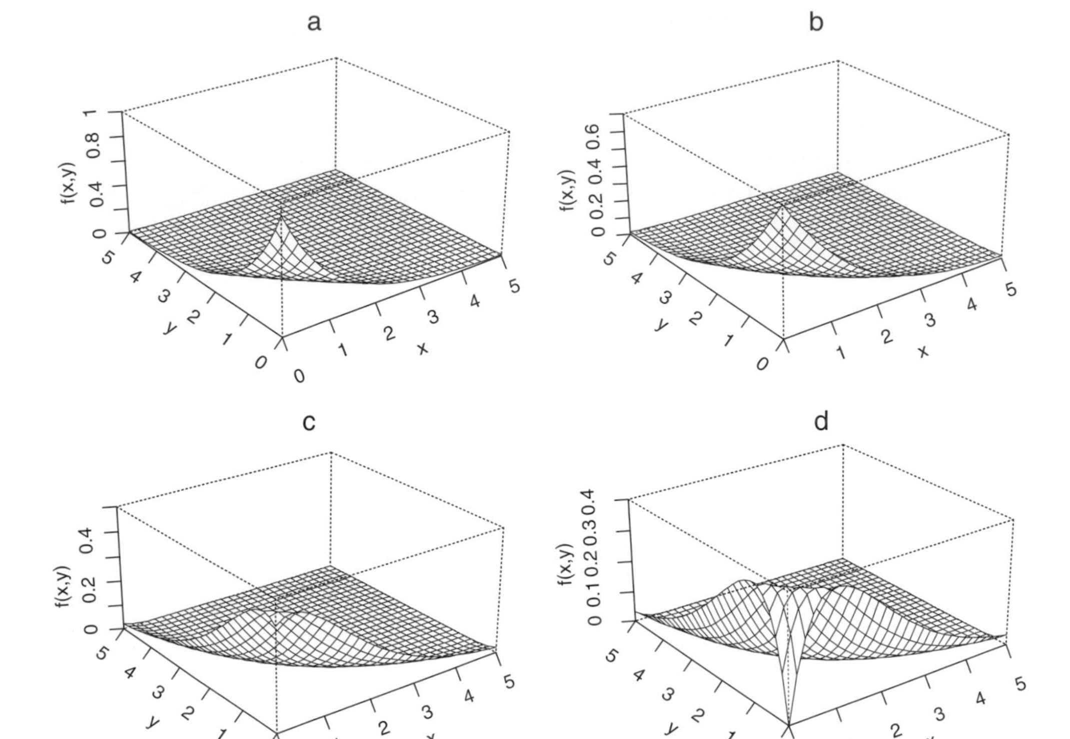

Fig. 21 Perspective plots of four bivariate exponential distributions (a) $\alpha = 0.1$; (b) $\alpha = 0.3$; (c) $\alpha = 0.6$; (d) $\alpha = 1.0$.

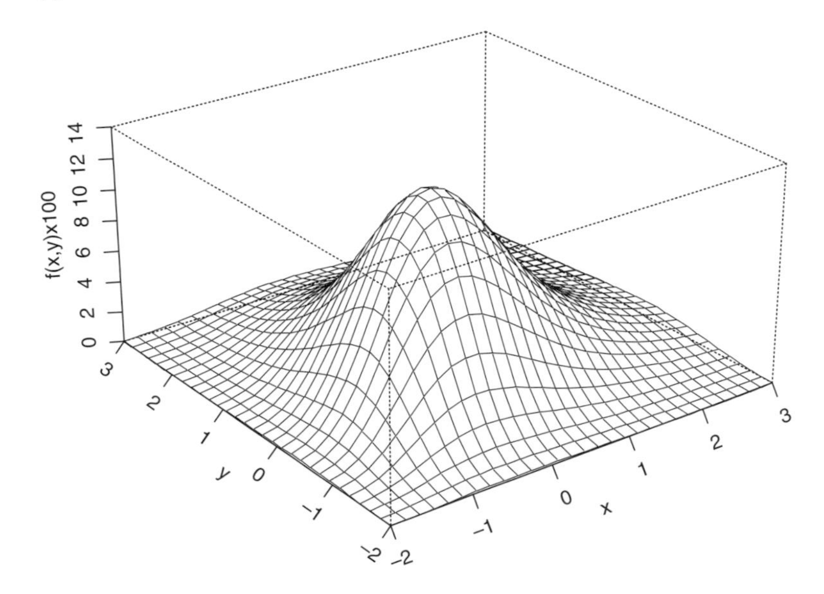

Fig. 22 Perspective plot of a bivariate Gumbel distribution with $\theta = 0.5$.

Bjerve, Petter Jakob (1914–2004): Born near Trondheim, Norway, Bjerve studied at Harvard University and the University of Chicago between 1947 and 1949. He then returned to his home country where he was appointed Director General of Statistics Norway, a position he held until 1980. Bjerve developed Statistics Norway's research particularly with regard to the national accounts and economic models. Bjerve was President of the International Statistical Institute from 1971 to 1974, and became a driving force behind the World Fertility Survey Programme. He died on January 12th, 2004.

Black, Arthur (1851–1893): Born in Brighton, UK, Black took a B.Sc degree of the University of London by private study, graduating in 1877. He earned a rather precarious living as an army coach and tutor in Brighton, while pursuing his mathematical and philosophical interests. He derived the *chi-squared distribution* as the limit to the *multinomial distribution* and independently discovered the *Poisson distribution.* Black's main aim was to quantify the theory of evolution. He died in January 1893.

Blinder–Oaxaca method: A method used for assessing the effect of the role of income on the racial wealth gap. The method is based on a decomposition of the mean interrace difference in wealth into the portion due to differences in the distribution of one or more explanatory variables and that due to differences in the conditional expectation function, and the parametric estimation of a wealth–earnings relationship by race. [*Yale University Economic Growth Center Discussion Paper No. 873*, 2003, pp. 1–11.]

Blinding: A procedure used in *clinical trials* to avoid the possible bias that might be introduced if the patient and/or doctor knew which treatment the patient is receiving. If neither the patient nor doctor are aware of which treatment has been given the trial is termed *double-blind.* If only one of the patient or doctor is unaware, the trial is called *single-blind.* Clinical trials should use the maximum degree of blindness that is possible, although in some areas, for example, surgery, it is often impossible for an investigation to be double-blind. [*Clinical Trials in Psychiatry*, 2nd edition, 2008, B. S. Everitt and S. Wesseley, Wiley, Chichester.]

Bliss, Chester Ittner (1899–1979): Born in Springfield, Ohio, Bliss studied entomology at Ohio State University, obtaining a Ph.D. in 1926. In the early 1930s after moving to London he collaborated with *Fisher*, and did work on *probit analysis.* After a brief period at the Institute of Plant Protection in Leningrad, Bliss returned to the United States becoming a biometrician at the Connecticut Agricultural Experimental Station where he remained until his retirement in 1971. He played a major role in founding the International Biometric Society. The author of over 130 papers on various aspects of *bioassay*, Bliss died in 1979.

BLKL algorithm: An algorithm for constructing *D-optimal designs* with specified block size. [*Optimum Experimental Designs*, 1992, A. C. Atkinson and A. N. Donev, Oxford Science Publications, Oxford.]

Block: A term used in experimental design to refer to a homogeneous grouping of experimental units (often subjects) designed to enable the experimenter to isolate and, if necessary, eliminate, variability due to extraneous causes. See also **randomized block design**.

Block, Borges and Savits model: A model used for recurrent events in reliability and engineering settings. In this model a system (or component) is put on test at time zero. On the system's failure at some time t, it undergoes a perfect repair with probability $p(t)$ or an imperfect repair with probability $q(t) = 1 - p(t)$. A perfect repair reverts the system's effective age to zero, whereas a minimal repair leaves the system's effective age unchanged from that at failure. [*Journal of Applied Probability*, 1985, **22**, 370–85.]

Block clustering: A method of *cluster analysis* in which the usual multivariate data matrix is partitioned into homogeneous rectangular blocks after reordering the rows and columns. The objectives of the analysis are to identify blocks or clusters of similar data values, to identify clusters of similar rows and columns, and to explore the relationship between the marginal (i.e. rows and column) clusters and the data blocks. [*Cluster Analysis*, 4th edition, 2002, B. S. Everitt, S. Landau, M. Leese, Arnold, London.]

Blocking: The grouping of plots, or experimental units, into blocks or groups, of homogeneous units. Treatments are then assigned at random to the plots within the blocks. The basic goal is to remove block-to-block variation from experimental error. Treatment comparisons are then made within the blocks on plots that are as nearly alike as possible.

Block randomization: A random allocation procedure used to keep the numbers of subjects in the different groups of a *clinical trial* closely balanced at all times. For example, if subjects are considered in sets of four at a time, there are six ways in which two treatments (A and B) can be allocated so that two subjects receive A and two receive B, namely

1.AABB	4.BBAA
2.ABAB	5.BABA
3.ABBA	6.BAAB

If only these six combinations are used for allocating treatments to each block of four subjects, the numbers in the two treatment groups can never, at any time, differ by more than two. See also **biased coin method** and **minimization**. [SMR Chapter 15.]

Blomqvist model: A model for assessing the relationship between severity (level) and rate of change (slope) in a *longitudinal study*. [*Journal of the American Statistical Association*, 1977, **72**, 746–9.]

BLUE: Abbreviation for **best linear unbiased estimator**.

Blunder index: A measure of the number of 'gross' errors made in a laboratory and detected by an external quality assessment exercise.

BLUP: Abbreviation for **best linear unbiased predictor**.

BMA: Abbreviation for **Bayesian model averaging**.

BMDP: A large and powerful statistical software package that allows the routine application of many statistical methods including *Student's t-tests*, *factor analysis*, and *analysis of variance*. Modules for data entry and data management are also available and the package is supported by a number of well-written manuals. Now largely taken over by SPSS. [Statistical Solutions Ltd., 8 South Bank, Crosse's Green, Cork, Ireland.]

BMI: Abbreviation for **body mass index**.

Body mass index (BMI): An index defined as weight in kilograms divided by the square of height in metres. An individual with a value less than 20 is generally regarded as underweight, while one with a value over 25 is overweight. [*Archives of Disease in Childhood*, 1995, **73**, 25–9.]

Bolshev, Login Nikolaevich (1922–1978): Born in Moscow, USSR, Bolshev graduated in 1951 from the Mechanical-Mathematical Faculty of Moscow State University and in 1966 was awarded a Doctorate of Science Degree by the Scientific Council of the prestigious Steklov Mathematical Institute where he then worked for his research on transformations of random variables. Bolshev also made significant contributions to statistics in the areas of parameter estimation, testing for *outliers* and epidemiology and demography. He died on August 29th 1978 in Moscow.

Bonferroni, Carlo Emilio (1892–1960): Born in Bergamo, Italy, Bonferroni studied mathematics at the University of Turin. In 1932 he was appointed to the Chair of Mathematics of

Finance at the University of Bari, and in 1933 moved to the University of Florence where he remained until his death on 18 August 1960. Bonferroni contributed to actuarial mathematics and economics but is most remembered among statisticians for the *Bonferroni correction*.

Bonferroni correction: A procedure for guarding against an increase in the probability of a type I error when performing multiple significance tests. To maintain the probability of a type I error at some selected value α, each of the m tests to be performed is judged against a significance level, α/m. For a small number of simultaneous tests (up to five) this method provides a simple and acceptable answer to the problem of multiple testing. It is however highly conservative and not recommended if large numbers of tests are to be applied, when one of the many other *multiple comparison procedures* available is generally preferable. See also **least significant difference test, Newman–Keuls test, Scheffé's test** and **Simes modified Bonferroni procedure**. [SMR Chapter 9.]

Boole's inequality: The following inequality for probabilities:

$$\sum_{i=1}^{k} \Pr(B_i) \geq \Pr(\cup_{i=1}^{k} B_i)$$

where the B_i are events which are not necessarily mutually exclusive or *exhaustive*. [KA1 Chapter 8.]

Boosting: A term used for a procedure which increases the performance of *allocation rules* in classification problems, by assigning a weight to each observation in the *training set* which reflects its importance. [*Statistical Pattern Recognition*, 1999, A. Webb, Arnold, London.]

Bootstrap: A data-based *simulation* method for statistical inference which can be used to study the variability of estimated characteristics of the probability distribution of a set of observations and provide *confidence intervals* for parameters in situations where these are difficult or impossible to derive in the usual way. (The use of the term bootstrap derives from the phrase 'to pull oneself up by one's bootstraps'.) The basic idea of the procedure involves *sampling with replacement* to produce random samples of size n from the original data, $x_1, x_2, \ldots, x_n$; each of these is known as a *bootstrap sample* and each provides an estimate of the parameter of interest. Repeating the process a large number of times provides the required information on the variability of the estimator and an approximate 95% confidence interval can, for example, be derived from the 2.5% and 97.5% *quantiles* of the replicate values. See also **jackknife**. [KA1 Chapter 10.]

Bootstrap sample: See **bootstrap**.

Borel–Tanner distribution: The probability distribution of the number of customers (N) served in a queueing system in which arrivals have a *Poisson distribution* with parameter λ and a constant server time, given that the length of the queue at the initial time is r. Given by

$$\Pr(N = n) = \frac{r}{(n-r)!} n^{n-r-1} e^{-\lambda n} \lambda^{n-r}, \qquad n = r, r+1, \ldots$$

[*Biometrika*, 1951, **38**, 383–92.]

Borrowing effect: A term used when abnormally low *standardized mortality rates* for one or more causes of death may be a reflection of an increase in the *proportional mortality rates* for other causes of death. For example, in a study of vinyl chloride workers, the overall proportional mortality rate for cancer indicated approximately a 50% excess, as compared with cancer death rates in the male US population. (One interpretation of this is a possible deficit of non-cancer deaths due to the *healthy worker effect*). Because the overall proportional mortality rate must by definition be equal to unity, a deficit in one type of mortality must entail a 'borrowing' from other causes. [*Statistics in Medicine*, 1985, **5**, 61–71.]

Bortkiewicz, Ladislaus von (1868–1931): Born in St Petersburg, Bortkiewicz graduated from the Faculty of Law at its university. Studied in Göttingen under *Lexis* eventually becoming Associate Professor at the University of Berlin in 1901. In 1920 he became a full Professor of Economics and Statistics. Bortkiewicz worked in the areas of classical economics, population statistics, acturial science and probability theory. Probably the first statistician to fit the *Poisson distribution* to the well-known data set detailing the number of soldiers killed by horse kicks per year in the Prussian army corps. Bortkiewicz died on 15 July 1931 in Berlin.

Bose, Raj Chandra (1901–1987): Born in Hoshangabad, India, Bose was educated at Punjab University, Delhi University and Calcutta University reading pure and applied mathematics. In 1932 he joined the Indian Statistical Institute under *Mahalanobis* where he worked on geometrical methods in multivariate analysis. Later he applied geometrical methods to the construction of experimental designs, in particular to *balanced incomplete block designs*. Bose was Head of the Department of Statistics at Calcutta University from 1945 to 1949, when he moved to the Department of Statistics at the University of North Carolina, USA where he remained until his retirement in 1971. He then accepted a chair at Colorado State University at Fort Collins from which he retired in 1980. He died on 30 October 1987 in Fort Collins, Colorado, USA.

Boundary estimation: The problem involved with estimating the boundary between two regions with different distributions when the data consists of independent observations taken at the nodes of a grid. The problem arises in epidemiology, forestry, marine science, meteorology and geology. [*Pattern Recognition*, 1995, **28**, 1599–609.]

Bowker, Albert Hosmer (1919–2008): Born in Wincendon, Massachusetts Bowker earned his bachelor's degree in mathematics at the Massachusetts Institute of Technology in 1941 and a doctorate in statistics at Columbia University in 1949. He became an assistant professor of mathematics and statistics at Stanford University in 1947 and was chair of its statistics department from 1948 until 1959. Bowker was President of the American Statistical Association in 1964 and he received the Shewhart Medal for pioneering work in applying mathematical statistics. He became chancellor of the University of California, Berkeley in 1971 staying in the post until 1980 a difficult period of reductions in state funding, student resentment about the Vietnam war and salary and hiring freezes. Bowker died in Portola Valley, California in January, 2008.

Bowker's test for symmetry: A test that can be applied to *square contingency tables*, to assess the hypothesis that the probability of being in cell i, j is equal to the probability of being in cell j, i. Under this hypothesis, the expected frequencies in both cells are $(n_{ij} + n_{ji})/2$ where n_{ij} and n_{ji} are the corresponding observed frequencies. The test statistic is

$$X^2 = \sum_{i<j} \frac{(n_{ij} - n_{ji})^2}{n_{ij} + n_{ji}}$$

Under the hypothesis of symmetry, X^2 has approximately a *chi-squared distribution* with $c(c-1)/2$ degrees of freedom, where c is the number of rows of the table (and the number of columns). In the case of a *two-by-two contingency table* the procedure is equivalent to *McNemar's test*. [*The Analysis of Contingency Tables*, 2nd edition, 1992, B. S. Everitt, Chapman and Hall/CRC Press, London.]

Bowley, Arthur Lyon (1869–1957): Born in Bristol, England, Bowley read mathematics at Cambridge. His first appointment was as a mathematics teacher at St. John's School, Leatherhead, Surrey where he worked from 1893 to 1899. It was during this period that his interest in statistics developed. Joined the newly established London School of Economics (LSE) in 1895 to give evening courses in statistics. In 1919 Bowley was elected to a newly-established Chair of Statistics at LSE, a post he held until his retirement in 1936. He made important contributions to the application of sampling techniques to economic and social surveys and produced a famous textbook, *Elements of Statistics*, which was first published in 1910 and went through seven editions up to 1937. Bowley was President of the Royal Statistical Society in 1938–1940, and received the Society's Guy medals in silver in 1895 and in gold in 1935. He died on 21 January 1957 in Haslemere, Surrey, England.

Box-and-whisker plot: A graphical method of displaying the important characteristics of a set of observations. The display is based on the *five-number summary* of the data with the 'box' part covering the *inter-quartile range*, and the 'whiskers' extending to include all but *outside observations*, these being indicated separately. Often particularly useful for comparing the characteristics of different samples as shown in Fig. 23. [SMR Chapter 2.]

Box-Behnken designs: A class of experimental designs that are constructed by combining two-level factorial designs with *balanced incomplete block designs*. [*Technometrics*, 1995, **37**, 399–410.]

Box-counting method: A method for estimating the *fractal dimension* of self-similar patterns in space which consists of plotting the number of *pixels* which intersect the pattern under consideration, versus length of the pixel unit. [*Fractal Geometry*, 1990, K. Falconer, Wiley, New York.]

Box–Cox transformation: A family of data transformations designed to achieve normality and given by

$$
\begin{aligned}
y &= (x^{\lambda} - 1)/\lambda, && \lambda \neq 0 \\
y &= \ln x, && \lambda = 0
\end{aligned}
$$

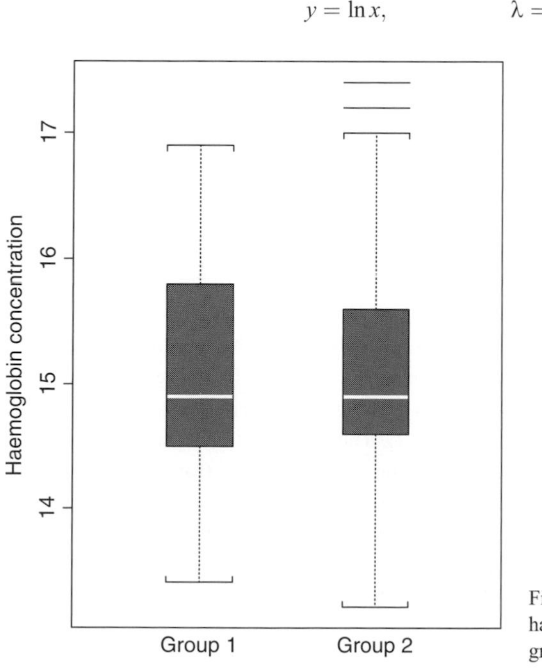

Fig. 23 Box-and-whisker plot of haemoglobin concentration for two groups of men.

Maximum likelihood estimation can be used to estimate a suitable value of λ for a particular variable. See also **Taylor's power law** and **Box–Tidwell transformation**. [ARA Chapter 11.]

Box–Jenkins models: Synonym for **autoregressive moving average models**.

Box–Müller transformation: A method of generating random variables from a normal distribution by using variables from a *uniform distribution* in the interval (0,1). Specifically if U_1 and U_2 are random uniform (0,1) variates then

$$X = (-2\ln U_1)^{\frac{1}{2}} \cos 2\pi U_2$$
$$Y = (-2\ln U_1)^{\frac{1}{2}} \sin 2\pi U_2$$

are independent standard normal variables. Often used for generating data from a normal distribution when using *simulation*. [KA1 Chapter 9.]

Box–Pierce test: A test to determine whether *time series* data are simply a *white noise sequence* and that therefore the observations are not related. The test statistic is

$$Q_m = n(n+2)\sum_{k=1}^{m} r_k^2/(n-k)$$

where r_k, $k = 1, 2, \ldots, m$, are the values of the *autocorrelations* up to lag m and n is the number of observations in the series. If the data are a white noise series then Q_m has a *chi-squared distribution* with m degrees of freedom. The test is valid only if m is very much less than n. [*Biometrika*, 1979, **66**, 1–22.]

Boxplot: Synonym for **box-and-whisker plot**.

Box's test: A test for assessing the equality of the variances in a number of populations that is less sensitive to departures from normality than *Bartlett's test*. See also **Hartley's test**.

Box–Tidwell transformation: See **fractional polynomials**.

Bradley, Ralph (1923–2001): Born in Smith Falls, Ontario, Canada, Bradley received an honours degree in mathematics and physics from Queen's University, Canada in 1944. After war service he took an M.A. in mathematics and statistics in 1946 and in 1949 completed his Ph.D. studies at the University of North Carolina at Chapel Hill. While working at Virginia Polytechnic Institute he collaborated with Milton Terry on a statistical test for paired comparisons, the *Bradley–Terry model*. Bradley also made contributions to nonparametric statistics and multivariate methods. He died on 30 October 2001 in Athens, Georgia, USA.

Bradley–Terry model: A model for experiments in which responses are pairwise rankings of treatments (a *paired comparison experiment*), which assigns probabilities to each of $\binom{t}{2}$ pairs among t treatments in terms of $t-1$ parameters, $\pi_1, \pi_2, \ldots, \pi_t$, $\pi_i \geq 0, i = 1, \ldots, t$ associated with treatments $T_1, \ldots, T_t$. These parameters represent relative selection probabilities for the treatments subject to the constraints $\pi_i \geq 0$ and $\sum_{i=1}^{t} \pi_i = 1$. The probability that treatment T_i is preferred over treatment T_j in a single comparison is $\pi_i/(\pi_i + \pi_j)$. Maximum likelihood estimators for the πs can be found from observations on the number of times treatment T_i is preferred to treatment T_j in a number of comparisons of the two treatments. [*Biometrics*, 1976, **32**, 213–32.]

Branch-and-bound algorithm: Synonym for **leaps-and-bounds algorithm**.

Branching process: A process involving a sequence of random variables $Y_1, Y_2, \ldots, Y_n$ where Y_k is the number of particles in existence at time k and the process of creation and annihilation of particles is as follows: each particle, independently of the other particles and of the previous history of the process, is transformed into j particles with probability $p_j, j = 0, 1, \ldots, M$. Used to describe the evolution of populations of individuals, which reproduce independently. A particularly interesting case is that in which each particle either vanishes with probability q or divides into two with probability $p, p+q=1$. In this case it can be shown that $p_{ij} = \Pr(Y_{k+1} = j | Y_k = i)$ is given by

$$p_{ij} = \binom{i}{j/2} p^{\frac{j}{2}} q^{(i-j/2)}, \quad j = 0, \ldots, 2i$$
$$= 0 \text{ in all other cases}$$

[*Branching Processes with Biological Applications*, 1975, P. Jagers, Wiley, New York.]

Brass, William (1921–99): Born in Edinburgh, Brass studied at Edinburgh University, first from 1940 to 1943 and then again, after serving in World War II, from 1946 to 1947. In 1948 he joined the East African Statistical Department as a statistician and later became its Director. It was during his seven years in these posts that Brass developed his ideas on both the collection and analysis of demographic data. In 1955 he joined the Department of Statistics in Aberdeen University where he remained for the next nine years, years in which, in collaboration with several others, he produced *The Demography of Tropical Africa*, which remained the definitive work on the subject for the next decade. From 1965 until his retirement in 1988 he worked at the London School of Tropical Medicine and Hygiene. Brass made major contributions to the study of the biosocial determinants of fertility and mortality, population forecasting and the evaluation of family planning programmes.

Breakdown point: A measure of the insensitivity of an estimator to multiple *outliers* in the data. Roughly it is given by the smallest fraction of data contamination needed to cause an arbitrarily large change in the estimate. [*Computational Statistics*, 1996, **11**, 137–46.]

Breiman, Leo (1928–2005): Breiman was born in New York City but moved to California when he was five years old. He graduated from Roosevelt High School in 1945 and in 1949 earned a degree in physics from the California Institute of Technology. Breiman then joined Columbia University intending to study philosophy, but changed to mathematics, receiving a master's degree in 1950. In 1954 he was awarded a Ph.D. from the University of California, Berkeley, and then began to teach probability theory at UCLA. Several years later he resigned from the university and earned a living as a statistical consultant before rejoining the university world at Berkeley where he established the Statistical Computing Facility. Breiman's main work in statistics was in the area of computationally intensive *multivariate analysis* and he is best known for his contribution to *classification and regression trees.* Leo Breiman died on 5 July 2005.

Breslow–Day test: A test of the null hypothesis of homogeneity of the *odds ratio* across a series of *2 × 2 contingency tables.* [*Statistical Methods in Cancer Research, I The Analysis of Case Control Studies.*, 1980, N. E. Breslow and N. E. Day, IARC, Lyon.]

Breslow's estimator: An estimator of the *cumulative hazard function* in *survival analysis.* [*Biometrics*, 1994, **50**, 1142–5.]

Breusch-Pagan test: A test of whether the error term in a regression model is heteroscedastic as a linear function of the explanatory variables. [*Econometrica*, 1979, **47**, 1287–1294.]

Bridge estimators: Methods that find an estimate of the parameters $\boldsymbol{\beta}$ in the regression model $y_i = \mathbf{x'}_i\boldsymbol{\beta} + \epsilon_i, i = 1, \ldots n$ by minimizing the penalized least squares objective function

$$L_n(\boldsymbol{\beta}) = \sum_{i=1}^{n}(y_i - \mathbf{x}'_i\boldsymbol{\beta})^2 + \lambda_n \sum_{j=1}^{q} |\beta_j|^\gamma$$

where q is the number of variables and λ_n is a penalty parameter. Includes two special cases; when $\gamma = 2$ the bridge estimator becomes the *ridge estimator* and when $\gamma = 1$ it becomes the *lasso.* [*Journal of Computational and Graphical Statistics*, 1998, **7**, 397–416.]

Bridging clinical trials: *Clinical trials* designed to evaluate whether a proposed dose of a particular drug for use in one populations, for example, children, gives similar pharmacokinetic levels, or has similar effects on a surrogate marker as an established effective dose of the same drug used in another population, for example, adults. [*Biometrics*, 2008, **64**, 1117–1125.]

Brier score: The *mean squared error* for binary data, measuring the difference between a predicted probability of an event and its occurrence expressed as 0 or 1 depending on whether or not the event has occurred. [*Monthly Weather Review*, 1950, **75**, 1–3.]

Broadband smoothing: A smoothing technique for *power spectra* which allows a trade-off between resolution and variance at different frequencies. [*Clinical Science*, 1999, **97**, 129–139.]

Broken-stick model: Essentially a synonym for **spline function**.

Brown–Forsythe test: A procedure for assessing whether a set of population variances are equal, based on applying an *F-test* to the following values calculated for a sample of observations from each of the populations;

$$z_{ij} = |x_{ij} - m_i|$$

where x_{ij} is the jth observation in the ith group and m_i is the median of the ith group. See also **Box's test, Bartlett's test** and **Hartley's test**. [*Journal of Statistical Computation and Simulation*, 1997, **56**, 353–72.]

Brownian motion: A *stochastic process*, X_t, with *state space* the real numbers, satisfying

- $X_0 = 0$;
- for any $s_1 \le t_1 \le s_2 \le t_2 \le \cdots \le s_n \le t_n$, the random variables $X_{t_1} - X_{s_1}, \ldots, X_{t_n} - X_{s_n}$ are independent;
- for any $s < t$, the random variable $X_t - X_s$ has a normal distribution with mean 0 and variance $(t - s)\sigma^2$.

[*Stochastic Modelling of Scientific Data*, 1995, P. Guttorp, Chapman and Hall/CRC Press, London.]

BRR: Abbreviation for **balanced repeated replication**.

Brushing scatterplots: An interactive computer graphics technique sometimes useful for exploring *multivariate data.* All possible two-variable scatterplots are displayed, and the points falling in a defined area of one of them (the 'brush') are highlighted in all. See also **dynamic graphics** and **tour**. [*Dynamic Graphics for Statistics*, 1987, W. S. Cleveland and M. E. McGill, Wadsworth, Belmont, California.]

Bubble plot: A method for displaying observations which involve three variable values. Two of the variables are used to form a scatter diagram and values of the third variable are represented by circles with differing radii centred at the appropriate position. An example of such a plot for variables age and weekly time spent looking after car with extroversion represented by

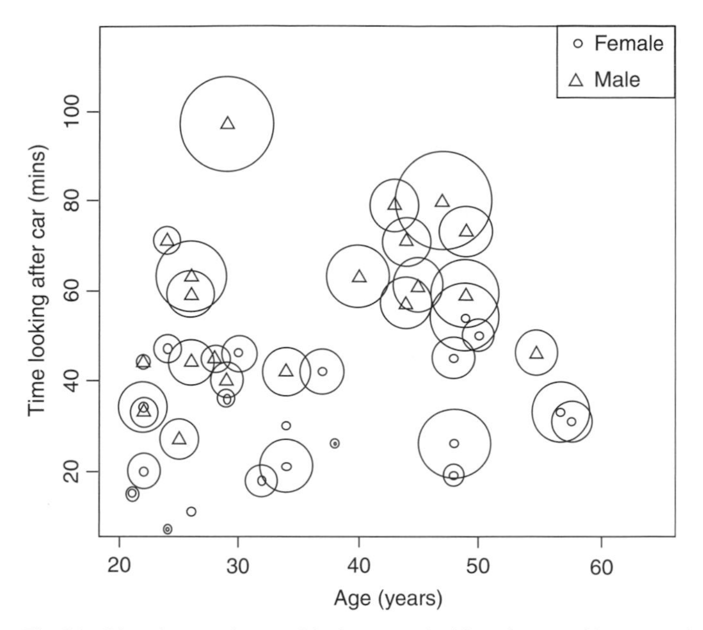

Fig. 24 Plot of age against weekly time spent looking after car with extroversion shown by 'bubbles' and gender also indicated.

the 'bubble' is shown in Fig. 24. Plot also shows gender of each individual. [*Modern Medical Statistics*, 2003, B. S. Everitt, Arnold, London.]

Buckley–James method: An iterative procedure for obtaining estimates of regression coefficients in models involving a response variable that is right-censored. [*Statistics in Medicine*, 1992, **11**, 1871 80.]

Buehler confidence bound: An upper confidence bound for a measure of reliability/maintainability of a series system in which component repair is assumed to restore the system to its original status. [*Journal of the American Statistical Association*, 1957, **52**, 482–93.]

Buffon's needle problem: A problem proposed and solved by Comte de Buffon in 1777 which involves determining the probability, p, that a needle of length l will intersect a line when thrown at random onto a horizontal plane that is ruled with parallel lines a distance $a > l$ apart. The solution is

$$p = \frac{2l}{\pi a}$$

Can be used to evaluate π since if the needle is tossed N times independently and $y_i = 1$ if it intersects a line on the ith toss and 0 otherwise then

$$\frac{2N}{y_1 + \cdots + y_N} \approx \pi$$

In 1850 R. Wolf (an astronomer in Zurich) threw a needle 5000 times and obtained the value 3.1596 for π. [*American Mathematical Monthly*, 1979, **81**, 26–9.]

BUGS: *Bayesian inference Using Gibbs Sampling*, a software package that provides a means of analysing complex models from a description of their structure provided by the user and automatically deriving the expressions for the Gibbs sampling needed to obtain the posterior marginal distributions. The software is capable of handling a wide range of problems, including hierarchical random effects and measurement errors in *generalized linear models*, latent variable and mixture models, and various forms of missing and censored data. [http://www.mrc-bsu.cam.ac.uk/bugs]

Bump hunting: A colourful term for the examination of frequency distributions for local maxima or modes that might be indicative of separate groups of subjects. See also **bimodal distribution, finite mixture distribution** and **dip test**. [*Journal of Computational and Graphic Statistics*, 2001, **10**, 713–29.]

Burgess method: See **point scoring**.

Burn-in: A term used in *Markov chain Monte Carlo methods* to indicate the necessary period of iterations before samples from the required distribution are achieved. [*Markov Chain Monte Carlo in Practice*, 1996, eds. W. R. Gilks, S. Richardson, and D. J. Spiegelhalter, Chapman and Hall/CRC Press, London.]

Burr distributions: A family of probability distributions which is very flexible and can have a wide range of shapes. The distribution involves two parameters ϕ and λ and is given explicitly as

$$f(y) = \frac{1}{\lambda}\frac{\phi y^{\phi-1}}{(1+y^{\phi})^{\frac{1}{\lambda}+1}} \quad y>0, \phi>0, \lambda>0$$

[*Encyclopedia of Statistical Sciences*, 2006, eds. S. Kotz, C. B. Read, N. Balakrishnan and B. Vidakovic, Wiley, New York.]

Bursty phenomena: A term sometimes applied to phenomena that exhibit occasional unusually large observations. [*Solid State Communications*, 2008, **145**, 545–550.]

Burt matrix: A matrix used in *correspondence analysis* when applied to *contingency tables* with more than two dimensions. [MV1 Chapter 5.]

Butler statistic: A statistic that can be used to test whether a distribution is symmetric about some point. [*Annals of Mathematical Statistics*, 1969, **40**, 2209–10.]

BW statistic: See **Cliff and Ord's BW statistic**.

Byar, D. P. (1938–1991): Born in Lockland, Ohio, Byar graduated from Harvard Medical School in 1964. Working at the Armed Forces Institute of Pathology, Byar became increasingly interested in the sources of variation in laboratory experiments and the statistical methods for coping with them. After joining the National Cancer Institute in 1968 he became a leading methodologist in *clinical trials*. Byar died on 5 August 1991 in Washington DC.

Byte: A unit of information, as used in digital computers, equals eight *bits*.

C

C: A low level programming language developed at Bell Laboratories. Widely used for software development.

C++: A general purpose programming language which is a superset of C.

CACE: Abbreviation for **complier average causal effect**.

Calendar plot: A method of describing compliance for individual patients in a *clinical trial*, where the number of tablets taken per day are set in a calendar-like-form – see Fig. 25. See also **chronology plot**. [*Statistics in Medicine*, 1997, **16**, 1653–64.]

Calendarization: A generic term for *benchmarking*.

Calibration: A process that enables a series of easily obtainable but inaccurate measurements of some quantity of interest to be used to provide more precise estimates of the required values. Suppose, for example, there is a well-established, accurate method of measuring the concentration of a given chemical compound, but that it is too expensive and/or cumbersome for routine use. A cheap and easy to apply alternative is developed that is, however, known to be imprecise and possibly subject to bias. By using both methods over a range of concentrations of the compound, and applying regression analysis to the values from the cheap method and the corresponding values from the accurate method, a *calibration curve* can be constructed that may, in future applications, be used to read off estimates of the required concentration from the values given by the less involved, inaccurate method. [KA2 Chapter 28.]

Calibration curve: See **calibration**.

Caliper matching: See **matching**.

Campion, Sir Harry (1905–1996): Born in Worsley, Lancashire, Campion studied at Manchester University where, in 1933, he became Robert Ottley Reader in Statistics. In the Second World War he became Head of the newly created Central Statistical Office remaining in this position until his retirement in 1967. Campion was responsible for drafting the Statistics of Trade Act 1947 and in establishing a new group in the civil service – the Statistician class. He was President of the Royal Statistical Society from 1957 to 1959 and President of the International Statistical Institute from 1963 to 1967. Awarded the CBE in 1945 and a knighthood in 1957.

Canberra metric: A *dissimilarity coefficient* given by

$$d_{ij} = \sum_{k=1}^{q} |x_{ik} - x_{jk}|/(x_{ik} + x_{jk})$$

where $x_{ik}, x_{jk}, k = 1, \ldots, q$ are the observations on q variables for individuals i and j. [MV1 Chapter 3.]

	Mon	Tue	Wed	Thr	Fri	Sat	Sun
						0	1
3	1	1	2	1	0	1	1
10	1	1	1	1	1	0	0
17	0	1	1	2	0	2	0
24	1	1	1	0	1	0	0
31	1						

Fig. 25 Calendar plot of number of tablets taken per day. (Reproduced from *Statistics in Medicine* with permission of the publishers John Wiley.)

CAN estimator: An estimator of a parameter that is consistent and asymptotically normal.

Canonical correlation analysis: A method of analysis for investigating the relationship between two groups of variables, by finding linear functions of one of the sets of variables that maximally correlate with linear functions of the variables in the other set. In many respects the method can be viewed as an extension of *multiple regression* to situations involving more than a single response variable. Alternatively it can be considered as analogous to *principal components analysis* except that a correlation rather than a variance is maximized. A simple example of where this type of technique might be of interest is when the results of tests for, say, reading speed (x_1), reading power (x_2), arithmetical speed (y_1), and arithmetical power (y_4), are available from a sample of school children, and the question of interest is whether or not reading ability (measured by x_1 and x_2) is related to arithmetical ability (as measured by y_1 and y_2). [MV1 Chapter 4.]

Canonical discriminant functions: See **discriminant analysis**.

Canonical variates: See **discriminant analysis**.

Cantelli, Francesco Paolo (1875–1966): Born in Palermo, Italy, Cantelli graduated from the University of Palermo with a degree in mathematics in 1899. From 1903 until 1923 he worked as an actuary at the Instituti di Previdenza della Cassa Depositi e Prestiti where he carried out research into financial and actuarial mathematics. Later Cantelli made major contributions to probability theory particularly to stochastic convergence. He died on July 21st, 1966 in Rome.

Capture–recapture sampling: An alternative approach to a census for estimating population size, which operates by sampling the population several times, identifying individuals which appear more than once. First used by *Laplace* to estimate the population of France, this approach received its main impetus in the context of estimating the size of wildlife populations. An initial sample is obtained and the individuals in that sample marked or otherwise identified. A second sample is, subsequently, independently obtained, and it is noted how many individuals in that sample are marked. If the second sample is representative of the population as a whole, then the sample proportion of marked individuals should be about the same as the corresponding population proportion. From this relationship the total number of individuals in the population can be estimated. Specifically if X individuals are 'captured', marked and released and y individuals then independently captured of which x are marked, then the estimator of population size (sometimes known as the *Petersen estimator*) is

$$\hat{N} = \frac{y}{x}X$$

with variance given by

$$\text{var}(\hat{N}) = \frac{Xy(X-x)(y-x)}{x^3}$$

The estimator does not have finite expectation since x can take the value zero. A modified version, *Chapman's estimator*, adds one to the frequency of animals caught in both samples (x) with the resulting population size estimator

$$\hat{N} = \frac{y+1}{x+1}(X+1) - 1$$

[KA1 Chapter 10.]

CAR: Abbreviation for **coarsening at random**.

Cardano, Girolamo (1501–1576): Born in Milan, Italy Cardano studied medicine in Padua and for a time practised as a country doctor. But he was better known as a gambler particularly with dice. His writings about gambling with dice contain the first expression of the mathematical concept of probability and Cardano clearly states that the six sides of an honest die are equally likely and introduces chance as the ratio of the number of favourable cases to the number of equally possible cases. Girolamo Cardano, physician, mathematician and eccentric died insane at the age of 75.

Cardiord distribution: A probability distribution, $f(\theta)$, for a *circular random variable*, θ, given by

$$f(\theta) = \frac{1}{2\pi}\{1 + 2\rho\cos(\theta - \mu)\}, \qquad 0 \leq \theta \leq 2\pi,\ 0 \leq \rho \leq \frac{1}{2}$$

[*Statistical Analysis of Circular Data*, 1995, N. I. Fisher, Cambridge University Press, Cambridge.]

Carryover effects: See **crossover design**.

CART: Abbreviation for **classification and regression tree technique**.

Cartogram: A diagram in which descriptive statistical information is displayed on a geographical map by means of shading, different symbols or in some other perhaps more subtle way. An example of the latter is given in Fig. 26. See also **disease mapping**. [*Statistics in Medicine*, 1988, **7**, 491–506.]

Cascaded parameters: A group of parameters that are interlinked and where selecting a value for the first parameter affects the choice available in the subsequent parameters. This approach brings important advantages to parameter estimation in the presence of *nuisance parameters*. [*Computational Statistics*, 2007, **22**, 335–351.]

Case-cohort study: A research design in *epidemiology* that involves sampling of controls at the outset of the study who are to be compared with cases from the cohort. The design is generally used when the cohort can be followed for disease outcome, but it is too expensive to collect and process information on all study subjects. See also **case-control study** and **nested case-control study**. [*Environmental Health Perspectives*, 1994, **102**, 53–56.]

Case-control study: A traditional case-control study is a common research design in *epidemiology* where the exposures to risk factors for cases (individuals getting a disease) are compared to exposures for controls (individuals not getting a disease). The design is retrospective in the sense that a sample of controls is sampled from subjects who are disease free at the end of the observation period. The primary advantage of this design is that exposure information needs to be gathered for only a small proportion of cohort members, thereby considerably reducing the data collection costs. A disadvantage is the potential for *recall bias*. See also **retrospective design, nested case-control study** and **case-cohort study**. [*Case-Control Studies: Design, Conduct, Analysis*, 1982, J. J. Schlesselman, Oxford University Press, Oxford.]

Fig. 26 A cartogram showing population density in the UK (used with permission of Dr. B. Hennig).

Case series: A series of reports on the condition of individual patients made by the treating physician. Such reports may be helpful and informative for rare diseases, for example, in the early days of the occurrence of AIDS, but it has to be remembered that in essence they are nothing more than anecdotal evidence dignified by the case series label. [*Clinical Trials in Psychiatry*, 2nd edition, 2008, B. S. Everitt and S.Wessely, Wiley, Chichester.]

Catastrophe theory: A theory of how small, continuous changes in independent variables can have sudden, discontinuous effects on dependent variables. Examples include the sudden collapse of a bridge under slowly mounting pressure, and the freezing of water when temperature is gradually decreased. Developed and popularized in the 1970s, catastrophe theory has, after a period of criticism, now become well established in physics, chemistry and biology. [*Physica Series A*, 2000, **285**, 199–210.]

Categorical variable: A variable that gives the appropriate label of an observation after allocation to one of several possible categories, for example, respiratory status: terrible, poor, fair, good, excellent, or blood group: A, B, AB or O. Respiratory status is an example of an ordered categorical variable or *ordinal variable* whereas blood type is an example of an unordered categorical variable. See also **binary variable** and **continuous variable**.

Categorizing continous variables: A practice that involves the conversion of continuous variables into a series of categories, that is common in medical research. The rationale is partly statistical (avoidance of certain assumptions about the nature of the data) and partly that clinicians are often happier when categorizing individuals. In general there are no statistical advantages in such a procedure. [*British Journal of Cancer*, 1991, **64**, 975.]

Cauchy, Augustin-Louis (1789–1857): Born in Paris, Cauchy studied to become an Engineer but ill health forced him to retire and teach mathematics at the École Polytechnique. Founder of the theory of functions of a complex variable, Cauchy also made considerable contributions to the theory of estimation and the classical linear model. He died on 22 May 1857 in Sceaux, France.

Cauchy distribution: The probability distribution, $f(x)$, given by

$$f(x) = \frac{\beta}{\pi[\beta^2 + (x - \alpha)^2]} \qquad -\infty \leq x \leq \infty, \beta > 0$$

where α is a *location parameter* (median) and beta β a *scale parameter. Moments* and *cumulants* of the distribution do not exist. The distribution is unimodal and symmetric about α with much heavier tails than the normal distribution (see Fig. 27). The upper and lower quartiles of the distribution are $\alpha \pm \beta$. Named after *Augustin-Louis Cauchy* (1789–1857). [STD Chapter 7.]

Cauchy integral: The integral of a function, $f(x)$, from a to b defined in terms of the sum

$$S_n = f(a)(x_1 - a) + f(x_1)(x_2 - x_1) + \cdots + f(x_n)(b - x_n)$$

where the range of the integral (a, b) is divided at points $x_1, x_2, \ldots, x_n$. It may be shown that under certain conditions (such as the continuity of $f(x)$ in the range) the sum tends to a limit as the length of the intervals tends to zero, independently of where the dividing points are drawn or the way in which the tendency to zero proceeds. This limit is the required integral. See also **Riemann–Stieltjes integral**. [KA1 Chapter 1.]

Cauchy–Schwarz inequality: The following inequality for the integrals of functions, $f(x)$, and $g(x)$, whose squares are integral

$$\left\{\int [f(x)g(x)]\mathrm{d}x\right\}^2 \leq \left\{\int [f(x)]^2 \mathrm{d}x\right\}\left\{\int [g(x)]^2 \mathrm{d}x\right\}$$

In statistics this leads to the following inequality for expected values of two random variables x and y with finite second moments

$$[E(xy)]^2 \leq E(x^2)E(y^2)$$

The result can be used to show that the correlation coefficient, ρ, satisfies the inequality $\rho^2 \leq 1$. [KA1 Chapter 2.]

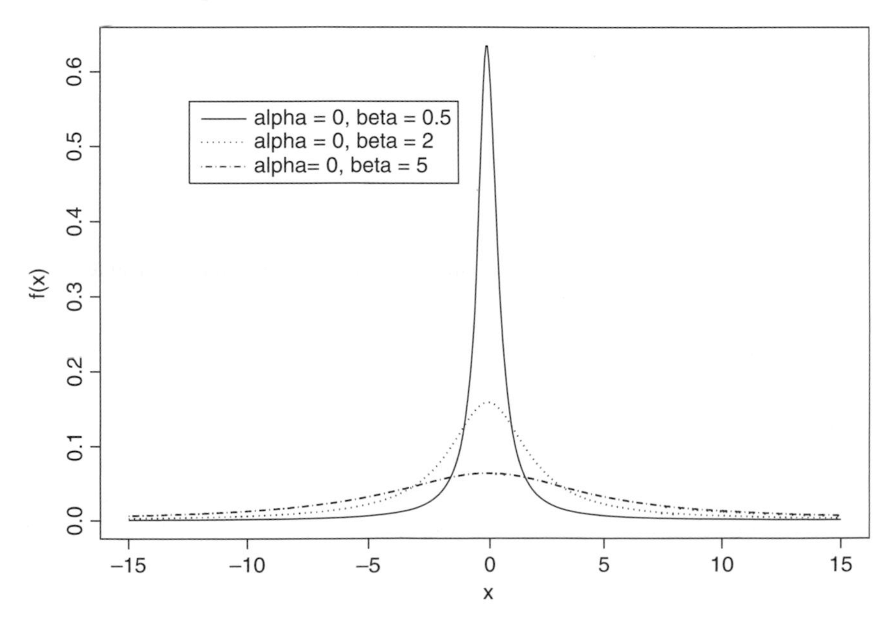

Fig. 27 Cauchy distributions for various parameter values.

Causality: The relating of causes to the effects they produce. Many investigations in medicine seek to establish causal links between events, for example, that receiving treatment A causes patients to live longer than taking treatment B. In general the strongest claims to have established causality come from data collected in experimental studies. Relationships established in *observational studies* may be very suggestive of a causal link but are always open to alternative explanations. [SMR Chapter 5.]

Cause specific death rate: A death rate calculated for people dying from a particular disease. For example, the following are the rates per 1000 people for three disease classes for developed and developing countries in 1985.

	C1	C2	C3
Developed	0.5	4.5	2.0
Developing	4.5	1.5	0.6

C1 = Infectious and parasitic diseases
C2 = Circulatory diseases
C3 = Cancer

See also **crude annual death rate** and **age-specific death rate**.

CCF: Abbreviation for **common cause failure**.

Ceiling effect: A term used to describe what happens when many subjects in a study have scores on a variable that are at or near the possible upper limit ('ceiling'). Such an effect may cause problems for some types of analysis because it reduces the possible amount of variation in the variable. The converse, or *floor effect*, causes similar problems. [*Arthritis and Rheumatism*, 1995, **38**, 1055.]

Cellular proliferation models: Models used to describe the growth of cell populations. One example is the *deterministic model*

$$N(t) = N(t_0)e^{\upsilon(t-t_0)}$$

where $N(t)$ is the number of cells in the population at time t, t_0 is an initial time and υ represents the difference between a constant birth rate and a constant death rate. Often also viewed as a *stochastic process* in which $N(t)$ is considered to be a random variable. [*Investigative Ophthalmology and Visual Science*, 1986, **27**, 1085–94.]

Censored observations: An observation x_i on some variable of interest is said to be censored if it is known only that $x_i \leq L_i$ (*left-censored*) or $x_i \geq U_i$ (*right-censored*) where L_i and U_i are fixed values. Such observations arise most frequently in studies where the main response variable is time until a particular event occurs (for example, time to death) when at the completion of the study, the event of interest has not happened to a number of subjects. See also **interval censored data, singly censored data, doubly censored data** and **non-informative censoring**. [SMR Chapter 2.]

Censored regression models: A general class of models for analysing truncated and censored data where the range of the dependent variable is constrained. A typical case of censoring occurs when the dependent variable has a number of its values concentrated at a limiting value, say zero. [*Statistics in Medicine*, 1984, **3**, 143–52.]

Census: A study that aims to observe every member of a population. The fundamental purpose of the population census is to provide the facts essential to government policy-making, planning and administration. [SMP Chapter 5.]

Centering: A procedure used in an effort to avoid *multicollinearity* problems when both linear, and higher order terms are included in a model. In *multilevel models* covariates are often cluster-mean centred to facilitate interpretation of corresponding regression parameters. [*Applied Longitudinal Analysis*, 2004, G.M Fitzmaurice, N.M. Laird and J.H. Ware, Wiley, New York.]

Centile: Synonym for **percentile**.

Centile reference charts: Charts used in medicine to observe clinical measurements on individual patients in the context of population values. If the population centile corresponding to the subject's value is atypical (i.e. far from the 50% value) this may indicate an underlying pathological condition. The chart can also provide a background with which to compare the measurement as it changes over time. An example is given in Fig. 28. [*Statistics in Medicine*, 1996, **15**, 2657–68.]

Centralized database: A *database* held and maintained in a central location, particularly in a *multicentre study.*

Central limit theorem: If a random variable Y has population mean μ and population variance σ^2, then the sample mean, $\bar{y}$, based on n observations, has an approximate normal distribution with mean μ and variance σ^2/n, for sufficiently large n. The theorem occupies an important place in statistical theory. [KA1 Chapter 8.]

Central range: The range within which the central 90% of values of a set of observations lie. [SMR Chapter 3.]

Central tendency: A property of the distribution of a variable usually measured by statistics such as the mean, median and mode.

Centroid method: A method of *factor analysis* widely used before the advent of high-speed computers but now only of historical interest. [*Psychological Review*, 1931, **38**, 406–27.]

CEP: Abbreviation for **circular error probable**.

CER: Abbreviation for **cost-effectiveness ratio**.

CERES plot: Abbreviation for **combining conditional expectations and residuals plot**.

CFA: Abbreviation for **confirmatory factor analysis**.

CHAID: Abbreviation for **chi-squared automated interaction detector**.

Chain-binomial models: Models arising in the mathematical theory of infectious diseases, that postulate that at any stage in an *epidemic* there are a certain number of infected and susceptibles, and that it is reasonable to suppose that the latter will yield a fresh crop of cases at the next stage, the number of new cases having a *binomial distribution.* This results in a 'chain' of binomial distributions, the actual probability of a new infection at any stage depending on the numbers of infectives and susceptibles at the previous stage. [*Stochastic Analysis and Applications*, 1995, **13**, 355–60.]

Chaining: A phenomenon often encountered in the application of *single linkage clustering* which relates to the tendency of the method to incorporate intermediate points between distinct clusters into an existing cluster rather than initiate a new one. [MV2 Chapter 10.]

Chain-of-events data: Data on a succession of events that can only occur in a prescribed order. One goal in the analysis of this type of data is to determine the distribution of times between successive events. [*Biometrics*, 1999, **55**, 179–87.]

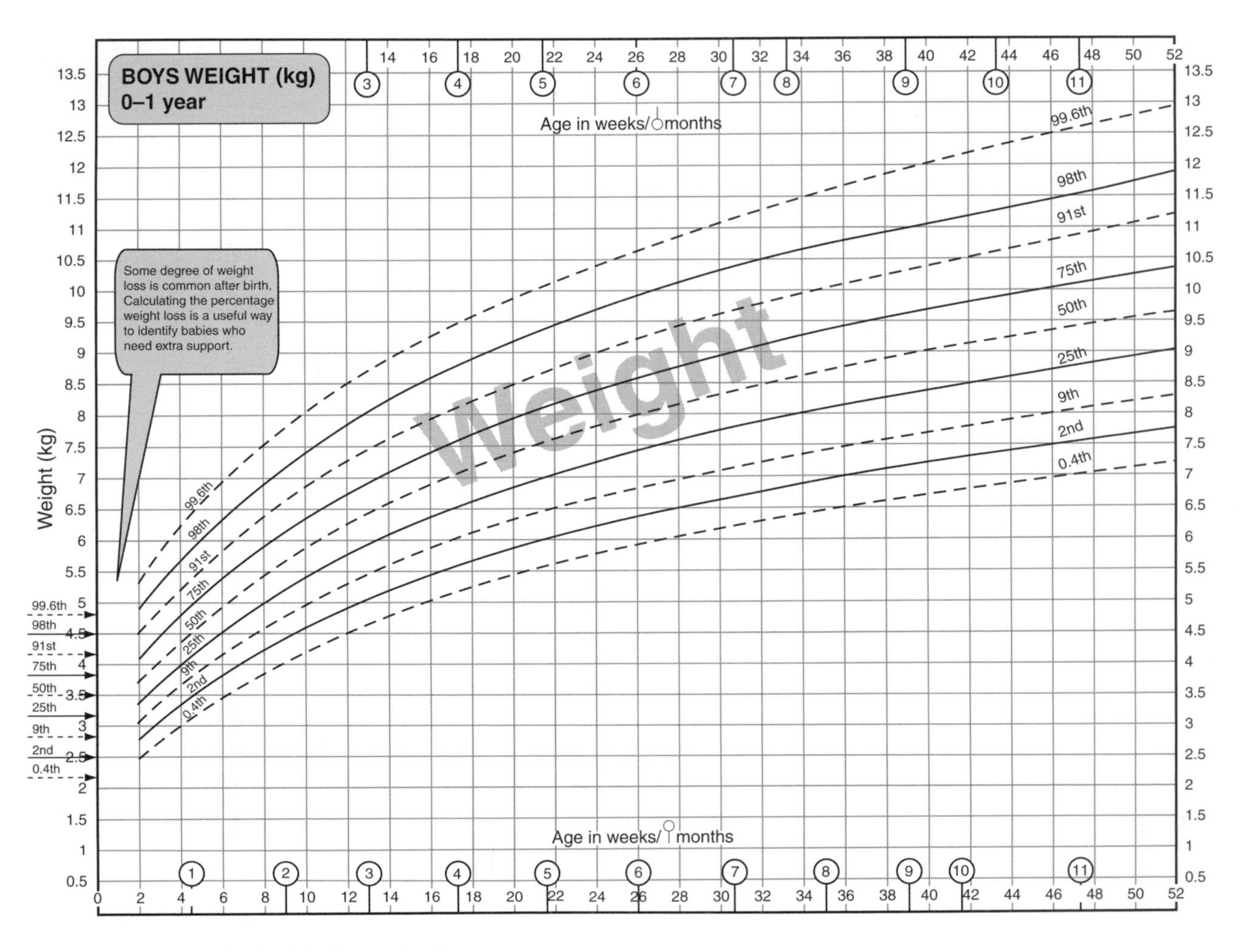

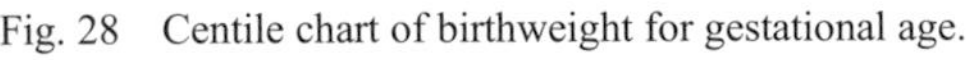
Fig. 28 Centile chart of birthweight for gestational age.

Chains of infection: A description of the course of an infection among a set of individuals. The susceptibles infected by direct contact with the introductory cases are said to make up the first generation of cases; the susceptibles infected by direct contact with the first generation are said to make up the second generation and so on. The enumeration of the number of cases in each generation is called an *epidemic chain*. Thus the sequence 1–2–1–0 denotes a chain consisting of one introductory case, two first generation cases, one second generation case and no cases in later generations. [*Occupational and Environmental Medicine*, 2004, **61**, 96–102.]

Chalmers, Thomas Clark (1917–1995): Born in Forest Hills, New York, Chalmers graduated from Columbia University College of Physicians and Surgeons in 1943. After entering private practice he became concerned over the lack of knowledge on the efficacy of accepted medical therapies, and eventually became a leading advocate for *clinical trials*, and later for *meta-analysis* setting up a meta-analysis consultancy company at the age of 75. In a distinguished research and teaching career, Chalmers was President and Dean of the Mount Sinai Medical Center and School of Medicine in New York City from 1973 to 1983. He died on 27 December 1995, in Hanover, New Hampshire.

Champernowne, David Gawen (1912–2000): Born in Oxford, Champernowne studied mathematics at King's College, Cambridge, later switching to economics, and gaining first class honours in both. Before World War II he worked at the London School of Economics and then at Cambridge where he demonstrated that the evolution of an income and wealth distribution could be represented by a Markovian model of income mobility. During the war he worked at the Ministry of Aircraft Production, and at the end of the war became Director of the Oxford Institute of Statistics. In 1948 he was made Professor of Statistics at Oxford, and carried out work on the application of *Bayesian analysis* to *autoregressive series*. In 1958 Champernowne moved to Cambridge and continued research into the theory of capital and the measurement of economic inequality. He died on 22 August 2000.

Chance events: According to Cicero these are events that occurred or will occur in ways that are uncertain-events that may happen, may not happen, or may happen in some other way. Cicero's characterization of such events is close to one dictionary definition of chance namely, 'the incalculable element in existence that renders events unpredictable'. But for Leucippus (circa 450 B.C.) the operation of chance was associated with some hidden cause and such a view was held widely in the middle ages where chance had no place in the universe, rather all events were though of as predetermined by God or by extrinsic causes determined by God. But in the 20th century the deterministic view of the universe was overthrown by the successful development of quantum mechanics with its central tenet being that it is impossible to predict exactly the outcome of an atomic (or molecular) system and where this uncertainty is not due to measurement error or experimental clumsiness but is a fundamental aspect of the basic physical laws themselves. [*Chance Rules*, 2nd edition, 2008, B. S. Everitt, Springer, New York.]

Change point problems: Problems with chronologically ordered data collected over a period of time during which there is known (or suspected) to have been a change in the underlying data generation process. Interest then lies in, retrospectively, making inferences about the time or position in the sequence that the change occurred. A famous example is the Lindisfarne scribes data in which a count is made of the occurrences of a particular type of pronoun ending observed in 13 chronologically ordered medieval manuscripts believed to be the work of more than one author. A plot of the data (see Fig. 29) shows strong evidence of a change point. A simple example of a possible model for such a problem is the following;

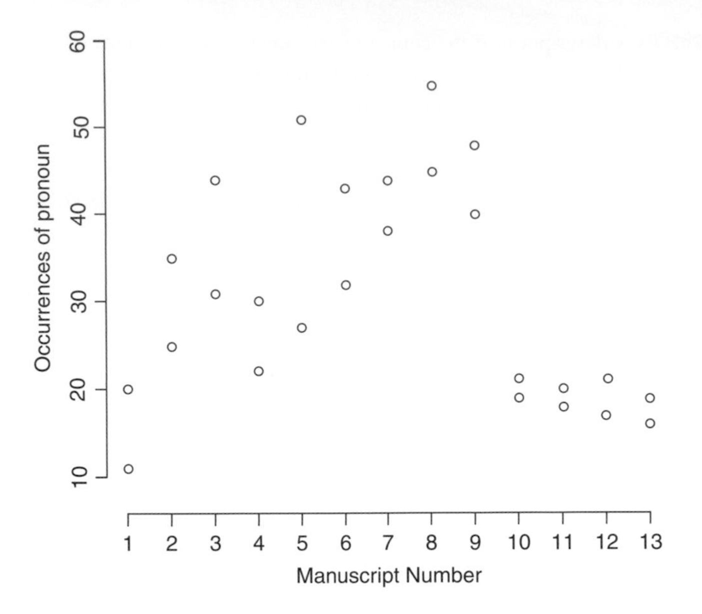

Fig. 29 A plot of the Lindisfarne scribes data indicating a clear change point.

$$y_i = \alpha_1 + \beta_1 x_i + e_i, \qquad i = 1, \ldots, \tau$$
$$y_i = \alpha_2 + \beta_2 x_i + e_i, \qquad i = \tau + 1, \ldots, T$$

Interest would centre on estimating the parameters in the model particularly the change point, τ. [*Statistics in Medicine*, 1983, **2**, 141–6.]

Change scores: The change in an outcome of interest from pre-treatment to post-treatment. Often used as a response variable in *longitudinal studies*. See **Lord's paradox**. [*Sociological Methodology*, 1990, **20**, 93–114.]

Chaos: Apparently random behaviour exhibited by a *deterministic model*. [*Chaos*, 1987, J. Gleick, Sphere Books Ltd, London.]

Chapman–Kolmogoroff equation: See **Markov chain**.

Chapman's estimator: See **capture–recapture sampling**.

Characteristic function: A function, $\phi(t)$, derived from a probability distribution, $f(x)$, as

$$\phi(t) = \int_{-\infty}^{\infty} e^{itx} f(x) \mathrm{d}x$$

where $i^2 = -1$ and t is real. The function is of great theoretical importance and under certain general conditions determines and is completely determined by the probability distribution. If $\phi(t)$ is expanded in powers of t then the rth central *moment*, μ_r', is equal to the coefficient of $(it)^r/r!$ in the expansion. Thus the characteristic function is also a *moment generating function*. [KA1 Chapter 4.]

Characteristic root: Synonym for **eigenvalue**.

Characteristic vector: Synonym for **eigenvector**.

Chebyshev–Hermite polynomial: A function, $H_r(x)$ defined by the identity

$$(-D)^r \alpha(x) = H_r(x)\alpha(x)$$

where $\alpha(x) = (1/\sqrt{2\pi})e^{-\frac{1}{2}x^2}$ and $D = d/dx$. It is easy to show that $H_r(x)$ is the coefficient of $t^r/r!$ in the expansion of $\exp(tx - \frac{1}{2}t^2)$ and is given explicitly by

$$H_r(x) = x^r - \frac{r^{[2]}}{2.1!}x^{r-2} + \frac{r^{[4]}}{2^2.2!}x^{r-4} - \frac{r^{[6]}}{2^3.3!}x^{r-6} + \cdots$$

where $r^{[i]} = r(r-1)\ldots(r-i+1)$. For example, $H_6(x)$ is

$$H_6(x) = x^6 - 15x^4 + 45x^2 - 15$$

The polynomials have an important orthogonal property namely that

$$\int_{-\infty}^{\infty} H_m(x)H_n(x)\alpha(x)dx = 0, \qquad m \neq n$$

$$= n!, \qquad m = n$$

Used in the *Gram–Charlier Type A series.* [KA1 Chapter 6.]

Chebyshev, Pafnuty Lvovich (1821–1894): Born in Okatovo, Russia, Chebyshev studied at Moscow University. He became professor at St Petersburg in 1860, where he founded the Petersburg mathematical school which influenced Russian mathematics for the rest of the century. Made important contributions to the theory of the distribution of prime numbers, but most remembered for his work in probability theory where he proved a number of fundamental limit theorems. Chebyshev died on 8 December 1894 in St Petersburg, Russia.

Chebyshev's inequality: A statement about the proportion of observations that fall within some number of standard deviations of the mean for any probability distribution. One version is that for a random variable, X

$$\text{Prob}\left[-k \leq \frac{x-\mu}{\sigma} \leq k\right] \leq 1 - \frac{1}{k^2}$$

where k is the number of standard deviations, σ, from the mean, μ. For example, the inequality states that at least 75% of the observations fall within two standard deviations of the mean. If the variable X can take on only positive values then the following, known as the *Markov inequality*, holds;

$$\text{Prob}[X \leq x] \leq 1 - \frac{\mu}{x}$$

[KA1 Chapter 3.]

Chemometrics: The application of mathematical and statistical methods to problems in Chemistry. [*Chemometrics: A Textbook*, 1988, D. L. Massart, B. G. M. Vandeginste, S. N. Deming, Y. Michotte and L. Kaufman, Elsevier, Amsterdam.]

Chernoff's faces: A technique for representing *multivariate data* graphically. Each observation is represented by a computer-generated face, the features of which are controlled by an observation's variable values. The collection of faces representing the set of observations may be useful in identifying groups of similar individuals, *outliers*, etc. See Fig. 30. See also **Andrews' plots** and **glyphs**. [MV1 Chapter 3.]

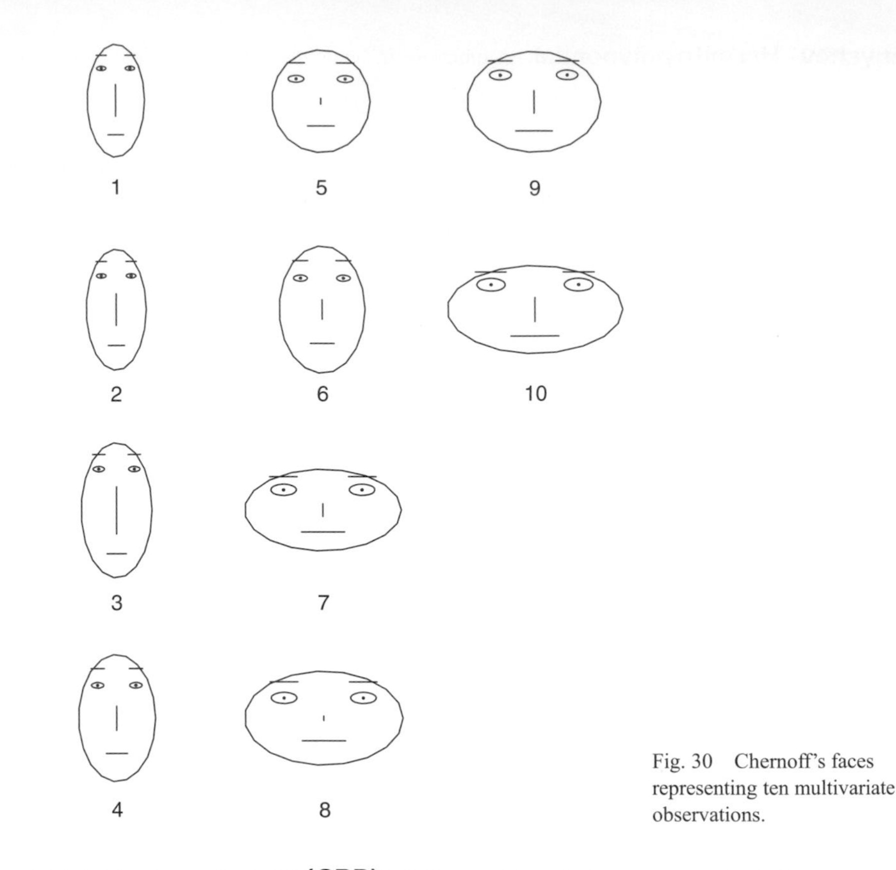

Fig. 30 Chernoff's faces representing ten multivariate observations.

Chinese restaurant process (CRP): A discrete-time *stochastic process* that at time n produces a partition of the integers $\{1, 2, \ldots, n\}$ and can be described in terms of a scheme by which n customers sit down in a Chinese restaurant with an infinite number of tables. (The terminology is inspired by the Chinese restaurants in San Francisco which appear to have an infinite seating capacity.) Customers walk in and sit down at some table, with the tables being chosen as follows;

1. The first customer always chooses the first table.
2. The mth customer chooses either the first unoccupied table with probability $\frac{\gamma}{m-1+\gamma}$ or occupied table i with probability $\frac{m_i}{m-1+\gamma}$ where m_i is the number of customers already seated at table i and γ is a parameter of the process.

After n customers have sat down the seating plan gives a partition of the integers $\{1, 2, \ldots, n\}$. As an example consider the following possible arrangement of 10 customers;

Table 1: (1,3,8), Table 2: (2,5,9,10), Table 3: (4,6,7)

The probability of this arrangement is given by;

$$1 \times \frac{\gamma}{1+\gamma} \times \frac{1}{2+\gamma} \times \frac{\gamma}{3+\gamma} \times \frac{1}{4+\gamma} \times \frac{1}{5+\gamma} \times \frac{2}{6+\gamma} \times \frac{2}{7+\gamma} \times \frac{2}{8+\gamma} \times \frac{3}{9+\gamma}$$

The CRP has been used to represent uncertainty in the number of components in a *finite mixture distribution*. The distribution induced by a *Dirichlet process* can be generated incrementally using the Chinese restaurant process. [*Statistica Sinica*, 2003, **13**, 1211–1235.]

Chi-bar squared distribution: A term used for a mixture of *chi-squared distributions* that is used in the simultaneous modelling of the *marginal distributions* and the association between two categorical variables. [*Journal of the American Statistical Association*, 2001, **96**, 1497–1505.]

Chi-plot: An auxiliary display to the *scatterplot* in which independence is manifested in a characteristic way. The plot provides a graph which has characteristic patterns depending on whether the variates are (1) independent, (2) have some degree of monotone relationship, (3) have more complex dependence structure. The plot depends on the data only through the values of their ranks. The plot is a scatterplot of the pairs (λ_i, χ_i), $|\lambda_i| < 4\{\frac{1}{n-1} - 0.5\}^2$, where

$$\lambda_i = 4S_i \max\{(F_i - 0.5)^2, (G_i - 0.5)^2\}$$
$$\chi_i = (H_i - F_i G_i)/\{F_i(1 - F_i)G_i(1 - F_i)\}^{\frac{1}{2}}$$

and where

$$H_i = \sum_{j \neq i} I(x_j \leq x_i, y_j \leq y_i)/(n-1)$$
$$F_i = \sum_{j \neq i} I(x_j \leq x_i)/(n-1)$$
$$G_i = \sum_{j \neq i} I(y_j \leq y_i)/(n-1)$$
$$S_i = \text{sign}\{(F_i - 0.5)(G_i - 0.5)\}$$

with $(x_1, y_1) \ldots (x_n, y_n)$ being the observed sample values, and $I(A)$ being the indicator function of the event A. Example plots are shown in Figure 31. Part (a) shows the situation in which x and y are independent, Part (b) in which they have a correlation of 0.6. In each case the left hand plot is a simple scatterplot of the data, and the right hand is the corresponding chi-plot. [*The American Statistician*, 2001, **55**, 233–239.]

Chi-squared automated interaction detector (CHAID): Essentially an *automatic interaction detector* for binary target variables. [*Multivariable Analysis*, 1996, A. R. Feinstein, Yale University Press, New Haven.]

Chi-squared distance: A distance measure for categorical variables that is central to *correspondence analysis*. Similar to *Euclidean distance* but effectively compensates for the different levels of occurrence of the categories. [MV1 Chapter 5.]

Chi-squared distribution: The probability distribution, $f(x)$, of a random variable defined as the sum of squares of a number (ν) of independent standard normal variables and given by

$$f(x) = \frac{1}{2^{\nu/2}\Gamma(\nu/2)} x^{(\nu-2)/2} e^{-x/2}, \quad x > 0$$

The *shape parameter*, ν, is usually known as the degrees of freedom of the distribution. This distribution arises in many areas of statistics, for example, assessing the goodness-of-fit of models, particularly those fitted to *contingency tables*. The mean of the distribution is ν and its variance is 2ν. See also **non-central chi-squared distribution**. [STD Chapter 8.]

Chi-squared probability plot: A procedure for testing whether a set of *multivariate data* have a *multivariate normal distribution*. The ordered generalized distances

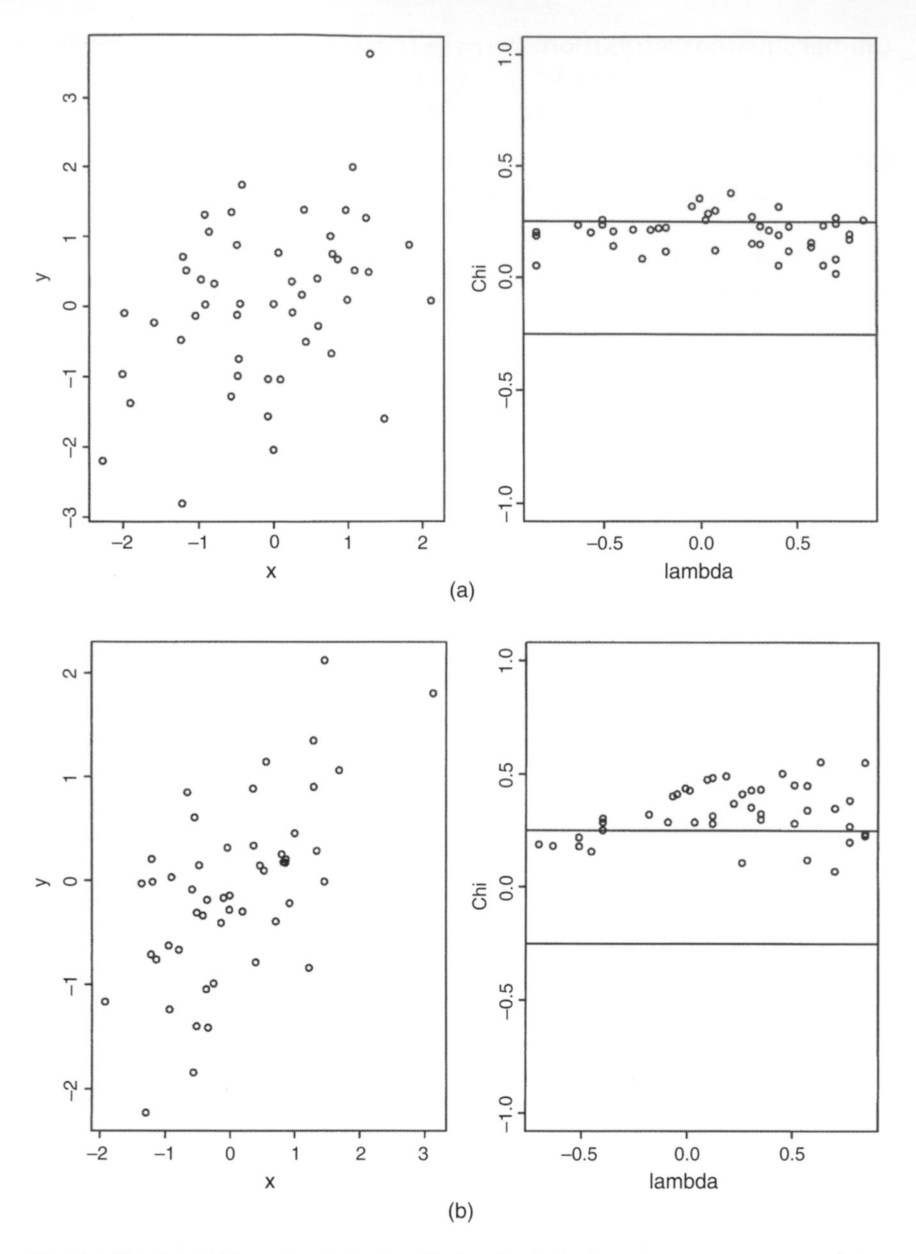

Fig. 31 Chi-plot. (a) Uncorrelated situation. (b) Correlated situation.

$$d_{(i)} = (\mathbf{x}_i - \bar{\mathbf{x}})'\mathbf{S}^{-1}(\mathbf{x}_i - \bar{\mathbf{x}})$$

where $\mathbf{x}_i, i = 1, \ldots, n$, are multivariate observations involving q variables, $\bar{\mathbf{x}}$ is the sample mean vector and $\mathbf{S}$ the sample *variance–covariance matrix*, are plotted against the quantiles of a *chi-squared distribution* with q degrees of freedom. Deviations from multivariate normality are indicated by depatures from a straight line in the plot. An example of such a plot appears in Fig. 32. See also **quantile–quantile plot**. [*Principles of Multivariate Analysis*, 2nd edition, 2000, W. J. Krzanowski, Oxford Science Publications, Oxford.]

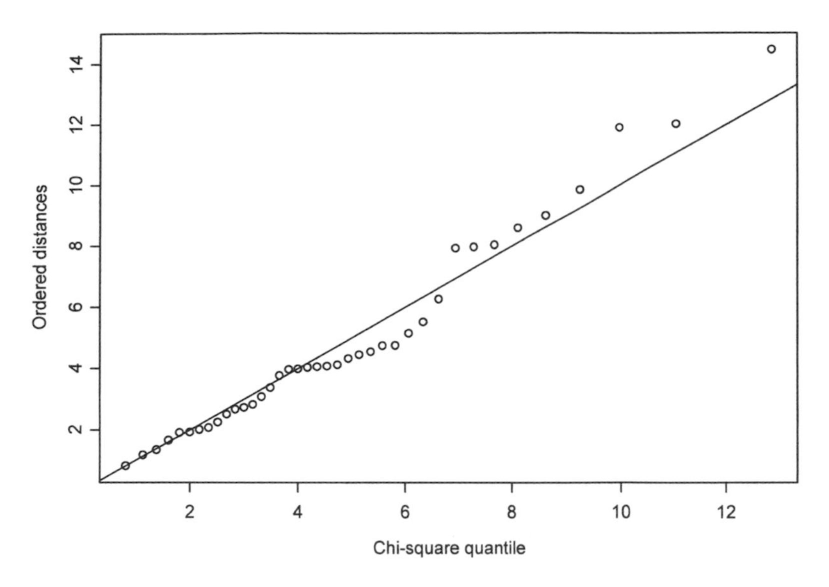

Fig. 32 Chi-squared probability plot indicating data do not have a multivariate normal distribution.

Chi-squared statistic: A statistic having, at least approximately, a *chi-squared distribution*. An example is the test statistic used to assess the independence of the two variables forming a *contingency table*

$$X^2 = \sum_{i=1}^{r}\sum_{j=1}^{c}(O_i - E_i)^2/E_i$$

where O_i represents an observed frequency and E_i the expected frequency under independence. Under the hypothesis of independence X^2 has, approximately, a chi-squared distribution with $(r-1)(c-1)$ degrees of freedom. [SMR Chapter 10.]

Chi-squared test for trend: A test applied to a *two-dimensional contingency table* in which one variable has two categories and the other has k ordered categories, to assess whether there is a difference in the trend of the proportions in the two groups. The result of using the ordering in this way is a test that is more powerful than using the *chi-squared statistic* to test for independence. [SMR Chapter 10.]

Choi-Williams distribution: A *time-frequency distribution* useful when high time and frequency resolutions are needed. See also **Wigner-Ville distribution**. [*Proceedings of the IEEE*, 1989, **77**, 941–981.]

Choleski decomposition: The decomposition of a *symmetric matrix*, **A** (which is not a *singular matrix*), into the form

$$\mathbf{A} = \mathbf{LL}'$$

where **L** is a *lower triangular matrix*. Widely used for solving linear equations and matrix inversion. [MV1 Appendix A.]

Chow test: A test of the equality of two independent sets of regression coefficients under the assumption of normally distributed errors. [*Biometrical Journal*, 1996, **38**, 819–28.]

Christmas tree boundaries: An adjustment to the stopping rule in a sequential *clinical trial* for the gaps between the 'looks'. [*The Design and Analysis of Sequential Clinical Trials*, 1997, J. Whitehead, Wiley, Chichester.]

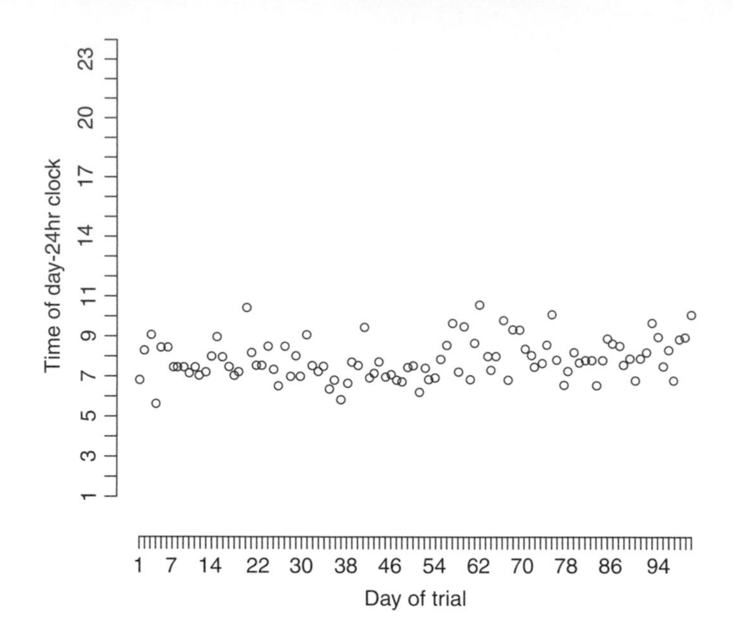

Fig. 33 Chronology plot of times that a tablet is taken in a clinical trial. (Reproduced from *Statistics in Medicine* with permission of the publishers John Wiley.)

Chronology plot: A method of describing *compliance* for individual patients in a *clinical trial*, where the times that a tablet are taken are depicted over the study period (see Fig. 33). See also **calendar plot**. [*Statistics in Medicine*, 1997, **16**, 1653–64.]

Chronomedicine: The study of the mechanisms underlying variability in *circadian* and other rhythms found in human beings. [*Biological Rhythms in Clinical and Laboratory Medicine*, 1992, Y. Touitou and E. Haus, Springer-Verlag, Berlin.]

Chuprov, Alexander Alexandrovich (1874–1926): Born in Mosal'sk, Russia Chuprov graduated from Moscow University in 1896 with a dissertation on probability theory as a basis for theoretical statistics. He then left for Germany to study political economy. In 1902 Chuprov took a position in the newly formed Economics Section of the St. Petersburg Polytechnic Institute and published work that emphasised the logical and mathematical approach to statistics which had considerable influence in Russia. Chuprov died on April 19th, 1926 in Geneva, Switzerland.

Circadian variation: The variation that takes place in variables such as blood pressure and body temperature over a 24 hour period. Most living organisms experience such variation which corresponds to the day/night cycle caused by the Earth's rotation about its own axis. [SMR Chapter 7.]

Circular data: Observations on a *circular random variable.*

Circular distribution: A probability distribution, $f(\theta)$, of a *circular random variable*, θ which ranges from 0 to 2π so that the probability may be regarded as distributed around the circumference of a circle. The function f is periodic with period 2π so that $f(\theta + 2\pi) = f(\theta)$. An example is the *von Mises distribution.* See also **cardiord distribution**. [KA1 Chapter 5.]

Circular probable error: An important measure of accuracy for problems of directing projectiles at targets, which is the bivariate version of a 50% quantile point. Defined explicitly as the value

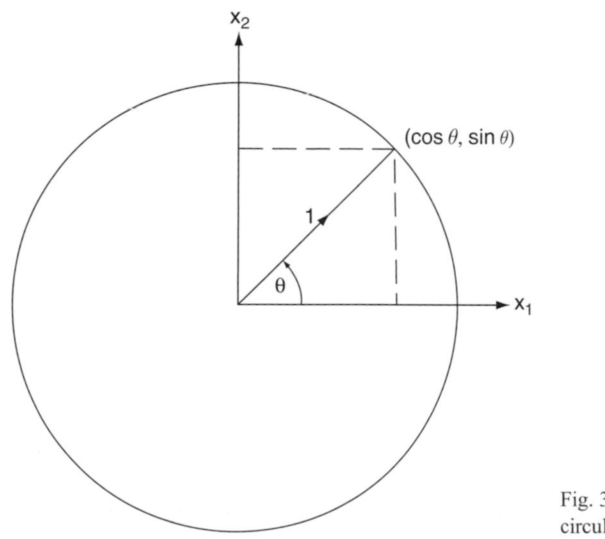

Fig. 34 Diagram illustrating a circular random variable.

R such that on average half of a group of projectiles will fall within the circle of radius R about the target point. [*Journal of the Royal Statistical Society, Series B*, 1960, **22**, 176–87.]

Circular random variable: An angular measure confined to be on the unit circle. Figure 34 shows a representation of such a variable. See also **cardiord distribution** and **von Mises distribution**. [*Multivariate Analysis*, 1979, K. V. Mardia, J. T. Kent and J. M. Bibby, Academic Press, London.]

City-block distance: A *distance measure* occasionally used in *cluster analysis* and given by

$$d_{ij} = \sum_{k=1}^{q} |x_{ik} - x_{jk}|$$

where q is the number of variables and $x_{ik}, x_{jk}, k = 1, \ldots, q$ are the observations on individuals i and j. [MV2 Chapter 10.]

Class frequency: The number of observations in a class interval of the observed frequency distribution of a variable.

Classical scaling: A form of *multidimensional scaling* in which the required coordinate values are found from the *eigenvectors* of a matrix of inner products. [MV1 Chapter 5.]

Classical statistics: Synonym for **frequentist inference**.

Classification and regression tree technique (CART): An alternative to *multiple regression* and associated techniques for determining subsets of explanatory variables most important for the prediction of the response variable. Rather than fitting a model to the sample data, a *tree* structure is generated by dividing the sample recursively into a number of groups, each division being chosen so as to maximize some measure of the difference in the response variable in the resulting two groups. The resulting structure often provides easier interpretation than a regression equation, as those variables most important for prediction can be quickly identified. Additionally this approach does not require distributional assumptions and is also more resistant to the effects of *outliers*. At each stage the sample is split on the basis of a variable, x_i, according to the answers to such questions as 'Is $x_i \leq c$' (univariate split), is '$\sum a_i x_i \leq c$' (linear function split) and 'does $x_i \in A$' (if x_i is a categorical variable).

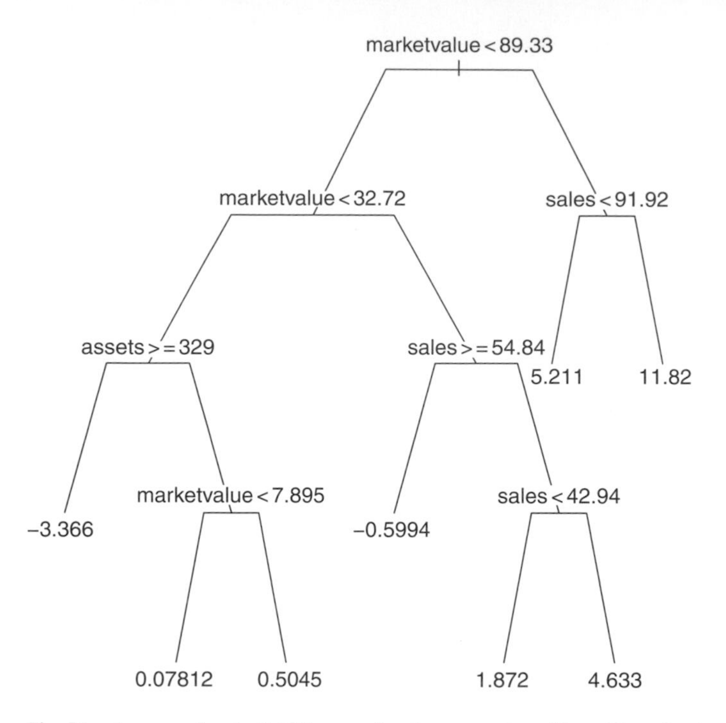

Fig. 35 An example of a CART regression for companies with profits as the dependent variable and assets, market value and sales as explanatory variables.

An illustration of an application of this method is shown in Fig. 35. See also **automatic interaction detector**. [MV2 Chapter 9.]

Classification matrix: A term often used in *discriminant analysis* for the matrix summarizing the results obtained from the derived classification rule, and obtained by crosstabulating observed against predicted group membership. Contains counts of correct classifications on the main diagonal and incorrect classifications elsewhere. [MV2 Chapter 9.]

Classification rule: See **discriminant analysis**.

Classification techniques: A generic term used for both *cluster analysis* methods and *discriminant methods* although more widely applied to the former. [MV1 Chapter 1.]

Class intervals: The intervals of the frequency distribution of a set of observations.

Clemmesen's hook: A phenomenon sometimes observed when interpreting parameter estimates from *age–period–cohort models*, where rates increase to some maximum, but then fall back slightly before continuing their upward trend. [*Anticancer Research*, 1995, **15**, 511–515.]

Cliff and Ord's BW statistic: A measure of the degree to which the presence of some factor in an area (or time period) increases the chances that this factor will be found in a nearby area. Defined explicitly as

$$\mathrm{BW} = \sum\sum \delta_{ij}(x_i - x_j)^2$$

where $x_i = 1$ if the ith area has the characteristic and zero otherwise and $\delta_{ij} = 1$ if areas i and j are adjacent and zero otherwise. See also **adjacency matrix and Moran's I**. [*Spatial*

Processes: Models, Inference and Applications, 1980, A. D. Cliff and J. K. Ord, Pion, London.]

Clinical epidemiology: The application of epidemiological methods to the study of clinical phenomena, particularly diagnosis, treatment decisions and outcomes. [*Clinical Epidemiology: The Essentials*, 3rd edition, 1996, R. H. Fletcher, S. W. Fletcher and E. H. Wagner, Williams and Wilkins, Baltimore.]

Clinical priors: See **prior distribution**.

Clinical trials: Medical experiments designed to evaluate which (if any) of two or more treatments is the more effective. It is based on one of the oldest principles of scientific investigation, namely that new information is obtained from a comparison of alternative states. The three main components of a clinical trial are:

- Comparison of a group of patients given the treatment under investigation (the treatment group) with another group of patients given either an older or standard treatment, in one exists, or an 'inert treatment' generally known as a *placebo* (the *control group*). Some trials may involve more than two groups.
- A method of assigning patients to the treatment and control groups.
- A measure of outcome, i.e., a response variable.

One of the most important aspects of a clinical trial is the question of how patients should be allocated to the treatment and control group. The objective in allocation is that the treatment group and control group should be alike in all respects except the treatment received. As a result the clinical trial is more likely to provide an unbiased estimate of the difference between the two treatments. The most appropriate (perhaps only appropriate) method of allocation is randomization leading to *randomized clinical trials*, the gold standard for treatment assessment. [*Clinical Trials in Psychiatry*, 2nd edn, 2008, B. S. Everitt and S. Wessely, Wiley, Chichester.]

Clinical vs statistical significance: The distinction between results in terms of their possible clinical importance rather than simply in terms of their statistical significance. With large samples, for example, very small differences that have little or no clinical importance may turn out to be statistically significant. The practical implications of any finding in a medical investigation must be judged on clinical as well as statistical grounds. [SMR Chapter 15.]

Clopper–Pearson interval: A *confidence interval* for the probability of success in a series of n independent repeated *Bernoulli trials*. For large sample sizes a $100(1-\alpha/2)\%$ interval is given by

$$\hat{p} - z_{\alpha/2}\left[\frac{\hat{p}(1-\hat{p})}{n}\right]^{\frac{1}{2}} \leq p \leq \hat{p} + z_{\alpha/2}\left[\frac{\hat{p}(1-\hat{p})}{n}\right]^{\frac{1}{2}}$$

where $z_{\alpha/2}$ is the appropriate standard normal deviate and $\hat{p} = B/n$ where B is the observed number of successes. See also **binomial test**. [NSM Chapter 2.]

Closed sequential design: See **sequential analysis**.

ClustanGraphics: A specialized software package that contains advanced graphical facilities for *cluster analysis*. [www.clustan.com]

Cluster: See **disease cluster**.

Cluster analysis: A set of methods for constructing a (hopefully) sensible and informative classification of an initially unclassified set of data, using the variable values observed on each

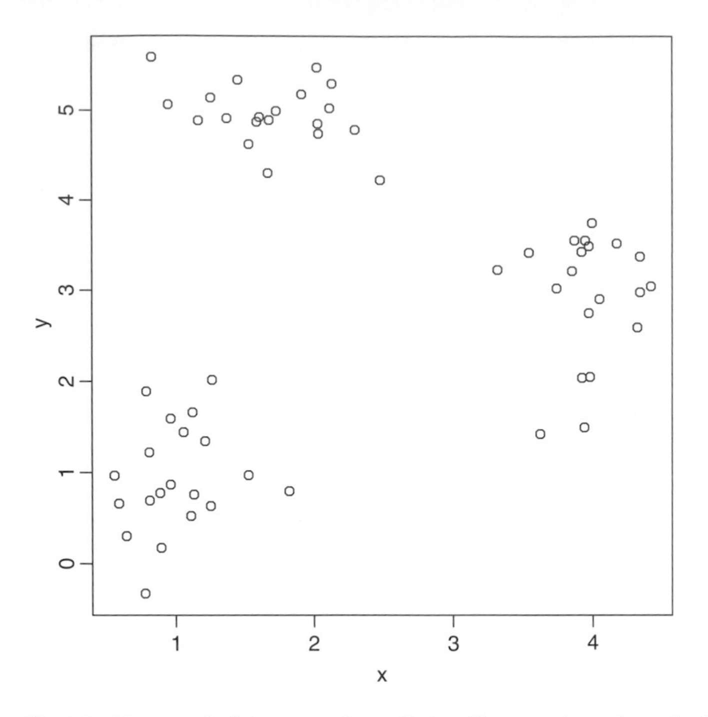

Fig. 36 Cluster analysis by eye can be applied to this scatterplot to detect the three distinct clusters.

individual. Essentially all such methods try to imitate what the eye–brain system does so well in two dimensions; in the scatterplot shown at Fig. 36, for example, it is very simple to detect the presence of three clusters without making the meaning of the term ‘ cluster’ explicit. See also **agglomerative hierarchical clustering**, ***K*-means clustering**, and **finite mixture densities**. [MV2 Chapter 10.]

Clustered binary data: See **clustered data**.

Clustered data: A term applied to both data in which the *sampling units* are grouped into clusters sharing some common feature, for example, animal litters, families or geographical regions, and *longitudinal data* in which a cluster is defined by a set of repeated measures on the same unit. A distinguishing feature of such data is that they tend to exhibit intracluster correlation, and their analysis needs to address this correlation to reach valid conclusions. Methods of analysis that ignore the correlations tend to be inadequate. In particular they are likely to give estimates of standard errors that are too low. When the observations have a normal distribution, *random effects models* and *mixed effects models* may be used. When the observations are binary, giving rise to *clustered binary data*, suitable methods are *mixed-effects logistic regression* and the *generalized estimating equation* approach. See also **multilevel models**. [*Statistics in Medicine*, 1992, **11**, 67–100.]

Cluster randomization: The random allocation of groups or clusters of individuals in the formation of treatment groups. Although not as statistically efficient as individual randomization, the procedure frequently offers important economic, feasibility or ethical advantages. Analysis of the resulting data may need to account for possible intracluster correlation (see **clustered data**). [*Design and Analysis of Cluster Randomization Trials in Health Research*, 2000, A. Donner and N. Klar, Arnold, London.]

Cluster sampling: A method of sampling in which the members of a population are arranged in groups (the 'clusters'). A number of clusters are selected at random and those chosen are then subsampled. The clusters generally consist of natural groupings, for example, families, hospitals, schools, etc. See also **random sample, area sampling, stratified random sampling** and **quota sample**. [KA1 Chapter 5.]

Cluster-specific models: Synonym for **subject-specific models**.

C_{max}: A measure traditionally used to compare treatments in *bioequivalence trials*. The measure is simply the highest recorded response value for a subject. See also **area under curve, response feature analysis** and **T_{max}**. [*Pharmaceutical Research*, 1995, **92**, 1634–41.]

Coale–Trussell fertility model: A model used to describe the variation in the age pattern of human fertility, which is given by

$$R_{ia} = n_a M_i e^{m_i \upsilon_a}$$

where R_{ia} is the expected marital fertility rate, for the ath age of the ith population, n_a is the standard age pattern of natural fertility, υ_a is the typical age-specific deviation of controlled fertility from natural fertility, and M_i and m_i measure the ith population's fertility level and control. The model states that marital fertility is the product of natural fertility, $n_a M_i$, and fertility control, $\exp(m_i \upsilon_a)$ [*Journal of Mathematical Biology*, 1983, **18**, 201–11.]

Coarse data: A term sometimes used when data are neither entirely missing nor perfectly present. A common situation where this occurs is when the data are subject to rounding; others correspond to *digit preference* and *age heaping*. [*Biometrics*, 1993, **49**, 1099–1109.]

Coarsening at random (CAR): The most general form of randomly grouped, censored or missing data, for which the coarsening mechanism can be ignored when making likelihood-based inference. See also *missing at random*, which is a special case. [*Annals of Statistics*, 1991, **19**, 2244–2253.]

Cobb–Douglas distribution: A name often used in the economics literature as an alternative for the *lognormal distribution*.

Cochran, William Gemmell (1909–1980): Born in Rutherglen, a surburb of Glasgow, Cochran read mathematics at Glasgow University and in 1931 went to Cambridge to study statistics for a Ph.D. While at Cambridge he published what was later to become known as *Cochran's Theorem*. Joined Rothamsted in 1934 where he worked with *Frank Yates* on various aspects of experimental design and sampling. Moved to the Iowa State College of Agriculture, Ames in 1939 where he continued to apply sound experimental techniques in agriculture and biology. In 1950, in collaboration with *Gertrude Cox*, published what quickly became the standard text book on experimental design, *Experimental Designs*. Became head of the Department of Statistics at Harvard in 1957, from where he eventually retired in 1976. President of the Biometric Society in 1954–1955 and Vice-president of the American Association for the Advancement of Science in 1966. Cochran died on 29 March 1980 in Orleans, Massachusetts.

Cochran-Armitage test: A test of independence in a *contingency table* for one binary and one ordinal variable. It has more power than the standard chi-squared test of independence because the alternative hypothesis is a linear trend in the probability of the binary variable instead of simply *any* association between the two variables. [*Biometrics*, 1955, **11**, 375–386.]

Cochran's C-test: A test that the variances of a number of populations are equal. The test statistic is

$$C = \frac{s^2_{\max}}{\sum_{i=1}^{g} s_i^2}$$

where g is the number of populations and $s_i^2, i = 1, \ldots, g$ are the variances of samples from each, of which $s^2_{\max}$ is the largest. Tables of critical values are available. See also **Bartlett's test, Box's test** and **Hartley's test**. [*Simultaneous Statistical Inference*, 1966, R. G. Miller, McGraw-Hill, New York.]

Cochran's Q-test: A procedure for assessing the hypothesis of no inter-observer bias in situations where a number of raters judge the presence or absence of some characteristic on a number of subjects. Essentially a generalized *McNemar's test.* The test statistic is given by

$$Q = \frac{r(r-1)\sum_{j=1}^{r}(y_{.j} - y_{..}/r)^2}{ry_{..} - \sum_{i=1}^{n} y_{i.}^2}$$

where $y_{ij} = 1$ if the i th patient is judged by the jth rater to have the characteristic present and 0 otherwise, $y_{i.}$ is the total number of raters who judge the ith subject to have the characteristic, $y_{.j}$ is the total number of subjects the jth rater judges as having the characteristic present, $y_{..}$ is the total number of 'present' judgements made, n is the number of subjects and r the number of raters. If the hypothesis of no inter-observer bias is true, Q, has approximately, a *chi-squared distribution* with $r - 1$ degrees of freedom. [*Biometrika*, 1950, **37**, 256–66.]

Cochran's Theorem: Let $\mathbf{x}$ be a vector of q independent standardized normal variables and let the sum of squares $Q = \mathbf{x}'\mathbf{x}$ be decomposed into k quadratic forms $Q_i = \mathbf{x}'\mathbf{A}_i\mathbf{x}$ with ranks r_i, i.e.

$$\sum_{i=1}^{k} \mathbf{x}'\mathbf{A}_i\mathbf{x} = \mathbf{x}'\mathbf{I}\mathbf{x}$$

Then any one of the following three conditions implies the other two;

- The ranks r_i of the Q_i add to that of Q.
- Each of the Q_i has a *chi-squared distribution.*
- Each Q_i is independent of every other.

[KA1 Chapter 15.]

Cochrane, Archibald Leman (1909–1988): Cochrane studied natural sciences in Cambridge and psychoanalysis in Vienna. In the 1940s he entered the field of *epidemiology* and then later became an enthusiastic advocate of *clinical trials.* His greatest contribution to medical research was to motivate the *Cochrane collaboration.* Cochrane died on 18 June 1988 in Dorset in the United Kingdom.

Cochrane collaboration: An international network of individuals committed to preparing , maintaining and disseminating systematic reviews of the effects of health care. The collaboration is guided by six principles: collaboration, building on people's existing enthusiasm and interests, minimizing unnecessary duplication, avoiding bias, keeping evidence up to data and ensuring access to the evidence. Most concerned with the evidence from *randomized clinical trials.* See also **evidence-based medicine**. [http://cochrane.co.uk/]

Cochrane–Orcutt procedure: A method for obtaining parameter estimates for a regression model in which the error terms follow an *autoregressive model* of order one, i.e. a model

$$y_t = \boldsymbol{\beta}'\mathbf{x}_t + \upsilon_t \qquad t = 1, 2, \ldots, n$$

where $\upsilon_t = \alpha\upsilon_{t-1} + \epsilon_t$ and y_t are the dependent values measured at time t, $\mathbf{x}_t$ is a vector of explanatory variables, $\boldsymbol{\beta}$ is a vector of regression coefficients and the ϵ_t are assumed independent normally distributed variables with mean zero and variance σ^2. Writing

$y_t^* = y_t - \alpha y_{t-1}$ and $\mathbf{x}_t^* = \mathbf{x}_t - \alpha \mathbf{x}_{t-1}$ an estimate of $\boldsymbol{\beta}$ can be obtained from *ordinary least squares* of y_t^* and $\mathbf{x}_t^*$. The unknown parameter α can be calculated from the *residuals* from the regression, leading to an iterative process in which a new set of transformed variables are calculated and thus a new set of regression estimates and so on until convergence. [*Communications in Statistics*, 1993, **22**, 1315–33.]

Codominance: The relationship between the *genotype* at a locus and a *phenotype* that it influences. If individuals with heterozygote (for example, AB) genotype is phenotypically different from individuals with either homozygous genotypes (AA and BB), then the genotype-phenotype relationship is said to be codominant. [*Statistics in Human Genetics*, 1998, P. Sham, Arnold, London.]

Coefficient of alienation: A name sometimes used for $1 - r^2$, where r is the estimated value of the correlation coefficient of two random variables. See also **coefficient of determination**.

Coefficient of concordance: A coefficient used to assess the *agreement* among m raters ranking n individuals according to some specific characteristic. Calculated as

$$W = 12S/[m^2(n^3 - n)]$$

where S is the sum of squares of the differences between the total of the ranks assigned to each individual and the value $m(n+1)/2$. W can vary from 0 to 1 with the value 1 indicating perfect agreement. [*Quick Statistics*, 1981, P. Sprent, Penguin Books, London.]

Coefficient of determination: The square of the correlation coefficient between two variables. Gives the proportion of the variation in one variable that is accounted for by the other. [ARA Chapter 7.]

Coefficient of racial likeness (CRL): Developed by *Karl Pearson* for measuring resemblances between two samples of skulls of various origins. The coefficient is defined for two samples I and J as:

$$\text{CRL} = \left[\frac{1}{p}\sum_{k=1}^{p}\left\{\frac{(\overline{X}_{Ik} - \overline{X}_{Jk})^2}{(s_{Ik}^2/n_I) + (s_{Jk}^2/n_J)}\right\}^{\frac{1}{2}}\right] - \frac{2}{p}$$

where $\overline{X}_{Ik}$ stands for the sample mean of the kth variable for sample I and s_{Ik}^2 stands for the variance of this variable. The number of variables is p. [*American Journal of Physical Anthropology*, 2005, **23**, 101–109.]

Coefficient of variation: A measure of spread for a set of data defined as

$$100 \times \text{standard deviation}/\text{mean}$$

Originally proposed as a way of comparing the variability in different distributions, but found to be sensitive to errors in the mean. [KA1 Chapter 2.]

Coefficient sign prediction methods: Methods used for predicting the signs of the coefficients in a regression model; this is useful when the main purpose in selecting a model is interpretation, because the sign of a coefficient is often of primary importance for this task. [*Journal of the Royal Statistical Society, Series B*, 2007, **69**, 447–461.]

Coherence: A term most commonly used in respect of the strength of association of two *time series*. Figure 37, for example, shows daily mortality and SO_2 concentration time series in London during the winter months of 1958, with an obvious question as to whether the pollution was in any way affecting mortality. The relationship is usually measured by the time series analogue of the correlation coefficient, although the association structure is likely to be more

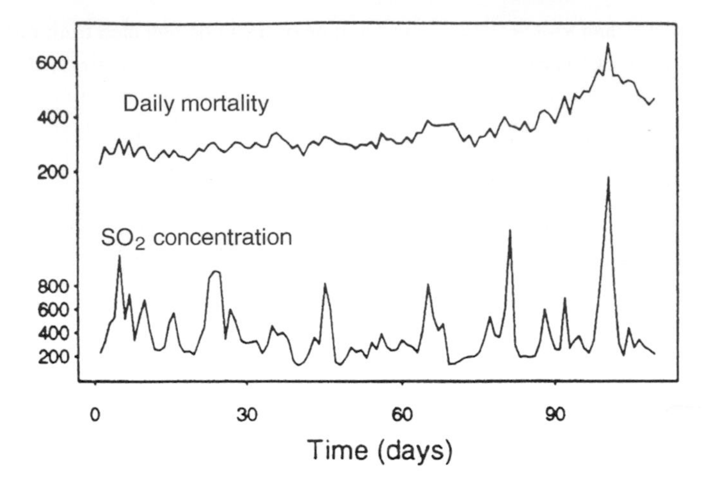

Fig. 37 Time series for daily mortality and sulphur dioxide concentration in London during the winter months of 1958.

complex and may include a leading or lagging relationship; the measure is no longer a single number but a function, $\rho(\omega)$, of frequency ω. Defined explicitly in terms of the Fourier transforms $f_x(\omega)$, $f_y(\omega)$ and $f_{xy}(\omega)$ of the *autocovariance functions* of X_t and Y_t and their *cross-covariance function* as

$$\rho^2(\omega) = \frac{|f_{xy}(\omega)|^2}{f_x(\omega)f_y(\omega)}$$

The squared coherence is the proportion of variability of the component of Y_t with frequency ω explained by the corresponding X_t component. This corresponds to the interpretation of the squared *multiple correlation coefficient* in *multiple regression*. See also **multiple time series**. [TMS Chapter 8.]

Cohort: See **cohort study**.

Cohort component method: A widely used method of forecasting the age- and sex-specific population to future years, in which the initial population is stratified by age and sex and projections are generated by application of survival ratios and birth rates, followed by an additive adjustment for net migration. The method is widely used because it provides detailed information on an area's future population, births, deaths, and migrants by age, sex and race, information that is uesful for many areas of planning and public administration. [*Demography*, 1977, **14**, 363–8.]

Cohort study: An investigation in which a group of individuals (the *cohort*) is identified and followed prospectively, perhaps for many years, and their subsequent medical history recorded. The cohort may be subdivided at the onset into groups with different characteristics, for example, exposed and not exposed to some risk factor, and at some later stage a comparison made of the incidence of a particular disease in each group. See also **prospective study**. [SMR Chapter 5.]

Coincidences: Surprising concurrence of events, perceived as meaningfully related, with no apparent causal connection. Such events abound in everyday life and are often the source of some amazement. As pointed out by *Fisher*, however, 'the one chance in a million will undoubtedly occur, with no less and no more than its appropriate frequency, however surprised we may be that it should occur to us'. See also **synchronicity** [*Chance Rules*, 2nd edn, 2008, B. S. Everitt, Springer, New York.]

Cointegration: A vector of *nonstationary time series* is said to be cointegrated if a linear combination of the individual series is stationary. Facilitates valid testing of the hypothesis that there is a relationship between the nonstationary series. [*Econometrica*, 1987, **55**, 251–276.]

Collapsing categories: A procedure often applied to *contingency tables* in which two or more row or column categories are combined, in many cases so as to yield a reduced table in which there are a larger number of observations in particular cells. Not to be recommended in general since it can lead to misleading conclusions. See also **Simpson's paradox**. [*Multivariable Analysis*, 1996, A. R. Feinstein, New York University Press, New Haven.]

Collective risk models: Models applied to insurance portfolios which do not make direct reference to the risk characteristics of the individual members of the portfolio when describing the aggregate claims experience of the whole portfolio itself. To model the total claims of a collection of risks over a fixed period of time in the future, the collective approach incorporates both claim frequency and claim severity components into the probability distribution of the aggregate. [*Encyclopedia of Quantitative Risk Analysis and Assessment, Volume 1*, 2008, eds. E. L. Melnick and B. S. Everitt, Wiley, Chichester.]

Collector's problem: A problem that derives from schemes in which packets of a particular brand of tea, cereal etc., are sold with cards, coupons or other tokens. There are say n different cards that make up a complete set, each packet is sold with one card, and *a priori* this card is equally likely to be any one of the n cards in the set. Of principal interest is the distribution of N, the number of packets that a typical customer must buy to obtain a complete set. It can be shown that the *cumulative probability distribution* of N, $F_N(r, n)$ is given by

$$F_N(r,n) = \sum_{t=0}^{n} \frac{(-1)^t n!(n-t)^r}{n^r t!(n-t)!}, \quad r \geq n$$

The expected value of N is

$$E(N) = n \sum_{t=1}^{n} \frac{1}{t}$$

[*The American Statistician*, 1998, **52**, 175–80.]

Collinearity: Synonym for **multicollinearity**.

Collision test: A test for randomness in a sequence of digits. [*The Art of Computer Programming*, 2nd edition, 1998, D. K. Knuth, Addison-Wesley, Reading, Maine.]

Combining conditional expectations and residuals plot (CERES): A generalization of the partial residual plot that is often useful for identifying curvature in regression models. If, for example, it is suspected that for one of the explanatory variables, x_i, a term in x_i^2 is needed, then this term is added to the usual residuals from the regression model and the result is plotted against x_i. Figure 38 shows an example from studying the relationship between pollution (sulphur dioxide level) and annual rainfall and wind speed in a number of cities in the USA. The plot for wind speed showing the linear and locally weighted fits indicates that a quadratic term for this variable might be needed in the model. [*Linear Regression Analysis*, 2003, G. A. F. Seber and A. J. Lee, Wiley, New York]

Combining p-values: The combining of a number of p-values obtained from testing hypotheses that are all related to a common question. In a *clinical trial*, for example, there may be several

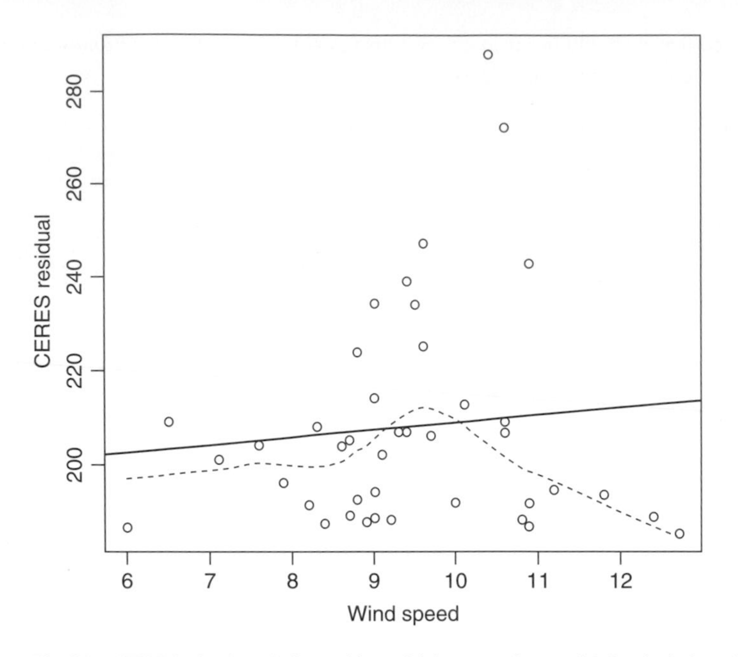

Fig. 38 CERES plot for wind speed in multiple regression model that includes wind speed and precipitation with the dependent variable being sulphur dioxide concentration.

outcome measures that are used in trying to determine whether or not a treatment is effective; combining the p-values from the tests on each outcome may give stronger evidence than is obtained from considering each p-value individually. The most common method of combining p-values is the *Fisher combination test* where the combined test statistic is $C_F = -2\sum_{i=1}^{k}\log(p_i)$ where $p_i, i = 1, \ldots k$, are the separate p-values; C_F is tested as a chi-square with $2k$ degrees of freedom. See also **meta-analysis**. [*Computational Statistics and Data Analysis*, 2004, **47**, 467–485.]

Comingling analysis: Essentially a synonym for the fitting of **finite mixture distributions**.

Commensurate variables: Variables that are on the same scale or expressed in the same units, for example, systolic and diastolic blood pressure.

Common cause failures (CCF): Simultaneous failures of multiple components due to a common cause. A cause can be external to the components, or it can be a single failure that propagates to other components in a cascade. Even if such events are rare compared to single failures of a component, they can dominate system unreliability or unavailability. A variety of models for predicting CCFs have been suggested. [*Reliability Engineering and System Safety*, 2005, **90**, 186–195.]

Common factor: See **factor analysis**.

Common factor variance: A term used in *factor analysis* for the proportion of the variance of a *manifest variable* that is attributed to the *common factors*. [MV2 Chapter 12.]

Common principal components analysis: A generalization of *principal components analysis* to several groups. The model for this form of analysis assumes that the *variance–covariance matrices*, $\mathbf{\Sigma}_i$ of k populations have identical *eigenvectors*, so that the same orthogonal matrix diagonalizes all $\mathbf{\Sigma}_i$ simultaneously. A modification of the model, *partial common* *principal components*, assumes that only r out of q eigenvectors are common to all Σ_i while the remaining $q - r$ eigenvectors are specific in each group. [MV1 Chapter 4.]

Communality: Synonym for **common factor variance**.

Community controls: See **control group**.

Community intervention study: An *intervention study* in which the experimental unit to be randomized to different treatments is not an individual patient or subject but a group of people, for example, a school, or a factory. See also **cluster randomization**. [*Statistics in Medicine*, 1996, **15**, 1069–92.]

Co-morbidity: The potential co-occurrence of two disorders in the same individual, family etc. [*Statistics in Medicine*, 1995, **14**, 721–33.]

Comparative calibration: The statistical methodology used to assess the calibration of a number of instruments, each designed to measure the same characteristic on a common group of individuals. Unlike a normal calibration exercise, the true underlying quantity measured is unobservable. [*Statistics in Medicine*, 1997, **16**, 1889–95.]

Comparative exposure rate: A measure of association for use in a *matched case–control study*, defined as the ratio of the number of case–control pairs, where the case has greater exposure to the risk factor under investigation, to the number where the control has greater exposure. In simple cases the measure is equivalent to the *odds ratio* or a weighted combination of odds ratios. In more general cases the measure can be used to assess association when an odds ratio computation is not feasible. [*Statistics in Medicine*, 1994, **13**, 245–60.]

Comparative trial: Synonym for **controlled trial**.

Comparison group: Synonym for **control group**.

Comparison wise error rate: Synonym for **per-comparison error rate**.

Compartment models: Models for the concentration or level of material of interest over time. A simple example is the *washout model* or *one compartment model* given by

$$E[Y(t)] = \alpha + \beta \exp(-\gamma t), \;\; (\gamma > 0)$$

where $Y(t)$ is the concentration at time $t \geq 0$ [*Compartmental Analysis in Biology in Medicine*, 1972, J. A. Jacquez, Elsevier, New York.]

Compensatory equalization: A process applied in some *clinical trials* and intervention studies, in which comparison groups not given the perceived preferred treatment are provided with compensations that make these comparison groups more equal than originally planned. [*Critically Evaluating the Role of Experiments in Program Evaluation*, 1994, ed. K. J. Conrad, Jossey-Bass, San Francisco.]

Competing risks: A term used particularly in *survival analysis* to indicate that the event of interest (for example, death), may occur from more than one cause. For example, in a study of smoking as a risk factor for lung cancer, a subject who dies of coronary heart disease is no longer at risk of lung cancer. Consequently coronary heart disease is a competing risk in this situation. [*Statistics in Medicine*, 1993, **12**, 737–52.]

Complementary events: *Mutually exclusive* events A and B for which

$$\Pr(A) + \Pr(B) = 1$$

where Pr denotes probability.

Complementary log–log transformation: A transformation of a proportion, p, that is often a useful alternative to the *logistic transformation*. It is given by

$$y = \ln[-\ln(1-p)]$$

This function transforms a probability in the range (0,1) to a value in $(-\infty, \infty)$, but unlike the logistic and *probit transformation* it is not symmetric about the value $p = 0.5$. In the context of a *bioassay*, this transformation can be derived by supposing that the tolerances of individuals have the *Gumbel distribution*. Very similar to the logistic transformation when p is small. [*Modelling Survival Data in Medical Research*, 2nd edn, 2003, D. Collett, Chapman and Hall/CRC, London.]

Complete case analysis: An analysis that uses only individuals who have a complete set of measurements. An individual with one or more missing values is not included in the analysis. When there are many individuals with missing values this approach can reduce the effective sample size considerably. In some circumstances ignoring the individuals with missing values can bias an analysis. See also **available case analysis** and **missing values**. [*Journal of the American Statistical Association*, 1992, **87**, 1227–37.]

Complete ennumeration: Synonym for **census**.

Complete estimator: A weighted combination of two (or more) component estimators. Mainly used in sample survey work. [*Journal of the American Statistical Association*, 1963, **58**, 454–67.]

Complete linkage cluster analysis: An *agglomerative hierarchical clustering method* in which the distance between two clusters is defined as the greatest distance between a member of one cluster and a member of the other. The between group distance measure used is illustrated in Fig. 39. [MV2 Chapter 10.]

Completely randomized design: An experimental design in which the treatments are allocated to the experimental units purely on a chance basis.

Completeness: A term applied to a statistic t when there is only one function of that statistic that can have a given expected value. If, for example, the one function of t is an unbiased estimator of a certain function of a parameter, θ, no other function of t will be. The concept confers a uniqueness property upon an estimator. [*Kendall's Advanced Theory of Statistics*, Volume 2A, 6th ed., 1999, A. Stuart, K. Ord and S. Arnold, Arnold, London.]

Complete spatial randomness: A *Poisson process* in the plane for which:

- the number of events *N(A)* in any region *A* follows a *Poisson distribution* with mean $\lambda|A|$;
- given *N(A)*=*n*, the events in *A* form an independent random sample from the uniform distribution on *A*.

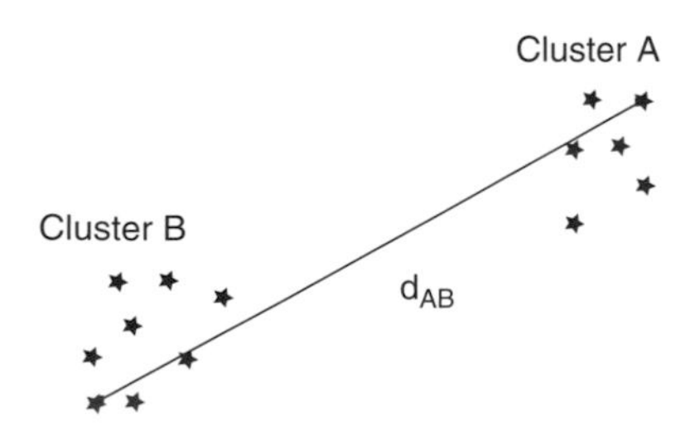

Fig. 39 Complete linkage distance for two clusters A and B.

Here $|A|$ denotes the area of A, and λ is the mean number of events per unit area. Often used as the standard against which to compare an observed spatial pattern. [*Spatial Data Analysis by Example*, Volume 1, 1985, G. Upton and B. Fingleton, Wiley, New York.]

Complex Bingham distributions: An exponential family with a *sufficient statistic* which is quadratic in the data. Such distributions provide tractable models for landmark-based shape analysis. [*Journal of the Royal Statistical Society, Series B*, 1994, **56**, 285–299.]

Complex survey data: Sample survey data obtained by using *cluster sampling*, unequal sampling probabilities and stratification instead of *simple random sampling*. [*Analysis of Complex Surveys*, 1989, C. J. Skinner, D. Holt and T. M. F. Smith, eds., Wiley, New York.]

Compliance: The extent to which participants in a *clinical trial* follow the trial protocol, for example, following both the intervention regimen and trial procedures (clinical visits, laboratory procedures and filling out forms). For clinical trial investigators it is an inescapable fact of life that the participants in their trials often make life difficult by missing appointments, forgetting to take their prescribed treatment from time to time, or not taking it at all but pretending to do so. Because poor participant compliance can adversely affect the outcome of a trial, it is important to use methods both to improve and monitor the level of compliance. See also **complier average causal effect** and **intention-to-treat analysis**. [*Clinical Trials in Psychiatry*, 2nd edn, 2008, B. S. Everitt and S. Wessely, Wiley.]

Complier average causal effect (CACE): The treatment effect among true compliers in a *clinical trial*. For a suitable response variable, the CACE is given by the mean difference in outcome between compliers in the treatment group and those controls who would have complied with treatment had they been randomized to the treatment group. The CACE may be viewed as a measure of 'efficacy' as opposed to 'effectiveness'. [*Review of Economic Studies*, 1997, **64**, 555–74.]

Component bar chart: A *bar chart* that shows the component parts of the aggregate represented by the total length of the bar. The component parts are shown as sectors of the bar with lengths in proportion to their relative size. Shading or colour can be used to enhance the display. Figure 40 gives an example.

Component-plus-residual plot: Synonym for **partial-residual plot**.

Composite estimators: Estimators that are a weighted combination of two or more component estimators. Often used in sample survey work. [*American Journal of Physical Anthropology*, 2007, **133**, 1028–1034.]

Composite hypothesis: A hypothesis that specifies more than a single value for a parameter. For example, the hypothesis that the mean of a population is greater than some value.

Composite indicators: Indicators of multidimensional concepts, for example, sustainability, single market policy and globalization, which cannot be captured by a single indicator. Such indicators (or indices) are formed from combining a number of sub-indicators on the basis of an underlying model of the multidimensional concept that is to be measured. [*Handbook on Constructing Composite Indicators: Methodology and User Guide*, 2008, M. Nardo, M.M Saisana, A. Saltelli and S. Tarantola, OECD publication.]

Composite likelihoods: Pseudo-likelihoods constructed by pooling *likelihood* components, with each component corresponding to a marginal or conditional event. Such likelihoods are used to counter the problem of large computational demands produced in particular by the need to evaluate integrals in many dimensions. [*Biometrika*, 2005, **92**, 519–528.]

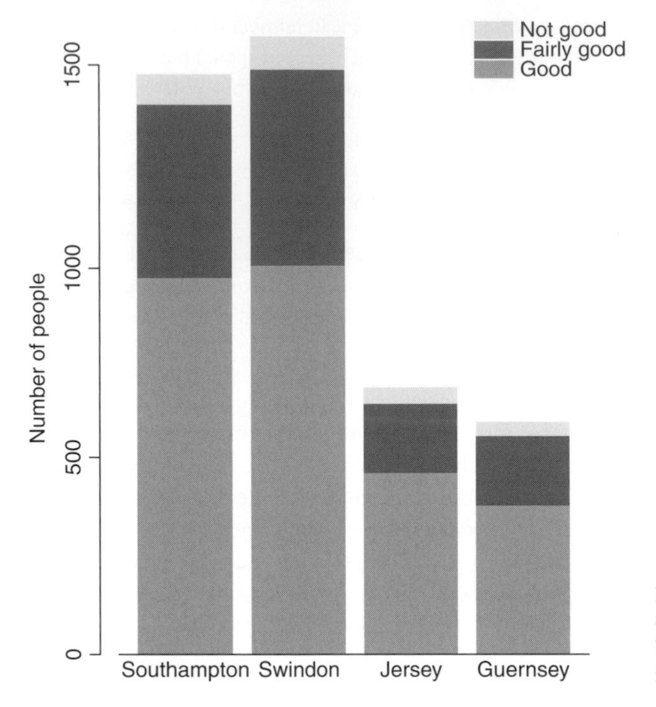

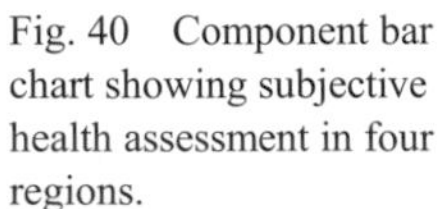
Fig. 40 Component bar chart showing subjective health assessment in four regions.

Composite sampling: A procedure whereby a collection of multiple sample units are combined in their entirety or in part, to form a new sample. One or more subsequent measurements are taken on the combined sample, and the information on the sample units is lost. An example is *Dorfman's scheme*. Because composite samples mask the respondent's identity their use may improve rates of test participation in senstive areas such as testing for HIV. [*Technometrics*, 1980, **22**, 179–86.]

Compositional data: A set of observations, $\mathbf{x}_1, \mathbf{x}_2, \ldots, \mathbf{x}_n$ for which each element of $\mathbf{x}_i$ is a proportion and the elements of $\mathbf{x}_i$ are constrained to sum to one. For example, a number of blood samples might be analysed and the proportion, in each, of a number of chemical elements recorded. Or, in geology, percentage weight composition of rock samples in terms of constituent oxides might be recorded. The appropriate *sample space* for such data is a simplex and much of the analysis of this type of data involves the *Dirichlet distribution*. [MV2 Chapter 15.]

Compound binomial distribution: Synonym for **beta binomial distribution**

Compound decision problem: The simultaneous consideration of n decision problems that have an identical formal structure. [*Annals of Statistics*, 1993, **21**, 736–745.]

Compound distribution: A type of probability distribution arising when a parameter of a distribution is itself a random variable with a corresponding probability distribution. See, for example, *beta-binomial distribution*. See also **contagious distribution**. [KA1 Chapter 5.]

Compound symmetry: The property possessed by a *variance–covariance matrix* of a set of *multivariate data* when its main diagonal elements are equal to one another, and additionally its off-diagonal elements are also equal. Consequently the matrix has the general form;

$$\Sigma = \begin{pmatrix} \sigma^2 & \rho\sigma^2 & \rho\sigma^2 & \cdots \rho\sigma^2 \\ \rho\sigma^2 & \sigma^2 & \cdots & \rho\sigma^2 \\ \vdots & \vdots & & \\ \rho\sigma^2 & \rho\sigma^2 & \cdots & \sigma^2 \end{pmatrix}$$

where ρ is the assumed common correlation coefficient of the measures. Of most importance in the analysis of *longitudinal data* since it is the correlation structure assumed by the *random intercept model* often used to analyse such data. See also **Mauchly test**. [MV2 Chapter 13.]

Computational complexity: The number of operations of the predominant type in an algorithm. Investigations of how computational complexity increases with the size of a problem are important in keeping computational time to a minimum. [*Communications of the ACM*, 1983, **26**, 400–8.]

Computer-aided diagnosis: Computer programs designed to support clinical decision making. In general, such systems are based on the repeated application of *Bayes' theorem*. In some cases a reasoning strategy is implemented that enables the programs to conduct clinically pertinent dialogue and explain their decisions. Such programs have been developed in a variety of diagnostic areas, for example, the diagnosis of dyspepsia and of acute abdominal pain. See also **expert system**. [*Biostatistics, A Methodology for the Health Sciences*, 2nd edn, 2004, G. Van Belle, L. D. Fisher P. J. Heagerty and T. S. Lumley, Wiley, New York.]

Computer algebra: Computer packages that permit programming using mathematical expressions. See also **Maple** and **Mathematica**.

Computer-assisted interviews: A method of interviewing subjects in which the interviewer reads the question from a computer screen instead of a printed page, and uses the keyboard to enter the answer. Skip patterns (i.e. 'if so-and-so, go to Question such-and-such') are built into the program, so that the screen automatically displays the appropriate question. Checks can be built in and an immediate warning given if a reply lies outside an acceptable range or is inconsistent with previous replies; revision of previous replies is permitted, with automatic return to the current question. The responses are entered directly on to the computer record, avoiding the need for subsequent coding and data entry. The program can make automatic selection of subjects who require additional procedures, such as special tests, supplementary questionnaires, or follow-up visits. [*Journal of Official Statistics*, 1995, **11**, 415–31; *American Journal of Epidemiology*, 1999, **149**, 950–4.]

Computer-intensive methods: Statistical methods that require almost identical computations on the data repeated many, many times. The term computer intensive is, of course, a relative quality and often the required 'intensive' computations may take only a few seconds or minutes on even a small PC. An example of such an approach is the *bootstrap*. See also **jackknife**. [MV1 Chapter 1.]

Computer virus: A computer program designed to sabotage by carrying out unwanted and often damaging operations. Viruses can be transmitted via disks or over *networks*. A number of procedures are available that provide protection against the problem.

Concentration matrix: A term sometimes used for the inverse of the *variance–covariance matrix* of a *multivariate normal distribution*.

Concentration measure: A measure, C, of the dispersion of a categorical random variable, Y, that assumes the integral values j, $1 \leq j \leq s$ with probability p_j, given by

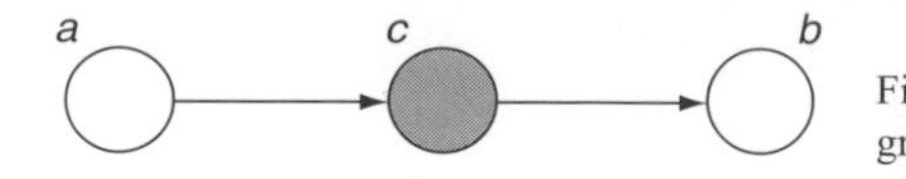

Fig. 41 An example of a conditional independence graph.

$$C = 1 - \sum_{j=1}^{s} p_j^2$$

See also **entropy measure**. [*Agricultural Biological Forum,* 2003, **6**, 134–40.]

Concomitant variables: Synonym for **covariates**.

Concordant mutations test: A statistical test used in cancer studies to determine whether or not a diagnosed second primary tumour is biologically independent of the original primary tumour. The test compares patterns of allelic losses at candidate genetic loci. It is a conditional test, an adaptation of *Fisher's exact test,* which requires no knowledge of the marginal mutation probabilities. [*Biometrics*, 2007, **63**, 522–530.]

Conditional distribution: The probability distribution of a random variable (or the *joint distribution* of several variables) when the values of one or more other random variables are held fixed. For example, in a *bivariate normal distribution* for random variables X and Y the conditional distribution of Y given X is normal with mean $\mu_2 + \rho\sigma_2\sigma_1^{-1}(x - \mu)$ and variance $\sigma_2^2(1 - \rho^2)$. [KA1 Chapter 8.]

Conditional independence graph: A graph that displays relationships between variables such that if two variables, *a* and *b*, are connected only via a third variable, *c*, then *a* and *b* are conditionally independent given c. A simple example is shown in Figure 41 [*Markov Chain Monte Carlo in Practice*, 1996, W. R. Gilks, S. Richardson and D. J. Spiegelhalter, Chapman and Hall/CRC Press, London.]

Conditional likelihood: The likelihood of the data given the *sufficient statistics* for a set of *nuisance parameters.* [GLM Chapter 4.]

Conditional inference: An approach to inference that states that inference about a parameter, θ, in a model, $f(y;\theta)$ should be conditional on any *ancillary statistic* for θ. This approach has caused a great deal of discussion in the literature on the foundations of statistics. [*Biometrika*, 1992, **79**, 247–259.]

Conditional logistic regression: A form of *logistic regression* designed to work with clustered data, such as data involving matched pairs of subjects, in which subject-specific fixed effects are used to take account of the matching. In order to take account of the matching in the analysis of the observed data a *conditional likelihood* is constructed where conditioning is on the sum of the responses in the matched pair, this being a *sufficient statistic* for the subject-specific effect. [*Modelling Binary Data*, 2nd edition, 2003, D. Collett, Chapman and Hall/CRC, London.]

Conditional mortality rate: Synonym for **hazard function**.

Conditional probability: The probability that an event occurs given the outcome of some other event. Usually written, $\Pr(A|B)$. For example, the probability of a person being colour blind given that the person is male is about 0.1, and the corresponding probability given that the person is female is approximately 0.0001. It is not, of course, necessary that $\Pr(A|B)=\Pr(B|A)$; the probability of having spots given that a patient has measles, for example, is very high, the probability of measles given that a patient has spots is, however, much less. If $\Pr(A|B)=\Pr$

(A) then the events A and B are said to be independent. See also **Bayes' Theorem**. [KA1 Chapter 8.]

Condition number: The ratio of the largest *eigenvalue* to the smallest eigenvalue of a matrix. Provides a measure of the sensitivity of the solution from a regression analysis to small changes in either the explanatory variables or the response variable. A large value indicates possible problems with the solution caused perhaps by *collinearity.* [ARA Chapter 10.]

Conference matrix: An $(n + 1) \times (n + 1)$ matrix C satisfying

$$\mathbf{C}'\mathbf{C} = \mathbf{C}\mathbf{C}' = n\mathbf{I}$$
$$c_{ii} = 0 \; i = 1, \ldots, n + 1$$
$$c_{ij} \in \{-1, 1\} \; i \neq j$$

The name derives from an application to telephone conference networks. [*Biometrika*, 1995, **82**, 589–602.]

Confidence interval: A range of values, calculated from the sample observations, that is believed, with a particular probability, to contain the true parameter value. A 95% confidence interval, for example, implies that were the estimation process repeated again and again, then 95% of the calculated intervals would be expected to contain the true parameter value. Note that the stated probability level refers to properties of the interval and not to the parameter itself which is not considered a random variable (although see, **Bayesian inference** and **Bayesian confidence interval)**. [KA2 Chapter 20.]

Confidence profile method: A Bayesian approach (see *Bayesian inference*) to *meta-analysis* in which the information in each piece of evidence is captured in a *likelihood function* which is then used along with an assumed *prior distribution* to produce a *posterior distribution.* The method can be used where the available evidence involves a variety of experimental designs, types of outcomes and effect measures, and uncertainty about biases. [*Bayesian Approaches to Clinical Trials and Health Care Evaluation*, 2004, D. J. Spiegelhalter, K. R. Abrams and J. P. Myles, Wiley, Chichester.]

Confirmatory data analysis: A term often used for model fitting and inferential statistical procedures to distinguish them from the methods of *exploratory data analysis.*

Confirmatory factor analysis: See **factor analysis**.

Confounding: A process observed in some *factorial designs* in which it is impossible to differentiate between some main effects or interactions, on the basis of the particular design used. In essence the *contrast* that measures one of the effects is exactly the same as the contrast that measures the other. The two effects are usually referred to as *aliases*. The term is also used in observational studies to highlight that a measured association found between an exposure and an outcome may not represent a causal effect because there may exist variables that are associated with both exposure and outcome. [SMR Chapter 5.]

Confusion matrix: Synonym for **misclassification matrix**.

Congruential methods: Methods for generating random numbers based on a fundamental congruence relationship, which may be expressed as the following recursive formula

$$n_{i+1} \equiv an_i + c(\text{mod } m)$$

where n_i, a, c and m are all non-negative integers. Given an initial starting value n_0, a constant multiplier a, and an additive constant C then the equation above yields a congruence relationship (modulo m) for any value for i over the sequence $\{n_1, n_2, \ldots, n_i, \ldots\}$.

From the integers in the sequence $\{n_i\}$, rational numbers in the unit interval (0, 1) can be obtained by forming the sequence $\{r_i\} = \{n_i/m\}$. Frequency tests and serial tests, as well as other tests of randomness, when applied to sequences generated by the method indicate that the numbers are uncorrelated and uniformly distributed, but although its statistical behaviour is generally good, in a few cases it is completely unacceptable. [*Handbook of Parametric and Nonparametric Statistical Procedures*, 3rd edition, 2007, D. J. Sheskin, Chapman and Hall/CRC, Boca Raton.]

Conjoint analysis: A method used primarily in market reasearch which is similar in many respects to *multidimensional scaling*. The method attempts to assign values to the levels of each attribute, so that the resulting values attached to the stimuli match, as closely as possible, the input evaluations provided by the respondents. [*Marketing Research-State of the Art Perspectives*, 2000, C. Chakrapani, ed., American Marketing Association, Chicago.]

Conjugate prior: A *prior distribution* for samples from a particular probability distribution such that the *posterior distribution* at each stage of sampling is of the same family, regardless of the values observed in the sample. For example, the family of *beta distributions* is conjugate for samples from a *binomial distribution*, and the family of *gamma distributions* is conjugate for samples from the *exponential distribution*. [KA1 Chapter 8.]

Conover test: A *distribution free method* for the equality of variance of two populations that can be used when the populations have different location parameters. The *asymptotic relative efficiency* of the test compared to the *F-test* for normal distributions is 76%. See also **Ansari–Bradley test** and **Klotz test**. [*Biostatistics: A Methodology for the Health Sciences*, 2nd edition, 2004, G. Van Belle, L. D. Fisher, P. J. Heagerty and T. S. Lumley, Wiley, New York.]

Conservative and non-conservative tests: Terms usually encountered in discussions of *multiple comparison tests*. Non-conservative tests provide poor control over the *per-experiment error rate*. Conservative tests on the other hand, may limit the *per-comparison error rate* to unecessarily low values, and tend to have low *power* unless the sample size is large. [*Biometrika*, 1988, **75**, 149–152.]

Consistency: An estimator is said to be consistent if it converges to its *estimand* as the sample size tends to infinity. [KA2 Chapter 17.]

Consistency checks: Checks built into the collection of a set of observations to assess their internal consistency. For example, data on age might be collected directly and also by asking about date of birth.

Consolidated Standards for Reporting Trials (CONSORT) statement: A protocol for reporting the results of *clinical trials*. The core contribution of the statement consists of a flow diagram (see Fig. 42) and a checklist. The flow diagram enables reviewers and readers to quickly grasp how many eligible participants were randomly assigned to each arm of the trial. [*Journal of the American Medical Association*, 1996, **276**, 637–9.]

CONSORT statement: See **Consolidated Standards for Reporting Trials (CONSORT) statement**.

Consumer price index (CPI): A measure of the changes in prices paid by urban consumers for the goods and services they purchase. Essentially, it measures the purchasing power of consumers' money by comparing what a sample or 'market basket' of goods and services costs today with what the same goods would have cost at an earlier date. [*The Economic Theory of Price Indexes*, 1972, F. M. Fisher and K. Schell, Academic Press, New York.]

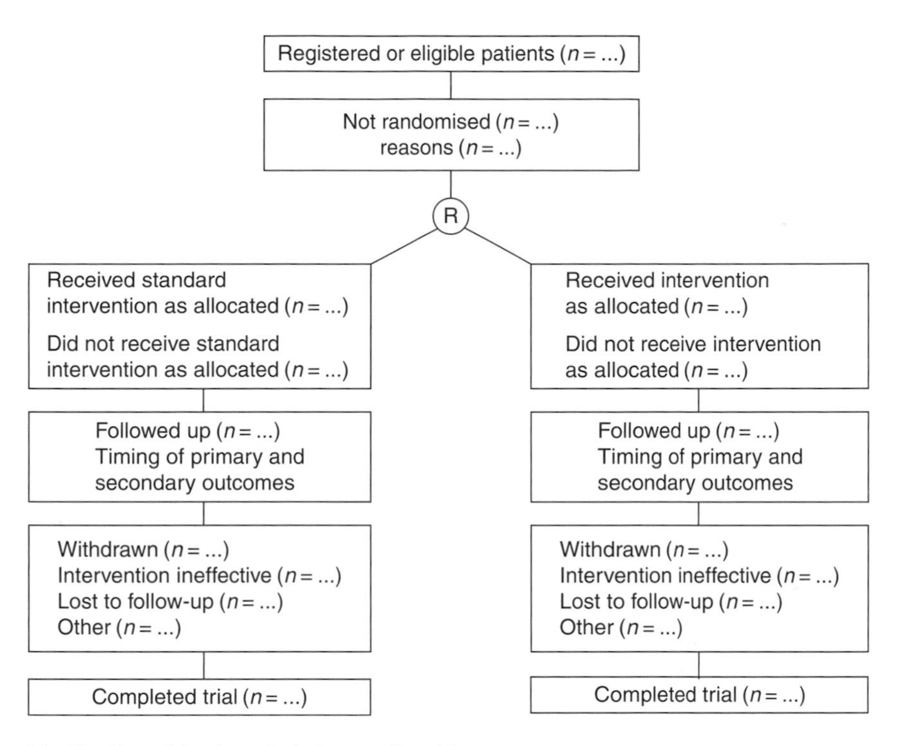

Fig. 42 Consolidated standards for reporting trials.

Contagious distribution: A term used for the probability distribution of the sum of a number (N) of random variables, particularly when N is also a random variable. For example, if $X_1, X_2, \ldots, X_N$ are variables with a *Bernoulli distribution* and N is a variable having a *Poisson distribution* with mean λ, then the sum S_N given by

$$S_N = X_1 + X_2 + \cdots + X_N$$

can be shown to have a Poisson distribution with mean λp where $p = \Pr(X_i = 1)$. See also **compound distribution**. [KA1 Chapter 5.]

Contaminated normal distribution: A term sometimes used for a *finite mixture distribution* of two normal distributions with the same mean but different variances. Such distributions have often been used in *Monte Carlo studies.* [*Transformation and Weighting in Regression*, 1988, R. J. Carroll and D. Ruppert, Chapman and Hall/CRC Press, London.]

Contingency coefficient: A measure of association, C, of the two variables forming a *two-dimensional contingency table*, given by

$$C = \sqrt{\frac{X^2}{X^2 + N}}$$

where X^2 is the usual *chi-squared statistic* for testing the independence of the two variables and N is the sample size. See also **phi-coefficient, Sakoda coefficient** and **Tschuprov coefficient**. [*The Analysis of Contingency Tables*, 2nd edition, 1992, B. S. Everitt, Chapman and Hall/CRC Press, London.]

Contingency tables: The tables arising when observations on a number of categorical variables are cross-classified. Entries in each cell are the number of individuals with the corresponding combination of variable values. Most common are *two-dimensional tables* involving two categorical variables, an example of which is shown below.

Retarded activity amongst psychiatric patients

	Affectives	Schizo	Neurotics	Total
Retarded activity	12	13	5	30
No retarded activity	18	17	25	60
Total	30	30	30	90

The analysis of such two-dimensional tables generally involves testing for the independence of the two variables using the familiar *chi-squared statistic*. Three- and higher-dimensional tables are now routinely analysed using *log-linear models*. [*The Analysis of Contingency Tables*, 2nd edition, 1992, B. S. Everitt, Chapman and Hall/CRC Press, London.]

Continual reassessment method: An approach that applies *Bayesian inference* to determining the maximum tolerated dose in a *phase I trial*. The method begins by assuming a *logistic regression* model for the dose–toxicity relationship and a *prior distribution* for the parameters. After each patient's toxicity result becomes available the *posterior distribution* of the parameters is recomputed and used to estimate the probability of toxicity at each of a series of dose levels. [*Statistics in Medicine*, 1995, **14**, 1149–62.]

Continuity correction: See **Yates' correction**.

Continuous proportion: Proportions of a continuum such as time or volume; for example, proportions of time spent in different conditions by a subject, or the proportions of different minerals in an ore deposit. [*Biometrika*, 1982, **69**, 197–203.]

Continuous screen design: See **screening studies**.

Continuous time Markov chain: See **Markov chain**.

Continuous time stochastic process: See **stochastic process**.

Continuous variable: A measurement not restricted to particular values except in so far as this is constrained by the accuracy of the measuring instrument. Common examples include weight, height, temperature, and blood pressure. For such a variable equal sized differences on different parts of the scale are equivalent. See also **categorical variable, discrete variable, ordinal variable** and **semi-continuous variable**.

Continuum regression: Regression of a response variable, y, on that linear combination t_γ of explanatory variables which maximizes $r^2(y, t)\text{var}(t)^\gamma$. The parameter γ is usually chosen by cross-validated optimization over the predictors $\mathbf{b}'_\gamma\mathbf{x}$. Introduced as an alternative to *ordinary least squares* to deal with situations in which the explanatory variables are nearly collinear, the method trades variance for bias. See also **principal components regression, partial least squares** and **ridge regression**. [*Journal of the Royal Statistical Society, Series B*, 1996, **58**, 703–10.]

Contour plot: A topographical map drawn from data involving observations on three variables. One variable is represented on the horizontal axis and a second variable is represented on the vertical axis. The third variable is represented by isolines (lines of constant value). These plots are often helpful in data analysis, especially when searching for maxima or minima in such data. The plots are most often used to display graphically *bivariate distributions* in which case the third variable is value of the probability density function corresponding to the values of the two variables. An alternative method of displaying the same material is provided by the *perspective plot* in which the values on the third variable are represented

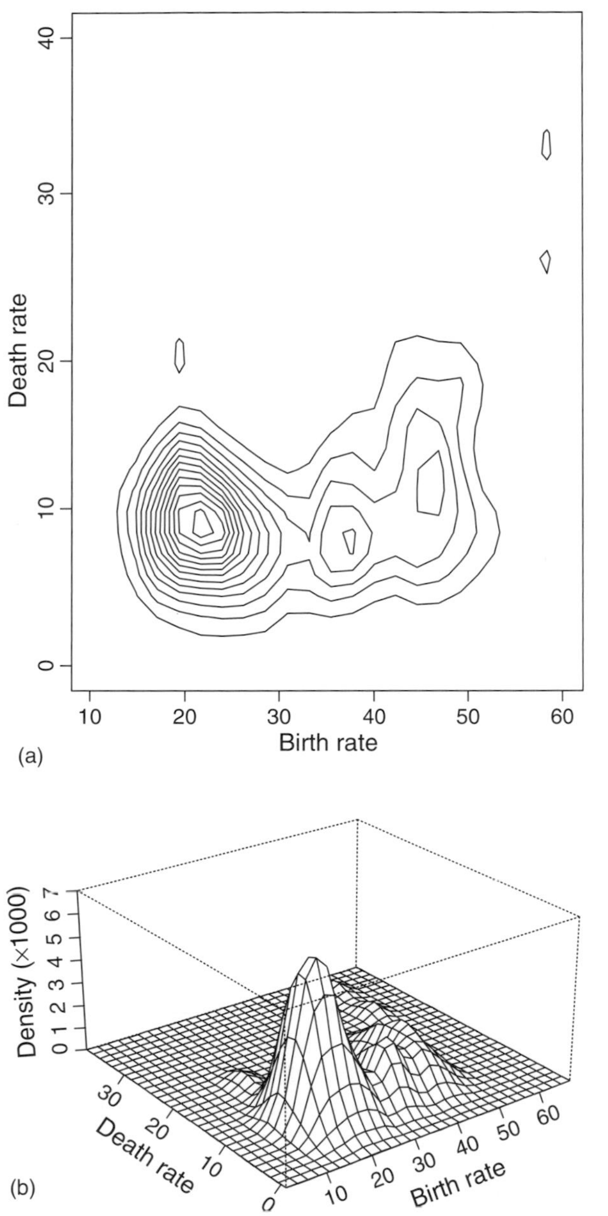

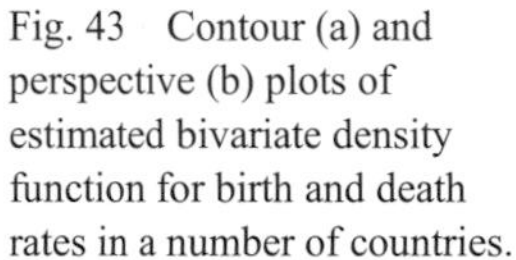

Fig. 43 Contour (a) and perspective (b) plots of estimated bivariate density function for birth and death rates in a number of countries.

by a series of lines constructed to give a three-dimensional view of the data. Figure 43 gives examples of these plots using birth and death rate data from a number of countries with a kernel density estimator used to calculate the bivariate distribution.

Contrast: A linear function of parameters or statistics in which the coefficients sum to zero. Most often encountered in the context of *analysis of variance*, in which the coefficients sum to zero (sometimes called effect coding). For example, in an application involving say three treatment groups (with means x_{T_1}, x_{T_2} and x_{T_3}) and a control group (with mean x_C), the following is the contrast for comparing the mean of the control group to the average of the treatment groups;

$$x_C - \tfrac{1}{3}x_{T_1} - \tfrac{1}{3}x_{T_2} - \tfrac{1}{3}x_{T_3}$$

See also **Helmert contrast** and **orthogonal contrast**. [*The Analysis of Variance*, 1959, H. Scheffé, Wiley, London.]

Control chart: See **quality control procedures**.

Control group: In experimental studies, a collection of individuals to which the experimental procedure of interest is not applied. In observational studies, most often used for a collection of individuals not subjected to the risk factor under investigation. In many medical studies the controls are drawn from the same clinical source as the cases to ensure that they represent the same catchment population and are subject to the same selective factors. These would be termed, *hospital controls*. An alternative is to use controls taken from the population from which the cases are drawn (*community controls*). The latter is suitable only if the source population is well defined and the cases are representative of the cases in this population. [SMR Chapter 15.]

Controlled trial: A *Phase III clinical trial* in which an experimental treatment is compared with a control treatment, the latter being either the current standard treatment or a placebo. [SMR Chapter 15.]

Control statistics: Statistics calculated from sample values $x_1, x_2, \ldots, x_n$ which elicit information about some characteristic of a process which is being monitored. The sample mean, for example, is often used to monitor the mean level of a process, and the sample variance its imprecision. See also **cusum** and **quality control procedures**.

Convenience sample: A non-random sample chosen because it is easy to collect, for example, people who walk by in the street. Such a sample is unlikely to be representative of the population of interest.

Convergence in probability: Convergence of the probability of a sequence of random variables to a value.

Convex hull: The vertices of the smallest convex polyhedron in variable space within or on which all the data points lie. An example is shown in Fig. 44. [MV1 Chapter 6.]

Convex hull trimming: A procedure that can be applied to a set of *bivariate data* to allow *robust estimation* of Pearson's product moment correlation coefficient. The points defining the *convex hull* of the observations, are deleted before the correlation coefficient is calculated. The major attraction of this method is that it eliminates isolated *outliers* without disturbing the general shape of the bivariate distribution. [*Interpreting Multivariate Data*, 1981, edited by V. Barnett, Wiley, Chichester.]

Convolution: An integral (or sum) used to obtain the probability distribution of the sum of two or more random variables. [KA1 Chapter 4.]

Conway–Maxwell–Poisson distribution: A generalization of the *Poisson distribution*, that has thicker or thinner tails than the Poisson distribution, which is included as a special case. The distribution is defined over positive integers and is flexible in representing a variety of shapes and in modelling *overdispersion*. [*Journal of the Royal Statistical Society, Series C*, 2005, **54**, 127–142.]

Cook's distance: An *influence statistic* designed to measure the shift in the estimated parameter vector, $\hat{\boldsymbol{\beta}}$, from fitting a regression model when a particular observation is omitted. It is a combined measure of the impact of that observation on all regression coefficients. The statistic is defined for the ith observation as

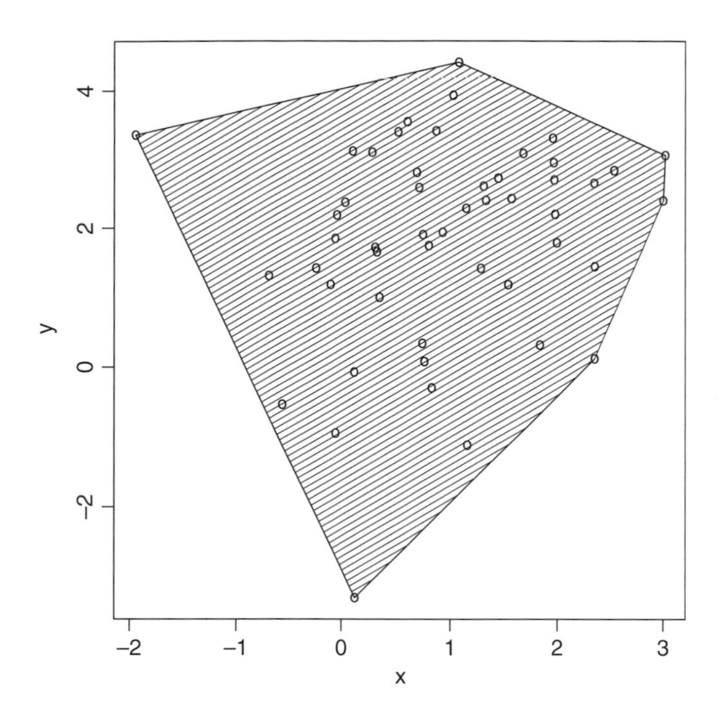

Fig. 44 An example of the convex hull of a set of bivariate data.

$$D_i = \frac{r_i^2}{\text{tr}(\mathbf{H})} \frac{h_i}{1 - h_i}$$

where r_i is the *standardized residual* of the ith observation and h_i is the ith diagonal element of the *hat matrix*, **H** arising from the regression analysis. Values of the statistic greater than one suggest that the corresponding observation has undue influence on the estimated regression coefficients. See also **COVRATIO, DFBETA** and **DFFIT**. [ARA Chapter 10.]

Cooperative study: A term sometimes used for *multicentre study*.

Cophenetic correlation: The correlation between the observed values in a *similarity matrix* or *dissimilarity matrix* and the corresponding fusion levels in the *dendrogram* obtained by applying an *agglomerative hierarchical clustering method* to the matrix. Used as a measure of how well the clustering matches the data. [*Cluster Analysis*, 4th edition, 2001, B. S. Everitt, S. Landau and M. Leese, Arnold, London.]

Coplot: A powerful visualization tool for studying how a response depends on an explanatory variable given the values of other explanatory variables. The plot consists of a number of panels one of which (the 'given' panel) shows the values of a particular explanatory variable divided into a number of intervals, while the others (the 'dependence' panels) show the *scatterplots* of the response variable and another explanatory variable corresponding to each interval in the given panel. The plot is examined by moving from left to right through the intervals in the given panel, while simultaneously moving from left to right and then from bottom to top through the dependence panels. The example shown (Fig. 45) involves the relationship between packed cell volume and white blood cell count for given haemoglobin concentration. [*Visualizing Data*, 1993, W. S. Cleveland, AT&T Bell Labs, New Jersey.]

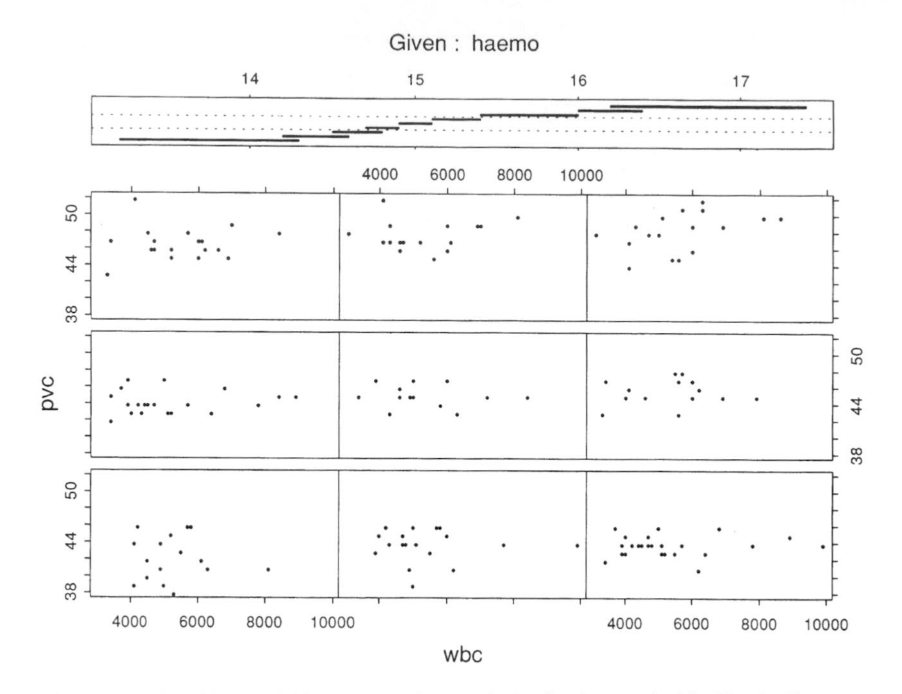

Fig. 45 Coplot of haemoglobin concentration; reached cell volume and white blood cell count.

Copulas: Invariant transformations to combine marginal probability functions to form multivariate distributions motivated by the need to enlarge the class of multivariate distributions beyond the *multivariate normal distribution* and its related functions such as the *multivariate Student's t-distribution* and the *Wishart distribution*. An example is *Frank's family of bivariate distributions*. (The word 'copula' comes from Latin and means to connect or join.) Quintessentially copulas are measures of the dependent structure of the marginal distributions and they have been used to model *correlated risks*, joint default probabilities in credit portfolios and groups of individuals that are exposed to similar economic and physical environments. Also used in *frailty models* for *survival data*. [*An Introduction to Copulas*, 1998, R. B. Nelson, Springer-Verlag, New York.]

Cornfield, Jerome (1912–1979): Cornfield studied at New York University where he graduated in history in 1933. Later he took a number of courses in statistics at the US Department of Agriculture. From 1935 to 1947 Cornfield was a statistician with the Bureau of Labour Statistics and then moved to the National Institutes of Health. In 1958 he was invited to succeed *William Cochrane* as Chairman of the Department of Biostatistics in the School of Hygiene and Public Health of the Johns Hopkins University. Cornfield devoted the major portion of his career to the development and application of statistical theory to the biomedical sciences, and was perhaps the most influential statistician in this area from the 1950s until his death. He was involved in many major issues, for example smoking and lung cancer, polio vaccines and risk factors for cardiovascular disease. Cornfield died on 17 September 1979 in Henden, Virginia.

Cornish, Edmund Alfred (1909–1973): Cornish graduated in Agricultural Science at the University of Melbourne. After becoming interested in statistics he then completed three years of mathematical studies at the University of Adelaide and in 1937 spent a year at University College, London studying with *Fisher*. On returning to Australia he was eventually appointed Professor of Mathematical Statistics at the University of Adelaide in 1960. Cornish made important contributions to the analysis of complex designs with missing values and *fiducial theory.*

Correlated binary data: Synonym for **clustered binary data**.

Correlated failure times: Data that occur when failure times are recorded which are dependent. Such data can arise in many contexts, for example, in epidemiological cohort studies in which the ages of disease occurrence are recorded among members of a sample of families; in animal experiments where treatments are applied to samples of littermates and in *clinical trials* in which individuals are followed for the occurrence of multiple events. See also **bivariate survival data**. [*Journal of the Royal Statistical Society, Series B*, 2003, **65**, 643–61.]

Correlated risks: The simultaneous occurrence of many losses from a single event. Earthquakes, for example, produce highly correlated losses with many homes in the affected area being damaged and destroyed; of particular importance in insurance. [*The Geneva Papers on Risk and Insurance Theory*, 1992, **17**, 35–60.]

Correlated samples *t*-test: Synonym for **matched pairs *t*-test**.

Correlation: A general term for interdependence between pairs of variables. See also **association**.

Correlation coefficient: An index that quantifies the linear relationship between a pair of variables. In a *bivariate normal distribution*, for example, the parameter, ρ. An estimator of ρ obtained from n sample values of the two variables of interest, $(x_1, y_1), (x_2, y_2), \ldots, (x_n, y_n)$, is *Pearson's product moment correlation coefficient*, r, given by

$$r = \frac{\sum_{i=1}^{n}(x_i - \bar{x})(y_i - \bar{y})}{\sqrt{\sum_{i=1}^{n}(x_i - \bar{x})^2 \sum_{i=1}^{n}(y_i - \bar{y})^2}}$$

The coefficient takes values between -1 and 1, with the sign indicating the direction of the relationship and the numerical magnitude its strength. Values of -1 or 1 indicate that the sample values fall on a straight line. A value of zero indicates the lack of any linear relationship between the two variables. See also **Spearman's rank correlation, intraclass correlation** and **Kendall's tau statistics**. [SMR Chapter 11.]

Correlation coefficient distribution: The probability distribution, $f(r)$, of Pearson's product moment correlation coefficient when n pairs of observations are sampled from a *bivariate normal distribution* with correlation parameter, ρ. Given by

$$f(r) = \frac{(1-\rho^2)^{(n-1)/2}(1-r^2)^{(n-4)/2}}{\sqrt{\pi}\Gamma(\frac{1}{2}(n-1))\Gamma(\frac{1}{2}n-1)} \sum_{j=o}^{\infty} \frac{[\Gamma(\frac{1}{2}(n-1+j))]^2}{j!}(2\rho r)^j, \qquad -1 \le r \le 1$$

[*Continuous Univariate Distributions*, Volume 2, 2nd edition, 1995, N. L. Johnson, S. Kotz and N. Balakrishnan, Wiley, New York.]

Correlation matrix: A square, *symmetric matrix* with rows and columns corresponding to variables, in which the off–diagonal elements are correlations between pairs of variables, and elements on the main diagonal are unity. An example for measures of muscle and body fat is as follows:

Correlation matrix for muscle, skin and body fat data

$$R = \begin{array}{c} \\ V1 \\ V2 \\ V3 \\ V4 \end{array} \begin{array}{c} \begin{array}{cccc} V1 & V2 & V3 & V4 \end{array} \\ \begin{pmatrix} 1.00 & 0.92 & 0.46 & 0.84 \\ 0.92 & 1.00 & 0.08 & 0.88 \\ 0.46 & 0.08 & 1.00 & 0.14 \\ 0.84 & 0.88 & 0.14 & 1.00 \end{pmatrix} \end{array}$$

V1 = tricep(thickness mm), V2 = thigh(circumference mm),
V3 = midarm(circumference mm), V4 = bodyfat(%).
[MV1 Chapter 1.]

Correlation matrix distribution: The joint distribution of the sample correlations from a set of data from a *multivariate normal distribution* when all the population correlations are zero is given by

$$f(\mathbf{R}) = \frac{|\mathbf{R}|^{\frac{1}{2}(n-p-2)}\left\{\Gamma\left[\frac{1}{2}(n-1)\right]\right\}^p}{\pi^{\frac{1}{4}p(p-1)}\prod_{j=1}^{p}\Gamma\left\{\frac{1}{2}(n-j)\right\}}$$

where $\mathbf{R}$ is the sample correlation matrix, n is the sample size and p is the number of variables. The moments of the correlation matrix determinant are given by

$$E(|\mathbf{R}|^t) = \frac{\left\{\Gamma\left[\frac{1}{2}(n-1)\right]\right\}^p}{\left\{\Gamma\left[\frac{1}{2}(n-1)+t\right]\right\}^p} \frac{\prod_{j=1}^{p}\Gamma\left[\frac{1}{2}(n-j)+t\right]}{\prod_{j=1}^{p}\Gamma\left[\frac{1}{2}(n-j)\right]}$$

[KA1]

Correlogram: See **autocorrelation**.

Correspondence analysis: A method for displaying the relationships between categorical variables in a type of scatterplot diagram. For two such variables displayed in the form of a *contingency table*, for example, a set of coordinate values representing the row and column categories are derived. A small number of these derived coordinate values (usually two) are then used to allow the table to be displayed graphically. In the resulting diagram *Euclidean distances* approximate *chi-squared distances* between row and column categories. The coordinates are analogous to those resulting from a *principal components analysis* of continuous variables, except that they involve a partition of a chi-squared statistic rather than the total variance. Such an analysis of a contingency table allows a visual examination of any structure or pattern in the data, and often acts as a useful supplement to more formal inferential analyses. Figure 46 arises from applying the method to the following table.

Eye Colour			**Hair Colour**		
	Fair (hf)	Red (hr)	Medium (hm)	Dark (hd)	Black (hb)
Light (EL)	688	116	584	188	4
Blue (EB)	326	38	241	110	3
Medium (EM)	343	84	909	412	26
Dark (ED)	98	48	403	681	81

[MV1 Chapter 5.]

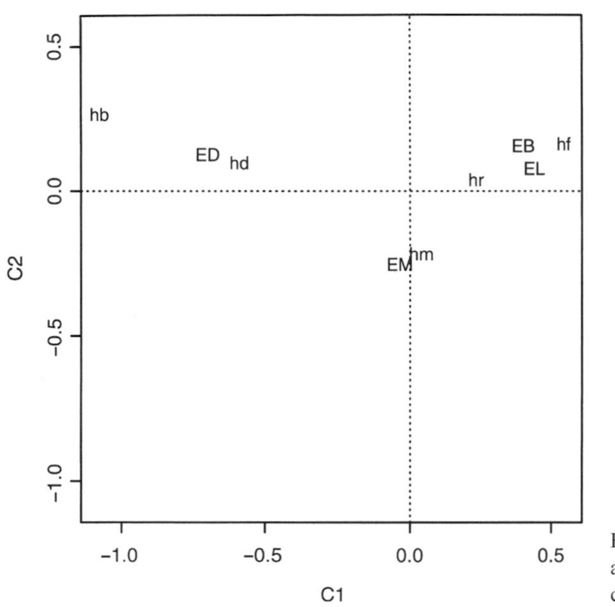

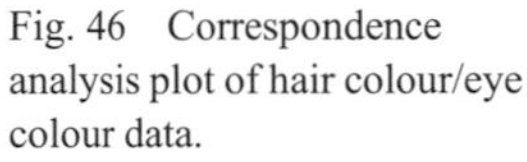
Fig. 46 Correspondence analysis plot of hair colour/eye colour data.

Cosine distribution: Synonym for **cardiord distribution**.

Cosinor analysis: The analysis of biological rhythm data, that is data with *circadian variation*, generally by fitting a single sinusoidal regression function having a known period of 24 hours, together with independent and identically distributed error terms. [*Statistics in Medicine*, 1987, **6**, 167–84.]

Cospectrum: See **multiple time series**.

Cost-benefit analysis: A technique where health benefits are valued in monetary units to facilitate comparisons between different programmes of health care. The main practical problem with this approach is getting agreement in estimating money values for health outcomes. [*Cost-Benefit Analysis*, 1971, E. J. Mishan, Allen and Unwin, London.]

Cost-effectiveness analysis: A method used to evaluate the outcomes and costs of an intervention, for example, one being tested in a *clinical trial.* The aim is to allow decisions to be made between various competing treatments or courses of action. The results of such an analysis are generally summarized in a series of cost-effectiveness ratios. [*Journal of Rheumatology*, 1995, **22**, 1403–7.]

Cost-effectiveness ratio (CER): The ratio of the difference in cost between a test and standard health programme to the difference in benefits. Generally used as a summary statistic to compare competing health care programmes relative to their cost and benefit. [*Statistics in Medicine*, 2001, **20**, 1469–77.]

Cost of living extremely well index: An index that tries to track the price fluctuations of items that are affordable only to those of very substantial means. The index is used to provide a barometer of economic forces at the top end of the market. The index includes 42 goods and services, including a kilogram of beluga malossal caviar and a face lift. [*Measurement Theory and Practice*, 2004, D. J. Hand, Arnold, London.]

Count data: Data obtained by counting the number of occurrences of particular events rather than by taking measurements on some scale.

Counter-matching: An approach to selecting controls in *nested case-control studies*, in which a covariate is known on all cohort members, and controls are sampled to yield covariate-stratified case-control sets. This approach has been shown to be generally efficient relative to matched case-control designs for studying interaction in the case of a rare risk factor X and an uncorrelated risk factor Y. [*Biometrika*, 1995, **82**, 69–79.]

Counternull value: An *effect size* that is just as well supported by the data as the 'zero effect' null hypothesis, i.e. the counternull value if used to replace the usual 'no difference' null hypothesis would result in the same p-value. When the data are drawn from a distribution that is symmetrical about its mean the counternull value is exactly twice the observed effect size. Reporting the counternull value in addition to the p-value has been suggested to avoid claiming that failure to reject the null hypothesis at a certain significance level implies that the effect size is zero. For most statisticians and applied researchers however, the *confidence interval* is a more useful safeguard in avoiding such a claim. [*Handbook of Experimental Psychology*, 2002, H. E. Pashler and S. S. Stevens, Wiley, Chichester.]

Counting process: A *stochastic process* $\{N(t), t \geq 0\}$ in which $N(t)$ represents the total number of 'events' that have occurred up to time t. The $N(t)$ in such a process must satisfy;

- $N(t) \geq 0$,
- $N(t)$ is integer valued,
- If $s < t$ then $N(s) \leq N(t)$.

For $s < t$, $N(t) - N(s)$ equals the number of events that have occurred in the interval $(s,t]$. [*Journal of the Royal Statistical Society, Series B*, 1996, **58**, 751–62.]

Courant–Fisher minimax theorem: This theorem states that for two quadratic forms, $\mathbf{X}'\mathbf{AX}$ and $\mathbf{X}'\mathbf{BX}$, assuming that $\mathbf{B}$ is positive definite, then

$$\lambda_S \leq \frac{\mathbf{X}'\mathbf{AX}}{\mathbf{X}'\mathbf{BX}} \leq \lambda_L$$

where λ_L and λ_S are the largest and smallest relative *eigenvalues* respectively of $\mathbf{A}$ and $\mathbf{B}$. [*IEEE Transactions on Pattern Analysis and Machine Intelligence*, 2000, **22**, 504–25.]

Covariance: The expected value of the product of the deviations of two random variables, x and y, from their respective means, μ_x and μ_y, i.e.

$$\text{cov}(x, y) = E(x - \mu_x)(y - \mu_y)$$

The corresponding sample statistic is

$$c_{xy} = \frac{1}{n}\sum_{i=1}^{n}(x_i - \bar{x})(y_i - \bar{y})$$

where $(x_i, y_i),\ i = 1, \ldots, n$ are the sample values on the two variables and $\bar{x}$ and $\bar{y}$ their respective means. See also **variance–covariance matrix** and **correlation coefficient**. [MV1 Chapter 2.]

Covariance inflation criterion: A procedure for model selection in regression analysis. [*Journal of the Royal Statistical Society, Series B*, 1999, 529–46.]

Covariance matrix: See **variance–covariance matrix**.

Covariance-regularized regression: A family of methods for prediction in *high-dimensional data* sets that uses a shrunken estimate of the inverse covariance matrix of the variables to achieve more accurate prediction. An estimate of the inverse covariance matrix is obtained by maximizing the *log-likelihood* of the data, under a multivariate normal model, subject to a penalty. *Ridge regression* and the *lasso* are special cases of this approach. [*Journal of the Royal Statistical Society, Series B*, 2009, **71**, 615–636.]

Covariance structure models: Synonym for **structural equation models**.

Covariates: Often used simply as an alternative name for explanatory variables, but perhaps more specifically to refer to variables that are not of primary interest in an investigation, but are measured because it is believed that they are likely to affect the response variable and consequently need to be included in analyses and model building. See also **analysis of covariance**.

COVRATIO: An *influence statistic* that measures the impact of an observation on the *variance–covariance matrix* of the estimated regression coefficients in a regression analysis. For the ith observation the statistic is given by

$$\text{COVRATIO}_i = \frac{\det\left(s^2_{(-i)}[\mathbf{X}'_{(-i)}\mathbf{X}_{(-i)}]^{-1}\right)}{\det(s^2[\mathbf{X}'\mathbf{X}]^{-1})}$$

where s^2 is the residual mean square from a regression analysis with all observations, $\mathbf{X}$ is the matrix appearing in the usual formulation of *multiple regression* and $s^2_{(-i)}$ and $\mathbf{X}_{(-i)}$ are the corresponding terms from a regression analysis with the ith observation omitted. Values outside the limits $1 \pm 3(\text{tr}(\mathbf{H})/n)$ where $\mathbf{H}$ is the *hat matrix* can be considered extreme for purposes of identifying influential observations. See also **Cook's distance, DFBETA, DFFIT**. [ARA Chapter 10.]

Cowles' algorithm: A hybrid *Metropolis-Hastings, Gibbs sampling algorithm* which overcomes problems associated with small candidate point probabilities. [*Statistics and Computing*, 1996, **6**, 101–11.]

Cox–Aalen model: A model for *survival data* in which some covariates are believed to have multiplicative effects on the *hazard function*, whereas others have effects which are better described as additive. [*Biometrics*, 2003, **59**, 1033–45.]

Cox, Gertrude Mary (1900–1978): Born in Dayton, Iowa, Gertrude Cox first intended to become a deaconess in the Methodist Episcopal Church. In 1925, however, she entered Iowa State College, Ames and took a first degree in mathematics in 1929, and the first MS degree in statistics to be awarded in Ames in 1931. Worked on psychological statistics at Berkeley for the next two years, before returning to Ames to join the newly formed Statistical Laboratory where she first met *Fisher* who was spending six weeks at the college as a visiting professor. In 1940 Gertrude Cox became Head of the Department of Experimental Statistics at North Carolina State College, Raleigh. After the war she became increasingly involved in the research problems of government agencies and industrial concerns. Joint authored the standard work on experimental design, *Experimental Designs*, with *William Cochran* in 1950. Gertrude Cox died on 17 October 1978.

Cox–Mantel test: A *distribution free method* for comparing two *survival curves*. Assuming $t_{(1)} < t_{(2)} < \cdots < t_{(k)}$ to be the distinct survival times in the two groups, the test statistic is

$$C = U/\sqrt{I}$$

where

$$U = r_2 - \sum_{i=1}^{k} m_{(i)} A_{(i)}$$

$$I = \sum_{i=1}^{k} \frac{m_i(r_{(i)} - m_{(i)})}{r_{(i)} - 1} A_{(i)}(1 - A_{(i)})$$

In these formulae, r_2 is the number of deaths in the second group, $m_{(i)}$ the number of *survival times* equal to $t_{(i)}$, $r_{(i)}$ the total number of individuals who died or were censored at time $t_{(i)}$, and $A_{(i)}$ is the proportion of these individuals in group two. If the survival experience of the two groups is the same then C has a standard normal distribution. [*Statistical Methods for Survival Data*, 3rd edn, E. T. Lee and J. W. Wang, Wiley, New York.]

Cox–Plackett model: A *logistic regression* model for the marginal probabilities from a *2 × 2 cross-over trial* with a binary outcome measure. [*Design and Analysis of Cross-Over trials*, 2nd edition, 2003, B. Jones and M. G. Kenward, Chapman and Hall/CRC Press, London.]

Cox–Snell residuals: Residuals widely used in the analysis of *survival time* data and defined as

$$r_i = -\ln \hat{S}_i(t_i)$$

where $\hat{S}_i(t_i)$ is the estimated *survival function* of the *i*th individual at the observed survival time of t_i. If the correct model has been fitted then these residuals will be n observations from an *exponential distribution* with mean one. [*Statistics in Medicine*, 1995, **14**, 1785–96.]

Cox–Spjøtvoll method: A method for partitioning treatment means in analysis of variance into a number of homogeneous groups consistent with the data. [*Biometrika*, 1986, **73**, 91–104.]

Coxian-2 distribution: The distribution of a random variable X such that

$$X = X_1 + X_2 \text{ with probability } b$$
$$X = X_1 \text{ with probability } 1 - b$$

where X_1 and X_2 are independent random variables having *exponential distributions* with different means. [*Scandinavian Journal of Statistics*, 1996, **23**, 419–41.]

Cox's proportional hazards model: A method that allows the *hazard function* to be modelled on a set of explanatory variables without making restrictive assumptions about the dependence of the hazard function on time. The model involved is

$$\ln h(t) = \ln \alpha(t) + \beta_1 x_1 + \beta_2 x_2 + \cdots + \beta_q x_q$$

where $x_1, x_2, \ldots, x_q$ are the explanatory variables of interest, and $h(t)$ the hazard function. The so-called *baseline hazard function*, $\alpha(t)$, is an arbitrary function of time. For any two individuals at any point in time the ratio of the hazard functions is a constant. Because the baseline hazard function, $\alpha(t)$, does not have to be specified explicitly, the procedure is essentially a *semi-parametric regression*. Estimates of the parameters in the model, i.e. $\beta_1, \beta_2, \ldots, \beta_p$, are usually obtained by maximizing the *partial likelihood*, and depend only on the order in which events occur, not on the exact times of their occurrence. See also **frailty** and **cumulative hazard function**. [SMR Chapter 13.]

Cox's test of randomness: A test that a sequence of events is random in time against the alternative that there exists a trend in the rate of occurrence. The test statistic is

$$m = \sum_{i=1}^{n} t_i/nT$$

where n events occur at times $t_1, t_2, \ldots, t_n$ during the time interval (0, T). Under the null hypothesis m has an *Irwin–Hall distribution* with mean $\frac{1}{2}$ and variance $\frac{1}{12n}$. As n increases the distribution of m under the null hypothesis approaches normality very rapidly and the normal approximation can be used safely for $n \geq 20$. [*Journal of the Royal Statistical Society, Series B*, 1955, **17**, 129–57.]

Craig's theorem: A theorem concerning the independence of *quadratic forms* in normal variables, which is given explicitly as:

> For $\mathbf{x}$ having a *multivariate normal distribution* with mean vector $\boldsymbol{\mu}$ and *variance–covariance matrix* $\boldsymbol{\Sigma}$, then $\mathbf{x}'\mathbf{A}\mathbf{x}$ and $\mathbf{x}'\mathbf{B}\mathbf{x}$ are stochastically independent if and only if $\mathbf{A}\boldsymbol{\Sigma}\mathbf{B} = \mathbf{0}$. [*The American Statistician*, 1995, **49**, 59–62.]

Cramér, Harald (1893–1985): Born in Stockholm, Sweden, Cramér studied chemistry and mathematics at Stockholm University. Later his interests turned more to mathematics and he obtained a Ph.D. degree in 1917 with a thesis on Dirichlet series. In 1929 he was appointed to a newly created professorship in actuarial mathematics and mathematical statistics. During the next 20 years he made important contributions to *central limit theorems*, *characteristic functions* and to mathematical statistics in general. Cramér died 5 October 1985 in Stockholm.

Cramér–Rao inequality: See **Cramér–Rao lower bound**.

Cramér–Rao lower bound: A lower bound to the variance of an estimator of a parameter that arises from the *Cramér–Rao inequality* given by

$$\text{var}(t) \geq -\{\tau'(\theta)\}^2 / E\left(\frac{\partial^2 \log L}{\partial \theta^2}\right)$$

where t is an unbiased estimator of some function of θ say $\tau(\theta)$, τ' is the derivative of τ with respect to θ and L is the relevant *likelihood*. In the case when $\tau(\theta) = \theta$, then $\tau'(\theta) = 1$, so for an unbiased estimator of θ

$$\text{var}(t) \geq 1/I$$

where I is the value of *Fisher's information*. [KA2 Chapter 17.]

Cramér's V: A measure of association for the two variables forming a *two-dimensional contingency table*. Related to the *phi-coefficient*, ϕ, but applicable to tables larger than 2 × 2. The coefficient is given by

$$\sqrt{\left\{\frac{\phi^2}{\min[(r-1)(c-1)]}\right\}}$$

where r is the number of rows of the table and c is the number of columns. See also **contingency coefficient, Sakoda coefficient** and **Tschuprov coefficient**. [*The Analysis of Contingency Tables*, 2nd edition, 1993, B. S. Everitt, Edward Arnold, London.]

Cramér–von Mises statistic: A goodness-of-fit statistic for testing the hypothesis that the cumulative probability distribution of a random variable take some particular form, F_0. If $x_1, x_2, \ldots, x_n$ denote a random sample, then the statistic U is given by

$$U = \int_{-\infty}^{\infty} \{F_n(x) - F_0(x)\}^2 \, \mathrm{d}F_0(x)$$

where $F_n(x)$ is the sample empirical cumulative distribution. [*Journal of the American Statistical Association*, 1974, **69**, 730–7.]

Cramér–von Mises test: A test of whether a set of observations arise from a normal distribution. The test statistic is

$$W = \sum_{i=1}^{n} \left[z_i - \frac{(2i-1)^2}{2n} \right] + \frac{1}{12n}$$

where the z_i are found from the ordered sample values $x_{(1)} \leq x_{(2)} \leq \cdots \leq x_{(n)}$ as

$$z_i = \int_{-\infty}^{x_{(i)}} \frac{1}{\sqrt{2\pi}} e^{-\frac{1}{2}x^2} \, \mathrm{d}x$$

Critical values of W can be found in many sets of statistical tables. [*Journal of the Royal Statistical Society*, Series B, 1996, **58**, 221–34.]

Craps test: A test for assessing the quality of random number generators. [*Random Number Generation and Monte Carlo Methods*, 1998, J. E. Gentle, Springer-Verlag, New York.]

Credible region: Synonym for **Bayesian confidence interval**.

Credibility theory: A class of techniques actuaries used to assign premiums to individual policy-holders in a heterogeneous portfolio. [*Journal of Actuarial Practice*, 1998, **6**, 5–62.]

Credit scoring: The process of determining how likely an applicant for credit is to default with repayments. Methods based on *discriminant analysis* are frequently employed to construct rules which can be helpful in deciding whether or not credit should be offered to an applicant. [*The Statistician*, 1996, **45**, 77–95.]

Creedy and Martin generalized gamma distribution: A probability distribution, $f(x)$, given by

$$f(x) = \exp\{\theta_1 \log x + \theta_2 x + \theta_3 x^2 + \theta_4 x^3 - \eta\}, \quad x > 0$$

The normalizing constant η needs to be determined numerically. Includes many well-known distributions as special cases. For example, $\theta_1 = \theta_3 = \theta_4 = 0$ corresponds to the *exponential distribution* and $\theta_3 = \theta_4 = 0$ to the *gamma distribution*. [*Communication in Statistics – Theory and Methods*, 1996, **25**, 1825–36.]

Cressie–Read statistic: A *goodness-of-fit statistic* which is, in some senses, somewhere between the *chi-squared statistic* and the *likelihood ratio*, and takes advantage of the desirable properties of both. [*Journal of the Royal Statistical Society, Series B*, 1979, **41**, 54–64.]

Criss-cross design: Synonym for **strip-plot design**.

Criteria of optimality: Criteria for choosing between competing experimental designs. The most common such criteria are based on the *eigenvalues* $\lambda_1, \ldots, \lambda_p$ of the matrix $\mathbf{X}'\mathbf{X}$ where $\mathbf{X}$ is the relevant *design matrix*. In terms of these eigenvalues three of the most useful criteria are:

A-optimality (*A-optimal designs*): Minimize the sum of the variances of the parameter estimates

$$\min\left\{\sum_{i=1}^{p}\frac{1}{\lambda_i}\right\}$$

D-optimality (*D-optimal designs*): Minimize the *generalized variance* of the parameter estimates

$$\min\left\{\prod_{i=1}^{p}\frac{1}{\lambda_i}\right\}$$

E-optimality (*E-optimal designs*): Minimize the variance of the least well estimated contrast

$$\min\left\{\max\frac{1}{\lambda_i}\right\}$$

All three criteria can be regarded as special cases of choosing designs to minimize

$$\left(\frac{1}{p}\sum_{i=1}^{p}\frac{1}{\lambda_i}\right)^{\frac{1}{k}} \qquad (0 \le k < \infty)$$

For A-, D- and E-optimality the values of k are 1, 0 and ∞, respectively. See also **response surface methodology**. [*Optimum Experimental Design*, 1992, A. C. Atkinson and A. N. Donev, Oxford University Press, Oxford.]

Critical region: The values of a *test statistic* that lead to rejection of a null hypothesis. The size of the critical region is the probability of obtaining an outcome belonging to this region when the null hypothesis is true, i.e. the probability of a type I error. Some typical critical regions are shown in Fig. 47. See also **acceptance region**. [KA2 Chapter 21.]

Critical value: The value with which a statistic calculated from sample data is compared in order to decide whether a null hypothesis should be rejected. The value is related to the particular significance level chosen. [KA2 Chapter 21.]

CRL: Abbreviation for **coefficient of racial likeness**.

Cronbach's alpha: An index of the internal consistency of a set of measurements, in particular psychological test. If the test consists of n items and an individual's score is the total answered correctly, then the coefficient is given specifically by

$$\alpha = \frac{n}{n-1}\left[1 - \frac{1}{\sigma^2}\sum_{i=1}^{n}\sigma_i^2\right]$$

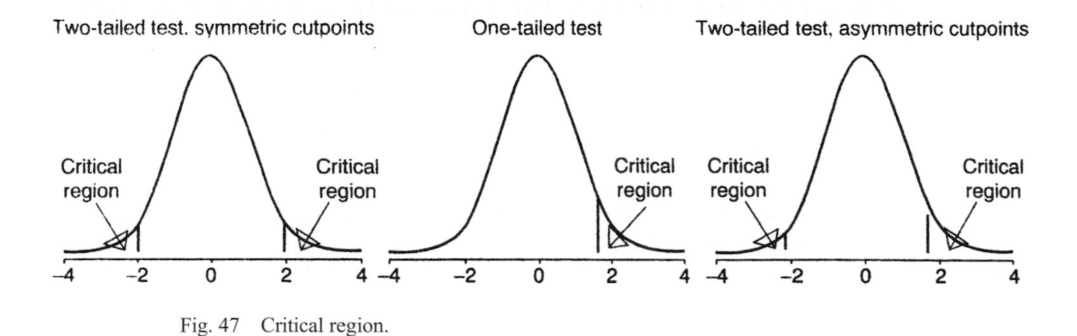

Fig. 47 Critical region.

where σ^2 is the variance of the total scores and σ_i^2 is the variance of the set of 0,1 scores representing correct and incorrect answers on item i. The theoretical range of the coefficient is 0 to 1. Suggested guidelines for interpretation are < 0.60 unacceptable, 0.60–0.65 undesirable, 0.65–0.70 minimally acceptable, 0.70–0.80 respectable, 0.80–0.90 very good, and > 0.90 consider shortening the scale by reducing the number of items. [*Statistical Evaluation of Measurement Errors: Design and Analysis of Reliability Studies*, 2004, G. Dunn, Arnold, London.]

Cross-correlation function: See **multiple time series**.

Cross-covariance function: See **multiple time series**.

Crossed treatments: Two or more treatments that are used in sequence (as in a *crossover design*) or in combination (as in a *factorial design*).

Crossover design: A type of *longitudinal study* in which subjects receive different treatments on different occasions. Random allocation is used to determine the order in which the treatments are received. The simplest such design involves two groups of subjects, one of which receives each of two treatments, A and B, in the order AB, while the other receives them in the reverse order. This is known as a *two-by-two crossover design*. Since the treatment comparison is 'within-subject' rather than 'between-subject', it is likely to require fewer subjects to achieve a given *power*. The analysis of such designs is not necessarily straightforward because of the possibility of *carryover effects*, that is residual effects of the treatment received on the first occasion that remain present into the second occasion. An attempt to minimize this problem is often made by including a *wash-out period* between the two treatment occasions. Some authorities have suggested that this type of design should only be used if such carryover effects can be ruled out *a priori*. Crossover designs are only applicable to chronic conditions for which short-term relief of symptoms is the goal rather than a cure. See also **three-period crossover designs**. [SMR Chapter 15.]

Crossover rate: The proportion of patients in a *clinical trial* transferring from the treatment decided by an initial random allocation to an alternative one.

Cross-sectional study: A study not involving the passing of time. All information is collected at the same time and subjects are contacted only once. Many surveys are of this type. The temporal sequence of cause and effect cannot be addressed in such a study, but it may be suggestive of an association that should be investigated more fully by, for example, a *prospective study*. [SMR Chapter 5.]

Cross-spectral density: See **multiple time series**.

Cross-validation: The division of data into two approximately equal sized subsets, one of which is used to estimate the parameters in some model of interest, and the second is used to assess whether the model with these parameter values fits adequately. See also **bootstrap** and **jackknife**. [MV2 Chapter 9.]

CRP: Abbreviation for **Chinese restaurant process**.

Crude annual death rate: The total deaths during a year divided by the total midyear population. To avoid many decimal places, it is customary to multiply death rates by 100 000 and express the results as deaths per 100 000 population. See also **age-specific death rates** and **cause specific death rates**. In 2005 the and annual death rate per 100 000 population ranged from 242 in Kuwait to 2936 in Botswana.

Crude risk: Synonym for **incidence rate**.

Cube law: A law supposedly applicable to voting behaviour which has a history of several decades. It may be stated thus:

> Consider a two-party system and suppose that the representatives of the two parties are elected according to a single member district system. Then the ratio of the number of representatives selected is approximately equal to the third power of the ratio of the national votes obtained by the two parties. [*Journal of the American Statistical Association*, 1970, **65**, 1213–19.]

Cubic spline: See **spline functions**.

Cultural assortment: See **assortative mating**.

Cumulant generating function: See **cumulants**.

Cumulants: A set of descriptive constants that, like *moments*, are useful for characterizing a probability distribution but have properties which are often more useful from a theoretical viewpoint. Formally the cumulants, $\kappa_1, \kappa_2, \ldots, \kappa_r$ are defined in terms of moments by the following identity in t:

$$\exp\left(\sum_{r=1}^{\infty} \kappa_r t^r / r!\right) = \sum_{r=0}^{\infty} \mu_r' t^r / r!$$

κ_r is the coefficient of $(it)^r / r!$ in $\log \phi(t)$ where $\phi(t)$ is the *characteristic function* of a random variable. The function $\psi(t) = \log \phi(t)$ is known as the *cumulant generating function*. The relationships between the first three cumulants and first four central moments are as follows:

$$\begin{aligned}
\mu_1' &= \kappa_1 \\
\mu_2' &= \kappa_2 + \kappa_1^2 \\
\mu_3' &= \kappa_3 + 3\kappa_2\kappa_1 + \kappa_1^3 \\
\mu_4' &= \kappa_4 + 4\kappa_3\kappa_1 + 6\kappa_2\kappa_1^2 - \kappa_1^4
\end{aligned}$$

[KA1 Chapter 3.]

Cumulative distribution function: A distribution giving the probability that a random variable is less than given values. For grouped data the given values correspond to the class boundaries.

Cumulative frequency distribution: The tabulation of a sample of observations in terms of numbers falling below particular values. The empirical equivalent of the *cumulative probability distribution*. An example of such a tabulation is shown below. [SMR Chapter 2.]

Hormone assay values (nmol/l)

Class limits	Cumulative frequency
75–79	1
80–84	3
85–89	8
90–94	17
95–99	27
100–104	34
105–109	38
110–114	40
≥ 115	41

Cumulative hazard function: A function, $H(t)$, used in the analysis of data involving *survival times* and given by

$$H(t) = \int_0^t h(u)\mathrm{d}u$$

where $h(t)$ is the *hazard function*. Can also be written in terms of the *survival function*, $S(t)$, as $H(t) = -\ln S(t)$. [*Modelling Survival Data in Medical Research*, 2nd edition, 2003, D. Collett, Chapman and Hall/CRC Press, London.]

Cumulative probability distribution: See **probability distribution**.

Cure models: Models for the analysis of *survival times*, or time to event, data in which it is expected that a fraction of the subjects will not experience the event of interest. In a clinical setting, this often corresponds to the assumption that a fraction of patients treated for a disease will be cured whereas the rest will experience a recurrence. Commonly such models involve the fitting of *finite mixture distributions*. [*Statistics in Medicine*, 1987, **6**, 483–489.]

Current status data: Current status data arise in *survival analysis* if observations are limited to indicators of whether or not the event of interest has occurred at the time the sample is collected. Hence, only the current status of each unit with respect to event occurrence is observed. [*Demography*, 1986, **23**, 607–620.]

Curse of dimensionality: A phrase first uttered by one of the witches in Macbeth. Now used to describe the exponential increase in number of possible locations in a multivariate space as dimensionality increases. Thus a single binary variable has two possible values, a 10-dimensional binary vector has over a thousand possible values and a 20-dimensional binary vector over a million possible values. This implies that sample sizes must increase exponentially with dimension in order to maintain a constant average sample size in the cells of the space. Another consequence is that, for a *multivariate normal distribution*, the vast bulk of the probability lies far from the centre if the dimensionality is large. [*Econometrica*, 1997, **65**, 487–516.]

Curvature measures: Diagnostic tools used to assess the closeness to linearity of a *non-linear model*. They measure the deviation of the so-called *expectation surface* from a plane with uniform grid. The expectation surface is the set of points in the space described by a prospective model, where each point is the expected value of the response variable based on a set of values for the parameters. [*Applied Statistics*, 1994, **43**, 477–88.]

Curve registration: The process of aligning important features or characteristics of curves and images by smooth, order-preserving nonlinear transformations (called *warping functions*), of the argument or domain over which the curve or, the image may be defined. [*Functional Data Analysis*, 2005, J. O. Ramsey and B. W. Silverman, Springer-Verlag, New York.]

Cusum: A procedure for investigating the influence of time even when it is not part of the design of a study. For a series $X_1, X_2, \ldots, X_n$, the cusum series is defined as

$$S_i = \sum_{j=1}^{i} (X_j - X_0)$$

where X_0 is a reference level representing an initial or normal value for the data. Depending on the application, X_0 may be chosen to be the mean of the series, the mean of the first few observations or some value justified by theory. If the true mean is X_0 and there is no time

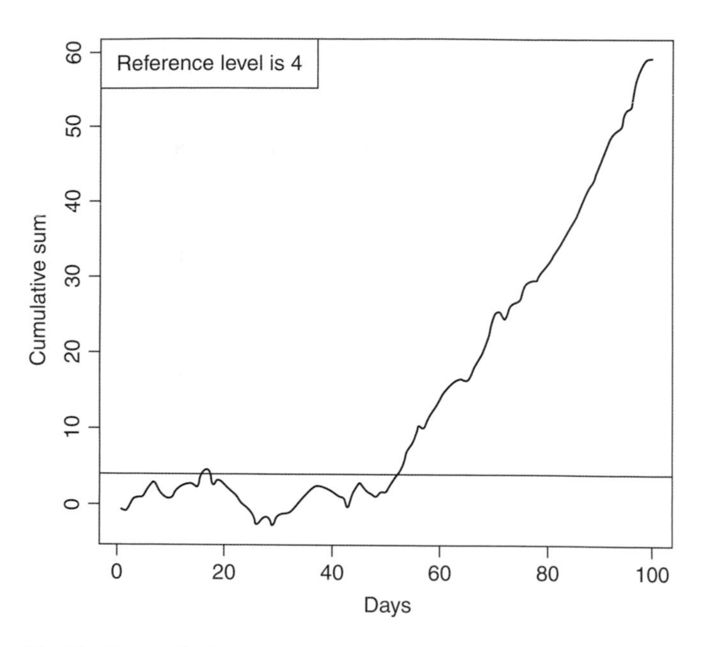

Fig. 48 Cusum chart.

trend then the cusum is basically flat. A change in level of the raw data over time appears as a change in the slope of the cusum. An example is shown in Fig. 48. See also **exponentially weighted moving average control chart**. [*Journal of Quality Techniques*, 1975, **7**, 183–92.]

Cuthbert, Daniel (1905–1997): Cuthbert attended MIT as an undergraduate, taking English and History along with engineering. He received a BS degree in chemical engineering in 1925 and an MS degree in the same subject in 1926. After a year in Berlin teaching Physics he returned to the US as an instructor at Cambridge School, Kendall Green, Maine. In the 1940s he read Fisher's *Statistical Methods for Research Workers* and began a career in statistics. Cuthbert made substantial contributions to the planning of experiments particularly in an individual setting. In 1972 he was elected an Honorary Fellow of the Royal Statistical Society. Cuthbert died in New York City on 8 August 1997.

Cuzick's trend test: A *distribution free method* for testing the trend in a measured variable across a series of ordered groups. The test statistic is, for a sample of N subjects, given by

$$T = \sum_{i=1}^{N} Z_i r_i$$

where $Z_i (i = 1, \ldots, N)$ is the group index for subject i (this may be one of the numbers $1, \ldots, G$ arranged in some natural order, where G is the number of groups, or, for example, a measure of exposure for the group), and r_i is the rank of the ith subject's observation in the combined sample. Under the null hypothesis that there is no trend across groups, the mean (μ) and variance (σ^2) of T are given by

$$\mu = N(N+1)E(Z)/2$$
$$\sigma^2 = N^2(N+1)V(Z)/12$$

where $E(Z)$ and $V(Z)$ are the calculated mean and variance of the Z values. [*Statistics in Medicine*, 1990, **9**, 829–34.]

Cycle: A term used when referring to *time series* for a periodic movement of the series. The *period* is the time it takes for one complete up-and-down and down-and-up movement. [*Cycles, the Mysterious Forces that Trigger Events*, 1971, E. R. Dewey, Hawthorn Books, New York.]

Cycle hunt analysis: A procedure for clustering variables in *multivariate data*, that forms clusters by performing one or other of the following three operations:

- combining two variables, neither of which belongs to any existing cluster,
- adding to an existing cluster a variable not previously in any cluster,
- combining two clusters to form a larger cluster.

Can be used as an alternative to *factor analysis*. See also **cluster analysis**. [*Multivariate Behavioral Research*, 1970, **5**, 101–16.]

Cycle plot: A graphical method for studying the behaviour of seasonal *time series*. In such a plot, the January values of the seasonal component are graphed for successive years, then the February values are graphed, and so forth. For each monthly subseries the mean of the values is represented by a horizontal line. The graph allows an assessment of the overall pattern of the seasonal change, as portrayed by the horizontal mean lines, as well as the behaviour of each monthly subseries. Since all of the latter are on the same graph it is readily seen whether the change in any subseries is large or small compared with that in the overall pattern of the seasonal component. Such a plot is shown in Fig. 49. [*Visualizing Data*, 1993, W. S. Cleveland, Hobart Press, Murray Hill, New Jersey.]

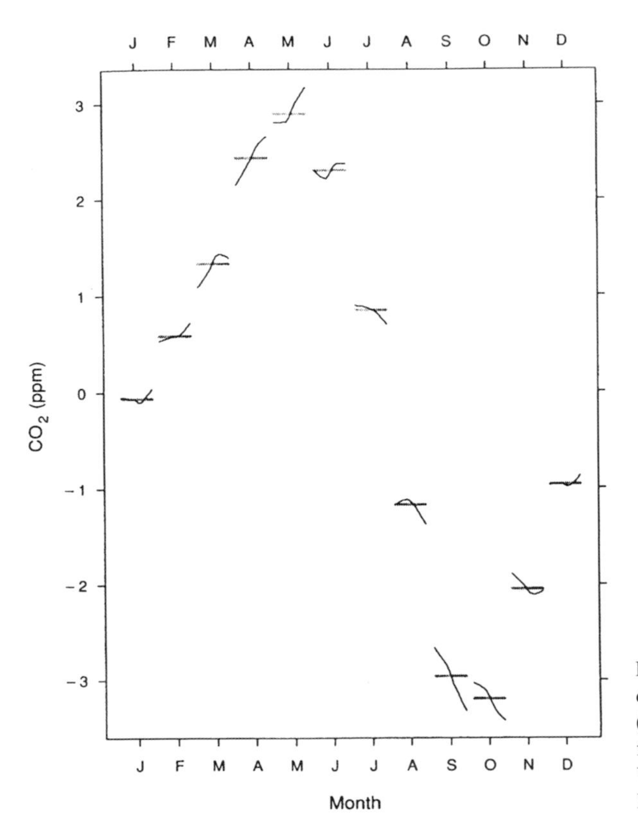

Fig. 49 Cycle plot of carbon dioxide concentrations. (Taken with permission from *Visualizing Data*, 1993, W. S. Cleveland, Hobart Press, Murray Hill, New Jersey.)

Cyclic designs: *Incomplete block designs* consisting of a set of blocks obtained by cyclic development of an initial block. For example, suppose a design for seven treatments using three blocks is required, the $\binom{7}{3}$ distinct blocks can be set out in a number of cyclic sets generated from different initial blocks, e.g. from (0,1,3)

0 1 2 3 4 5 6
1 2 3 4 5 6 0
3 4 5 6 0 1 2

[*Statistical Design and Analysis of Experiments*, 1971, P. W. M. John, MacMillan, New York.]

Cyclic variation: The systematic and repeatable variation of some variable over time. Most people's blood pressure, for example, shows such variation over a 24 hour period, being lowest at night and highest during the morning. Such *circadian variation* is also seen in many hormone levels. [SMR Chapter 14.]

Czekanowski coefficient: A *dissimilarity coefficient*, d_{ij}, for two individuals i and j each having scores, $\mathbf{x}_i' = [x_{i1}, x_{i2}, \ldots, x_{iq}]$ and $\mathbf{x}_j' = [x_{j1}, x_{j2}, \ldots, x_{jq}]$ on q variables, which is given by

$$d_{ij} = 1 - \frac{2\sum_{k=1}^{q} \min(x_{ik}, x_{jk})}{\sum_{k=1}^{q}(x_{ik} + x_{jk})}$$

[MV1 Chapter 3.]

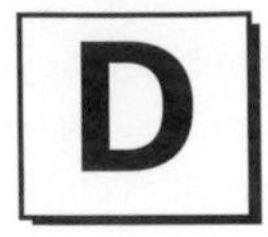

DAG: See **directed acyclic graph**.

D'Agostino's test: A test based on ordered sample values $x_{(1)} \leq x_{(2)} \leq \cdots \leq x_{(n)}$ with mean $\bar{x}$, used to assess whether the observations arise from a normal distribution. The *test statistic* is

$$D = \frac{\sum_{i=1}^{n} \{i - \frac{1}{2}(n+1)\} x_{(i)}}{n\sqrt{n \sum_{i=1}^{n} (x_{(i)} - \bar{x})^2}}$$

Appropriate for testing departures from normality due to *skewness.* Tables of critical values are available. [*Communication in Statistics: Theory and Methods*, 1994, **23**, 45–47.]

DALYs: Abbreviation for **disability adjusted life years**.

Daniels, Henry (1912–2000): Daniels studied at the Universities of Edinburgh and Cambridge, and was first employed at the Wool Industries Research Association in Leeds. This environment allowed Daniels to apply both his mathematical skills to the strength of bundles of threads and his mechanical bent to inventing apparatus for the fibre measurement laboratory. In 1947 he joined the statistical laboratory at the University of Cambridge and in 1957 was appointed as the first Professor of Mathematical Statistics at the University of Birmingham. He remained in Birmingham until his retirement in 1978 and then returned to live in Cambridge. Daniels was a major figure in the development of statistical theory in the 20th Century and was President of the Royal Statistical Society in 1974–1975. He was awarded the Guy medal of the Royal Statistical Society in bronze in 1957 and in gold in 1984. In 1980 Daniels was elected as a Fellow of the Royal Society. Daniels was a expert watch-repairer and in 1984 was created a Liveryman of the Worshipful Company of Clockmakers in recognition of his contribution to watch design. Daniels died on 16 April 2000 whilst attending a statistics conference at Gregynog, Powys, Wales.

Darling test: A test that a set of random variables arise from an *exponential distribution.* If $x_1, x_2, \ldots, x_n$ are the n sample values the test statistic is

$$K_m = \frac{\sum_{i=1}^{n} (x_i - \bar{x})^2}{\bar{x}^2}$$

where $\bar{x}$ is the mean of the sample. Asymptotically K_m can be shown to have mean (μ) and variance (σ^2) given by

$$\mu = \frac{n(n-1)}{n+1}$$

$$\sigma^2 = \frac{4n^4(n-1)}{(n+1)^2(n+2)(n+3)}$$

so that $z = (K_m - \mu)/\sigma$ has asymptotically a standard normal distribution under the exponential distribution hypothesis. [*Journal of Statistical Planning and Inference*, 1994, **39**, 399–424.]

Dantzig selector: An approach to variable selection with *high-dimensional data* that is similar to the *lasso*. [*Annals of Statistics*, 2007, **35**, 2313–2404.]

Darmois-Skitovitch's theorem: If X_k (k = 1,2, ... n) are independent random variables and L_1 and L_2 are defined as

$$L_1 = \sum_{k=1}^{n} b_k X_k L_2 = \sum_{k=1}^{n} c_k X_x$$

where b_k and c_k (k = 1,2 ..., n) are nonzero real coefficients, then if L_1 and L_2 are independent, X_1 ...... X_k all have normal distributions. [*Bulletin of the International Statistical Institute*, 1951, **23** (II), 79–82.]

Data: See **data set**.

Data archives: Collections of data that are suitably indexed and are accessible to be utilized by researchers aiming to perform secondary data analysis. Examples are the Economic and Social Research Council Data Archive held at the University of Essex in the UK, the ICPSR in Michigan (http://www.icpsr.umich.edu/) and StatLib (http://lib.stat.cmu.edu/datasets/)

Data augmentation: A scheme for augmenting observed data so as to make it more easy to analyse. A simple example is the estimation of *missing values* to balance a *factorial design* with different numbers of observations in each cell. The term is most often used, however, in respect of an iterative procedure, the *data augmentation algorithm* common in the computation of the *posterior distribution* in *Bayesian inference*. The basic idea behind this algorithm is to augment the observed data y by a quantity z which is usually referred to as latent data. It is assumed that given both y and z it is possible to calculate or sample from the augmented data posterior distribution $p(\theta|y, z)$. To obtain the observed posterior $p(\theta|y)$, multiple values (imputations) of z from the predictive distribution $p(z|y)$ are generated and the average of $p(\theta|y, z)$ over the imputations calculated. Because $p(z|y)$ depends on $p(\theta|y)$ an iterative algorithm for calculating $p(\theta|y)$ results. Specifically, given the current approximation $g_i(\theta)$ to the observed posterior $p(\theta|y)$ the algorithm specifies:

- generate a sample $z^{(1)}, \ldots, z^{(m)}$ from the current approximation to the predictive distribution $p(z|y)$;
- update the approximation to $p(\theta|y)$ to be the mixture of augmented posteriors of θ given the augmented data from the step above, i.e.

$$g_{i+1}(\theta) = \frac{1}{m} \sum_{j=1}^{m} p(\theta|z^{(j)}, y)$$

The two steps are then iterated. See also **EM algorithm** and **Markov chain Monte Carlo methods**. [*Analysis of Incomplete Multivariate Data*, 1997, J.L. Schafer, Chapman and Hall/CRC Press, London.]

Data augmentation algorithm: See **data augmentation**.

Database: A structured collection of data that is organized in such a way that it may be accessed easily by a wide variety of applications programs. Large clinical databases are becoming increasingly available to clinical and policy researchers; they are generally used for two purposes; to facilitate health care delivery, and for research. An example of such a database is that provided by the US Health Care Financing Administration which contains information about all Medicare patients' hospitalizations, surgical procedures and office visits. [SMR Chapter 6.]

Database management system: A computer system organized for the systematic management of a large structured collection of information, that can be used for storage, modification and retrieval of data.

Data depth: A quantitative measurement of how central a point is with respect to a data set, and used to measure the 'depth' of 'outlyingness' of a given multivariate sample with respect to its underlying distribution. Several such measures have been proposed, for example, *convex hull trimming* and the *bivariate Oja median*. Desirable properties of data depth measures are *affine invariance* and *robustness*. [*Annals of Statistics*, 1999, **27**, 783–858.]

Data dredging: A term used to describe comparisons made within a data set not specifically prescribed prior to the start of the study. See also **data mining** and **subgroup analysis**. [SMR Chapter 6.]

Data editing: The action of removing format errors and keying errors from data.

Data fusion: The act of combining data from heterogeneous sources with the intent of extracting information that would not be available for any single source in isolation. An example is the combination of different satellite images to facilitate identification and tracking of objects. The term is also used to mean the combination of statistically heterogeneous samples to construct a new sample that can be regarded as having come from an unobserved joint distribution of interest, i.e. the act of inferring a joint distribution when one only has information about the marginal distributions. See also **record linkage** and **copulas**. [*Journal of Marketing Research*, 2006, **43**, 1–22.]

Data generating mechanism (DGM): A term sometimes used for the statistical model which is assumed to have generated a dataset.

Data intrusion simulation: A method of estimating the probability that a data intruder who has matched an arbitrary population unit against a sample unit in a target *microdata* file has done so correctly. [*Statistical Journal of the United Nations Economics Commission for Europe*, 2001, **18**, 383–391.]

Data matrix: See **multivariate data**.

Data mining: The non-trivial extraction of implicit, previously unknown and potentially useful information from data, particularly *high-dimensional data*, using *pattern recognition*, *artificial intelligence* and *machine learning*, and the presentation of the information extracted in a form that is easily comprehensible to humans. Significant biological discoveries are now often made by combining data mining methods with traditional laboratory techniques; an example is the discovery of novel regulatory regions for heat shock genes in *C. elegans* made by mining vast amounts of gene expression and sequence data for significant patterns. [*Principles of Data Mining*, 2001, D.J. Hand, H. Mannila and P. Smyth, MIT Press.]

Data monitoring committees (DMC): Committees to monitor accumulating data from *clinical trials*. Such committees have major responsibilities for ensuring the continuing safety of trial participants, relevance of the trial question, appropriateness of the treatment protocol and integrity and quality of the accumulating data. The committees should be multidisciplinary, and should always include individuals with relevant clinical and statistical expertise. [*Data*

Monitoring Committees in Clinical Trials: A Practical Perspective, S. S. Ellenberg, T. R. Fleming and D. L. DeMets, 2002, Wiley, Chichester.]

Data perturbation: See **statistical disclosure limitation**.

Data reduction: The process of summarizing large amounts of data by forming frequency distributions, histograms, scatter diagrams, etc., and calculating statistics such as means, variances and correlation coefficients. The term is also used when obtaining a low-dimensional representation of *multivariate data* by procedures such as *principal components analysis* and *factor analysis.* [*Data Reduction and Error Analysis for the Physical Sciences*, 1991, P. R. Bevington, D. K. Robinson, McGraw-Hill.]

Data science: A term intended to unify statistics, data analysis and related methods. Consists of three phases, design for data, collection of data and analysis of data. [*Data Science, Classification and Related Methods*, 1998, C. Hayashi *et al.* eds., Springer, Tokyo.]

Data screening: The initial assessment of a set of observations to see whether or not they appear to satisfy the assumptions of the methods to be used in their analysis. Techniques which highlight possible *outliers*, or, for example, departures from normality, such as a *normal probability plot*, are important in this phase of an investigation. See also **initial data analysis**. [SMR Chapter 6.]

Data set: A general term for observations and measurements collected during any type of scientific investigation.

Data smoothing algorithms: Procedures for extracting a pattern in a sequence of observations when this is obscured by noise. Basically any such technique separates the original series into a smooth sequence and a residual sequence (commonly called the 'rough'). For example, a smoother can separate seasonal fluctuations from briefer events such as identifiable peaks and random noise. A simple example of such a procedure is the *moving average*; a more complex one is *locally weighted regression.* See also **Kalman filter** and **spline function**.

Data squashing: An approach to reducing the size of very large data sets in which the data are first 'binned' and then statistics such as the mean and variance/covariance are computed on each bin. These statistics are then used to generate a new sample in each bin to construct a reduced data set with similar statistical properties to the original one. [*Graphics of Large Data Sets*, 2006, A. Unwin, M. Theus, and H. Hofmann, Springer, New York.]

Data swapping: See **statistical disclosure limitation**.

Data tilting: A term applied to techniques for adjusting the empirical distribution by altering the data weights from their usual uniform values, i.e., n^{-1} where n is the sample size, to multinomial weights, p_i for the ith data point. Often used in the analysis of *time series.* [*Journal of the Royal Statistical Society, Series B*, 2003, **65**, 425–442.]

Data theory: Data theory is concerned with how observations are transformed into data that can be analyzed. Data are hence viewed as theory laden in the sense that observations can be given widely different interpretations, none of which are necessitated by the observations themselves. [*Data Theory and Dimensional Anaysis*, 1991, W. G. Jacoby, Sage, Newbury Park.]

Data vizualization: Interpretable graphical representations of abstract data and their relationships. See also **statistical graphics**. [*Vizualization Handbook*, 2004, C. Hansen and C. R. Johnson, Academic Press, Orlando, Florida.]

David, Florence Nightingale (1909–1993): Born near Leominster, Florence David obtained a first degree in mathematics from Bedford College for Women in 1931. Originally applied to become an actuary but had the offer of a post withdrawn when it was discovered that the 'F.N. David' who had applied was a women. Worked with *Karl Pearson* at University College, London and was awarded a doctorate in 1938. Also worked closely with *Jerzy Neyman* both in the United Kingdom and later in Berkeley. During the next 22 years she published eight books and over 80 papers. In 1962 David became Professor of Statistics at University College, London and in 1967 left England to accept a position at the University of California at Riverside where she established the Department of Statistics. She retired in 1977.

Davies–Quade test: A *distribution free method* that tests the hypothesis that the common underlying probability distribution of a sample of observations is symmetric about an unknown median. [NSM Chapter 3.]

Death rate: See **crude death rate**.

Debugging: The process of locating and correcting errors in a computer routine or of isolating and eliminating malfunctions of a computer itself.

Deciles: The values of a variable that divide its probability distribution or its frequency distribution into ten equal parts.

Decision function: See **decision theory**.

Decision theory: A unified approach to all problems of estimation, prediction and hypothesis testing. It is based on the concept of a *decision function*, which tells the experimenter how to conduct the statistical aspects of an experiment and what action to take for each possible outcome. Choosing a decision function requires a *loss function* to be defined which assigns numerical values to making good or bad decisions. Explicitly a general loss function is denoted as $L(d, \theta)$ expressing how bad it would be to make decision d if the parameter value were θ. A *quadratic loss function*, for example, is defined as

$$L(d, \theta) = (d - \theta)^2$$

and a *bilinear loss function* as

$$L(d, \theta) = a(d - \theta) \text{ if } d \leq \theta$$
$$L(d, \theta) = b(\theta - d) \text{ if } d \geq \theta$$

where a and b are positive constants. [KA2 Chapter 31.]

Decision tree: A graphic representation of the alternatives in a decision making problem that summarizes all the possibilities foreseen by the decision maker. For example, suppose we are given the following problem.

A physician must choose between two treatments. The patient is known to have one of two diseases but the diagnosis is not certain. A thorough examination of the patient was not able to resolve the diagnostic uncertainty. The best that can be said is that the probability that the patient has disease A is p.

A simple decision tree for the problem is shown in Fig. 50. [KA2 Chapter 31.]

Deep models: A term used for those models applied in *screening studies* that incorporate hypotheses about the disease process that generates the observed events. The aim of such models is to attempt an understanding of the underlying disease dynamics. See also **surface models**. [*Statistical Methods in Medical Research*, 1995, **4**, 3–17.]

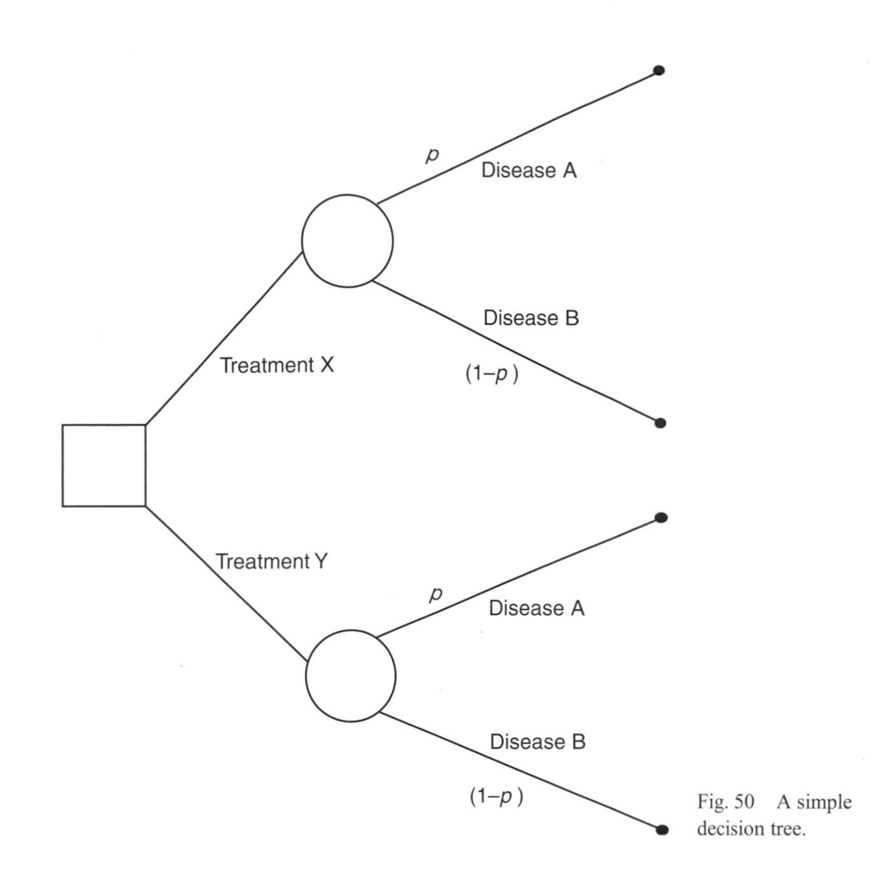

Fig. 50 A simple decision tree.

De Finetti, Bruno (1906–1985): Born in Innsbruck, Austria, De Finetti studied mathematics at the University of Milan, graduating in 1927. He became an actuary and then worked at the National Institute of Statistics in Rome later becoming a professor at the university. De Finetti is now recognized as a leading probability theorist for whom the sole interpretation of probability was a number describing the belief of a person in the truth of a proposition. He coined the aphorism 'probability does not exist', meaning that it has no reality outside an individual's perception of the world. A major contributor to *subjective probability* and *Bayesian inference*, De Finetti died on 20 July 1985 in Rome.

DeFries–Fulker analysis: A class of regression models that can be used to provide possible estimates of the fundamental biometrical genetic constructs, heritability and shared or common environment. See also **ACE model**. [*Annual Review of Psychology*, 1991, **42**, 161–90.]

Degenerate distributions: Special cases of probability distributions in which a random variable's distribution is concentrated at only one point. For example, a discrete uniform distribution when $k = 1$. Such distributions play an important role in *queuing theory*. [*A Primer on Statistical Distributions*, 2003, N. Balakrishnan and V.B. Neizorow, Wiley, New York.]

Degrees of freedom: An elusive concept that occurs throughout statistics. Essentially the term means the number of independent units of information in a sample relevant to the estimation of a parameter or calculation of a statistic. For example, in a *two-by-two contingency table* with a given set of marginal totals, only one of the four cell frequencies is free and the table has therefore a single degree of freedom. In many cases the term corresponds to the number of parameters in a model. Also used to refer to a parameter of various families of distributions, for example, *Student's t-distribution* and the *F- distribution*. [SMR Chapter 9.]

Delay distribution: The probability distribution of the delay in reporting an event. Particularly important in AIDS research, since AIDS surveillance data needs to be appropriately corrected for reporting delay before they can be used to reflect the current AIDS incidence. See also **back-projection**. [*Philosophical Transactions of the Royal Society of London, Series B*, 1989, **325**, 135–45.]

Delayed entry: In *survival analysis* this refers to situations where units do not come under observation when they become at risk for the event of interest but instead after a delay. [*Survival Analysis: State of the Art*, 1992, eds J. P. Klein and P. K. Goel, Springer, New York.]

Delta(δ) method: A procedure that uses the *Taylor series expansion* of a function of one or more random variables to obtain approximations to the expected value of the function and to its variance. For example, writing a variable x as $x = \mu + \epsilon$ where $E(x) = \mu$ and $E(\epsilon) = 0$, Taylor's expansion gives

$$f(x) = f(\mu) + \epsilon \frac{df(x)}{dx}|_{x=\mu} + \frac{\epsilon^2}{2}\frac{d^2f(x)}{dx^2}|_{x=\mu} + \cdots$$

If terms involving ϵ^2, ϵ^3, etc. are assumed to be negligible then

$$f(x) \approx f(\mu) + (x - \mu)f'(\mu)$$

So that

$$\text{var}[f(x)] \approx [f'(\mu)]^2\text{var}(x)$$

So if $f(x) = \ln x$ then $\text{var}(\ln x) = (1/\mu^2)\text{var}(x)$. [*The American Statistician*, 1992, **46**, 27–29.]

Deming, Edwards (1900–1993): Born in Sioux City, Iowa, Deming graduated from the University of Wyoming in 1921 in electrical engineering, received an MS in mathematics and physics from the University of Colorado in 1925 and a Ph.D. in mathematics and physics from Yale University in 1928. He became aware of early work on *quality control procedures* while working at the Hawthorne plant of the Western Electric Company in Chicago. Deming's interest in statistics grew in the early 1930s and, in 1939, he joined the US Bureau of the Census. During World War II, Deming was responsible for a vast programme throughout the USA teaching the use of sampling plans and control charts but it was in Japan in the 1950s that Deming's ideas about industrial production as a single system involving both the suppliers and manufacturers all aimed at satisfying customer need were put into action on a national scale. In 1960 Deming received Japan's Second Order Medal of the Sacred Treasure and became a national hero. He died on 20 December 1993.

Demography: The study of human populations with respect to their size, structure and dynamics, by statistical methods. [*Demography: Measuring and Modeling Population Processes*, S. H. Preston, P. Heiveline and M. Guillot, 2000, Wiley, New York.]

De Moivre, Abraham (1667–1754): Born in Vitry, France, de Moivre came to England in c. 1686 to avoid religious persecution as a Protestant and earned his living at first as a travelling teacher of mathematics, and later in life sitting daily in Slaughter's Coffee House in Long Acre, at the beck and call of gamblers, who paid him a small sum for calculating odds. A close friend of Isaac Newton, de Moivre reached the normal curve as the limit to the skew binomial and gave the correct measure of dispersion $\sqrt{np(1-p)}$. Also considered the

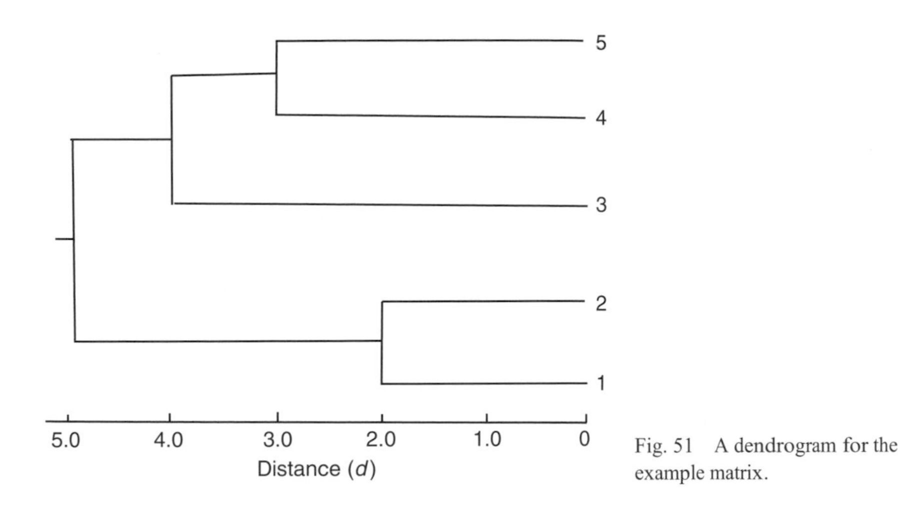

Fig. 51 A dendrogram for the example matrix.

concept of independence and arrived at a reasonable definition. His principal work, *The Doctrine of Chance*, which was on probability theory, was published in 1718. Just before his death in 1754 the French Academy elected him a foreign associate of the Academy of Science.

De Moivre–Laplace theorem: This theorem states that if x is a random variable having the *binomial distribution* with parameters n and p, then the *asymptotic distribution* of x is a normal distribution with mean np and variance $np(1-p)$. See also **normal approximation**. [KA1 Chapter 5.]

Dendrogram: A term usually encountered in the application of *agglomerative hierarchical clustering methods*, where it refers to the 'tree-like' diagram illustrating the series of steps taken by the method in proceeding from n single member 'clusters' to a single group containing all n individuals. The example shown (Fig. 51) arises from applying *single linkage clustering* to the following matrix of *Euclidean distances* between five points:

$$\mathbf{D} = \begin{pmatrix} 0.0 & & & & \\ 2.0 & 0.0 & & & \\ 6.0 & 5.0 & 0.0 & & \\ 10.0 & 9.0 & 4.0 & 0.0 & \\ 9.0 & 8.0 & 5.0 & 3.0 & 0.0 \end{pmatrix}$$

[MV1 Chapter 1.]

Density estimation: Procedures for estimating probability distributions without assuming any particular functional form. Constructing a histogram is perhaps the simplest example of such estimation, and *kernel density estimators* provide a more sophisticated approach. Density estimates can give valuable indication of such features as *skewness* and *multimodality* in the data. [*Density Estimation in Statistics and Data Analysis*, 1986, B.W. Silverman, Chapman and Hall/CRC Press, London.]

Density function: See **probability density**.

Density ratio model: A semiparametric model for testing the relative treatment effect between two populations based on a random sample from each. The model specifies that the log-likelihood of the ratio of two unknown densities is linear in some parameters. [*Statistical Modelling*, 2007, **7**, 155–173.]

Density sampling: A method of sampling controls in a *case-control study* which can reduce *bias* from changes in the prevalence of exposure during the course of a study. Controls are samples from the population at risk at the times of incidence of each case. [*American Journal of Epidemiology*, 1976, **103**, 226–35.]

Denton method: A widely used method for *benchmarking* a *time series* to annual benchmarks while preserving as far as possible the month-to-month movement of the original series. [*International Statistical Review*, 1994, **62** 365–77.]

Dependent variable: See **response variable**.

Deprivation indices: Socioeconomic indices constructed from variables such as social class or housing tenure and used in an examination of the relationship between deprivation and ill-health. [*Journal of Epidemiology and Community Health*, 1995, **49**, S3–S8.]

Descriptive statistics: A general term for methods of summarizing and tabulating data that make their main features more transparent. For example, calculating means and variances and plotting histograms. See also **exploratory data analysis** and **initial data analysis**.

Design-based inference: Statistical inference for parameters of a finite population where variability is due to hypothetical replications of the sampling design, which is sometimes complex (see **complex survey data**). Often contrasted to *model-based inference.* [*Canadian Journal of Forest Research*, 1998, **88**, 1429–1447.]

Design effect: The ratio of the variance of an estimator under the particular sampling design used in a study to its variance at equivalent sample size under simple random sampling without replacement. [*Survey Sampling*, 1995, L. Kish, Wiley, New York.]

Design matrix: Used generally for a matrix that specifies a statistical model for a set of observations. For example, in a *one-way design* with three observations in one group, two observations in a second group and a single observation in the third group, and where the model is

$$y_{ij} = \mu + \alpha_i + \epsilon_{ij}$$

the design matrix, **X** is

$$\mathbf{X} = \begin{pmatrix} 1 & 1 & 0 & 0 \\ 1 & 1 & 0 & 0 \\ 1 & 1 & 0 & 0 \\ 1 & 0 & 1 & 0 \\ 1 & 0 & 1 & 0 \\ 1 & 0 & 0 & 1 \end{pmatrix}$$

Using this matrix the model for all the observations can be conveniently expressed in matrix form as

$$\mathbf{y} = \mathbf{X}\boldsymbol{\beta} + \boldsymbol{\epsilon}$$

where $\mathbf{y}' = [y_{11}, y_{12}, y_{13}, y_{21}, y_{22}, y_{31}]$, $\boldsymbol{\beta}' = [\mu, \alpha_1, \alpha_2, \alpha_3]$ and $\boldsymbol{\epsilon}' = [\epsilon_{11}, \epsilon_{12}, \epsilon_{13}, \epsilon_{21}, \epsilon_{22}, \epsilon_{31}]$. Also used specifically for the matrix **X** in designed industrial experiments which specify the chosen values of the explanatory variables; these are often selected using one or other *criteria of optimatily.* See also **multiple regression**.

Design regions: Regions relevant to an experiment which are defined by specification of intervals of interest on the explanatory variables. For quantitative variables the most common region is that corresponding to lower and upper limits for the explanatory variables, which depend upon the physical limitations of the system and upon the range of values thought by the experimenter to be of interest. [*Journal of the Royal Statistical Society, Series B*, 1996, **58**, 59–76.]

Design rotatability: A term used in applications of *response surface methodology* for the requirement that the quality of the derived predictor of future response values is roughly the same throughout the region of interest. More formally a *rotatable design* is one for which $N\,\mathrm{var}(\hat{y}(\mathbf{x}))/\sigma^2$ has the same value at any two locations that are the same distance from the design centre. [*Journal of the Royal Statistical Society, Series B*, 1996, **58**, 59–76.]

Design set: Synonym for **training set**.

Detection bias: See **ascertainment bias**.

Detection limits: A term used to denote low-level data that cannot be distinguished from a zero concentration. Higher limits are also used to denote data with low, nonzero concentrations that are too imprecise to report as distinct numbers. See also **low-dose extrapolation**. [*Environmental Science and Technology*, 1988, **22**, 856–861.]

Determinant: A value associated with a *square matrix* that represents sums and products of its elements. For example, if the matrix is

$$\mathbf{A} = \begin{pmatrix} a & b \\ c & d \end{pmatrix}$$

then the determinant of **A** (conventionally written as det (**A**) or |**A**|) is given by

$$ad - bc$$

Deterministic model: One that contains no random or probabilistic elements. See also **random model**.

DETMAX: An algorithm for constructing exact *D-optimal designs*. [*Technometrics*, 1980, **22**, 301–13.]

Detrending: A term used in the analysis of *time series* data for the process of calculating a *trend* in some way and then subtracting the trend values from those of the original series. Often needed to achieve *stationarity* before fitting models to times series. See also **differencing**. [*Journal of Applied Economics*, 2003, **18**, 271–89.]

Deviance: A measure of the extent to which a particular model differs from the *saturated model* for a data set. Defined explicitly in terms of the *likelihoods* of the two models as

$$D = -2[\ln L_c - \ln L_s]$$

where L_c and L_s are the likelihoods of the current model and the saturated model, respectively. Large values of d are encountered when L_c is small relative to L_s, indicating that the current model is a poor one. Small values of d are obtained in the reverse case. The deviance has asymptotically a *chi-squared distribution* with degrees of freedom equal to the difference in the number of parameters in the two models when the current model is correct. See also $\mathbf{G}^2$ and **likelihood ratio**. [GLM Chapter 2.]

Deviance information criterion (DIC): A goodness of fit measure similar to *Akaike's information criterion* which arises from consideration of the posterior expectation of the *deviance* as a measure of fit and the effective number of parameters as a measure of complexity. Widely used for comparing models in a Bayesian framework. [*Journal of Business and Economic Statistics*, 2004, **22**, 107–20.]

Deviance residuals: The signed square root of an observation's contribution to total model *deviance*. [*Ordinal Data Modelling*, 1999, V.E. Johnson and J.H. Albert, Springer, New York.]

Deviate: The value of a variable measured from some standard point of location, usually the mean.

DeWitt, Johan (1625–1672): Born in Dordrecht, Holland, DeWitt entered Leiden University at the age of 16 to study law. Contributed to actuarial science and economic statistics before becoming the most prominent Dutch statesman of the third quarter of the seventeenth century. DeWitt died in The Hague on 20 August 1672.

DF(df): Abbreviation for **degrees of freedom**.

DFBETA: An *influence statistic* which measures the impact of a particular observation, i, on a specific estimated regression coefficient, $\hat{\beta}_j$, in a *multiple regression*. The statistic is the standardized change in $\hat{\beta}_j$ when the ith observation is deleted from the analysis; it is defined explicitly as

$$\text{DFBETA}_{j(i)} = \frac{\hat{\beta}_j - \hat{\beta}_{j(i)}}{s_{(-i)}\sqrt{c_j}}$$

where $s_{(-i)}$ is the residual mean square obtained from the regression analysis with observation i omitted, and c_j is the $(j+1)$th diagonal element of $(\mathbf{X}'\mathbf{X})^{-1}$ with $\mathbf{X}$ being the matrix appearing in the usual formulation of this type of analysis. See also **Cook's distance, DFFITS** and **COVRATIO**. [ARA Chapter 10.]

DFFITS: An *influence statistic* that is closely related to *Cook's distance*, which measures the impact of an observation on the predicted response value of the observation obtained from a *multiple regression*. Defined explicitly as;

$$\text{DFFITS}_i = \frac{\hat{y}_i - \hat{y}_{i(-i)}}{s_{(-i)}\sqrt{h_i}}$$

where $\hat{y}_i$ is the predicted response value for the ith observation obtained in the usual way and $\hat{y}_{(-i)}$ is the corresponding value obtained when observation i is not used in estimating the regression coefficients; $s^2_{(-i)}$ is the residual mean square obtained from the regression analysis performed with the ith observation omitted. The relationship of this statistic to Cook's distance, D_i, is

$$D_i = (\text{DFFITS}_i)^2 \frac{s^2_{(-i)}}{\text{tr}(\mathbf{H})s^2}$$

where $\mathbf{H}$ is the *hat matrix* with diagonal elements h_i and s^2 is the residual sum of squares obtained from the regression analysis including all observations. Absolute values of the statistic larger than $2\sqrt{\text{tr}(\mathbf{H}/n)}$ indicate those observations that give most cause for concern. [SMR Chapter 9.]

DGM: Abbreviation for **data generating mechanism**.

Diagnostic key: A sequence of binary or polytomous tests applied sequentially in order to indentify the population of origin of a specimen. [*Biometrika*, 1975, **62**, 665–72.]

Diagnostics: Procedures for indentifying departures from assumptions when fitting statistical models. See, for example, **DFBETA** and **DFFITS**. [*Residuals and Influence in Regression*, 1994, R.D. Cook and S. Weisberg, Chapman and Hall/CRC Press, London.]

Diagnostic tests: Procedures used in clinical medicine and also in *epidemiology*, to screen for the presence or absence of a disease. In the simplest case the test will result in a positive (disease likely) or negative (disease unlikely) finding. Ideally, all those with the disease should be classified by the test as positive and all those without the disease as negative. Two indices of the performance of a test which measure how often such correct classifications occur are its *sensitivity* and *specificity.* See also **believe the positive rule, positive predictive value** and **negative predictive value**. [SMR Chapter 14.]

Diagonal matrix: A *square matrix* whose off-diagonal elements are all zero. For example,

$$\mathbf{D} = \begin{pmatrix} 10 & 0 & 0 \\ 0 & 5 & 0 \\ 0 & 0 & 3 \end{pmatrix}$$

Diary survey: A form of data collection in which respondents are asked to write information at regular intervals or soon after a particular event has occurred. [*Lancet*, 2008, **37**, 1519–1525.]

DIC: Abbreviation for **deviance information criterion**.

Dichotomous variable: Synonym for **binary variable**.

Dickey-Fuller test: Synonym for **unit-root test**.

Dieulefait, Carlos Eugenio (1901–1982): Born in Buenos Aires, Dieulefait graduated from the Universidad del Litoral in 1922. In 1930 he became first director of the Institute of Statistics established by the University of Buenos Aires. For the next 30 years, Dieulefait successfully developed statistics in Argentina while also making his own contributions to areas such as correlation theory and *multivariate analysis.* He died on 3 November 1982 in Rosario, Argentina.

Differences-in-differences estimator: An estimator of group by period interaction in a study in which subjects in two different groups are observed in two different periods. Typically one of the groups is a control group and the other group is given the treatment in the second period but not in the first. The differences-in-differences estimator of the causal effect of treatment is simply the difference in period means for the second group minus the difference in the period means for the first group. The differences-in-differences estimator is attractive because it can be viewed as resulting from sweeping out both subject-specific intercepts and period-specific intercepts from a model including such terms. It is a standard approach to program evaluation in economics. [*Review of Economic Studies*, 2005, **72**, 1–19.]

Differences vs totals plot: A graphical procedure most often used in the analysis of data from a *two-by-two crossover design.* For each subject the difference between the response variable values on each treatment are plotted against the total of the two treatment values. The two groups corresponding to the order in which the treatments were given are differentiated on the plot by different plotting symbols. A large shift between the groups in the horizontal direction implies a differential *carryover effect.* If this shift is small, then the shift between the groups in a vertical direction is a measure of the treatment effect. An example of this type of plot appears in Fig. 52. [*The Statistical Consultant in Action*, 1987, edited by D.J. Hand and B.S. Everitt, Cambridge University Press, Cambridge.]

Differencing: A simple approach to removing trends in *time series.* The first difference of a time series, $\{y_t\}$, is defined as the transformation

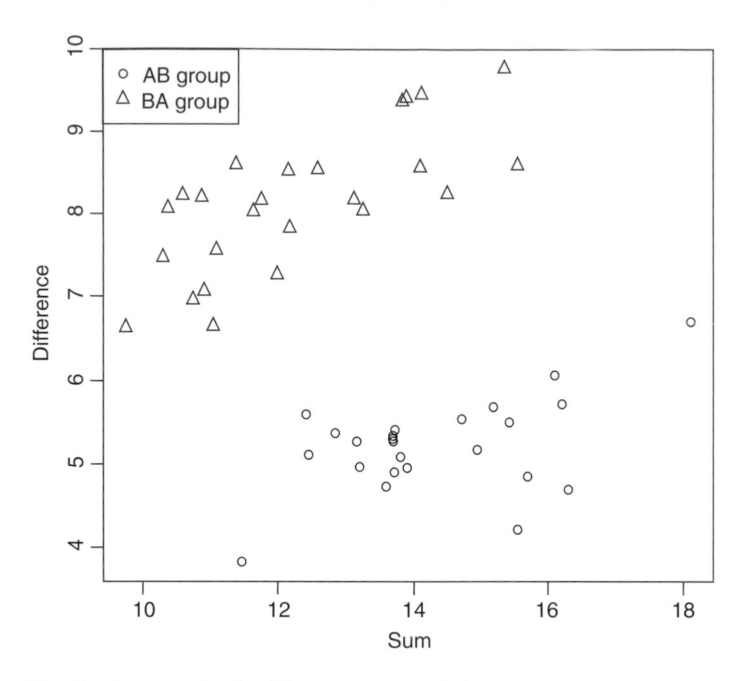

Fig. 52 An example of a difference versus total plot.

$$Dy_t = y_t - y_{t-1}$$

Higher-order differences are defined by repeated application. So, for example, the second difference, D^2y_t, is given by

$$D^2y_t = D(Dy_t) = Dy_t - Dy_{t-1} = y_t - 2y_{t-1} + y_{t-2}$$

Frequently used in applications to achieve a *stationarity* before fitting models. See also **backward shift operator** and **autoregressive integrated moving average models**.

Diggle–Kenward model for dropouts: A model applicable to *longitudinal data* in which the dropout process may give rise to *informative missing values*. Specifically if the study protocol specifies a common set of n measurement times for all subjects, and d is used to represent the subject's dropout time, with $D = d$ if the values corresponding to times $d, d+1, \ldots, n$ are missing and $D = n+1$ indicating that the subject did not drop out, then a statistical model involves the joint distribution of the observations y and d. This joint distribution can be written in two equivalent ways,

$$\begin{aligned} f(y, d) &= f(y)g(d|y) \\ &= g(d)f(y|d) \end{aligned}$$

Models derived from the first factorization are known as *selection models* and those derived from the second factorisation are called *pattern mixture models*. The Diggle–Kenward model is an example of the former which specifies a *multivariate normal distribution* for $f(y)$ and a *logistic regression* for $g(d|y)$. Explicitly if $p_t(y)$ denotes the *conditional probability* of dropout at time t, given $Y = y$ then,

$$\ln\left\{\frac{p_t(y)}{1 - p_t(y)}\right\} = \alpha_0 y_t + \sum_{k=1}^{r} \alpha_k y_{t-k}$$

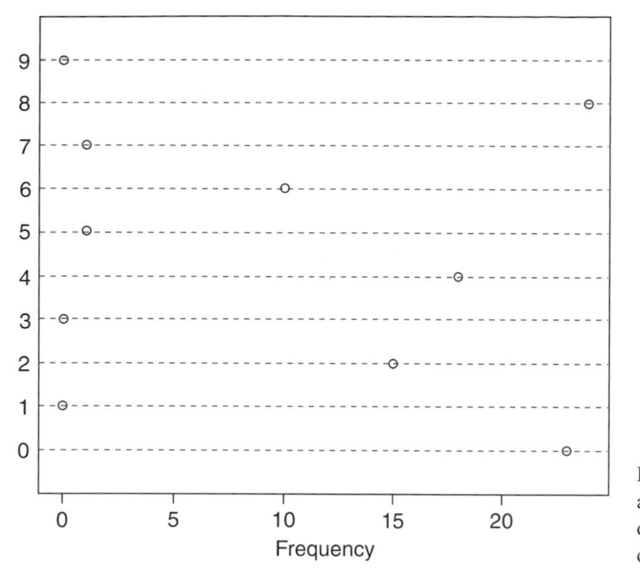

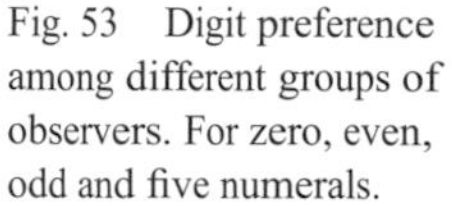
Fig. 53 Digit preference among different groups of observers. For zero, even, odd and five numerals.

When the dropout mechanism is informative the probability of dropout at time t can depend on the unobserved y_t. See also **missing values** and **selection models**. [*Analysis of Longitudinal Data*, 2nd edition, 2002, P. J. Diggle, P. J. Heagerty, K.-Y. Liang and S. L. Zeger, Oxford Science Publications, Oxford.]

Digit preference: The personal and often subconscious bias that frequently occurs in the recording of observations. Usually most obvious in the final recorded digit of a measurement. Figure 53 illustrates the phenomenon. [SMR Chapter 7.]

Digraph: Synonym for **directed graph**.

DIP test: A test for multimodality in a sample, based on the maximum difference, over all sample points, between the empirical distribution function and the unimodal distribution function that minimizes that maximum difference. [MV2 Chapter 10.]

Directed acyclic graph: Formal graphical representation of "causal diagrams" or "path diagrams" where relationships are directed (causal order specified for all relations) but acyclic (no feedback relations allowed). Plays an important role in effectively conveying an assumed causal model and determining which variables that should be controlled for in estimation of causal effects. Scc also **graph theory**. [*Journal of Epidemiology and Community Health*, 2008, **62**, 842–846.]

Directed deviance: Synonymous with **signed root transformation**.

Directed graph: See **graph theory**.

Directional data: Data where the observations are directions, points on the circumference of a circle in two dimensions (*circular data*) or on the surface of a sphere in three dimensions (*spherical variable*). In general directions may be visualized as points on the surface of a hypersphere. [*Multivariate Analysis*, 1979, K.V. Mardia, J. T. Kent and J. M. Bibby, Academic, New York.]

Directional neighbourhoods approach (DNA): A method for classifying pixels and reconstructing images from remotely sensed noisy data. The approach is partly Bayesian and partly data analytic and uses observational data to select an optimal, generally asymmetric, but relatively homogeneous neighbourhood for classifying pixels. The procedure involves two stages: a zero-neighbourhood pre-classification stage, followed by selection of the most homogeneous neighbourhood and then a final classification.

Direct matrix product: Synonym for **Kronecker product**.

Direct standardization: The process of adjusting a crude mortality or morbidity rate estimate for one or more variables, by using a known *reference population*. It might, for example, be required to compare cancer mortality rates of single and married women with adjustment being made for the age distribution of the two groups, which is very likely to differ with the married women being older. *Age-specific mortality rates* derived from each of the two groups would be applied to the population age distribution to yield mortality rates that could be directly compared. See also **indirect standardization**. [*Statistics in Medicine*, 1993, **12**, 3–12.]

Dirichlet distribution: The multivariate version of the *beta distribution*. Given by

$$f(x_1, x_2, \ldots, x_q) = \frac{\Gamma(\nu_1 + \cdots + \nu_q)}{\Gamma(\nu_1)\cdots\Gamma(\nu_q)} x_1^{\nu_1 - 1} \cdots x_q^{\nu_q - 1}$$

where

$$0 \leq x_i < 1, \ \sum_{i=1}^{q} x_i = 1, \ \nu_i > 0, \ \sum_{i=1}^{q} \nu_i = \nu_0$$

The expected value of x_i is ν_i/ν_0 and its variance is

$$\frac{\nu_i(\nu_0 - \nu_i)}{\nu_0^2(\nu_0 + 1)}$$

The covariance of x_i and x_j is

$$\frac{-\nu_i\nu_j}{\nu_0^2(\nu_0 + 1)}$$

[STD Chapter 10.]

Dirichlet process: A distribution over distributions in the sense that each draw from the process is itself a distribution. The name Dirichlet process is due to the fact that the finite dimensional marginal distributions of the process follows the *Dirichlet distribution*. Commonly used as a *prior distribution* in *nonparametric Bayesian models*, particularly in *Dirichlet process mixture models*. Called nonparametric because, although the distributions drawn from a Dirichlet process are discrete, they cannot be described using a finite number of parameters. The prior distribution induced by a Dirichlet process can be generated incrementally using the *Chinese restaurant process*. See also **Pitman-Yor process**. [*Annals of Statistics*, 1973, **1**, 209–230.]

Dirichlet process mixture models: A nonparametric *Bayesian inference* approach to using *finite mixture distributions* for modelling data suspected of containing distinct groups of observations; this approach does not require the number of mixture components to be known in advance. The basic idea is that the *Dirichlet process* induces a *prior distribution* over partitions of the data which can then be combined with a prior distribution over parameters and likelihood. The distribution over partitions can be generated incrementally using the *Chinese restaurant process*. [*Annals of Statistics*, 1974, **2**, 1152–1174.]

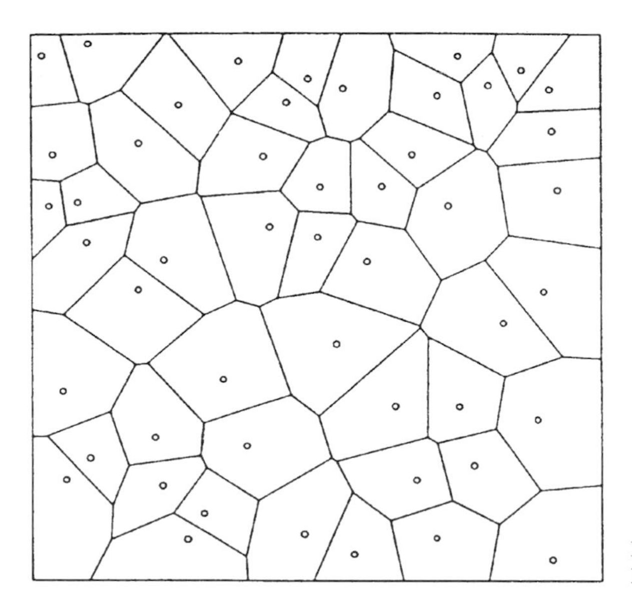

Fig. 54 An example of Dirichlet tessellation.

Dirichlet tessellation: A construction for events that occur in some planar region a, consisting of a series of 'territories' each of which consists of that part of a closer to a particular event x_i than to any other event x_j. An example is shown in Fig. 54. [*Pattern Recognition and Neural Networks*, 1996, B.D. Ripley, Cambridge University Press, Cambridge.]

Disability adjusted life years (DALYs): An attempt to measure the suffering caused by an illness, that takes into account both the years of potential life lost due to premature mortality as well as the years lost due to a disease or health condition. One DALY represents the equivalent of the loss of one year of full health. Useful in comparisons across diseases and in setting national and international health priorities. See also **healthy life expectancy**. [*Journal of Health Economics*, 1997, **16**, 685–702.]

Disclosure risk: The risk of being able to identify a respondent's confidential information in a data set. Several approaches have been proposed to measure the disclosure risk some of which concentrate on the risk per individual record and others of which involve a global measure for the entire data file. See also **data intrusion simulation**. [*Statistics and Computing*, 2003, **13**, 343–354.]

Discrete rectangular distribution: A probability distribution for which any one of a finite set of equally spaced values is equally likely. The distribution is given by

$$\Pr(X = a + jh) = \frac{1}{n+1} \quad j = 0, 1, \ldots, n$$

so that the random variable x can take any one of the equally spaced values $a, a+h, \ldots, a+nh$. As $n \to \infty$ and $h \to 0$, with $nh = b - a$ the distribution tends to a *uniform distribution* over (a, b). The *standard discrete rectangular distribution* has $a = 0$ and $h = 1/n$ so that x takes values $0, 1/n, 2/n, \ldots, 1$. [*Univariate Discrete Distributions*, 3rd edn, 2005, N. L. Johnson, A. W. Kemp and S. Kotz, Wiley, New York.]

Discrete time Markov chain: See **Markov chain**.

Discrete time stochastic process: See **stochastic process**.

Discrete uniform distribution: A probability distribution for a discrete random variable that takes on k distinct values $x_1, x_2, \ldots, x_k$ with equal probabilities where k is a positive integer. See also **lattice distributions**. [*A Primer on Statistical Distributions*, 2003, N. Balakrishnan and V.B. Neizorow, Wiley, New York.]

Discrete variables: Variables having only integer values, for example, number of births, number of pregnancies, number of teeth extracted, etc. [SMR Chapter 2.]

Discrete wavelet transform (DWT): The calculation of the coefficients of the *wavelet series approximation* for a discrete signal $f_1, f_2, \ldots, f_n$ of finite extent. Essentially maps the vector $\mathbf{f}' = [f_1, f_2, \ldots, f_n]$ to a vector of n *wavelet transform coefficients*. [*IEEE Transactions on Pattern Analysis and Machine Intelligence*, 1989, **11**, 674–93.]

Discriminant analysis: A term that covers a large number of techniques for the analysis of multivariate data that have in common the aim to assess whether or not a set of variables distinguish or discriminate between two (or more) groups of individuals. In medicine, for example, such methods are generally applied to the problem of using optimally the results from a number of tests or the observations of a number of symptoms to make a diagnosis that can only be confirmed perhaps by post-mortem examination. In the two group case the most commonly used method is *Fisher's linear discriminant function*, in which a linear function of the variables giving maximal separation between the groups is determined. This results in a *classification rule* (often also known as an *allocation rule*) that may be used to assign a new patient to one of the two groups. The derivation of this linear function assumes that the *variance–covariance matrices* of the two groups are the same. If they are not then a *quadratic discriminant function* may be necessary to distinguish between the groups. Such a function contains powers and cross-products of variables. The sample of observations from which the discriminant function is derived is often known as the *training set*. When more than two groups are involved (all with the same variance–covariance matrix) then it is possible to determine several linear functions of the variables for separating them. In general the number of such functions that can be derived is the smaller of q and $g - 1$ where q is the number of variables and g the number of groups. The collection of linear functions for discrimination are known as *canonical discriminant functions* or often simply as *canonical variates*. See also **error rate estimation** and **regularised discriminant analysis**. [MV2 Chapter 9.]

Discrimination information: Synonymous with **Kullback-Leibler information**.

Disease clusters: An unusual aggregation of health events, real or perceived. The events may be grouped in a particular area or in some short period of time, or they may occur among a certain group of people, for example, those having a particular occupation. The significance of studying such clusters as a means of determining the origins of public health problems has long been recognized. In 1850, for example, the Broad Street pump in London was identified as a major source of cholera by plotting cases on a map and noting the cluster around the well. More recently, recognition of clusters of relatively rare kinds of pneumonia and tumours among young homosexual men led to the identification of acquired immunodeficiency syndrome (AIDS) and eventually to the discovery of the human immunodeficiency virus (HIV). See also **scan statistic**. [*Statistics in Medicine*, 1995, **14**, 799–810.]

Disease mapping: The process of displaying the geographical variability of disease on maps using different colours, shading, etc. The idea is not new, but the advent of computers and computer graphics has made it simpler to apply and it is now widely used in descriptive *epidemiology*, for example, to display morbidity or mortality information for an area. Figure 55 shows an example. Such mapping may involve absolute rates, relative rates, etc., and often the viewers impression of geographical variation in the data may vary quite

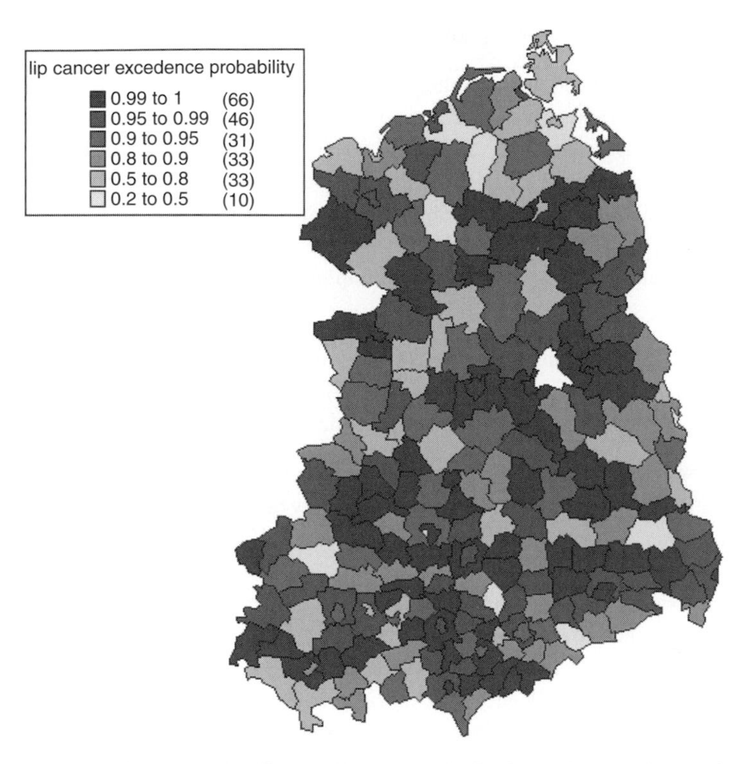

Fig. 55 Disease mapping illustrated by age-standardized mortality rates in part of Germany. (Reproduced from *Biometrics* by permission of the International Biometric Society.)

markedly according to the methodology used. [*Statistical Methods in Spatial Epidemiology*, 2nd edn, 2006, A. B. Lawson, Wiley, New York.]

Disease surveillance: A process that aims to use health and health-related data that precede diagnosis and/or confirmation to identify possible outbreaks of a disease, mobilize a rapid response and therefore reduce morbidity and mortality. Such a process provides essential information for control and response planning, helping to identify changes in incidence and affected group, thereby providing valuable additional time for public health interventions. [*The Lancet*, 1997, **349**, 794–795.]

Dispersion: The amount by which a set of observations deviate from their mean. When the values of a set of observations are close to their mean, the dispersion is less than when they are spread out widely from the mean. See also **variance**. [MV1 Chapter 1.]

Dispersion parameter: See **generalized linear models**.

Dissimilarity coefficient: A measure of the difference between two observations from (usually) a set of *multivariate data*. For two observations with identical variable values the dissimilarity is usually defined as zero. See also **metric inequality**. [MV1 Chapter 1.]

Dissimilarity matrix: A *square matrix* in which the elements on the main diagonal are zero, and the off-diagonal elements are *dissimilarity coefficients* of each pair of stimuli or objects of interest. Such matrices can either arise by calculations on the usual multivariate data matrix **X** or directly from experiments in which human subjects make judgements about the stimuli. The pattern or structure of this type of matrix is often investigated by applying some form of *multidimensional scaling*. [MV1 Chapter 4.]

Distance measure: See **metric inequality**.

Distance sampling: A method of sampling used in ecology for determining how many plants or animals are in a given fixed area. A set of randomly placed lines or points is established and distances measured to those objects detected by travelling the line or surveying the points. As a particular object is detected, its distance to the randomly chosen line or point is measured and on completion n objects have been detected and their associated distances recorded. Under certain assumptions unbiased estimates of density can be made from these distance data. See also **line-intersect sampling**. [*Canadian Field-Naturalist*, 1996, **110**, 642–8.]

Distributed database: A *database* that consists of a number of component parts which are situated at geographically separate locations. [*Principles of Distributed Database Systems*, 2nd edition, 1999, M.T. Ozsu, P. Valdunez, Prentice Hall.]

Distribution free methods: Statistical techniques of estimation and inference that are based on a function of the sample observations, the probability distribution of which does not depend on a complete specification of the probability distribution of the population from which the sample was drawn. Consequently the techniques are valid under relatively general assumptions about the underlying population. Often such methods involve only the ranks of the observations rather than the observations themselves. Examples are *Wilcoxon's signed rank test* and *Friedman's two way analysis of variance*. In many cases these tests are only marginally less powerful than their analogues which assume a particular population distribution (usually a normal distribution), even when that assumption is true. Also commonly known as *non-parametric methods* although the terms are not completely synonymous. [SMR Chapter 9.]

Distribution function: See **probability distribution**.

Divide-by-total models: Models for polytomous responses in which the probability of a particular category of the response is typically modelled as a multinomial *logistic function*. [*Psychometrika*, 1986, **51**, 567–77.]

Divisive clustering: *Cluster analysis* procedures that begin with all individuals in a single cluster, which is then successively divided by maximizing some particular measure of the separation of the two resulting clusters. Rarely used in practice. See also **agglomerative hierarchical clustering methods** and **K-means cluster analysis**. [MV2 Chapter 10.]

Dixon, Wilfred J. (1915–2008): Born in Portland, Oregon, USA, Dixon received a B.A. in mathematics from Oregon State College in 1938, an M.A. in mathematics from the University of Wisconsin in 1939 and a Ph.D in mathematical statistics from Princeton in 1944 under the supervision of *Samuel S. Wilks*. During the Second World War he was an operations analyst at Princeton University. After the war Dixon went to work at the University of Oregon and then in 1955 moved to the University of California at Los Angeles where he remained for the rest of his career. He made major contributions in several areas of statistics including *nonparametric statistics*, *experimental design* and *robust statistics*, but Dixon is probably most widely remembered for his pioneering efforts in the development of statistical software which culminated in the 1960s with the release of the Biomedical Computer Programs, later to become the *BMDP* Statistical Software. Dixon died on September 20th 2008.

Dixon test: A test for *outliers*. When the sample size, n, is less than or equal to seven, the test statistic is

$$r = \frac{y_{(1)} - y_{(2)}}{y_{(1)} - y_{(n)}}$$

where $y_{(1)}$ is the suspected outlier and is the smallest observation in the sample, $y_{(2)}$ is the next smallest and $y_{(n)}$ the largest observation. For $n > 7$, $y_{(3)}$ is used instead of $y_{(2)}$ and $y_{(n-1)}$ in place of $y_{(n)}$. Critical values are available in some statistical tables. [*Journal of Statistical Computation and Simulation*, 1997, **58**, 1–20.]

DMC: Abbreviation for **data monitoring committees**.

Doane's rule: A rule for calculating the number of classes to use when constructing a *histogram* and given by

$$\text{no. of classes} = \log_2 n + 1 + \log_2(1 + \hat{\gamma}\sqrt{n/6})$$

where n is the sample size and $\hat{\gamma}$ is an estimate of *kurtosis*. See also **Sturges' rule**. [*The American Statistician*, 1977, **30**, 181–183.]

Dodge, Harold (1893–1976): Born in the mill city of Lowell, Massachusetts, Dodge became one of the most important figures in the introduction of quality control and the development and introduction of *acceptance sampling*. He was the originator of the operating characteristic curve. Dodge's career was mainly spent at Bell Laboratories, and his contributions were recognized with the Shewhart medal of the American Society for Quality Control in 1950 and an Honorary Fellowship of the Royal Statistical Society in 1975. He died on 10 December 1976 at Mountain Lakes, New Jersey.

Dodge's continuous sampling plan: A procedure for monitoring continuous production processes. [*Annals of Mathematical Statistics*, 1943, **14**, 264–79.]

Doll, Sir Richard (1912–2005): Born in Hampton, England, Richard Doll studied medicine at St. Thomas's Hospital Medical School in London, graduating in 1937. From 1939 until 1945 he served in the Royal Army Medical Corps and in 1946 started work at the Medical Research Council. In 1951 Doll and *Bradford Hill* started a study that would eventually last for 50 years, asking all the doctors in Britain what they themselves smoked and then tracking them down over the years to see what they died of. The early results confirmed that smokers were much more likely to die of lung cancer than non-smokers, and the 10-year results showed that smoking killed far more people from other diseases than from lung cancer. The study has arguably helped to save millions of lives. In 1969 Doll was appointed Regius Professor of Medicine at Oxford and during the next ten years helped to develop one of the top medical schools in the world. He retired from administrative work in 1983 but continued his research, publishing the 50-year follow-up on the British Doctors' Study when he was 91 years old, on the 50th anniversary of the first publication from the study. Doll received many honours during his distinguished career including an OBE in 1956, a knighthood in 1971, becoming a Companion of Honour in 1996, the UN award for cancer research in 1962 and the Royal Medal from the Royal Society in 1986. He also received honorary degrees from 13 universities. Doll died in Oxford on 24 July 2005, aged 92.

Doob–Meyer decomposition: A theorem which shows that any *counting process* may be uniquely decomposed as the sum of a *martingale* and a predictable, right-continous process called the compensator, assuming certain mathematical conditions. [*Modelling Survival Data*, 2001, T.M. Therneau and P.M. Grambsch, Springer, New York.]

D-optimal design: See **criteria of optimality**.

Doran estimator: An estimator of the *missing values* in a *time series* for which monthly observations are available in a later period but only quarterly observations in an earlier period. [*Journal of the American Statistical Association*, 1974, **69**, 546–554.]

Dorfman–Berbaum–Metz method: An approach to analysing multireader *receiver operating characteristic curves* data, that applies an analysis of variance to pseudovalues of the ROC parameters computed by *jackknifing* cases separately for each reader–treatment combination. See also **Obuchowski and Rockette method**. [*Academic Radiology*, 2005, **12**, 1534–1541.]

Dorfman scheme: An approach to investigations designed to identify a particular medical condition in a large population, usually by means of a blood test, that may result in a considerable saving in the number of tests carried out. Instead of testing each person separately, blood samples from, say, k people are pooled and analysed together. If the test is negative, this one test clears k people. If the test is positive then each of the k individual blood samples must be tested separately, and in all $k + 1$ tests are required for these k people. If the probability of a positive test (p) is small, the scheme is likely to result in far fewer tests being necessary. For example, if $p = 0.01$, it can be shown that the value of k that minimizes the expected number of tests per person is 11, and that the expected number of tests is 0.2, resulting in 80% saving in the number of tests compared with testing each individual separately. [*Statistics in Medicine*, 1994, **22**, 2337–2343.]

Dose-ranging trial: A *clinical trial*, usually undertaken at a late stage in the development of a drug, to obtain information about the appropriate magnitude of initial and subsequent doses. Most common is the *parallel-dose design*, in which one group of subjects in given a placebo, and other groups different doses of the active treatment. [*Controlled Clinical Trials*, 1995, **16**, 319–30.]

Dose–response curve: A plot of the values of a response variable against corresponding values of dose of drug received, or level of exposure endured, etc. See Fig. 56. [*Hepatology*, 2001, **33**, 433–8.]

Dot plot: A more effective display than a number of other methods, for example, *pie charts* and *bar charts*, for displaying quantitative data which are labelled. An example of such a plot showing standardized mortality rates (SMR) for lung cancer for a number of occupational groups is given in Fig. 57.

Double-blind: See **blinding**.

Double-centred matrices: Matrices of numerical elements from which both row and column means have been subtracted. Such matrices occur in some forms of *multidimensional scaling*.

Double-count surveys: Surveys in which two observers search the sampled area for the species of interest. The presence of the two observers permits the calculation of a survey-specific correction for visibility bias. [*The Survey Kit*, 1995, A. Fink, ed., Sage, London.]

Double-dummy technique: A technique used in *clinical trials* when it is possible to make an acceptable placebo for an active treatment but not to make two active treatments identical. In this instance, the patients can be asked to take two sets of tablets throughout the trial: one representing treatment A (active or placebo) and one treatment B (active or placebo). Particularly useful in a *crossover trial*. [SMR Chapter 15.]

Double-exponential distribution: Synonym for **Laplace distribution**.

Double-masked: Synonym for **double-blind**.

Double reciprocal plot: Synonym for **Lineweaver–Burke plot**.

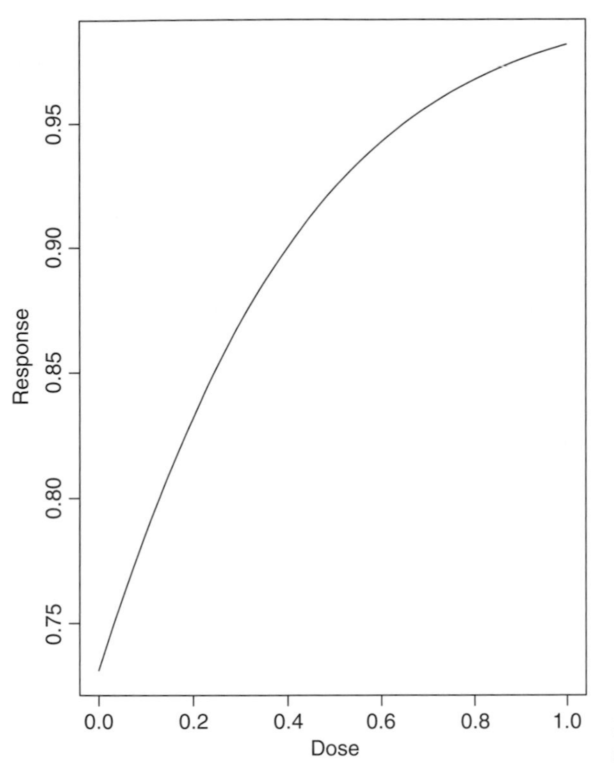

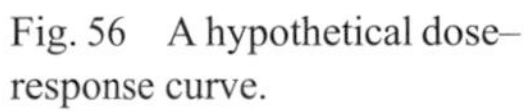
Fig. 56 A hypothetical dose–response curve.

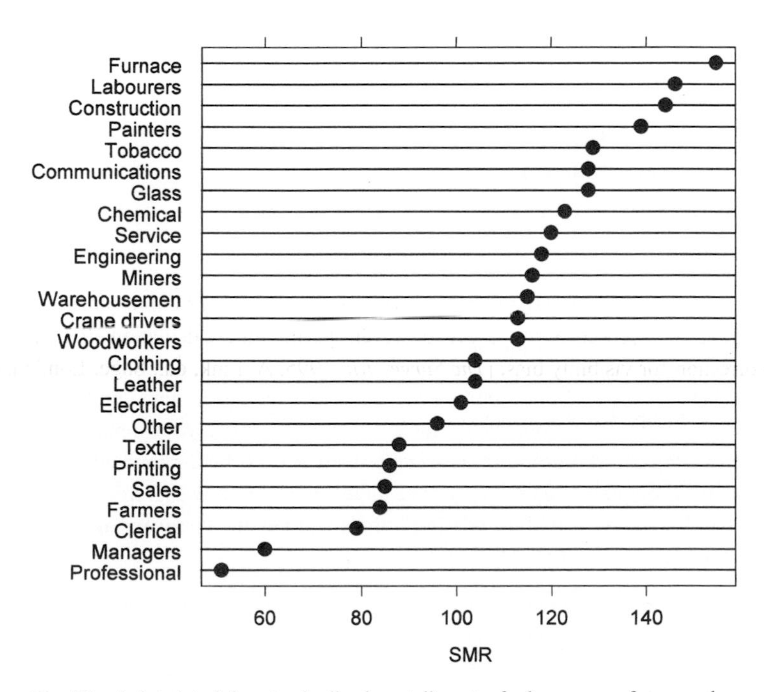

Fig. 57 A dot plot giving standardized mortality rates for lung cancer for several occupational groups.

Double sampling: A procedure in which initially a sample of subjects is selected for obtaining auxillary information only, and then a second sample is selected in which the variable of interest is observed in addition to the auxillary information. The second sample is often selected as a subsample of the first. The purpose of this type of sampling is to obtain better estimators by using the relationship between the auxillary variables and the variable of interest. See also **two-phase sampling**. [KA2 Chapter 24.]

Doubly censored data: Data involving *survival times* in which the time of the originating event and the failure event may both be *censored observations*. Such data can arise when the originating event is not directly observable but is detected via periodic *screening studies*. [*Statistics in Medicine,* 1992, **11**, 1569–78.]

Doubly multivariate data: A term sometimes used for the data collected in those *longitudinal studies* in which more than a single response variable is recorded for each subject on each occasion. For example, in a *clinical trial*, weight and blood pressure may be recorded for each patient on each planned visit.

Doubly ordered contingency tables: *Contingency tables* in which both the row and column categories follow a natural order. An example might be, drug toxicity ranging from mild to severe, against drug dose grouped into a number of classes. A further example from a more esoteric area is shown.

Cross-classification of whiskey for age and grade

	Grade		
Years/maturity	1	2	3
7	4	2	0
5	2	2	2
1	0	0	4

Downton, Frank (1925–1984): Downton studied mathematics at Imperial College London and after a succession of university teaching posts became Professor of Statistics at the University of Birmingham in 1970. Contributed to *reliability theory* and *queueing theory.* Downton died on 9 July 1984.

Dragstedt–Behrens estimator: An estimator of the *median effective dose* in *bioassay.* See also **Reed–Muench estimator**. [*Probit Analysis* 3rd ed., 1971, D.J. Finney, Cambridge Unversity Press, Cambridge.]

Draughtsman's plot: Synonym for **scatterplot matrix.**

Drift: A term used for the progressive change in assay results throughout an assay run.

Drift parameter: See **Brownian motion**.

Dropout: A subject who withdraws from a study for whatever reason, noncompliance, adverse side effects, moving away from the district, etc. In many cases the reason may not be known. The fate of subjects who drop out of an investigation must be determined whenever possible since the dropout mechanism may have implications for how data from the study should be analysed. See also **attrition, missing values** and **Diggle–Kenward model for dropouts**. [*Applied Statistics*, 1994, **43**, 49–94.]

Drug stability studies: Studies conducted in the pharmaceutical industry to measure the degradation of a new drug product or an old drug formulated or packaged in a new way. The main study objective is to estimate a drug's shelf life, defined as the time point where the 95% lower confidence limit for the regression line crosses the lowest acceptable limit for drug content according to the *Guidelines for Stability Testing*. [*Journal of Pharmaceutical and Biomedical Analysis*, 2005, **38**, 653–663.]

Dual scaling: Synonym for **correspondence analysis**.

Dual system estimates: Estimates which are based on a census and a post-enumeration survey, which try to overcome the problems that arise from the former in trying, but typically failing, to count everyone.

Dummy variables: The variables resulting from recoding categorical variables with more than two categories into a series of binary variables. Marital status, for example, if originally labelled 1 for married, 2 for single and 3 for divorced, widowed or separated, could be redefined in terms of two variables as follows

Variable 1: 1 if single, 0 otherwise;
Variable 2: 1 if divorced, widowed or separated, 0 otherwise;

For a married person both new variables would be zero. In general a categorical variable with k categories would be recoded in terms of $k-1$ dummy variables. Such recoding is used before *polychotomous variables* are used as explanatory variables in a regression analysis to avoid the unreasonable assumption that the original numerical codes for the categories, i.e. the values $1, 2, \ldots, k$, correspond to an *interval scale*. This procedure is often known as dummy coding [ARA Chapter 8.]

Duncan's test: A modified form of the *Newman–Keuls multiple comparison test*. [SMR Chapter 9.]

Dunnett's test: A *multiple comparison test* intended for comparing each of a number of treatments with a single control. [*Biostatistics: A Methodology for the Health Sciences*, 2nd edn, 2004, G. Van Belle, L.D. Fisher P. J. Heagerty and T. S. Lumley, Wiley, New York.]

Dunn's test: A *multiple comparison test* based on the *Bonferroni correction*. [*Biostatistics: A Methodology for the Health Sciences*, 2nd edn, 2004, G. Van Belle, L. D. Fisher P. J. Heagerty and T. S. Lumley, Wiley, New York.]

Duration dependence: The extent to which the *hazard function* of the event of interest in a survival analysis is rising or falling over time. An important point is that duration dependence may be 'spurious' due to unobserved heterogeneity or *frailty*. [*Review of Economic Studies*, 1982, **49**, 403–409.]

Duration time: The time that elapses before an *epidemic* ceases. [*Biometrika*, 1975, **62**, 477–482.]

Durbin–Watson test: A test that the residuals from a linear regression or *multiple regression* are independent. The test statistic is

$$D = \frac{\sum_{i=2}^{n}(r_i - r_{i-1})^2}{\sum_{i=1}^{n} r_i^2}$$

where $r_i = y_i - \hat{y}_i$ and y_i and $\hat{y}_i$ are, respectively, the observed and predicted values of the response variable for individual i. D becomes smaller as the *serial correlations* increase. Upper and lower critical values, D_U and D_L have been tabulated for different values of q (the number of explanatory variables) and n. If the observed value of d lies between these limits then the test is inconclusive. [TMS Chapter 3.]

Dutch book: A gamble that gives rise to certain loss, no matter what actually occurs. Used as a rhetorical device in *subjective probability* and *Bayesian statistics*. [*Betting on Theories*, 1992, P. Maher, Cambridge University Press, New York.]

Dvoretzky-Kiefer-Wolfowitz inequality: A prediction of how close an empirically determined distribution function will be to the assumed population distribution form which the empirical samples are taken. [*Annals of Mathematical Statistics*, 1956, **27**, 642–669.]

Dynamic allocation indices: Indices that give a priority for each project in a situation where it is necessary to optimize in a sequential manner the allocation of effort between a number of competing projects. The indices may change as more effort is allocated. [*Multiarmed Bandit Allocation*, 1989, J.C. Gittins, Wiley, Chichester.]

Dynamic graphics: Computer graphics for the exploration of *multivariate data* which allow the observations to be rotated and viewed from all directions. Particular sets of observations can be highlighted. Often useful for discovering structure or pattern, for example, the presence of clusters. See also **brushing scatterplots**. [*Dynamic Graphics for Statistics*, 1987, W.S. Cleveland and M.E. McGill, Wadsworth, Belmont, California.]

Dynamic panel data model: A term used in *econometrics* when an *autoregressive model* is specified for the response variable in a *panel study*.

Dynamic population: A population that gains and looses members. See also **fixed population**.

Dynamic population modelling: The application and analysis of population models that have changing vital rates. [*Dynamic Population Models*, 2006, R. Schoen, Springer, New York.]

E: Abbreviation for **expected value**.

Early detection program: Synonymous with **screening studies**.

EAST: A computer package for the design and analysis of group sequential *clinical trials*. See also **PEST**. [CYTEL Software Corporation, 675 Massachusetts Avenue, Cambridge, MA 02139, USA.]

Eberhardt's statistic: A statistic, A, for assessing whether a large number of small items within a region are distributed completely randomly within the region. The statistic is based on the *Euclidean distance*, X_j from each of m randomly selected sampling locations to the nearest item and is given explicitly by

$$A = m\sum_{i=1}^{m} X_i^2 / \left(\sum_{j=1}^{m} X_j\right)^2$$

[*Biometrika*, 1979, **66**, 73–79.]

EBM: Abbreviation for **evidence-based medicine**.

ECM algorithm: An extension of the *EM algorithm* that typically converges more slowly than EM in terms of iterations but can be faster in total computer time. The basic idea of the algorithm is to replace the M-step of each EM iteration with a sequence of $S > 1$ conditional or constrained maximization or CM-steps, each of which maximizes the expected complete-data *log-likelihood* found in the previous E-step subject to constraints on the parameter of interest, θ, where the collection of all constraints is such that the maximization is over the full parameter space of θ. Because the CM maximizations are over smaller dimensional spaces, often they are simpler, faster and more reliable than the corresponding full maximization called for in the M-step of the EM algorithm. See also **ECME algorithm**. [*Statistics in Medicine*, 1995, **14**, 747–68.]

ECME algorithm: The Expectation/Conditional Maximization Either algorithm which is a generalization of the *ECM algorithm* obtained by replacing some CM-steps of ECM which maximize the constrained expected complete-data *log-likelihood*, with steps that maximize the correspondingly constrained actual *likelihood*. The algorithm can have substantially faster convergence rate than either the *EM algorithm* or ECM measured using either the number of iterations or actual computer time. There are two reasons for this improvement. First, in some of ECME's maximization steps the actual likelihood is being conditionally maximized, rather than a current approximation to it as with EM and ECM. Secondly, ECME allows faster converging numerical methods to be used on only those constrained maximizations where they are most efficacious. [*Biometrika*, 1997, **84**, 269–81.]

Ecological fallacy: A term used when aggregated data (for example, aggregated over different regions) are analysed and the results assumed to apply to relationships at the individual level. In most cases analyses based on aggregated level means give conclusions very different from those that would be obtained from an analysis of unit level data. An example from the literature is a correlation coefficient of 0.11 between illiteracy and being foreign born calculated from person level data, compared with a value of −0.53 between percentage illiterate and percentage foreign born at the State level. [*Statistics in Medicine*, 1992, **11**, 1209–24.]

Ecological statistics: Procedures for studying the dynamics of natural communities and their relation to environmental variables.

Econometrics: The area of economics concerned with developing and applying quantitative or statistical methods to the study and elucidation of economic principles.

EDA: Abbreviation for **exploratory data analysis**.

ED50: Abbreviation for **median effective dose**.

Edgeworth, Francis Ysidro (1845–1926): Born in Edgeworthstown, Longford, Ireland, Edgeworth entered Trinity College, Dublin in 1862 and in 1867 went to Oxford University where he obtained a first in classics. He was called to the Bar in 1877. After leaving Oxford and while studying law, Edgeworth undertook a programme of self study in mathematics and in 1880 obtained a position as lecturer in logic at Kings College, London later becoming Tooke Professor of Political Economy. In 1891 he was elected Drummond Professor of Political Economy at Oxford and a Fellow of All Souls, where he remained for the rest of his life. In 1883 Edgeworth began publication of a sequence of articles devoted exclusively to probability and statistics in which he attempted to adapt the statistical methods of the theory of errors to the quantification of uncertainty in the social, particularly the economic sciences. In 1885 he read a paper to the Cambridge Philosophical Society which presented, through an extensive series of examples, an exposition and interpretation of significance tests for the comparison of means. Edgeworth died in London on 13 February 1926.

Edgeworth's form of the Type A series: An expression for representing a probability distribution, $f(x)$, in terms of *Chebyshev–Hermite polynomials*, H_r, given explicitly by

$$f(x) = \alpha(x)\left\{1 + \frac{\kappa_3}{6}H_3 + \frac{\kappa_4}{24}H_4 + \frac{\kappa_5}{120}H_5 + \frac{\kappa_6 + 10\kappa_3^2}{720}H_6\right\}$$

where κ_i are the *cumulants* of $f(x)$ and

$$\alpha(x) = \frac{1}{\sqrt{2\pi}}e^{-\frac{1}{2}x^2}$$

Essentially equivalent to the *Gram–Charlier Type A series*. [KA1 Chapter 6.]

Effect: Generally used for the change in a response variable produced by a change in one or more explanatory or factor variables.

Effect coding: See **contrast**.

Effective sample size: The sample size after dropouts, deaths and other specified exclusions from the original sample. [*Survey Sampling*, 1995, L. Kish, Wiley, New York.]

Effect size: Most commonly the difference between the control group and experimental group population means of a response variable divided by the assumed common population standard deviation. Estimated by the difference of the sample means in the two groups divided by a pooled estimate of the assumed common standard deviation. Often used in *meta-analysis*. See also **counternull-value**. [*Psychological Methods*, 2003, **8**, 434–447.]

Effect sparsity: A term used in industrial experimentation, where there is often a large set of candidate factors believed to have possible significant influence on the response of interest, but where it is reasonable to assume that only a small fraction are influential. [*Technometrics*, 1986, **28**, 11–18.]

Efficiency: A term applied in the context of comparing different methods of estimating the same parameter; the estimate with lowest variance being regarded as the most efficient. Also used when comparing competing experimental designs, with one design being more efficient than another if it can achieve the same precision with fewer resources. [KA2 Chapter 17.]

EGRET: Acronym for the Epidemiological, Graphics, Estimation and Testing program developed for the analysis of data from studies in *epidemiology*. Can be used for *logistic regression* and models may include *random effects* to allow *overdispersion* to be modelled. The *beta-binomial distribution* can also be fitted. [Statistics & Epidemiology Research Corporation, 909 Northeast 43rd Street, Suite 310, Seattle, Washington 98105, USA.]

Ehrenberg's equation: An equation linking the height and weight of children between the ages of 5 and 13 and given by

$$\log \bar{w} = 0.8\bar{h} + 0.4$$

where $\bar{w}$ is the mean weight in kilograms and $\bar{h}$ the mean height in metres. The relationship has been found to hold in England, Canada and France. [*Indian Journal of Medical Research*, 1998, **107**, 406–9.]

Eigenvalues: The roots, $\lambda_1, \lambda_2, \ldots, \lambda_q$ of the qth-order polynomial defined by

$$|\mathbf{A} - \lambda\mathbf{I}|$$

where $\mathbf{A}$ is a $q \times q$ square matrix and $\mathbf{I}$ is an identity matrix of order q. Associated with each root is a non-zero vector $\mathbf{z}_i$ satisfying

$$\mathbf{A}\mathbf{z}_i = \lambda_i\mathbf{z}_i$$

and $\mathbf{z}_i$ is known as an *eigenvector* of $\mathbf{A}$. Both eigenvalues and eigenvectors appear frequently in accounts of techniques for the analysis of multivariate data such as *principal components analysis* and *factor analysis*. In such methods, eigenvalues usually give the variance of a linear function of the variables, and the elements of the eigenvector define a linear function of the variables with a particular property. [MV1 Chapter 3.]

Eigenvector: See **eigenvalue**.

Eisenhart, Churchill (1913–1994): Born in Rochester, New York, Eisenhart received an A.B. degree in mathematical physics in 1934 and an A.M. degree in mathematics in 1935. He obtained a Ph.D from University College London in 1937, studying under *Egon Pearson*, *Jerzy Neyman* and *R. A. Fisher*. In 1946 Eisenhart joined the National Bureau of Standards and undertook pioneering work in introducing modern statistical methods in experimental work in the physical sciences. He was President of the American Statistical Association in 1971. Eisenhart died in Bethesda, Maryland on 25 June 1994.

Electronic mail: The use of computer systems to transfer messages between users; it is usual for messages to be held in a central store for retrieval at the user's convenience.

Elfving, Erik Gustav (1908–1984): Born in Helsinki, Elfving studied mathematics, physics and astronomy at university but after completing his doctoral thesis his interest turned to probability theory. In 1948 he was appointed Professor of Mathematics at the University of Helsinki from where he retired in 1975. Elfving worked in a variety of areas of theoretical statistics including *Markov chains* and *distribution free methods.* After his retirement Elfving wrote a monograph on the history of mathematics in Finland between 1828 and 1918, a period of Finland's autonomy under Russia. He died on 25 March 1984 in Helsinki.

Elliptically symmetric distributions: Multivariate probability distributions of the form,

$$f(\mathbf{x}) = |\Sigma|^{-\frac{1}{2}} g[(\mathbf{x} - \boldsymbol{\mu})'\Sigma^{-1}(\mathbf{x} - \boldsymbol{\mu})]$$

By varying the function g, distributions with longer or shorter tails than the normal can be obtained. [MV1 Chapter 2.]

Email: Abbreviation for **electronic mail**.

EM algorithm: A method for producing a sequence of parameter estimates that, under mild regularity conditions, converges to the *maximum likelihood estimator.* Of particular importance in the context of incomplete data problems. The algorithm consists of two steps, known as the E, or Expectation step and the M, or Maximization step. In the former, the expected value of the *log-likelihood* conditional on the observed data and the current estimates of the parameters, is found. In the M-step, this function is maximized to give updated parameter estimates that increase the *likelihood.* The two steps are alternated until convergence is achieved. The algorithm may, in some cases, be very slow to converge. See also **finite mixture distributions, imputation, ECM algorithm** and **ECME algorithm**. [KA2 Chapter 18.]

Empirical: Based on observation or experiment rather than deduction from basic laws or theory.

Empirical Bayes method: A procedure in which the *prior distribution* needed in the application of *Bayesian inference*, is determined from empirical evidence, namely the same data for which the *posterior distribution* is obtained. [*Empirical Bayes' Methods*, 1970, J. S. Maritz, Chapman and Hall/CRC Press, London.]

Empirical distribution function: A probability distribution function estimated directly from sample data without assuming an underlying algebraic form.

Empirical likelihood: An approach to using *likelihood* as the basis of estimation without the need to specify a parametric family for the data. Empirical likelihood can be viewed as an instance of *nonparametric maximum likelihood.* [*Empirical Likelihood*, 2000, A. B. Owen, Chapman and Hall/CRC, Boca Raton.]

Empirical logits: The *logistic transformation* of an observed proportion y_i/n_i, adjusted so that finite values are obtained when y_i is equal to either zero or n_i. Commonly 0.5 is added to both y_i and n_i. [*Modelling Binary Data*, 2nd edition, 2003, D. Collett, Chapman and Hall/CRC, Boca Raton.]

Empirical variogram: See **variogram**.

End-aversion bias: A term which refers to the reluctance of some people to use the extreme categories of a scale. See also **acquiescence bias**. [*Expert Review of Pharmacoeconomics and Outcomes Research*, 2002, **2**, 99–108.]

Endogenous variable: A term primarily used in econometrics to describe those variables which are an inherent part of a system. Typically refers to a covariate which is correlated with the error term in a regression model due to for instance omitted variables and measurement error. See also **exogeneous variable**.

Endpoint: A clearly defined outcome or event associated with an individual in a medical investigation. A simple example is the death of a patient. See also **Surrogate endpoint**.

Engel, Ernst (1821–1896): Born in Dresden, Germany, Engel studied mining engineering at the Mining Academy, Saxony from 1841 until 1845. Moving to Brussels he was influenced by the work of Adolphe Quetelet and in 1850 he became head of the newly established Royal Saxon Statistical Bureau in Dresden. Engel contributed to census techniques, economic statistics and the organization of official statistics. He died on December 8th, 1896 in Oberlossnitz, near Dresden, Germany.

Entropy: A measure of amount of information received or output by some system, usually given in *bits*. [MV1 Chapter 4.]

Entropy measure: A measure, H, of the dispersion of a categorical random variable, Y, that assumes the integral values j, $1 \leq j \leq s$ with probability p_j, given by

$$H = -\sum_{j=1}^{s} p_j \log p_j$$

See also **concentration measure**. [MV1 Chapter 4.]

Environmental statistics: Procedures for determining how quality of life is affected by the environment, in particular by such factors as air and water pollution, solid wastes, hazardous substances, foods and drugs. [*Environmental Statistics*, 2000, S. P. Millard, N. K. Neerchal, CRC Press, Boca Raton.]

E-optimal design: See **criteria of optimality**.

Epanechnikov kernel: See **kernel density estimation**.

Epidemic: The rapid development, spread or growth of a disease in a community or area. Statistical thinking has made significant contributions to the understanding of such phenomena. A recent example concerns the Acquired Immunodeficiency Syndrome (AIDS) where complex statistical methods have been used to estimate the number of infected individuals, the incubation period of the disease, the aetiology of the disease and monitoring and forecasting the course of the disease. Figure 58, for example, shows the annual numbers of new HIV infections in the US by risk group based on a deconvolution of AIDS incidence data. [*Methods in Observational Epidemiology*, 1986, J. L. Kelsey, W. D. Thompson and A. S. Evans, Oxford University Press, New York.]

Epidemic chain: See **chains of infection**.

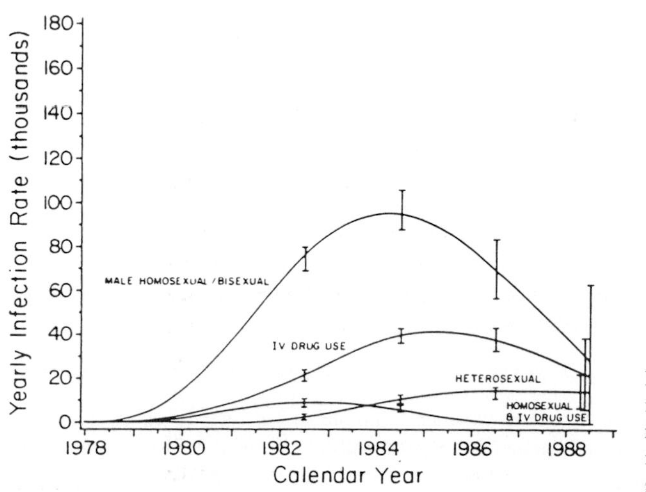

Fig. 58 Epidemic illustrated by annual numbers of new HIV infections in the US by risk group.

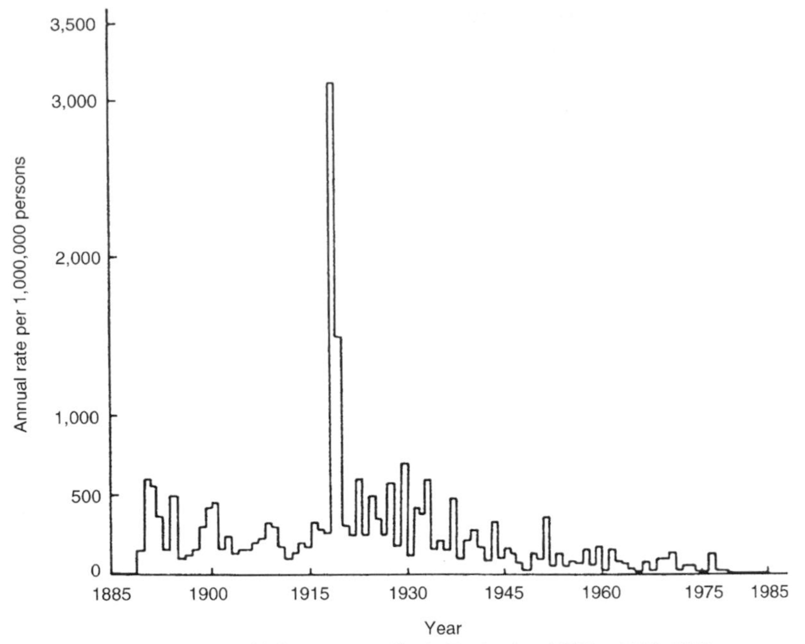

Fig. 59 Epidemic curve of influenza mortality in England and Wales 1885–1985.

Epidemic curve: A plot of the number of cases of a disease against time. A large and sudden rise corresponds to an *epidemic*. An example is shown in Fig. 59. [*Methods in Observational Epidemiology*, 1986, J. L. Kelsey, W. D. Thompson and A. S. Evans, Oxford University Press, New York.]

Epidemic model: A model for the spread of an *epidemic* in a population. Such models may be deterministic, spatial or stochastic. [*The Mathematical Theory of Infectious Diseases*, 1975, N. T. J. Bailey, Chapman and Hall/CRC Press, London.]

Epidemic threshold: The value above which the susceptible population size has to be for an *epidemic* to become established in a population. [*The Mathematical Theory of Infectious Diseases*, 1975, N. T. J. Bailey, Chapman and Hall/CRC Press, London.]

Epidemiology: The study of the distribution and size of disease problems in human populations, in particular to identify aetiological factors in the pathogenesis of disease and to provide the data essential for the management, evaluation and planning of services for the prevention, control and treatment of disease. See also **incidence, prevalence, prospective study** and **retrospective study**. [*An Introduction to Epidemiology*, 1983, M. Alderson, Macmillan, London.]

Epi Info: Free software for performing basic sample size calculations, developing a study questionnaire, performing statistical analyses, etc. The software can be downloaded from the web site, http://www.cdc.gov/epiinfo/.

EPILOG PLUS: Software for *epidemiology* and *clinical trials*. [Epicenter Software, PO Box 90073, Pasadena, CA 91109, USA.]

EPITOME: An epidemiological data analysis package developed at the National Cancer Institute in the US.

Epstein test: A test for assessing whether a sample of *survival times* arise from an *exponential distribution* with constant failure rate against the alternative that the failure rate is increasing. [NSM Chapter 11.]

Equipoise: A term applied to a state of genuine uncertainty about which experimental therapy is more effective, a state that needs to be held by clinicians running a randomized clinical trial to make the trial ethical. [*Randomization in Clinical Trials: Theory and Practice*, 2002, W. F. Rosenberger and J. M. Lachin, Wiley, New York.]

EQS: A software package for fitting *structural equation models*. See also **LISREL**. [Distributor is *Multivariate Software*: http://www.mvsoft.com/]

Ergodicity: A property of many time-dependent processes, such as certain *Markov chains*, namely that the eventual distribution of states of the system is independent of the initial state. [*Stochastic Modelling of Scientific Data*, 1995, P. Guttorp, Chapman and Hall/CRC Press, London.]

Erlang distribution: A *gamma distribution* in which the parameter γ takes integer values. Arises as the time to the νth event in a *Poisson process*. [STD Chapter 11.]

Error distribution: The probability distribution, $f(x)$, given by

$$f(x) = \frac{\exp[-|x-\alpha|^{2/\gamma}/2\beta]}{\beta^{\frac{1}{2}}2^{\gamma/2+1}\Gamma(1+\gamma/2)}, \quad -\infty<x<\infty, \beta>0, \gamma>0, -\infty<\alpha<\infty$$

α is a *location parameter*, β is a *scale parameter*, and γ a *shape parameter*. The mean, variance, *skewness* and *kurtosis* of the distribution are as follows:

$$\text{mean} = \alpha$$

$$\text{variance} = \frac{2^{\gamma}\beta^2\Gamma(3\gamma/2)}{\Gamma(\gamma/2)}$$

$$\text{skewness} = 0$$

$$\text{kurtosis} = \frac{\Gamma(5\gamma/2)\Gamma(\gamma/2)}{[\Gamma(3\gamma/2)]^2}$$

The distribution is symmetric, and for $\gamma > 1$ is *leptokurtic* and for $\gamma > 1$ is *platykurtic*. For $\alpha = 0$, $\beta = \gamma = 1$ the distribution corresponds to a standard normal distribution. [STD Chapter 12.]

Error mean square: See **mean squares**.

Error rate: The proportion of subjects misclassified by a classification rule, for example, one derived for a *discriminant analysis*. [MV2 Chapter 9.]

Error rate estimation: A term used for the estimation of the misclassification rate in *discriminant analysis*. Many techniques have been proposed for the two-group situation, but the multiple-group situation has very rarely been addressed. The simplest procedure is the *resubstitution method*, in which the *training data* are classified using the estimated classification rule and the proportion incorrectly placed used as the estimate of the misclassification rate. This method is known to have a large optimistic bias, but it has the advantage that it can be applied to multigroup problems with no modification. An alternative method is the *leave one out estimator*, in which each observation in turn is omitted from the data and the classification rule recomputed using the remaining data. The proportion incorrectly classified by the procedure will have reduced bias compared to the resubstitution method. This method can also be applied to the multigroup problem with no modification but it has a large variance. See also **b632 method**. [MV2 Chapter 9.]

Errors-in-variables problem: See **regression dilution** and **attenuation**.

Errors of classification: A term most often used in the context of *retrospective studies*, where it is recognized that a certain proportion of the controls may be at an early stage of disease and should have been diagnosed as cases. Additionally misdiagnosed cases might be included in the disease group. Both errors lead to underestimates of the *relative risk*.

Errors of the third kind: Giving the right answer to the wrong question! (Not to be confused with *Type III error*.)

Estimand: The quantity (usually a parameter) to be estimated using an estimator.

Estimate: See **estimation**.

Estimating functions: Functions of the data and the parameters of interest which can be used to conduct inference about the parameters when the full distribution of the observations is unknown. This approach has an advantage over methods based on the *likelihood* since it requires the specification of only a few *moments* of the random variable involved rather than the entire probability distribution. The most familiar example is the *quasi-score estimating function*

$$h(y, \boldsymbol{\theta}) = \frac{\partial \boldsymbol{\mu}'}{\partial \boldsymbol{\theta}} \mathbf{V}_{\boldsymbol{\mu}}^{-1}(y_{\boldsymbol{\mu}})$$

where $\boldsymbol{\mu}$ is the $n \times 1$ vector of mean responses, $\boldsymbol{\theta}$ is a $q \times 1$ vector of *regression coefficients* relating explanatory variables to $\boldsymbol{\mu}$ and $\mathbf{V}_{\boldsymbol{\mu}}$ denotes the $n \times n$ *variance–covariance matrix* of the responses. See also **quasi-likelihood** and **generalized estimating equations**. [*Scandinavian Journal of Statistics*, 2001, **28**, 99–112.]

Estimation: The process of providing a numerical value for a population parameter on the basis of information collected from a sample. If a single figure is calculated for the unknown parameter the process is called *point estimation*. If an interval is calculated which is likely to contain the parameter, then the procedure is called *interval estimation*. See also **least squares estimation, maximum likelihood estimation**, and **confidence interval**. [KA1 Chapter 9.]

Estimator: A statistic used to provide an estimate for a parameter. The sample mean, for example, is an unbiased estimator of the population mean.

Etiological fraction: Synonym for **attributable risk**.

Euclidean distance: For two observations $\mathbf{x}' = [x_1, x_2, \ldots, x_q]$ and $\mathbf{y}' = [y_1, y_2, \ldots, y_q]$ from a set of multivariate data, the *distance measure* given by

$$d_{\mathbf{xy}} = \sqrt{\sum_{i=1}^{q} (x_i - y_i)^2}$$

See also **Minkowski distance** and **city block distance**. [MV1, Chapter 3.]

EU model: A model used in investigations of the rate of success of *in vitro* fertilization (IVF), defined in terms of the following two parameters.

$$e = \text{Prob(of a viable embryo)}$$
$$u = \text{Prob(of a receptive uterus)}$$

Assuming the two events, viable embryro and receptive uterus, are independent, the probability of observing *j* implantations out of *i* transferred embryos is given by

$$P(j|i,u,e) = \binom{i}{j} ue^j(1-e)^{i-j}, \quad \text{for } j = 1,2,3,\ldots,i.$$

The probability of at least one implantation out of i transferred embryos is given by

$$P(j > 0|i,u,e) = u[1-(1-e)^i]$$

and the probability of no implantations by

$$P(j = 0|i,u,e) = 1 - u[1-(1-e)^i]$$

The parameters u and e can be estimated by *maximum likelihood estimation* from observations on the number of attempts at IVF with j implantations from i transferred. See also **Barrett and Marshall model for conception**. [*Biostatistics*, 2002, **3**, 361–77.]

Evaluable patients: The patients in a *clinical trial* regarded by the investigator as having satisfied certain conditions and, as a result, are retained for the purpose of analysis. Patients not satisfying the required condition are not included in the final analysis. [*Clinical Trials in Psychiatry*, 2nd edition, B. S. Everitt and S. Wessely, Wiley, Chichester.]

Event history data: Observations on a collection of individuals, each moving among a small number of states. Of primary interest are the times taken to move between the various states, which are often only incompletely observed because of some form of censoring. The simplest such data involves *survival times*. [*The Lancet Oncology*, 2007, **8**, 889–897.]

Evidence-based medicine (EBM): Described by its leading proponent as 'the conscientious, explicit, and judicious uses of current best evidence in making decisions about the care of individual patients, and integrating individual clinical experience with the best available external clinical evidence from systematic research'. [*Evidence Based Medicine*, 1996, **1**, 98–9.]

Excel: Software that started out as a spreadsheet aiming at manipulating tables of number for financial analysis, which has now grown into a more flexible package for working with all types of information. Particularly good features for managing and annotating data. [http://www.Microsoft.com]

Excess hazards model: A model for *survival data* in cases when what is of interest is to model the excess mortality that is the result of subtracting the expected mortality calculated from ordinary *life tables*. Essentially synonymous with *relative survival*. [*Lifetime Data Analysis*, 1998, **4**, 149–68.]

Excess risk: The difference between the risk for a population exposed to some risk factor and the risk in an otherwise identical population without the exposure. See also **attributable risk**. [*Case-Control Studies: Design, Conduct and Analysis*, 1982, J. J. Schlesselman, Oxford University Press, New York.]

Exchangeability: A term usually attributed to *Bruno De Finetti* and introduced in the context of *personal probability* to describe a particular sense in which quantities treated as random variables in a probability specification are thought to be similar. Formally it describes a property possessed by a set of random variables, $X_1, X_2, \ldots$ which, for all $n \geq 1$ and for every permutation of the n subscripts, the joint distributions of $X_{j_1}, X_{j_2}, \ldots, X_{j_n}$ are identical. Central to some aspects of *Bayesian inference* and *randomization tests*. [KA1 Chapter 12.]

Exhaustive: A term applied to events $B_1, B_2, \ldots, B_k$ for which $\cup_{i=1}^k B_i = \Omega$ where Ω is the *sample space*.

Exogeneous variable: A term primarily used in econometrics to describe those variables that impinge on a system from outside. Typically refers to a covariate which is uncorrelated with the error term in a regression model. See also **endogenous variable**. [MV2 Chapter 11.]

Expectation surface: See **curvature measures**.

Expected frequencies: A term usually encountered in the analysis of *contingency tables*. Such frequencies are estimates of the values to be expected under the hypothesis of interest. In a two-dimensional table, for example, the values under independence are calculated from the product of the appropriate row and column totals divided by the total number of observations. [*The Analysis of Contingency Tables*, 2nd edition, 1992, B. S. Everitt, Chapman and Hall/CRC Press, London.]

Expected value: The mean of a random variable, X, generally denoted as $E(X)$. If the variable is discrete with probability distribution, $\Pr(X = x)$, then $E(X) = \sum_x x \Pr(X = x)$. If the variable is continuous the summation is replaced by an integral. The expected value of a function of a random variable, $f(x)$, is defined similarly, i.e.

$$E(f(x)) = \int_u f(u)g(u)du$$

where $g(x)$ is the probability distribution of x. [KA1 Chapter 2.]

Experimental design: The arrangement and procedures used in an experimental study. Some general principles of good design are, simplicity, avoidance of bias, the use of random allocation for forming treatment groups, replication and adequate sample size. [*Modern Experimental Design*, 2007, T. P. Ryan, Wiley, New York.]

Experimental study: A general term for investigations in which the researcher can deliberately influence events, and investigate the effects of the intervention. *Clinical trials* and many animal studies fall under this heading.

Experimentwise error rate: Synonym for **per-experiment error rate**.

Expert systems: Computer programs designed to mimic the role of an expert human consultant. Such systems are able to cope with the complex problems of medical decision making because of their ability to manipulate symbolic rather than just numeric information and their use of judgemental or heuristic knowledge to construct intelligible solutions to diagnostic problems. Well known examples include the *MYCIN* system developed at Stanford University, and ABEL developed at MIT. See also **computer-aided diagnosis** and **statistical expert system**. [*Expert Systems: Principles and Programming*, 4th edn, 2004, J. C. Giarratano and E. D. Ripley, Course Technology.]

Explanatory analysis: A term sometimes used for the analysis of data from a *clinical trial* in which treatments A and B are to be compared under the assumption that patients remain on their assigned treatment throughout the trial. In contrast a *pragmatic analysis* of the trial would involve asking whether it would be better to start with A (with the intention of continuing this therapy if possible but the willingness to be flexible) or to start with B (and have the same intention). With the explanatory approach the aim is to acquire information on the true effect of treatment, while the pragmatic aim is to make a decision about the therapeutic strategy after taking into account the cost (e.g. withdrawal due to side effects) of administering treatment. See also **intention-to-treat analysis**. [*Statistics in Medicine*, 1988, **7**, 1179–86.]

Explanatory trials: A term sometimes used to describe *clinical trials* that are designed to explain how a treatment works. [*Statistics in Medicine*, 1988, **7**, 1179–86.]

Explanatory variables: The variables appearing on the right-hand side of the equations defining, for example, *multiple regression* or *logistic regression*, and which seek to predict or 'explain' the response variable. Also commonly known as the *independent variables*, although this is not to be recommended since they are rarely independent of one another. [ARA Chapter 1.]

Exploratory data analysis: An approach to data analysis that emphasizes the use of informal graphical procedures not based on prior assumptions about the structure of the data or on formal models for the data. The essence of this approach is that, broadly speaking, data are assumed to possess the following structure

$$\text{Data} = \text{Smooth} + \text{Rough}$$

where the 'Smooth' is the underlying regularity or pattern in the data. The objective of the exploratory approach is to separate the smooth from the 'Rough' with minimal use of formal mathematics or statistical methods. See also **initial data analysis**. [SMR Chapter 6.]

Exploratory factor analysis: See **factor analysis**.

Exponential autoregressive model: An *autoregressive model* in which the 'parameters' are made exponential functions of X_{t-1}, that is

$$X_t + (\phi_1 + \pi_1 e^{-\gamma X_{t-1}^2})X_{t-1} + \cdots + (\phi_k + \pi_k e^{-\gamma X_{t-1}^2})X_{t-k} = \epsilon_t$$

[*Applications of Time Series Analysis in Astronomy and Meteorology*, 1997, edited by T. Subba Rao, M. B. Priestley and O. Lissi, Chapman and Hall/CRC Press, London.]

Exponential distribution: The probability distribution $f(x)$ given by

$$f(x) = \lambda e^{-\lambda x}, \quad x > 0$$

The mean, variance, *skewness* and *kurtosis* of the distribution are as follows:

$$\text{mean} = 1/\lambda$$
$$\text{variance} = 1/\lambda^2$$
$$\text{skewness} = 2$$
$$\text{kurtosis} = 9$$

The distribution of intervals between independent consecutive random events, i.e. those following a *Poisson process*. [STD Chapter 13.]

Exponential family: A family of probability distributions of the form

$$f(x) = \exp\{a(\theta)b(x) + c(\theta) + d(x)\}$$

where θ is a parameter and a, b, c, d are known functions. Includes the normal distribution, *gamma distribution, binomial distribution* and *Poisson distribution* as special cases. The binomial distribution, for example, can be written in the form above as follows:

$$\binom{n}{x} p^x (1-p)^{n-x} = \binom{n}{x} \exp\{x \ln \frac{p}{1-p} + n \ln(1-p)\}$$

[STD Chapter 14.]

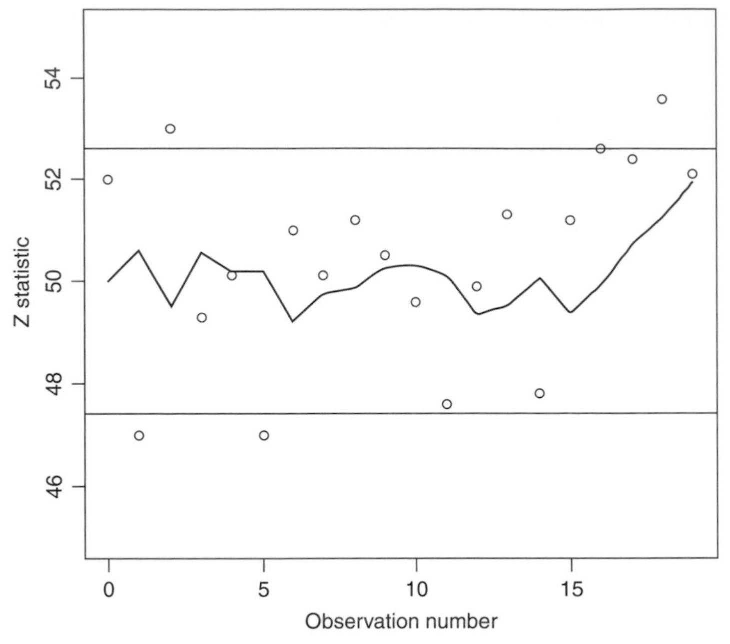

Fig. 60 Example of an exponentially weighted moving average control chart.

Exponentially weighted moving average control chart: An alternative form of *cusum* based on the statistic

$$Z_i = \lambda Y_i + (1 - \lambda) Z_{i-1} \qquad 0 < \lambda \leq 1$$

together with upper control and lower control limits. The starting value Z_0 is often taken to be the target value. The sequentially recorded observations, Y_i, can be individually observed values from the process, although they are often sample averages obtained from a designated sampling plan. The process is considered out of control and action should be taken whenever Z_i falls outside the range of the control limits. An example of such a chart is shown in Fig. 60. [*Journal of Quality Technology*, 1997, **29**, 41–8.]

Exponential order statistics model: A model that arises in the context of estimating the size of a closed population where individuals within the population can be identified only during some observation period of fixed length, say $T > 0$, and when the detection times are assumed to follow an *exponential distribution* with unknown rate λ. [*Journal of the American Statistical Association*, 2000, **95**, 1220–8.]

Exponential power distribution: Synonym for **error distribution**.

Exponential trend: A *trend*, usually in a *time series*, which is adequately described by an equation of the form $y = ab^x$.

Exposure effect: A measure of the impact of exposure on an outcome measure. Estimates are derived by contrasting the outcomes in an exposed population with outcomes in an unexposed population or in a population with a different level of exposure. See also **attributable risk** and **excess risk**.

Exposure factor: Synonym for **risk factor**.

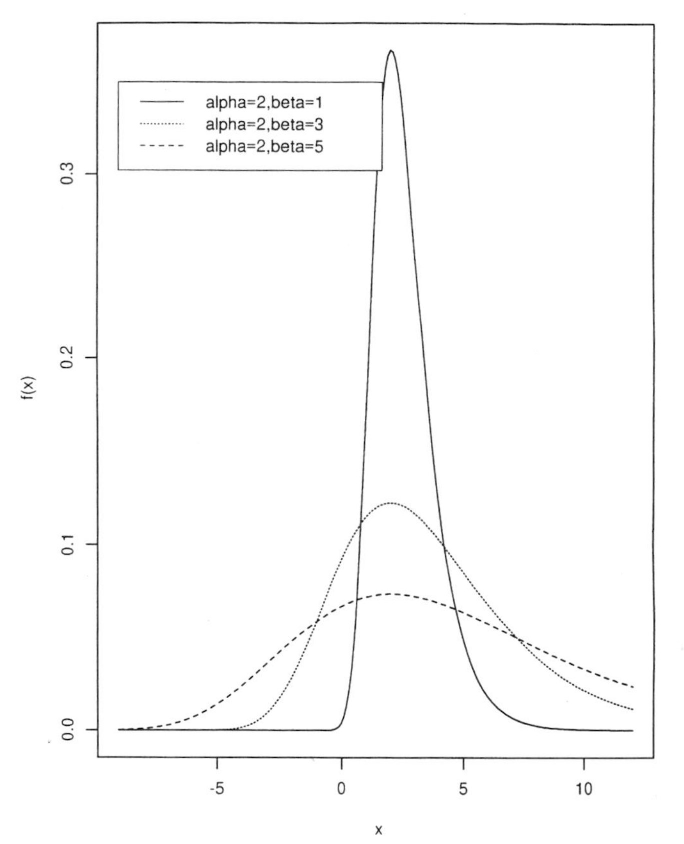

Fig. 61 Examples of extreme value distribution for $\alpha = 2$ and $\beta = 1, 3$ and 5.

Extra binomial variation: A form of *overdispersion*. [*Modelling Binary Data*, 2nd edition, 2003, D. Collett, Chapman and Hall/CRC Press, London.]

Extraim analysis: An analysis involving increasing the sample size in a *clinical trial* that ends without a clear conclusion, in an attempt to resolve the issue of say a group difference that is currently marginally non-significant. As in *interim analysis*, significance levels need to be suitably adjusted. [*Statistical Science*, 2004, **19**, 175–87.]

Extrapolation: The process of estimating from a data set those values lying beyond the range of the data. In regression analysis, for example, a value of the response variable may be estimated from the fitted equation for a new observation having values of the explanatory variables beyond the range of those used in deriving the equation. Often a dangerous procedure. [*Extrapolation: Theory and Practice*, 1991, C. Brezinski and M. R. Zaglia, North Holland.]

Extremal quotient: Defined for nonnegative observations as the ratio of the largest observation in a sample to the smallest observation. [*Health Services Research*, 1989, **24**, 665–684.]

Extreme standardized deviate: See **many-outlier detection procedures**.

Extreme value distribution: The probability distribution, $f(x)$, of the largest extreme given by

$$f(x) = \frac{1}{\beta}\exp[-(x-\alpha)/\beta]\exp\{-\exp[-(x-\alpha)/\beta]\}, \qquad -\infty<x<\infty, \beta>0$$

The *location parameter*, α is the mode and β is a *scale parameter*. The mean, variance *skewness* and *kurtosis* of the distribution are as follows:

$$\text{mean} = \alpha - \beta\Gamma'(1)$$
$$\text{variance} = \beta^2\pi^2/6$$
$$\text{skewness} = 1.139547$$
$$\text{kurtosis} = 5.4$$

where $\Gamma'(1) = -0.57721$ is the first derivative of the *gamma function*, $\Gamma(n)$ with respect to n at $n = 1$. The distribution of the maximum value of n-independent random variables with the same continuous distribution as n tends to infinity. Examples of this distribution are given in Fig. 61. See also **bivariate Gumbel distribution**. [STD Chapter 15.]

Extreme values: The largest and smallest variate values among a sample of observations. Important in many areas, for example flood levels of a river, wind speeds and snowfall. Statistical analysis of extreme values aims to model their occurrence and size with the aim of predicting future values. [KA1 Chapter 14.]

Extrinsic aliasing: See **aliasing**.

Eyeball test: Informal assessment of data simply by inspection and mental calculation allied with experience of the particular area from which the data arise.

Facets: See **generalizability theory**.

Factor: A term used in a variety of ways in statistics, but most commonly to refer to a categorical variable, with a small number of levels, under investigation in an experiment as a possible source of variation. Essentially simply a categorical explanatory variable. [SMR Chapter 5.]

Factor analysis: A procedure that postulates that the correlations or covariances between a set of observed variables, $\mathbf{x}' = [x_1, x_2, \ldots, x_q]$, arise from the relationship of these variables to a small number of underlying, unobservable, *latent variables*, usually known as the *common factors*, $\mathbf{f}' = [f_1, f_2, \ldots, f_k]$, where $k < q$. Explicitly the model used is

$$\mathbf{x} = \mathbf{\Lambda f} + \mathbf{e}$$

where

$$\mathbf{\Lambda} = \begin{pmatrix} \lambda_{11} & \lambda_{12} & \cdots & \lambda_{1k} \\ \lambda_{21} & \lambda_{22} & \cdots & \lambda_{2k} \\ \vdots & \vdots & \vdots & \vdots \\ \lambda_{q1} & \lambda_{q2} & \cdots & \lambda_{qk} \end{pmatrix}$$

contains the regression coefficients (usually known in this context as *factor loadings*) of the observed variables on the common factors. The matrix, $\mathbf{\Lambda}$, is known as the *loading matrix*. The elements of the vector $\mathbf{e}$ are known as *specific variates*. Assuming that the common factors are uncorrelated, in standardized form and also uncorrelated with the specific variates, the model implies that the variance–covariance matrix of the observed variables, $\mathbf{\Sigma}$, is of the form

$$\mathbf{\Sigma} = \mathbf{\Lambda\Lambda}' + \mathbf{\Psi}$$

where $\mathbf{\Psi}$ is a diagonal matrix containing the variances of the specific variates. A number of approaches are used to estimate the parameters in the model, i.e. the elements of $\mathbf{\Lambda}$ and $\mathbf{\Psi}$, including *principal factor analysis* and *maximum likelihood estimation*. After thc initial estimation phase an attempt is generally made to simplify the often difficult task of interpreting the derived factors using a process known as *factor rotation*. In general the aim is to produce a solution having what is known as *simple structure*, i.e. each common factor affects only a small number of the observed variables. Although based on a well-defined model the method is, in its initial stages at least, essentially exploratory and such *exploratory factor analysis* needs to be carefully differentiated from *confirmatory factor analysis* in which a prespecified set of common factors with some variables constrained to have zero loadings is tested for consistency with the correlations of the observed variables. See also **structural equation models** and **principal components analysis**. [MV2 Chapter 12.]

Factorial designs: Designs which allow two or more questions to be addressed in an investigation. The simplest factorial design is one in which each of two treatments or interventions are either present or absent, so that subjects are divided into four groups; those receiving neither treatment, those having only the first treatment, those having only the second treatment and those receiving both treatments. Such designs enable possible interactions between factors to be investigated. A very important special case of a factorial design is that where each of k factors of interest has only two levels; these are usually known as 2^k *factorial designs*. Particularly important in industrial applications of *response surface methodology*. A single replicate of a 2^k design is sometimes called an *unreplicated factorial*. [SMR Chapter 5.]

Factorial moment generating function: A function of a variable t which, when expanded formally as a power series in t, yields the *factorial moments* as coefficients of the respective powers. If $P(t)$ is the *probability generating function* of a discrete random variable, the factorial moment generating function is simply $P(1+t)$. [KA1 Chapter 3.]

Factorial moments: A type of *moment* used almost entirely for discrete random variables, or occasionally for continuous distributions grouped into intervals. For a discrete random variable, X, taking values $0, 1, 2, \ldots, \infty$, the jth factorial moment about the origin, $\mu'_{[j]}$ is defined as

$$\mu'_{[j]} = \sum_{r=0}^{\infty} r(r-1)\cdots(r-j+1)\Pr(X=r)$$

By direct expansion it is seen that

$$\mu'_{[1]} = \mu'_1$$
$$\mu'_{[2]} = \mu'_2 + \mu'_1$$
$$\mu'_{[3]} = \mu'_3 - 3\mu'_2 + 2\mu'_1$$

[KA1 Chapter 3.]

Factorization theorem: A theorem relating the structure of the *likelihood* to the concept of a *sufficient statistic*. Formally a necessary and sufficient condition that a statistic S be sufficient for a parameter θ is that the likelihood, $l(\theta; y)$ can be expressed in the form;

$$l(\theta; y) = m_1(S, \theta)m_2(y)$$

For example, if $Y_1, Y_2, \ldots, Y_n$ are independent random variables from a *Poisson distribution* with mean μ, the likelihood is given by;

$$l(\mu; y_1, y_2, \ldots, y_n) = \prod_{j=1}^{n} \frac{e^{-\mu}\mu^{y_j}}{y_j!}$$

which can be factorized into

$$l = e^{-n\mu}\mu^{\Sigma y_j}\left(\prod \frac{1}{y_j!}\right)$$

Consequently $S = \sum y_j$ is a sufficient statistic for μ. [KA2 Chapter 18.]

Factor loading: See **factor analysis**.

Factor rotation: Usually the final stage of an exploratory *factor analysis* in which the factors derived initially are transformed to make their interpretation simpler. In general the aim of the process is to make the common factors more clearly defined by increasing the size of large factor loadings and decreasing the size of those that are small. *Bipolar factors* are generally split into two separate parts, one corresponding to those variables with positive loadings and the other to those variables with negative loadings. Rotated factors can be constrained to be orthogonal but may also be allowed to be correlated or oblique if this aids in simplifying the resulting solution. See also **varimax rotation**. [MV2 Chapter 12.]

Factor scores: Values assigned to factors for individual sample units in a *factor analysis*. The most common approach is the "regression method". When the factors are viewed as random variables this corresponds to a *best linear unbiased predictor* and if factors are assumed to have normal distributions to *empirical Bayes prediction* (given model parameters). The *Bartlett method* is also sometimes used which corresponds to *maximum likelihood estimation* of factor scores (given model parameters) if the factors are viewed as fixed. [*British Journal of Statistical and Mathematical Psychology*, 2009, **62**, 569–582.]

Factor sparsity: A term used to describe the belief that in many industrial experiments to study large numbers of factors in small numbers of runs, only a few of the many factors studied will have major effects. [*Technometrics*, 1996, **38**, 303–13.]

Failure record data: Data in which time to failure of an item and an associated set of explanatory variables are observed only for those items that fail in some prespecified warranty period $(0, T^0]$. For all other items it is known only that $T_i > T_0$, and for these items no covariates are observed.

Failure time: Synonym for **survival time**.

Fair game: A game in which the entry cost or stake equals the expected gain. In a sequence of such games between two opponents the one with the larger capital has the greater chance of ruining his opponent. See also **random walk**.

False discovery rate (FDR): An approach to controlling the error rate in an exploratory analysis where large numbers of hypotheses are tested, but where the strict control that is provided by *multiple comparison procedures* controlling the *familywise error rate* is not required. Suppose m hypotheses are to be tested, of which m_0 relate to cases where the null hypothesis is true and the remaining $m - m_0$ relate to cases where the alternative hypothesis is true. The FDR is defined as the expected proportion of incorrectly rejected null hypotheses. Explicitly the FDR is given by

$$\text{FDR} = E\left(\frac{V}{R}|R>0\right)\Pr(R>0)$$

where V represents the number of true null hypotheses that are rejected and R is the total number of rejected hypotheses. Procedures that exercise control over the FDR guarantee that FDR $< \alpha$, for some fixed value of α. [*Journal of the Royal Statistical Society, Series B*, 1995, **57**, 289–300.]

False-negative rate: The proportion of cases in which a *diagnostic test* indicates disease absent in patients who have the disease. See also **false-positive rate**. [SMR Chapter 14.]

False-positive rate: The proportion of cases in which a *diagnostic test* indicates disease present in disease-free patients. See also **false-negative rate**. [SMR Chapter 14.]

Familial correlations: The correlation in some phenotypic trait between genetically related individuals. The magnitude of the correlation is dependent both on the heritability of the trait and on the degree of relatedness between the individuals. [*Statistics in Human Genetics*, 1998, P. Sham, Arnold, London.]

Familywise error rate: The probability of making any error in a given family of inferences. See also **per-comparison error rate** and **per-experiment error rate**. [*Statistics in Genetics and Molecular Biology*, 2004, **3**, article 15.]

Fan plot: A graphical procedure for determining the effect of one or more observations on the transformation parameter, λ, in the *Box-Cox transformation* of a response variable in a *multiple regression*. [*Chemometrics and Intelligent Laboratory Systems*, 2002, **60**, 87–100.]

Fan-spread model: A term sometimes applied to a model for explaining differences found between naturally occurring groups that are greater than those observed on some earlier occasion; under this model this effect is assumed to arise because individuals who are less unwell, less impaired, etc., and thus score higher initially, may have greater capacity for change or improvement over time. [*Educational and Psychological Measurement*, 1982, **42**, 759–762.]

Farr, William (1807–1883): Born in Kenley, Shropshire, Farr studied medicine in England from 1826 to 1828 and then in Paris in 1829. It was while in Paris that he became interested in medical statistics. Farr became the founder of the English national system of *vital statistics*. On the basis of national statistics he compiled *life tables* to be used in actuarial calculations. Farr was elected a Fellow of the Royal Society in 1855 and was President of the Royal Statistical Society from 1871 to 1873. He died on 14 April 1883.

Fast Fourier transformation (FFT): A method of calculating the Fourier transform of a set of observations $x_0, x_1, \ldots, x_{n-1}$, i.e. calculating $d(\omega_p)$ given by

$$d(\omega_p) = \sum_{t=0}^{n-1} x_t e^{i\omega_p t}, \quad p = 0, 1, 2, \ldots, n-1; \quad \omega_p = \frac{2\pi p}{n}$$

The number of computations required to calculate $\{d(\omega_p)\}$ is n^2 which can be very large. The FFT reduces the number of computations to $O(n \log_2 n)$, and operates in the following way.

Let $n = rs$ where r and s are integers. Let $t = rt_1 + t_0$, $t = 0, 1, 2, \ldots, n-1$; $t_1 = 0, 1, 2, \ldots, s-1$; $t_0 = 0, 1, 2, \ldots, r-1$. Further let $p = sp_1 + p_0$, $p_1 = 0, 1, 2, \ldots, r-1$, $p_0 = 0, 1, \ldots, s-1$. The FFT can now be written

$$\begin{aligned} d(\omega_p) &= \sum_{t=0}^{n-1} x_t e^{i\omega_p t} = \sum_{t_0=0}^{r-1} \sum_{t_1=0}^{s-1} x_{rt_1+t_0} e^{\frac{2\pi ip}{n}} (rt_1 + t_0) \\ &= \sum_{t_0=0}^{r-1} e^{\frac{2\pi ip}{n} t_0} a(p_0, t_0) \end{aligned}$$

where

$$a(p_0, t_0) = \sum_{t_1=0}^{s-1} x_{rt_1+t_0} e^{\frac{2\pi i t_1 p_0}{S}}$$

Calculation of $a(p_0, t_0)$ requires only s^2 operations, and $d(\omega_p)$ only rs^2. The evaluation of $d(\omega_p)$ reduces to the evaluation of $a(p_0, t_0)$ which is itself a Fourier transform. Following the same procedure, the computation of $a(p_0, t_0)$ can be reduced in a similar way. The procedures can be repeated until a single term is reached. [*Spectral Analysis and Time Series*, 1981, M. B. Priestley, Academic Press, New York.]

Father wavelet: See **wavelet functions**.

Fatigue models: Models for data obtained from observations of the failure of materials, principally metals, which occurs after an extended period of service time. See also **variable-stress accelerated life testing trials**. [*Methods for Statistical Analysis of Reliability and Life Data*, 1974, N. R. Mann, R. E. Schafer and N. D. Singpurwalla, Wiley, New York.]

F-distribution: The probability distribution of the ratio of two independent random variables, each having a *chi-squared distribution*, divided by their respective degrees of freedom. Widely used to assign P-values to *mean square ratios* in the *analysis of variance*. The form of the distribution function is that of a *beta distribution* with $\alpha = \nu_1/2$ and $\beta = \nu_2/2$ where ν_1 and ν_2 are the degrees of freedom of the numerator and denominator chi-squared variables, respectively. [STD Chapter 16.]

Feasible generalized least squares (FGLS) estimator: See **generalized least squares (GLS) estimator**.

Feasibility study: Essentially a synonym for **pilot study**.

Feed-forward network: See **artificial neural network**.

Feller, William (1906–1970): Born in Zagreb, Croatia, Feller was educated first at the University of Zagreb and then at the University of Göttingen from where he received a Ph.D. degree in 1926. After spending some time in Copenhagen and Stockholm, he moved to Providence, Rhode Island in 1939. Feller was appointed professor at Cornell University in 1945 and then in 1950 became Eugene Higgins Professor of Mathematics at Princeton University, a position he held until his death on 14 January 1970. Made significant contributions to probability theory and wrote a very popular two-volume text *An Introduction to Probability Theory and its Applications*. Feller was awarded a National Medal for Science in 1969.

FFT: Abbreviation for **fast Fourier transform**.

FGLS estimator: Abbreviation for **feasible generalized least squares (FGLS) estimator**.

Fibonacci distribution: The probability distribution of the number of observations needed to obtain a run of two successes in a series of *Bernoulli trials* with probability of success equal to a half. Given by

$$\Pr(X = x) = 2^{-x} F_{2,x-2}$$

where the $F_{2,x}$s are Fibonacci numbers. [*Fibonacci Quarterly*, 1973, **11**, 517–22.]

Fiducial inference: A problematic and enigmatic theory of inference introduced by *Fisher*, which extracts a probability distribution for a parameter on the basis of the data without having first assigned the parameter any *prior distribution*. The outcome of such an inference is not the acceptance or rejection of a hypothesis, not a decision, but a probability distribution of a special kind. [KA2 Chapter 20.]

Field plot: A term used in agricultural field experiments for the piece of experimental material to which one treatment is applied.

Fieller's theorem: A general result that enables *confidence intervals* to be calculated for the ratio of two random variables with normal distributions. [*Journal of the Royal Statistical Society, Series B*, 1954, **16**, 175–85.]

File drawer problem: The problem that studies are not uniformly likely to be published in scientific journals. There is evidence that statistical significance is a major determining factor of publication since some researchers may not submit a nonsignificant results for publication and editors may not publish nonsignificant results even if they are submitted.

A particular concern for the application of *meta-analysis* when the data come solely from the published scientific literature. [*Statistical Aspects of the Design and Analysis of Clinical Trials*, 2004, 2nd edn, B.S. Everitt and A. Pickles, ICP, London.]

Final-state data: A term often applied to data collected after the end of an outbreak of a disease and consisting of observations of whether or not each individual member of a household was ever infected at any time during the outbreak. [*Biometrics*, 1995, **51**, 956–68.]

Finite mixture distribution: A probability distribution that is a linear function of a number of component probability distributions. Such distributions are used to model populations thought to contain relatively distinct groups of observations. An early example of the application of such a distribution was that of Pearson in 1894 who applied the following mixture of two normal distributions to measurements made on a particular type of crab:

$$f(x) = pN(\mu_1, \sigma_1) + (1 - p)N(\mu_2, \sigma_2)$$

where p is the proportion of the first group in the population, μ_1, σ_1 are, respectively, the mean and standard deviation of the variable in the first group and μ_2, σ_2 are the corresponding values in the second group. An example of such a distribution is shown in Fig. 62. Mixtures of *multivariate normal distributions* are often used as models for *cluster analysis*. See also **NORMIX, contaminated normal distribution, bimodality** and **multimodality**. [MV2 Chapter 10.]

Finite population: A population of finite size.

Finite population correction: A term sometimes used to describe the extra factor in the variance of the sample mean when n sample values are drawn without replacement from a *finite population* of size N. This variance is given by

$$\text{var}(\bar{x}) = \frac{\sigma^2}{n}\left(1 - \frac{n}{N}\right)$$

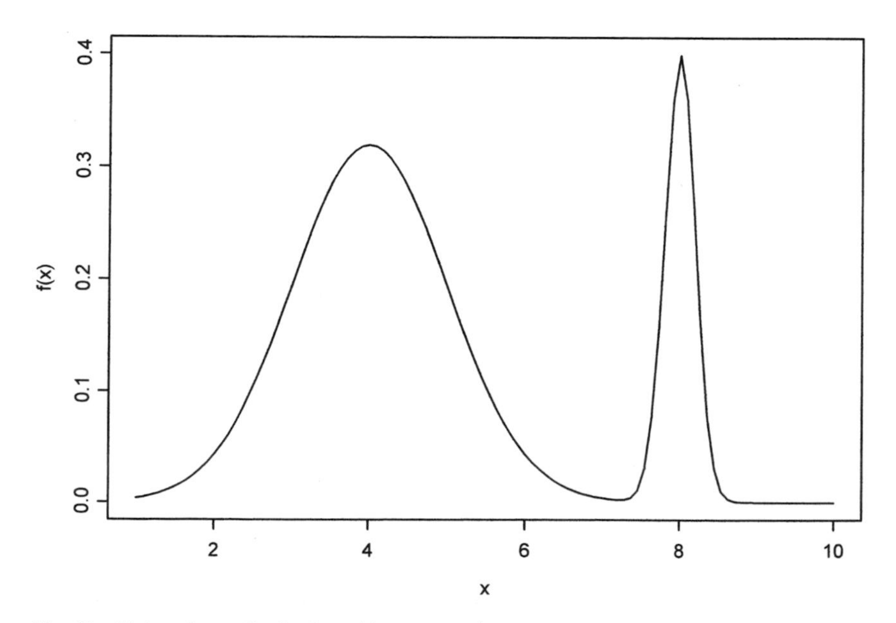

Fig. 62 Finite mixture distribution with two normal components.

the 'correction' term being $(1 - n/N)$. [*Sampling of Populations, Methods and Applications*, 4th edition, 2008, P. S. Levy and S. Lemeshow, Wiley, New York.]

First-order autoregressive model: See **autoregressive model**.

First-order Markov chain: See **Markov chain**.

First passage time: An important concept in the theory of *stochastic processes*, being the time, T, until the first instant that a system enters a state j given that it starts in state i. Even for simple models of the underlying process very few analytical results are available for first passage times. [*Fractals*, 2000, **8**, 139–45.]

Fisher combination test: See **combining p-values**.

Fisher, Irving (1867–1947): Born in Saugerties, New York, Fisher studied mathematics at Yale, where he became Professor of Economics in 1895. Most remembered for his work on *index numbers*, he was also an early user of correlation and regression. Fisher was President of the American Statistical Association in 1932. He died on 29 April 1947 in New York.

Fisher's exact test: An alternative procedure to use of the *chi-squared statistic* for assessing the independence of two variables forming a *two-by-two contingency table* particularly when the expected frequencies are small. The method consists of evaluating the sum of the probabilities associated with the observed table and all possible two-by-two tables that have the same row and column totals as the observed data but exhibit more extreme departure from independence. The probability of each table is calculated from the *hypergeometric distribution*. [*The Analysis of Contingency Tables*, 2nd edition, 1992, B. S. Everitt, Chapman and Hall/CRC Press, London.]

Fisher's *g* statistic: A statistic for assessing ordinates in a *periodogram* and given by

$$g = \frac{\max[I(\omega_p)]}{\sum_{p=1}^{[N/2]} I(\omega_p)}$$

where N is the length of the associated *time series*, and $I(\omega_p)$ is an ordinate of the periodogram. Under the hypothesis that the population ordinate is zero, the exact distribution of g can be derived. [*Applications of Time Series Analysis in Astronomy and Meteorology*, 1997, eds. T. Subba Rao, M. B. Priestley and O. Lissi, Chapman and Hall/CRC Press, London.]

Fisher's information: Essentially the amount of information supplied by data about an unknown parameter. Given by the quantity;

$$I = E\left(\frac{\partial \ln L}{\partial \theta}\right)^2$$

where L is the *likelihood* of a sample of n independent observations from a probability distribution, $f(x|\theta)$. In the case of a vector parameter, $\boldsymbol{\theta}' = [\theta_1, \ldots, \theta_s]$, *Fisher's information matrix* has elements i_{jk} given by

$$i_{jj} = E\left(\frac{\partial \ln L}{\partial \theta_j}\right)^2$$

$$i_{jk} = E\left(\frac{\partial \ln L}{\partial \theta_j \partial \theta_k}\right)$$

The *variance–covariance matrix* of the *maximum likelihood estimator* of $\boldsymbol{\theta}$ is the inverse of this matrix. [KA2 Chapter 18.]

Fisher's information matrix: See **Fisher's information.**

Fisher's linear discriminant function: A technique for classifying an object or indivdual into one of two possible groups, on the basis of a set of measurements $x_1, x_2, \ldots, x_q$. The essence of the approach is to seek a linear transformation, z, of the measurements, i.e.

$$z = a_1x_1 + a_2x_2 + \cdots + a_qx_q$$

such that the ratio of the between-group variance of z to its within-group variance is maximized. The solution for the coefficients $\mathbf{a}' = [a_1, \ldots, a_q]$ is

$$\mathbf{a} = \mathbf{S}^{-1}(\bar{\mathbf{x}}_1 - \bar{\mathbf{x}}_2)$$

where $\mathbf{S}$ is the pooled within groups *variance–covariance matrix* of the two groups and $\bar{\mathbf{x}}_1, \bar{\mathbf{x}}_2$ are the group mean vectors. [MV2 Chapter 9.]

Fisher's scoring method: An alternative to the *Newton–Raphson method* for optimization (finding the minimum or the maximum) of some function, which involves replacing the matrix of second derivatives of the function with respect to the parameters, by the corresponding matrix of expected values. This method will often converge from starting values further away from the optimum than will Newton–Raphson. [GLM Chapter 2.]

Fisher, Sir Ronald Aylmer (1890–1962): Born in East Finchley in Greater London, Fisher studied mathematics at Cambridge and became a statistician at the Rothamsted Agricultural Research Institute in 1919. By this time Fisher had already published papers introducing *maximum likelihood estimation* (although not yet by that name) and deriving the general sampling distribution of the correlation coefficient. At Rothamsted his work on data obtained in on-going field trials made him aware of the inadequacies of the arrangements of the experiments themselves and so led to the evolution of the science of experimental design and to the analysis of variance. From this work there emerged the crucial principle of randomization and one of the most important statistical texts of the 20th century, *Statistical Methods for Research Workers*, the first edition of which appeared in 1925 and now (1997) is in its ninth edition. Fisher continued to produce a succession of original research on a wide variety of statistical topics as well as on his related interests of genetics and evolutionary theory throughout his life. Perhaps Fisher's central contribution to 20th century science was his deepening of the understanding of uncertainty and of the many types of measurable uncertainty. He was Professor of Eugenics at University College, London from 1933 until 1943, and then became the Arthur Balfour Professor of Genetics at Cambridge University. After his retirement in 1957, Fisher moved to Australia. He died on 29 July 1962 in Adelaide.

Fisher's *z* transformation: A transformation of *Pearson's product moment correlation coefficient*, r, given by

$$z = \frac{1}{2}\ln\frac{1+r}{1-r}$$

The statistic z has a normal distribution with mean

$$\frac{1}{2}\ln\frac{1+\rho}{1-\rho}$$

where ρ is the population correlation value and variance $1/(n-3)$ where n is the sample size. The transformation may be used to test hypotheses and to construct *confidence intervals* for ρ. [SMR Chapter 11.]

Fishing expedition: Synonym for **data dredging**.

Fitted value: Usually used to refer to the value of the response variable as predicted by some estimated model.

Five-number summary: A method of summarizing a set of observations using the minimum value, the lower quartile, the median, upper quartile and maximum value. Forms the basis of the *box-and-whisker plot.* [*Exploratory Data Analysis*, 1977, J. W. Tukey, Addison-Wesley, Reading, MA.]

Fixed effects: The effects attributable to a finite set of levels of a factor that are of specific interest. For example, the investigator may wish to compare the effects of three particular drugs on a response variable. *Fixed effects models* are those that contain only factors with this type of effect. See also **multilevel models**.

Fixed effects model: See **fixed effects**.

Fixed population: A group of individuals defined by a common fixed characteristic, such as all men born in Essex in 1944. Membership of such a population does not change over time by immigration or emigration. [*Radiology*, 1982, **143**, 469–74.]

Fix–Neyman process: A stochastic model used to describe recovery, relapse, death and loss of patients in medical follow-up studies of cancer patients. [*Human Biology*, 1951, **23**, 205–41.]

Fleiss, Joseph (1937–2003): Born in Brooklyn, New York, Fleiss studied at Columbia College, graduating in 1959. In 1961 he received an M.S. degree in Biostatistics from the Columbia School of Public Health, and six years later completed his Ph.D. dissertation on 'Analysis of variance in assessing errors in interview data'. In 1975 Fleiss became Professor and Head of the Division of Biostatistics at the School of Public Health in Columbia University. He made influential contributions in the analysis of categorical data and *clinical trials* and wrote two classic biostatistical texts, *Statistical Methods for Rates and Proportions* and *The Design and Analysis of Clinical Experiments*. Fleiss died on 12 June 2003, in Ridgewood, New Jersey.

Flingner–Policello test: A robust rank test for the *Behrens–Fisher problem.* [NSM Chapter 4.]

Fligner–Wolfe test: A *distribution free method* designed for the setting where one of the treatments in a study corresponds to a control or baseline set of conditions and interest centres on assessing which, if any, of the treatments are better than the control. The test statistic arises from combining all N observations from the k groups and ordering them from least to greatest. If r_{ij} is the rank of the jth observation in the ith group the test statistic is given specifically by

$$FW = \sum_{j=2}^{k} \sum_{i=1}^{n_j} r_{ij}$$

the summations being over the non-control groups. Critical values of FW are available in appropriate tables, and a large-sample approximation can also be found. [NSM Chapter 6.]

Floor effect: See **ceiling effect**.

Flow-chart: A graphical display illustrating the interrelationships between the different components of a system. It acts as a convenient bridge between the conceptualization of a model and the construction of equations.

Folded log: A term sometimes used for half the *logistic transformation* of a proportion.

Folded normal distribution: The probability distribution of $Z = |X|$, where the random variable, X, has a normal distribution with zero mean and variance σ^2. Given specifically by

$$f(z) = \frac{1}{\sigma}\sqrt{\frac{2}{\pi}}e^{-z^2/2\sigma^2}$$

[*Handbook of the Normal Distribution*, 2nd edition, 1996, J. K. Patel and C. B. Read, Marcel Dekker, New York.]

Folded square root: A term sometimes used for the following transformation of a proportion p

$$\sqrt{2p} - \sqrt{2(1-p)}$$

Foldover design: A design used to augment a *fractional factorial design*, which is obtained by reversing the signs of all columns of the original *design matrix*. The original design runs are combined with the mirror-image foldover design runs, and the combination can then be used to estimate all the main effects clear of any two-factor interactions. [*Statistics and Probability Letters*, 2003, **3**, 245–250.]

Foldover frequency: Synonym for **Nyquist frequency**.

Flog: An unattractive synonym for **folded log**.

Focus groups: A research technique used to collect data through group interaction on a topic determined by the researcher. The goal in such groups is to learn the opinions and values of those involved with products and/or services from diverse points of view. [*American Journal of Sociology*, 1996, **22**, 129–52.]

Folk theorem of queueing behaviour: The queue you join moves the slowest.

Follmann's test: A one-sided modified *Hotelling's* T^2 test for multivariate normally distributed variables that combines ease of use, theoretically exact control of the significance level and good power to detect a large spectrum of alternative hypotheses. [*Journal of the American Statistical Association*, 1996, **91**, 854–861.]

Followback surveys: Surveys which use lists associated with vital statistics to sample individuals for further information. For example, the 1988 National Mortality Followback Survey sampled death certificates for 1986 decedents 25 years or older that were filed in the USA. Information was then sought from the next of kin or some other person familiar with the decedent, and from health care facilities used by the decedent in the last year of life. Information was obtained by emailed questionnaire, or by telephone or personal interview. Another example of such a survey is the 1988 National Maternal and Infant Health Survey, the live birth component of which sampled birth certificates of live births in the USA. Mothers corresponding to the sampled birth certificates were then mailed a questionnaire. The sample designs of this type of survey are usually quite simple, involving a stratified selection of individuals from a list frame. [*American Journal of Industrial Medicine*, 1995, **27**, 195–205.]

Follow-up: The process of locating research subjects or patients to determine whether or not some outcome of interest has occurred. [SMR Chapter 13.]

Follow-up plots: Plots for following up the frequency and patterns of *longitudinal data*. In one version the vertical axis consists of the time units of the study. Each subject is represented

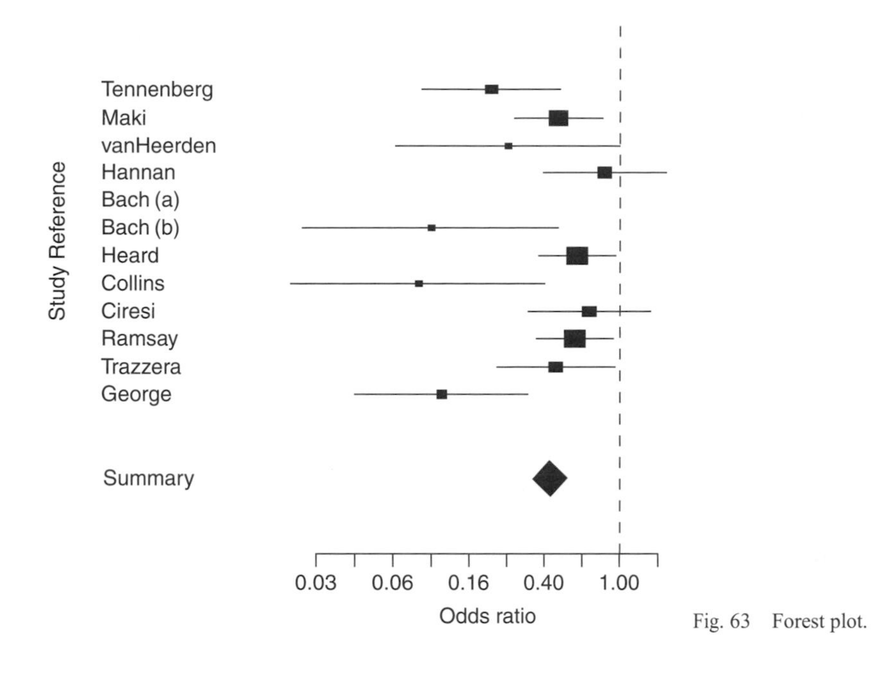

Fig. 63 Forest plot.

by a column, which is a vertical line proportional in length to the subject's total time in the system. [*The American Statistician*, 1995, **49**, 139–44.]

Force of mortality: Synonym for **hazard function**.

Forecast: The specific projection that an investigator believes is most likely to provide an accurate prediction of a future value of some process. Generally used in the context of the analysis of *time series*. Many different methods of forecasting are available, for example *autoregressive integrated moving average models*. [*Statistical Forecasting*, 1976, W.G. Gilchrist, Wiley, London.]

Forest plot: A name sometimes given to a type of diagram commonly used in *meta-analysis*, in which point estimates and confidence intervals are displayed for all studies included in the analysis. An example from a meta-analysis of clozapine v other drugs in the treatment of schizophrenia is shown in Fig. 63. A drawback of such a plot is that the viewer's eyes are often drawn to the least significant studies, because these have the widest confidence intervals and are graphically more imposing. [*Methods in Meta-Analysis*, 2000, A. J. Sutton, K. R. Abrams, D. R. Jones, T. A. Sheldon, and F. Song, Wiley, Chichester, UK.]

Forward-looking study: An alternative term for *prospective study*.

Forward selection procedure: See **selection methods in regression**.

Fourfold table: Synonym for **two-by-two contingency table**.

Fourier coefficients: See **Fourier series**.

Fourier series: A series used in the analysis of generally a periodic function into its constituent sine waves of different frequencies and amplitudes. The series is

$$\frac{1}{2}a_0 + \sum(a_n \cos nx + b_n \sin nx)$$

where the coefficients are chosen so that the series converges to the function of interest, *f*; these coefficients (the *Fourier coefficients*) are given by

$$a_n = \frac{1}{\pi}\int_{-\pi}^{\pi} f(x)\cos nx\,\mathrm{d}x$$
$$b_n = \frac{1}{\pi}\int_{-\pi}^{\pi} f(x)\sin nx\,\mathrm{d}x$$

for $n = 1, 2, 3, \ldots$ See also **fast Fourier transformation** and **wavelet analysis**. [TMS Chapter 7.]

Fourier transform: See fast **Fourier transformation**.

Fractal: A term used to describe a geometrical object that continues to exhibit detailed structure over a large range of scales. Snowflakes and coastlines are frequently quoted examples. A medical example is provided by electrocardiograms. [*The Fractal Geometry of Nature*, 1982, B. B. Mandelbrot, Freeman, San Francisco.]

Fractal dimension: A numerical measure of the degree of roughness of a *fractal*. Need not be a whole number, for example, the value for a typical coastline is between 1.15 and 1.25. [*The Fractal Geometry of Nature*, 1982, B.B. Mandelbrot, W.H. Freeman, San Francisco.]

Fractional factorial design: Designs in which information on main effects and low-order interactions are obtained by running only a fraction of the complete factorial experiment and assuming that particular high-order interactions are negligible. Among the most widely used type of designs in industry. See also **response surface methodology**. [*Experimental Designs*, 2nd edition, 1992, W. G. Cochran and G. M. Cox, Wiley, New York.]

Fractional polynomials: An extended family of curves which are often more useful than low- or high-order polynomials for modelling the often curved relationship between a response variable and one or more continuous covariates. Formally such a polynomial of degree *m* is defined as

$$\phi_m(X : \boldsymbol{\beta}, \mathbf{p}) = \beta_0 + \sum_{j=1}^{m} \beta_j X^{(p_j)}$$

where $\mathbf{p}' = [p_1, \ldots, p_m]$ is a real-valued vector of powers with $p_1 < p_2 < \cdots < p_m$ and $\boldsymbol{\beta}' = [\beta_0, \ldots, \beta_m]$ are real-valued coefficients. The round bracket notation signifies the *Box–Tidwell transformation*

$$X^{(p_j)} = X^{p_j} \quad \text{if } p_j \neq 0$$
$$= \ln X \quad \text{if } p_j = 0$$

So, for example, a fractional polynomial of degree 3 with powers (1,2,0) is of the form

$$\beta_0 + \beta_1 X + \beta_2 X^2 + \beta_3 \log(X)$$

[*International Journal of Epidemiology*, 1999, **28**, 964–974.]

Frailty: A term generally used for unobserved individual heterogeneity. Such variation is of major concern in medical statistics particularly in the analysis of *survival times* where *hazard functions* can be strongly influenced by selection effects operating in the population. There are a number of possible sources of this heterogeneity, the most obvious of which is that it reflects biological differences, so that, for example, some individuals are born with a weaker heart, or a genetic disposition for cancer. A further possibility is that the heterogeneity arises from the induced weaknesses that result from the stresses of life. Failure to take account of this type of variation may often obscure comparisons between groups, for example, by measures of *relative risk*. A simple model (*frailty model*) which attempts to allow for the variation between individuals is

$$\text{individual hazard function} = Z\lambda(t)$$

where Z is a quantity specific to an individual, considered as a *random variable* over the population of individuals, and $\lambda(t)$ is a basic rate. What is observed in a population for which such a model holds is not the individual hazard rate but the net result for a number of individuals with different values of Z. [*Statistics in Medicine*, 1988, **7**, 819–42.]

Frailty model: See **frailty**.

Frank's family of bivariate distributions: A class of *bivariate probability distributions* of the form

$$P(X \leq x, Y \leq y) = H_\alpha(x,y) = \log_\alpha\left\{1 + \frac{(\alpha^x - 1)(\alpha^y - 1)}{\alpha - 1}\right\}, \quad \alpha \neq 1$$

A perspective plot of such a distribution is shown in Fig. 64.

See also **copulas**. [*Biometrika*, 1987, **74**, 549–55.]

Frechet bounds: Bounds for any *bivariate distribution*, $H(x,y)$, and given by

$$\max\{F(x) + G(y) - 1, 0\} \leq H(x,y) \leq \min\{F(\text{x}), G(\text{y})\}$$

where $F(x)$ and $G(y)$ are the *marginal distributions*. The bounds are important in the development of *copulas*. [*Scandinavian Actuarial Journal*, 2000, **100**, 17–32.]

Freedman, David A. (1938–2008): Born in Montreal, Canada, Friedman received a first degree from McGill University in 1958 and a Ph.D from Princeton University in 1960. In 1961 Freedman joined the statistics department at the University of California, Berkeley where he remained for his entire career. Freedman made major contributions to both theoretical statistics, for example, the *bootstrap* and martingale inequalities and applied statistics, for example, the use of regression to analyse observational data. He also wrote about statistics for lawyers and judges. Freedman died in Berkeley on the 17th October, 2008.

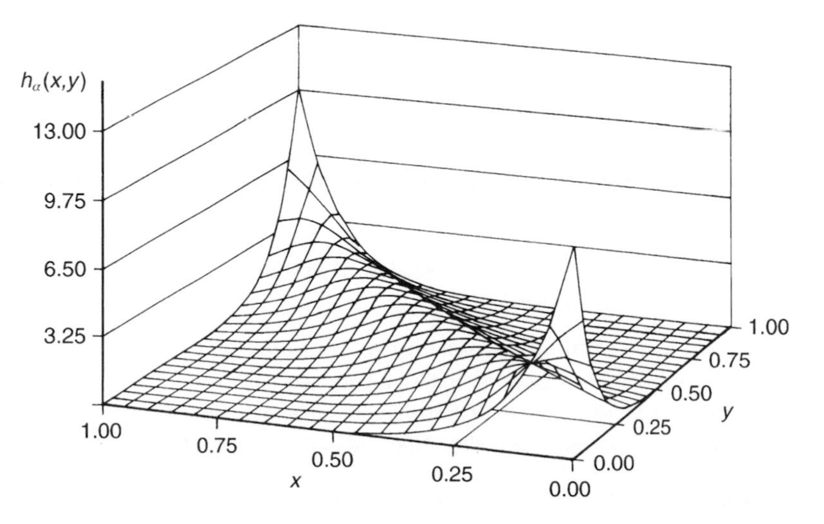

Fig. 64 An example of a bivariate distribution from Frank's family.

Freeman–Tukey test: A procedure for assessing the goodness-of-fit of some model for a set of data involving counts. The test statistic is

$$T = \sum_{i=1}^{k} \left(\sqrt{O_i} + \sqrt{O_i + 1} - \sqrt{4E_i + 1} \right)^2$$

where k is the number of categories, $O_i, i = 1, 2, \ldots, k$ the observed counts and $E_i, i = 1, 2, \ldots, k$, the expected frequencies under the assumed model. The statistic T has asymptotically a *chi-squared distribution* with $k - s - 1$ degrees of freedom, where s is the number of parameters in the model. See also **chi-squared statistic**, and **likelihood ratio**. [*Annals of Mathematical Statistics*, 1950, **21**, 607–11.]

Freeman–Tukey transformation: A transformation of a random variable, X, having a *Poisson distribution*, to the form $\sqrt{X} + \sqrt{X + 1}$ in order to stabilize its variance. [*Annals of Mathematical Statistics*, 1950, **21**, 607–11.]

Frequency distribution: The division of a sample of observations into a number of classes, together with the number of observations in each class. Acts as a useful summary of the main features of the data such as location, shape and spread. Essentially the empirical equivalent of the probability distribution. An example is as follows:

Hormone assay values(nmol/L)

Class limits	Observed frequency
75–79	1
80–84	2
85–89	5
90–94	9
95–99	10
100–104	7
105–109	4
110–114	2
≥ 115	1

See also **histogram**. [SMR Chapter 2.]

Frequency polygon: A diagram used to display graphically the values in a frequency distribution. The frequencies are graphed as ordinate against the class mid-points as abscissae. The points are then joined by a series of straight lines. Particularly useful in displaying a number of frequency distributions on the same diagram. An example is given in Fig. 65. [SMR Chapter 2.]

Frequentist inference: An approach to statistics based on a frequency view of probability in which it is assumed that it is possible to consider an infinite sequence of independent repetitions of the same statistical experiment. Significance tests, hypothesis tests and *likelihood* are the main tools associated with this form of inference. See also **Bayesian inference**. [KA2 Chapter 31.]

Friedman's two-way analysis of variance: A *distribution free method* that is the analogue of the *analysis of variance* for a design with two factors. Can be applied to data sets that do not meet the assumptions of the parametric approach, namely normality and homogeneity of variance. Uses only the ranks of the observations. [SMR Chapter 12.]

Friedman's urn model: A possible alternative to random allocation of patients to treatments in a *clinical trial* with K treatments, that avoids the possible problem of imbalance when

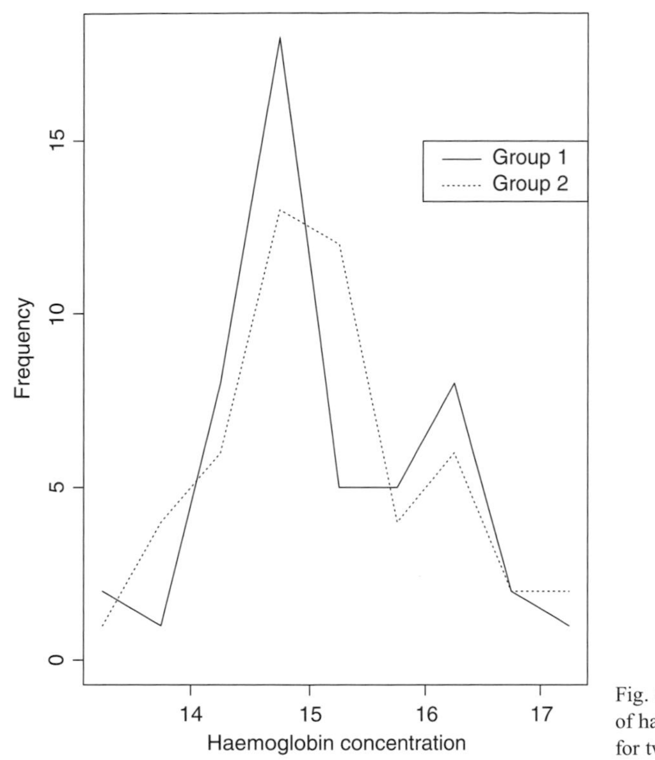

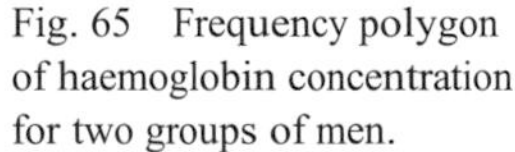

Fig. 65 Frequency polygon of haemoglobin concentration for two groups of men.

the number of available subjects is small. The model considers an urn containing balls of K different colours, and begins with w balls of colour k, $k = 1, \ldots, K$. A draw consists of the following operations

- select a ball at random from the urn,
- notice its colour k' and return the ball to the urn;
- add to the urn α more balls of colour k' and β more balls of each other colour k where $k \neq k'$.

Each time a subject is waiting to be assigned to a treatment, a ball is drawn at random from the urn; if its colour is k' then treatment k' is assigned. The values of w, α and β can be any reasonable nonnegative numbers. If β is large with respect to α then the scheme forces the trial to be balanced. The value of w determines the first few stages of the trial. If w is large, more randomness is introduced into the trial; otherwise more balance is enforced. [*Encyclopedia of Statistical Sciences*, 2006, eds. S. Kotz, C. B. Read, N. Balakrishnan and B. Vidakovic, Wiley, New York.]

Frindall, William Howard (1939–2009): Born in Epsom, Surrey, United Kingdom, Frindall attended Reigate Grammar School in Surrey and studied architecture at the Kingston School of Art. After National Service in the Royal Air Force, Frindall became scorer and statistician on the BBC radio program, Test Match Special in 1966 and continued in this role until his death, watching all 246 test matches held in England from 1966 to 2008. He introduced a scoring system for cricket matches, that was named after him and he was meticulously accurate. Frindall wrote a large number of books on cricket statistics. In 1998 Frindall was awarded the honorary degree of Doctor of Technology by Staffordshire University for his contributions to statistics and in 2004 he was awarded an MBE for services to cricket and broadcasting. 'Bill' Frindall, cricket scorer, statistician and broadcaster died in Swindon, UK on the 30th January, 2009.

Frisch, Ragnar Anton Killi (1895–1973): Born in 1895 in Oslo, Norway as the only son of a silversmith Frisch was expected to follow his father's trade and took steps in that direction including an apprenticeship. He studied economics at the University of Oslo because it was "the shortest and easiest study" available at the university, but remained involved in his father's business. After a couple of years studying in Paris and England, Frisch returned to lecture at Oslo in 1924 before leaving for the United States in 1930 visiting Yale and Minnesota. In 1931 Frisch became a full professor at the University of Oslo and founded the Rockefeller-funded Institute of Economics in 1932. Ragnar Frisch was a founding father of econometrics and editor of *Econometrica* for more than 20 years. He was awarded the first Nobel Memorial Prize in Economics in 1969 with Jan Tinbergen. Ragnar Frisch died in Oslo in 1973.

Froot: An unattractive synonym for **folded square root**.

F-test: A test for the equality of the variances of two populations having normal distributions, based on the ratio of the variances of a sample of observations taken from each. Most often encountered in the *analysis of variance*, where testing whether particular variances are the same also tests for the equality of a set of means. [SMR Chapter 9.]

F-to-enter: See **selection methods in regression**.

F-to-remove: See **selection methods in regression**.

Full model: Synonym for **saturated model**.

Functional data analysis: The analysis of data that are functions observed continuously, for example, functions of time. Essentially a collection of statistical techniques for answering questions like 'in what way do the curves in the sample differ?', using information on the curves such as slopes and curvature. [*Applied Statistics*, 1995, **44**, 12–30.]

Functional principal components analysis: A version of *principal components analysis* for data that may be considered as curves rather than the vectors of classical *multivariate analysis*. Denoting the observations $X_1(t), X_2(t), \ldots, X_n(t)$, where $X_i(t)$ is essentially a *time series* for individual i, the model assumed is that

$$X(t) = \mu(t) + \sum \gamma_\nu U_\nu(t)$$

where the principal component scores, γ_ν are uncorrelated variables with mean zero, and the principal component functions, $U_\nu(t)$ are scaled to satisfy $\int U_\nu^2 = 1$; these functions often have interesting physical meanings which aid in the interpretation of the data. [*Computational Statistics and Data Analysis*, 1997, **24**, 255–70.]

Functional relationship: The relationship between the 'true' values of variables, i.e. the values assuming that the variables were measured without error. See also **latent variables** and **structural equation models**.

Funnel plot: An informal method of assessing the effect of *publication bias*, usually in the context of a *meta-analysis*. The effect measures from each reported study are plotted on the x-axis against the corresponding sample sizes on the y-axis. Because of the nature of sampling variability this plot should, in the absence of publication bias, have the shape of a pyramid with a tapering 'funnel-like' peak. Publication bias will tend to skew the pyramid by selectively excluding studies with small or no significant effects. Such studies

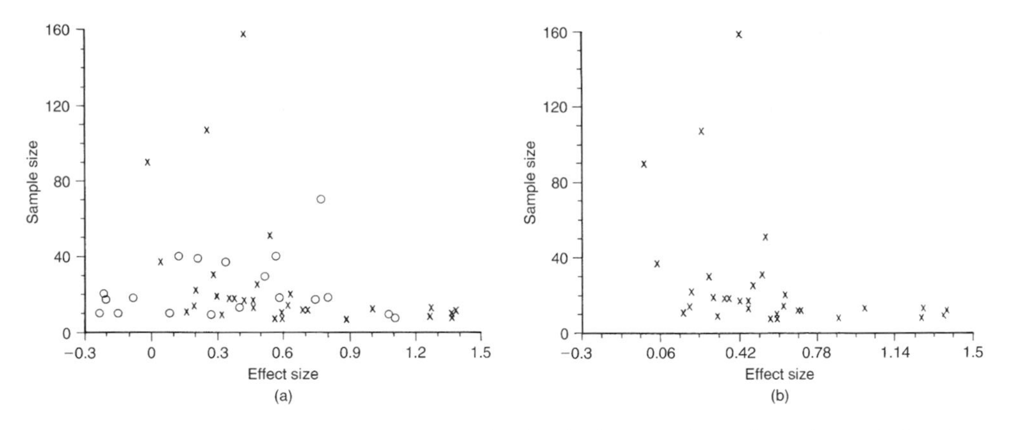

Fig. 66 Funnel plot of studies of psychoeducational progress for surgical patients: (a) all studies; (b) published studies only.

predominate when the sample sizes are small but are increasingly less common as the sample sizes increase. Therefore their absence removes part of the lower left-hand corner of the pyramid. This effect is illustrated in Fig. 66. [*Reproductive Health*, 2007, **4**, 1742–1748.]

FU-plots: Abbreviation for **follow-up plots**.

Future years of life lost: An alternative way of presenting data on mortality in a population, by using the difference between age at death and *life expectancy.* [*An Introduction to Epidemiology*, 1983, M. A. Alderson, Macmillan, London.]

Fuzzy set theory: A radically different approach to dealing with uncertainty than the traditional probabilistic and statistical methods. The essential feature of a fuzzy set is a membership function that assigns a grade of membership between 0 and 1 to each member of the set. Mathematically a membership function of a fuzzy set A is a mapping from a space χ to the unit interval $m_A : \chi \rightarrow [0, 1]$. Because memberships take their values in the unit interval, it is tempting to think of them as probabilities; however, memberships do not follow the laws of probability and it is possible to allow an object to simultaneously hold nonzero degrees of membership in sets traditionally considered mutually exclusive. Methods derived from the theory have been proposed as alternatives to traditional statistical methods in areas such as quality control, linear regression and forecasting, although they have not met with universal acceptance and a number of statisticians have commented that they have found no solution using such an approach that could not have been achieved as least as effectively using probability and statistics. See also **grade of membership model**. [*Theory of Fuzzy Subsets*, 1975, M. Kaufman, Academic Press, New York.]

G^2: Symbol for the *goodness-of-fit test statistic* based on the *likelihood ratio*, often used when using *log-linear models*. Specifically given by

$$G^2 = 2\sum O \ln(O/E)$$

where O and E denote observed and expected frequencies. Also used more generally to denote *deviance*.

Gabor regression: An approach to the modelling of time–frequency surfaces that consists of a Bayesian regularization scheme in which *prior distributions* over the time–frequency coefficients are constructed to favour both smoothness of the estimated function and sparseness of the coefficient representation. [*Journal of the Royal Statistical Society, Series B*, 2004, **66**, 575–89.]

Gain: Synonym for **power transfer function**.

Galbraith plot: A graphical method for identifying *outliers* in a *meta-analysis*. The standardized effect size is plotted against precision (the reciprocal of the standard error). If the studies are homogeneous, they should be distributed within ±2 standard errors of the regression line through the origin. [*Research Methodology*, 1999, edited by H. J. Ader and G. J. Mellenbergh, Sage, London.]

Galton, Sir Francis (1822–1911): Born in Birmingham, Galton studied medicine at London and Cambridge, but achieved no great distinction. Upon receiving his inheritance he eventually abandoned his studies to travel in North and South Africa in the period 1850-1852 and was given the gold medal of the Royal Geographical Society in 1853 in recognition of his achievements in exploring the then unknown area of Central South West Africa and establishing the existence of anticyclones. In the early 1860s he turned to meteorology where the first signs of his statistical interests and abilities emerged. His later interests ranged over psychology, anthropology, sociology, education and fingerprints but he remains best known for his studies of heredity and intelligence which eventually led to the controversial field he referred to as eugenics, the evolutionary doctrine that the condition of the human species could most effectively be improved through a scientifically directed process of controlled breeding. His first major work was *Heriditary Genius* published in 1869, in which he argued that mental characteristics are inherited in the same way as physical characteristics. This line of thought lead, in 1876, to the very first behavioural study of twins in an endeavour to distinguish between genetic and environmental influences. Galton applied somewhat naive regression methods to the heights of brothers in his book *Natural Inheritance* and in 1888 proposed the index of co-relation, later elaborated by his student *Karl Pearson* into the correlation coefficient. Galton was a cousin of Charles Darwin and was knighted in 1909. He died on 17 January 1911 in Surrey.

Galton–Watson process: A commonly used name for what is more properly called the *Bienaymé–Galton–Watson process*.

GAM: Abbreviation for **geographical analysis machine** and **generalized additive models**.

Gambler's fallacy: The belief that if an event has not happened for a long time it is bound to occur soon. [*Chance Rules*, 2nd edition, 2008, B. S. Everitt, Springer, New York.]

Gambler's ruin problem: A term applied to a game in which a player wins or loses a fixed amount with probabilities p and q ($p \neq q$). The player's initial capital is z and he plays an adversary with capital $a - z$. The game continues until the player's capital is reduced to zero or increased to a, i.e. until one of the two players is ruined. The probability of ruin for the player starting with capital z is q_z given by

$$q_z = \frac{(q/p)^a - (q/p)^z}{(q/p)^a - 1}$$

[KA2 Chapter 24.]

Gambling: The art of attempting to exchange something small and certain, for something large and uncertain. Gambling is big business; in the US, for example, it is at least a $40-billion-a-year industry. Popular forms of gambling are *national lotteries*, roulette and horse racing. For these (and other types of gambling) it is relatively easy to apply statistical methods to discover the chances of winning, but in no form of gambling does a statistical analysis increase your chance of winning. [*Chance Rules*, 2nd edn, 2008, B. S. Everitt, Springer, New York.]

Game theory: The branch of mathematics that deals with the theory of contests between two or more players under specified sets of rules. The subject assumes a statistical aspect when part of the game proceeds under a chance scheme. Game theory has a long history of being applied to security, beginning with military applications, and has also been used in the context of arms control. [*Simulation and Gaming*, 2003, **34**, 319–337.]

Gamma distribution: The probability distribution, $f(x)$, given by

$$f(x) = \left(\frac{x}{\beta}\right)^{\gamma-1} \frac{\exp(-x/\beta)}{\beta\Gamma(\gamma)}, \quad 0 \le x < \infty, \ \beta > 0, \gamma > 0$$

β is a *scale parameter* and γ a *shape parameter*. Examples of the distribution are shown in Fig. 67. The mean, variance, *skewness* and *kurtosis* of the distribution are as follows

$$\text{mean} = \beta\gamma$$
$$\text{variance} = \beta^2\gamma$$
$$\text{skewness} = 2\gamma^{-\frac{1}{2}}$$
$$\text{kurtosis} = 3 + \frac{6}{\gamma}$$

The distribution of $u = x/\beta$ is the *standard gamma distribution* with corresponding density function given by

$$f(u) = \frac{x^{\gamma-1}e^{-x}}{\Gamma(\gamma)}$$

[STD Chapter 18.]

Gamma function: The function Γ defined by

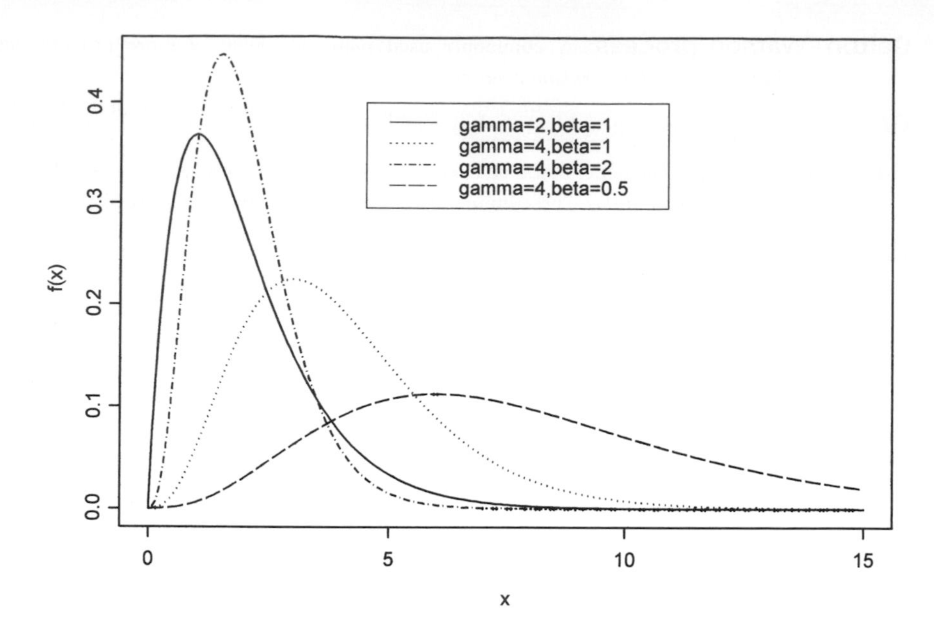

Fig. 67 Gamma distributions for a number of parameter values.

$$\Gamma(r) = \int_0^\infty t^{r-1}e^{-t}\mathrm{d}t$$

where $r > 0$ (r need not be an integer). The function is recursive satisfying the relationship

$$\Gamma(r+1) = r\Gamma(r)$$

The integral

$$\int_0^T t^{r-1}e^{-t}dt$$

is known as the *incomplete gamma function*.

Gap statistic: A statistic for estimating the number of clusters in applications of *cluster analysis*. Applicable to virtually any clustering method, but in terms of *K-means cluster analysis*, the statistic is defined specifically as

$$\mathrm{Gap}_n(k) = E_n^*[\log(W_k)] - \log(W_k)$$

where W_k is the pooled within-cluster sum of squares around the cluster means and E_n^* denotes expectation under a sample of size n from the reference distribution. The estimate of the number of clusters, $\hat{k}$, is the value of k maximizing $\mathrm{Gap}_n(k)$ after the sampling distribution has been accounted for. [*Journal of the Royal Statistical Society, Series B*, 2001, **63**, 411–23.]

Gap-straggler method: A procedure for partitioning of treatment means in a *one-way design* under the usual normal theory assumptions. [*Biometrics*, 1949, **5**, 99–114.]

Gap time: The time between two successive events in survival data in which each individual subject can potentially experience a series of events. An example is the time from the development of AIDS to death. [*Biometrika*, 1999, **86**, 59–70.]

Garbage in garbage out: A term that draws attention to the fact that sensible output only follows from sensible input. Specifically if the data is originally of dubious quality then so also will be the results.

Gardner, Martin (1940–1993): Gardner read mathematics at Durham University followed by a diploma in statistics at Cambridge. In 1971 he became Senior Lecturer in Medical Statistics in the Medical School of Southampton University. Gardner was one of the founders of the Medical Research Council's Environmental Epidemiology Unit. Worked on the geographical distribution of disease, and, in particular, on investigating possible links between radiation and the risk of childhood leukaemia. Gardner died on 22 January 1993 in Southampton.

GAUSS: A high level programming language with extensive facilities for the manipulation of matrices. [Aptech Systems, P.O. Box 250, Black Diamond, WA 98010, USA. Timberlake Consulting, Unit B3, Broomsley Business Park, Worsley Bridge Road, London SE26 5BN, UK.]

Gauss, Karl Friedrich (1777–1855): Born in Brunswick, Germany, Gauss was educated at the Universities of Göttingen and Helmstedt where he received a doctorate in 1799. He was a prodigy in mental calculation who made numerous contributions in mathematics and statistics. He wrote the first modern book on number theory and pioneered the application of mathematics to such areas as gravitation, magnetism and electricity–the unit of magnetic induction was named after him. In statistics Gauss' greatest contribution was the development of *least squares estimation* under the label 'the combination of observations'. He also applied the technique to the analysis of observational data, much of which he himself collected. The normal curve is also often attributed to Gauss and sometimes referred to as the Gaussian curve, but there is some doubt as to whether this is appropriate since there is considerable evidence that it is more properly due to *de Moivre*. Gauss died on 23 February 1855 in Göttingen, Germany.

Gaussian distribution: Synonym for **normal distribution**.

Gaussian Markov random field: A multivariate normal random vector that satisfies certain *conditional independence* assumptions. Can be viewed as a model framework that contains a wide range of statistical models, including models for images, *time-series*, *longitudinal data*, spatio-temporal processes, and graphical models. [*Gaussian Markov Random Fields: Theory and Applications*, 2005, H. Rue and L. Held, Chapman and Hall/CRC, Boca Raton.]

Gaussian process: A generalization of the normal distribution used to characterize functions. It is called a Gaussian process because it has Gaussian distributed finite dimensional marginal distributions. A main attraction of Gaussian processes is computational tractability. They are sometimes called Gaussian random fields and are popular in the application of *nonparametric Bayesian models*. [*Gaussian Processes for Machine Learning*, 2006, C. E. Rasmussen and C. K. I. Williams, MIT Press, Boston.]

Gaussian quadrature: An approach to approximating the integral of a function using a weighted sum of function values at specified points within the domain of integration. n-point Gaussian quadrature involves an optimal choice of quadrature points x_i and quadrature weights w_i for $i = 1, \ldots, n$ that yields exact results for polynomials of degree $2n$–1 or less. For instance, the Gaussian quadrature approximation of the integral $[-\infty, \infty]$ for a function f(x) becomes

$$\int_{-\infty}^{\infty} f(x)dx \approx \sum_{i=1}^{n} w_i f(x_i).$$

[*Generalized Latent Variable Modeling: Multievel, Longitudinal and Structural Equation Models*, 2004, A. Skrondal and S. Rabe-Hesketh, Chapman and Hall/CRC, Boca Raton.]

Gaussian random field: Synonymous with **Gaussian process**.

Gauss–Markov theorem: A theorem that states that if the error terms in a *multiple regression* have the same variance and are uncorrelated, then the estimators of the parameters in the model produced by *least squares estimation* are better (in the sense of having lower dispersion about the mean) than any other unbiased linear estimator. See also **best linear unbiased estimator**. [MV1 Chapter 7.]

Gauss-Newton method: A procedure for minimizing an objective function that is a sum of squares. The method is similar to the *Newton-Raphson method* but with the advantage of not requiring second derivatives of the function.

Geary's ratio: A test of normality, in which the test statistic is the ratio of the *mean deviation* of a sample of observations to the standard deviation of the sample. In samples from a normal distribution, G tends to $\sqrt{(2/\pi)}$ as n tends to infinity. Aims to detect departures from a *mesokurtic curve* in the parent population. [*Biometrika*, 1947, **34**, 209–42.]

GEE: Abbreviation for **generalized estimating equations**.

Gehan's generalized Wilcoxon test: A *distribution free method* for comparing the survival times of two groups of individuals. See also **Cox–Mantel test** and **log-rank test**. [*Statistics in Medicine*, 1989, **8**, 937–46.]

Geisser, Seymour (1929–2004): Born in New York City, Geisser graduated from the City College of New York in 1950. From New York he moved to the University of North Carolina to undertake his doctoral studies under the direction of *Harold Hotelling*. From 1955 to 1965 Geisser worked at the US National Institutes of Health as a statistician, and from 1960 to 1965 was also a Professorial Lecturer at George Washington University. He made important contributions to *multivariate analysis* and prediction. Geisser died on 11 March 2004.

Gene: A DNA sequence that performs a defined function, usually by coding for an amino acid sequence that forms a protein molecule. [*Statistics in Human Genetics*, 1998, P. Sham, Arnold, London.]

Gene–environment interaction: The interplay of genes and environment on, for example, the risk of disease. The term represents a step away from the argument as to whether nature or nurture is the predominant determinant of human traits, to developing a fuller understanding of how genetic makeup and history of environmental exposures work together to influence an individual's traits. [*The Lancet*, 2001, **358**, 1356–1360.]

Gene frequency estimation: The estimation of the frequency of an *allele* in a population from the *genotypes* of a sample of individuals. [*Statistics in Human Genetics*, 1998, P. Sham, Arnold, London.]

Gene mapping: The placing of *genes* onto their positions on chromosomes. It includes both the construction of marker maps and the localization of genes that confer susceptibility to disease. [*Statistics in Human Genetics*, 1998, P. Sham, Arnold, London.]

General Household Survey: A survey carried out in Great Britain on a continuous basis since 1971. Approximately 100 000 households are included in the sample each year. The main aim of the survey is to collect data on a range of topics including household and family information, vehicle ownership, employment and education. The information is used by government departments and other organizations for planning, policy and monitoring purposes.

General location model: A model for data containing both continuous and categorical variables. The categorical data are summarized by a *contingency table* and their *marginal distribution*, by a *multinomial distribution*. The continuous variables are assumed to have a *multivariate normal distribution* in which the means of the variables are allowed to vary from cell to cell of the contingency table, but with the *variance-covariance matrix* of the variables being common to all cells. When there is a single categorical variable with two categories the model becomes that assumed by *Fisher's linear discriminant analysis*. [*Annals of Statistics*, 1961, **32**, 448–65.]

Generalizability theory: A theory of measurement that recognizes that in any measurement situation there are multiple (in fact infinite) sources of variation (called *facets* in the theory), and that an important goal of measurement is to attempt to identify and measure *variance components* which are contributing error to an estimate. Strategies can then be implemented to reduce the influence of these sources on the measurement. [*Statistical Evaluation of Measurement Errors*, 2004, G. Dunn, Arnold, London.]

Generalized additive mixed models (GAMM): A class of models that uses additive non-parametric functions, for example, *splines*, to model covariate effects while accounting for *overdispersion* and correlation by adding *random effects* to the additive predictor. [*Journal of the Royal Statistical Society, Series B*, 1999, **61**, 381–400.]

Generalized additive models: Models which use smoothing techniques such as *locally weighted regression* to identify and represent possible non-linear relationships between the explanatory and response variables as an alternative to considering polynomial terms or searching for the appropriate transformations of both response and explanatory variables. With these models, the *link function* of the expected value of the response variable is modelled as the sum of a number of smooth functions of the explanatory variables rather than in terms of the explanatory variables themselves. See also **generalized linear models** and **smoothing**. [*Generalized Additive Models*, 1990, T. Hastie and R. Tibshirani, Chapman and Hall/CRC Press, London.]

Generalized distance: See **Mahalanobis** D^2.

Generalized estimating equations (GEE): Technically the multivariate analogue of *quasi-likelihood* with the same feature that it leads to consistent inferences about mean responses without requiring specific assumptions to be made about second and higher order *moments*. Most often used for likelihood-based inference on *longitudinal data* where the response variable cannot be assumed to be normally distributed. Simple models are used for within-subject correlation and a *working correlation matrix* is introduced into the model specification to accommodate these correlations. The procedure provides consistent estimates for the mean parameters even if the covariance structure is incorrectly specified. The method assumes that missing data are *missing completely at random*, otherwise the resulting parameter estimates are biased. An amended approach, *weighted generalized estimating equations*, is available which produces unbiased parameter estimates under the less stringent assumption that missing data are *missing at random*. See also **sandwich estimator**. [*Analysis of Longitudinal Data*, 2nd edition, 2002, P. J. Diggle, P. J. Heagerty, K.-Y. Liang and S. Zeger, Oxford Science Publications, Oxford.]

Generalized gamma distribution: Synonym for **Creedy and Martin generalized gamma distribution**.

Generalized least squares (GLS): An estimator for the regression parameter vector $\boldsymbol{\beta}$ in the multivariate linear regression model

$$\mathbf{y} = \mathbf{X}\boldsymbol{\beta} + \epsilon$$

where $\mathbf{y}$ is the vector of all n responses, $\mathbf{X}$ is a covariate matrix where the n covariate vectors are stacked, and $\boldsymbol{\epsilon}$ a residual error vector. For the general case where $\boldsymbol{\epsilon}$ has a *variance-covariance matrix* $\boldsymbol{\Sigma}$, the generalized least squares estimator of $\boldsymbol{\beta}$ is

$$\hat{\boldsymbol{\beta}}_{GLS} = (\mathbf{X}'\Sigma^{-1}\mathbf{X})^{-1}\mathbf{X}'\Sigma^{-1}\mathbf{Y}$$

In practice Σ is unknown and estimated by the sample covariance matrix, $\hat{\boldsymbol{\Sigma}}$ yielding the *feasible generalized least squares (FGLS) estimator*

$$\hat{\boldsymbol{\beta}}_{FGLS} = (\mathbf{X}'\hat{\Sigma}^{-1}\mathbf{X})^{-1}\mathbf{X}'\hat{\Sigma}^{-1}\mathbf{Y}$$

In the special case where $\Sigma = \sigma^2\mathbf{I}$, the *ordinary least squares (OLS)* estimator is obtained. [*Multivariate Analysis*, 1979, K. V. Mardia, J. T. Kent and J. M. Bibby, Academic, New York.]

Generalized linear mixed models (GLMM): *Generalized linear models* extended to include *random effects* in the linear predictor. See **multilevel models** and **mixed-effects logistic regression**.

Generalized linear models: A class of models that arise from a natural generalization of ordinary linear models. Here some function (the *link function*) of the expected value of the response variable is modelled as a linear combination of the explanatory variables, $x_1, x_2, \ldots, x_q$, i.e.

$$f(E(y)) = \beta_0 + \beta_1 x_1 + \beta_2 x_2 + \cdots + \beta_q x_q$$

where f is the link function. The other components of such models are a specification of the form of the variance of the response variable and of its probability distribution (some member of the *exponential family*). Particular types of model arise from this general formulation by specifying the appropriate link function, variance and distribution. For example, multiple regression corresponds to an identity link function, constant variance and a normal distribution. *Logistic regression* arises from a *logit* link function and a *binomial distribution*; here the variance of the response is related to its mean as, variance $=$ mean$(1 - (\text{mean}/n))$ where n is the number of observations. A *dispersion parameter* (often also known as a *scale factor*), can also be introduced to allow for a phenomenon such as *overdispersion*. For example, if the variance is greater than would be expected from a binomial distribution then it could be specified as ϕmean$(1 - (\text{mean}/n))$. In most applications of such models the scaling factor, ϕ, will be one. Estimates of the parameters in such models are generally found by *maximum likelihood estimation*. See also **GLIM, generalized additive models** and **generalized estimating equations**. [GLM] [*Generalized Latent Variable Modeling*, 2004, A. Skrondal, S. Rabe-Hesketh, Chapman and Hall/CRC Press, Boca Raton.]

Generalized method of moments (GMM): An estimation method popular in *econometrics* that generalizes the *method of moments* estimator. Essentially the same as what is known as *estimating functions* in statistics. The *maximum likelihood estimator* and the *instrumental variables estimator* are special cases. [*Generalized Method of Moments*, 2005, A. R. Hall, Oxford University Press, Oxford]

Generalized multinomial distribution: The joint distribution of n discrete variables $x_1, x_2, \ldots, x_n$ each having the same *marginal distribution*

$$\Pr(x = j) = p_j \quad (j = 0, 1, 2, \ldots, k)$$

and such that the correlation between two different xs has a specified value ρ. [*Journal of the Royal Statistical Society, Series B*, 1962, **24**, 530–4.]

Generalized odds ratio: Synonym for **Agresti's α**.

Generalized p-values: A procedure introduced to deal with those situations where it is difficult to derive a significance test because of the presence of *nuisance parameters*. The central idea is to construct an appropriate generalized *test statistics* and to define an extreme region consisting of all random samples for the testing problem that are as extreme as the observed sample. Some general conditions are assumed satisfied, in particular that the generalized test statistic is free of nuisance parameters. Given the observed sample the generalized p-value is defined to be the largest probability that a sample is in the extreme region under the null hypothesis. The procedure has been shown to provide small sample solutions for many hypothesis testing problems. [*Journal of the American Statistical Association*, 1989, **84**, 602–607.]

Generalized Poisson distribution: A probability distribution defined as follows:

$$\Pr(X = 0) = e^{-\lambda}(1 + \lambda\theta)$$
$$\Pr(X = 1) = \lambda e^{-\lambda}(1 - \theta)$$
$$\Pr(X = x) = \frac{\lambda^x e^{-\lambda}}{x!} (x \geq 2)$$

The distribution corresponds to a situation in which values of a random variable with a *Poisson distribution* are recorded correctly, except when the true value is unity, when there is a non-zero probability that it will be recorded as zero. [*Communication in Statistics: Theory and Methods*, 1984, **10**, 977–991.]

Generalized principal components analysis: A non-linear version of *principal components analysis* in which the aim is to determine the non-linear coordinate system that is most in agreement with the data configuration. For example, for *bivariate data*, y_1, y_2, if a quadratic coordinate system is sought, then as a first step, a variable z is defined as follows:

$$z = ay_1 + by_2 + cy_1y_2 + dy_1^2 + ey_2^2$$

with the coefficients being found so that the variance of z is a maximum among all such quadratic functions of y_1 and y_2. [*Methods of Statistical Data Analysis of Multivariate Observations*, R. Gnanadesikan, 2nd edition, 1997, Wiley, New York.]

Generalized Procrustes analysis: See **Procrustes analysis**.

Generalized variance: See **moments of the generalized variance**.

Genetic algorithms: Optimization procedures motivated by biological analogies. The primary idea is to try to mimic the 'survival of the fittest' rule of genetic mutation in the development of optimization algorithms. The process begins with a population of potential solutions to a problem and a way of measuring the fitness or value of each solution. A new generation of solutions is then produced by allowing existing solutions to 'mutate' (change a little) or cross over (two solutions combine to produce a new solution with aspects of both). The aim is to produce new generations of solutions that have higher values. [*IMA Journal of Mathematics Applied in Business and Industry*, 1997, **8**, 323–46.]

Genetic epidemiology: A science that deals with etiology, distribution, and control of disease in groups of relatives and with inherited causes of disease in populations. [*Outline of Genetic Epidemiology*, 1982, N. E. Morton, Karger, New York.]

Genetic heritability: The proportion of the trait variance that is due to genetic variation in a population. [*Statistics in Human Genetics*, 1998, P. Sham, Arnold, London.]

Genomics: The study of the structure, fuction and evolution of the deoxyribonucleic acid (DNA) or ribonucleic acid (RNA) sequences that comprise the genome of living organisms. Genomics is closely related (and almost synoynmous) to genetics; the former is more directly concerned with DNA structure, function and evolution whereas the latter emphasizes the consequences of genetic transmission for the distribution of heritable traits in families and in populations.

Genotype: The set of *alleles* present at one or more loci in an individual. [*Statistics in Human Genetics*, 1998, P. Sham, Arnold, London.]

Genotypic assortment: See **assortative mating**.

GENSTAT: A general purpose piece of statistical software for the management and analysis of data. The package incorporates a wide variety of data handling procedures and a wide range of statistical techniques including, regression analysis, *cluster analysis*, and *principal components analysis*. Its use as a sophisticated statistical programming language enables non-standard methods of analysis to be implemented relatively easily. [http://www.vsni.co.uk.]

Geographical analysis machine: A procedure designed to detect clusters of rare diseases in a particular region. Circles of fixed radii are created at each point of a square grid covering the study region. Neighbouring circles are allowed to overlap to some fixed extent and the number of cases of the disease within each circle counted. Significance tests are then performed based on the total number of cases and on the number of individuals at risk, both in total and in the circle in question, during a particular census year. See also **scan statistic**. [*Statistics in Medicine*, 1996, **15**, 1961–78.]

Geographical correlations: The correlations between variables measured as averages over geographical units. See also **ecological fallacy**. [*Environmental Research*, 2003, **92**, 78–84.]

Geographical information system (GIS): Software and hardware configurations through which digital georeferences are processed and displayed. Used to identify the geographic or spatial location of any known disease outbreak and, over time, follow its movements as well as changes in *incidence* and *prevalence*. [*Computers and Geoscience*, 1997, **23**, 371–85.]

Geometric distribution: The probability distribution of number of trials (N) before the first success in a sequence of *Bernoulli trials*. Specifically the distribution is given by

$$\Pr(N = n) = p(1-p)^{n-1}, \quad n = 1, 2, \ldots$$

where p is the probability of a success on each trial. The mean, variance, *skewness* and *kurtosis* of the distribution are as follows:

$$\text{mean} = 1/p$$
$$\text{variance} = (1-p)/p^2$$
$$\text{skewness} = (2-p)/(1-p)^{\frac{1}{2}}$$
$$\text{kurtosis} = 9 + p^2/(1-p)$$

[STD Chapter 19.]

Geometric mean: A measure of location, g, calculated from a set of observations $x_1, x_2, \ldots, x_n$ as

$$g = \left(\prod_{j=1}^{n} x_j \right)^{\frac{1}{n}}$$

The geometric mean is always less than or equal to the arithmetic mean. [SMR Chapter 3.]

Geostatistics: A body of methods useful for understanding and modelling the spatial variability in a process of interest. Central to these methods is the idea that measurements taken at locations close together tend to be more alike than values observed at locations farther apart. See also **kriging** and **variogram**. [*Practical Geostatistics*, 1979, I. Clark, Applied Science, London.]

G-estimator: Estimator for causal effects, for instance in *structural nested models*, based on the notion of *potential outcomes* (see *Neyman-Rubin causal framework*). Can be viewed as a generalization of *standardization*. [*Epidemiology*, 1992, **3**, 319–336.]

GGobi: Free software for interactive and dynamic graphics that can be operated using a command line interface or from a graphical user interface. [*Computational Statistics and Data Analysis*, 2003, **43**, 423–444.]

Gibbs sampling: See **Markov chain Monte Carlo methods**.

Gini concentration: A measure of spread, $V_G(X)$, for a variable X taking k categorical values and defined as

$$V_G(X) = \sum_{i=1}^{k} \pi_i(1 - \pi_i)$$

where $\pi_i = \Pr(X = i),\ i = 1, \ldots, k$. The statistic takes its minimum value of zero when X is least spread, i.e., when $\Pr(X=j)=1$ for some category j, and its maximum value $(k - 1)/k$ when X is most spread, i.e., when $\Pr(X = i) = 1/k$ for all i. [*Statistical Analysis of Categorical Data*, 1999, C. J. Lloyd, Wiley, New York.]

Gini, Corrado (1884–1965): Born in a small town in the region of Veneto, Italy, Gini studied jurisprudence at the University of Bologna before becoming interested in statistics. His thesis for his degree became, in 1908, the first of his eighty books, *Il sesso dal punto di vista statistico*. At the age of 26 Gini was already a university professor and throughout his life held chairs in the Universities of Cagliari, Padua and Rome. Founded two journals *Genus* and *Metron* and wrote over a thousand scientific papers in the areas of probability, demography and biometry. Elected an Honarary Fellow of the Royal Statistical Society in 1920, Gini was also a member of the Academia dei Lincei. He died on 13 March 1965.

Gini index: See **Lorenz curve**.

GIS: Abbreviation for **geographical information system**.

Gittins indices: Synonym for **dynamic allocation indices**.

Glejser test: A test for heteroscedasticity in the error terms of a regression analysis that involves regressing the absolute values of regression residuals for the sample on the values of the independent variable thought to covary with the variance of the error terms. See also **Goldfield–Quandt test**. [*Regression Analysis*, Volume 2, ed. M. S. Lewis-Beck, Sage Publications, London.]

GLIM: A software package particularly suited for fitting *generalized linear models* (the acronym stands for Generalized Linear Interactive Modelling), including *log-linear models*, *logistic models*, and models based on the *complementary log-log transformation*. A large number of GLIM macros are now available that can be used for a variety of non-standard statistical analyses. [NAG Ltd, Wilkinson House, Jordan Hill Road, Oxford OX2 8DR, UK; NAG Inc., 1400 Opus Place, Suite 200, Downers Grove, Illinois 60515-5702, USA.]

Glivenko-Cantelli theorem: A theorem stating that the empirical distribution function for a random variable x converges, for all x, almost surely to the true distribution function as the number of independent identically distributed observations increases. [*Annals of Probability*, 1987, **15**, 837–870.]

GLLAMM: A program that estimates generalized linear latent and mixed models by maximum likelihood. The models that can be fitted include *multi-level models, structural equation models* and *latent class models*. The response variables can be of mixed types including continuous, counts, survival times, dichotomous, and ordinal. [Details and manuals available at http://www.gllamm.org/]

GLM: Abbrevation for **generalized linear model**.

GLMM: Abbreviation for **generalized linear mixed models**.

GLS: Abbreviation for **generalized least squares.**

Glyphs: A general term for the graphics that are sometimes used to represent multivariate observations. Two examples are *Chernoff faces* and *snowflakes*.

Gnedenko, Boris (1912–1995): Born on 1 January 1912 in Simbirsk a town on the River Volga, Gnedenko studied at the University of Saratov. In 1934 he joined the Institute of Mathematics at Moscow State University and studied under Khinchin and later Kolmogorov. In 1938 he became associate professor in the Department of Mechanics and Mathematics. Gnedenko's main work was on various aspects of theoretical statistics particularly the limiting distribution of maxima of independent and identically distributed random variables. The first edition of what is widely regarded as his most important published work, *Limit Distributions for Sums of Independent Random Variables* appeared in 1949. In 1951 Gnedenko published *The Theory of Probability* which remained a popular account of the topic for students for over a decade. Later in his career Gnedenko took an interest in *reliability theory* and *quality control procedures* and played a role in the modernization of Soviet industry including the space programme. He also devoted much time to popularizing mathematics. Gnedenko died in Moscow on 27 December 1995.

Goelles, Josef (1929–2000): Goelles studied mathematics, physics and psychology at the University of Graz, and received a Ph.D in mathematics in 1964. He began his scientific career at the Technical University of Graz and in 1985 founded the Institute of Applied Statistics and Systems Analysis at Joanneum Research and chaired it until shortly before his death. He collaborated in a large number of joint projects with clinicians and biologists and tried hard to convince other colleagues to incorporate statistical reasoning into their own discipline.

Golden–Thompson inequality: An inequality relating to the *matrix exponential transformation* and given by

$$\text{trace}[\exp(\mathbf{A})\exp(\mathbf{B})] \geq \text{trace}[\exp(\mathbf{A}+\mathbf{B})]$$

with equality if and only if **A** and **B** commute. [*Linear Algebra and its Applications*, 1996, **243**, 105–12.]

Goldfeld–Quandt test: A test for heteroscedasticity in the error terms of a regression analysis that involves examining the monotonic relationship between an explanatory variable and the variance of the error term. See also **Glejser test**. [*Review of Agricultural Economics*, 2000, **22**, 421–37.]

Gold standard trials: A term usually retained for those *clinical trials* in which there is random allocation to treatments, a control group and double-blinding.

Golub, Gene Howard (1932–2007): Golub was born in Chicago in 1932. After taking his PhD at the University of Illinois he arrived at Stanford University in 1965. Gloub became a founding member of the Computer Science Department in 1962 and remained at Stanford throughout his career. Golub made fundamental contributions to computational linear algebra. In particular, he transformed the field of matrix computations, his most famous achievements being associated with the *singular value decomposition.* Golub served as president of the Society for Industrial and Applied Mathematics (SIAM) and was founding editor of two SIAM journals. He died at Stanford on November 16th 2007.

Gompertz curve: A curve used to describe the size of a population (y) as a function of time (t), where relative growth rate declines at a constant rate. Explicitly given by

$$y = ae^{-b^t}$$

[MV2 Chapter 13.]

Good, Irving John (1916–2009): Born in London, England, Good was a child prodigy in mathematics and won a mathematics scholarship to Jesus College, Cambridge from where he graduated in 1938 with a first. He obtained his in Ph.D in 1941 for a thesis on topological dimension. During the Second World War Good worked at Bletchley Park on deciphering German military and naval radio traffic. After the war he moved to the University of Manchester and then to Trinity College, Oxford. In 1967 Good moved to the USA as a professor of statistics at Virginia Tech where he remained until his retirement in 1994. Good made fundamental and far reaching contributions to several areas of statistics particularly *Bayesian statistics.* He died on April 5th, 2009.

Goodman and Kruskal measures of association: Measures of associations that are useful in the situation where two categorical variables cannot be assumed to be derived from perhaps unobservable continuous variables and where there is no natural ordering of interest. The rationale behind the measures is the question, 'how much does knowledge of the classification of one of the variables improve the ability to predict the classification on the other variable'. [*The Analysis of Contingency Tables*, 2nd edition, 1992, B. S. Everitt, Chapman and Hall/CRC Press, London.]

Goodness-of-fit statistics: Measures of the agreement between a set of sample observations and the corresponding values predicted from some model of interest. Many such measures have been suggested; see *chi-squared statistic, deviance, likelihood ratio,* G^2 and X^2. [SMR Chapter 8.]

Good's method: A procedure for combining independent tests of hypotheses. [*Probability and the Weighing of Evidence*, 1950, I. J. Good, Hafner, New York.]

Gossett, William Sealy (1876–1937): Born in Canterbury, England, Gossett obtained a degree in chemistry at Oxford before joining Guinness breweries in Dublin in 1899. He continued to work for Guinness for the next three decades. Practical problems in his work led him to seek exact error probabilities of statistics from small samples, a previously un-researched area. Spent the academic year 1906–1907 studying under *Karl Pearson* at University College, London. Writing under the pseudonym of 'Student' (as required by Guinness) his paper essentially introducing the *Student's t-distribution* was published in *Biometrika* in 1908. Gossett died in London on 16 October 1937.

Goulden, Cyril Harold (1897–1981): Born in Bridgend, Wales, Goulden's family moved to Canada and he graduated from the University of Saskatchewan in 1921. Between 1925 and 1930 he taught himself mathematics and statistical methods and began to apply methods developed by *Gosset* and *Pearson* in agricultural experiments as part of his job at the Dominion Rust Research Laboratory in Winnipeg. In 1930 Goulden went to work at Rothamsted Experimental Station and met *Fisher.* His book, *Methods of Statistical Analysis*, became a standard textbook in Canada and the USA. In 1958 he acted as President of the Biometrics Society. Goulden died on February 3rd, 1981 in Ottawa, Canada.

Gower's similarity coefficient: A *similarity coefficient* particularly suitable when the measurements contain both continuous variables and categorical variables. [MV2 Chapter 10.]

Grade of membership model: A general *distribution free method* for the clustering of multivariate data in which only categorical variables are involved. The model assumes that individuals can exhibit characteristics of more than one cluster, and that the state of an individual can be represented by a set of numerical quantities, each one corresponding to one of the clusters, that measure the 'strength' or grade of membership of the individual for the cluster. Estimation of these quantities and the other parameters in the model is undertaken by *maximum likelihood estimation.* See also **latent class analysis** and **fuzzy set theory**. [*Statistical Applications of Fuzzy Sets*, 1994, K. G. Manton, M. A. Woodbury and H. D. Tolley, Wiley, New York.]

Graduation: A term employed most often in the application of *actuarial statistics* to denote procedures by which a set of observed probabilities is adjusted to provide a suitable basis for inferences and further practical calculations to be made. [*The Analysis of Mortality and Other Actuarial Statistics*, 3rd edition, 1993, B. Benjamin and J. H. Pollard, Institute and Faculty of Actuaries, Oxford.]

Graeco-Latin square: An extension of a *Latin square* that allows for three extraneous sources of variation in an experiment. A three-by-three example of such a square is

$$\begin{array}{ccc} A\alpha & B\beta & C\gamma \\ B\gamma & C\alpha & A\beta \\ C\beta & A\gamma & B\alpha \end{array}$$

[*The Mathematics of Experimental Designs; Incomplete Block Designs and Latin Squares*, S. Vajda, 1967, Charles Griffin, London.]

Gram–Charlier Type A series: An expansion of a probability distribution, $f(x)$ in terms of *Chebyshev–Hermite polynomials*, $H_r(x)$. Given explicitly by

$$f(x) = \alpha(x)\left\{1 + \frac{1}{2}(\mu_2 - 1)H_2 + \frac{1}{6}\mu_3 H_3 + \frac{1}{24}(\mu_4 - 6\mu_2 + 3)H_4 + \cdots\right\}$$

where

$$\alpha(x) = \frac{1}{\sqrt{2\pi}}e^{-\frac{1}{2}x^2}$$

See also **Edgeworth's form of the Type A series**. [KA1 Chapter 6.]

Gramian matrix: A symmetric matrix, **A**, whose elements are real numbers and for which there exists a matrix **B** also consisting of real numbers, such that **BB′=A** or **B′B=A**. An example is a *correlation matrix.*

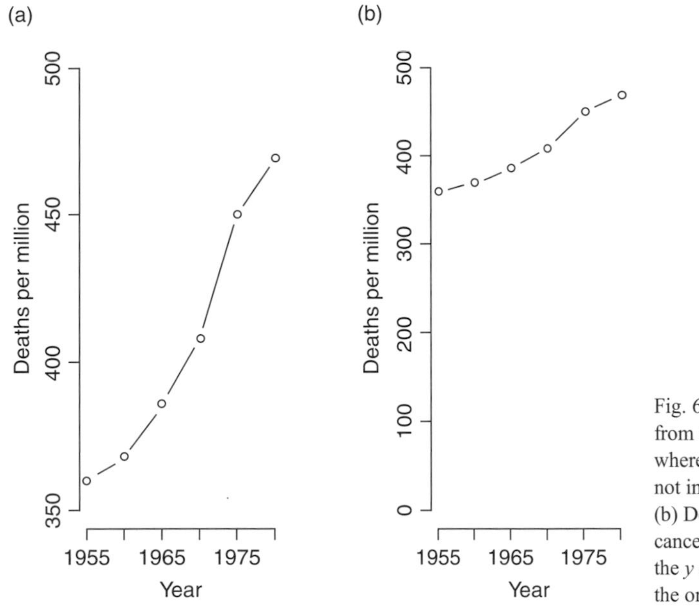

Fig. 68 (a) Death rates from cancer of the breast where the y-axis does not include the origin. (b) Death rates from cancer of the breast when the y axis does include the origin.

Grand mean: Mean of all the values in a grouped data set irrespective of groups.

Granger causality: A *time series* X is said to Granger-cause a time series Y if lagged values of X predict future values of Y, given the lagged values of Y. [*Econometrica*, 1969, **37**, 424–438.]

Granger, Sir Clive William John (1934–2009): Granger was born in Swansea, Wales, in 1934. He developed a strong interest in mathematics as a child but ambled into a career in statistics haphazardly. As a grammar school student in 1946, his teacher instructed him and his classmates to stand up one day and announce the careers they wanted. "I preferred to use mathematics in some practical fashion and thought that meteorology sounded promising," he wrote in his Nobel biography. "In those days I stuttered somewhat and when my turn came to stand up, I tried to say 'meteorology' but found I could not get the word out, so I just said 'statistics,' thereby determining my future path." Granger went on to the University of Nottingham, where he earned a bachelor's degree in mathematics in 1955 and a Ph.D. in statistics in 1959, publishing his first statistical paper on sunspot activity. He worked at Nottingham until the early 1970s when he moved to the University of California at San Diego, where he built one of the world's most prominent econometrics programs. Granger was awarded the Nobel Prize in economics in 2003 for his contributions to the analysis of *nonstationary time-series.*

Graphical deception: *Statistical graphics* that are not as honest as they should be. It is relatively easy to mislead the unwary with graphical material. For example, consider the plot of the death rate per million from cancer of the breast, for several periods over the last three decades, shown in Figure 68(a). The rate appears to show a rather alarming increase. However, when the data are replotted with the vertical scale beginning at zero, as shown in Figure 68(b), the increase in the breast cancer death rate is altogether less startling. This example illustrates that undue exaggeration or compression of the scales is best avoided when drawing graphs. For another example of a common graphical distortion see *lie factor.* [*Visual Display of Quantitative Information*, 1983, E. R. Tufte, Graphics Press, Cheshire, Connecticut.]

Graph theory: A branch of mathematics concerned with the properties of sets of points (vertices or nodes) some of which are connected by lines known as edges. A *directed graph* is one in which direction is associated with the edges and an *undirected graph* is one in which no direction is involved in the connections between points. A graph may be represented as an *adjacency matrix*. See also **conditional independence graph**. [*Graphs and Digraphs*, 1979, M. Behzad, G. Chartrand and L. Lesniak-Foster, Prindle, Weber and Schmidt, Boston.]

Graunt, John (1620–1674): The son of a city tradesman, Graunt is generally regarded as having laid the foundations of *demography* as a science with the publication of his *Natural and Political Observations Made Upon the Bills of Mortality* published in 1662. His most important contribution was the introduction of a rudimentary *life table*. Graunt died on 18 April, 1674 in London.

Greatest characteristic root test: Synonym for **Roy's largest root criterion**.

Greenberg, Bernard George (1919–1985): Born in New York City, Greenberg obtained a degree in mathematics from the City College of New York in 1939. Ten years later after a period of military service he obtained a Ph.D. from the North Carolina State University where he studied under *Hotelling*. Greenberg founded the Department of Biostatistics at the University of North Carolina and was a pioneer in the field of public health and medical research. He died on 24 November 1985 in Chapel Hill.

Greenhouse–Geisser correction: A method of adjusting the degrees of freedom of the within-subject *F-tests* in the *analysis of variance* of *longitudinal data* so as to allow for possible departures of the *variance–covariance matrix* of the measurements from the assumption of *sphericity*. If this condition holds for the data then the correction factor is one and the simple F-tests are valid. Departures from *sphericity* result in an estimated correction factor less than one, thus reducing the degrees of freedom of the relevant F-tests. See also **Huynh–Feldt correction** and **Mauchly test**. [MV2 Chapter 13.]

Greenhouse, Samuel (1918–2000): Greenhouse began his career at the National Institutes of Health in the National Cancer Institute. Later he become Chief of the Theoretical Statistics and Mathematics Section in the National Institute of Mental Health. After 1974 Greenhouse undertook a full time academic career at George Washington University. He was influential in the early development of the theory and practice of *clinical trials* and his work also included the evaluation of diagnostic tests and the analysis of *repeated measure designs*. Greenhouse died on 28 September 2000 in Rockville, MD, USA.

Greenwood, Major (1880–1949): Born in the East End of London, Greenwood studied medicine at University College, London and the London Hospital, but shortly after qualifying forsook clinical medicine and following a period of study with Karl Pearson was, in 1910, appointed statistician to the Lister Institute. Here he carried out statistical investigations into such diverse topics as the fatality of fractures and pneumonia in hospital practice, the epidemiology of plague and factors influencing rates of infant mortality. In 1919 he became Head of Medical Statistics in the newly created Ministry of Health where he remained until 1928, when he was appointed to the chair of Vital Statistics and Epidemiology at the London School of Hygiene and Tropical Medicine. Here he remained until his retirement in 1945. Greenwood was President of the Royal Statistical Society from 1934 to 1936 and was awarded their Guy medal in gold in 1945. He died on 5 October 1949.

Greenwood's formula: A formula giving the variance of the *product limit estimator* of a *survival function*, namely

$$\text{var}(\hat{S}(t)) = [\hat{S}(t)]^2 \sum_{j|t_{(j)} \le t} \frac{d_j}{r_j(r_j - d_j)}$$

where $\hat{S}(t)$ is the estimated *survival function* at time, t, $t_{(1)} < t_{(2)} < \cdots < t_{(n)}$ are the ordered, observed *survival times*, r_j is the number of individuals at risk at time $t_{(j)}$, and d_j is the number who experience the event of interest at time t_j. (Individuals censored at $t_{(j)}$ are included in r_j.) [*Modelling Survival Data in Medical Research*, 2nd edn, 2003, D. Collett, Chapman and Hall/CRC Press, London.]

Gripenberg estimator: A *distribution free* estimator for the *partial correlation* between two variables, X and Y, conditional on a third variable, Z. [*Journal of the American Statistical Association*, 1992, **87**, 546–51.]

Group average clustering: Synonym for **average linkage clustering**.

Group divisible design: An arrangement of $v = mn$ treatments in b blocks such that:

- each block contains k distinct treatments, $k < v$;
- each treatment is replicated r times;
- the treatments can be divided into m groups of n treatments each, any two treatments occurring together in λ_1 blocks if they belong to the same group and in λ_2 blocks if they belong to different groups. [*Biometrika*, 1976, **63**, 555–8.]

Grouped binary data: Observations on a binary variable tabulated in terms of the proportion of one of the two possible outcomes amongst patients or subjects who are, for example, the same diagnosis or same sex, etc. [*SORT*, 2004, **28**, 125–60.]

Grouped data: Data recorded as frequencies of observations in particular intervals.

Group matching: See **matching**.

Group sequential design: See **sequential design**.

Growth charts: Synonym for **centile reference charts**.

Growth curve: See **growth curve analysis**.

Growth curve analysis: A general term for methods dealing with the development of individuals over time. A classic example involves recordings made on a group of children, say, of height or weight at particular ages. A plot of the former against the latter for each child gives the individual's *growth curve*. Traditionally low-degree polynomials are fitted to such curves, and the resulting parameter estimates used for inferences such as comparisons between boy and girls. See also **multilevel models** and **random** effects. [*An Introduction to Latent Variable Growth Curve Modeling: Concepts, Issues and Applications*, 1999, S. C. Duncan and L. A. Strycker, Lawrence Erlbaum, Mahwah, NJ.]

Growth rate: A measure of population growth calculated as

$$\frac{\text{live births during the year} - \text{deaths during the year}}{\text{midyear population}} \times 100$$

Grubb's estimators: Estimators of the measuring precisions when two instruments or techniques are used to measure the same quantity. For example, if the two measurements are denoted by x_i and y_i for $i = 1,\ldots,$ n, we assume that

$$x_i = \tau_i + \epsilon_i$$
$$y_i = \tau_i + \delta_i$$

where τ_i is the correct unknown value of the ith quantity and ϵ_i and δ_i are measurement errors assumed to be independent, then Grubb's estimators are

$$\hat{V}(\tau_i) = \text{covariance}(x, y)$$
$$\hat{V}(\epsilon_i) = \text{variance}(x) - \text{covariance}(x, y)$$
$$\hat{V}(\delta_i) = \text{variance}(y) - \text{covariance}(x, y)$$

[*Technometrics*, 1973, **15**, 53–66.]

GTdistribution: A probability distribution, $f(x)$, related to *Student's t-distribution* and given by

$$f(x) = \frac{p}{2\sigma q^{1/p} B(p^{-1}, q)} \left[1 + \frac{|x|^p}{(q\sigma^p)}\right]^{q+p^{-1}}$$

where B is the *beta function.*

In $f(x)$, σ is a *scale parameter* while p and q control the shape of the distribution. Larger values of p and q are associated with lighter tails for the distribution. When $p = 2$ and $\sigma = \sqrt{2}\alpha$, this becomes Student's t-distribution with $2q$ degrees of freedom. [*Continuous Univariate Distributions*, Volume 2, 2nd edition, 1995, N. L. Johnson, S. Kotz and N. Balakrishnan, Wiley, New York.]

Guarantee time: See **two-parameter exponential distribution**.

Gumbel distribution: Synonym for **extreme value distribution**.

Gumbel, Emil Julius (1891–1966): Born in Munich, Germany, Gumbel obtained a Ph.D. in economics and mathematical statistics from the University of Munich in 1914. Between 1933 and 1940 he worked in France, first at the Institut Henri Poincaré in Paris, later at the University of Lyon. Gumbel made important contributions to the theory of *extreme values.*

Gupta's selection procedure: A method which, for given samples of size n from each of k normal populations, selects a subset of the k populations which contains the population with the largest mean with some minimum probability. [*Communications in Statistics: Theory and Methods*, 1997, **26**, 1291–1312.]

Guttman, Louis (1916–1987): Born in Brooklyn, New York, Guttman received his Ph.D in sociology in 1942 from the University of Minnesota with a thesis on the algebra of *factor analysis.* In 1947 he emigrated to Israel where he founded and directed the Israel Institute of Applied Social Research, later to be known as the Guttman Institute of Applied Social Research. From 1955 he also became Professor of Social and Psychological Assessment at the Hebrew University of Jerusalem. Guttman's greatest contributions to statistics were in the area of scaling theory and factor analysis. He died on October 25th, 1987 in Minneapolis, Minnesota, USA.

Guttman scale: A scale based on a set of *binary variables* which measure a one-dimensional *latent variable.* The basic idea is that k binary items are ordered such that a subject who answers "Yes" to item j ($< k$) will probably answer "Yes" to items $j + 1, \ldots, k$. See also **Cronbach's alpha**. [*Scale Development Theory and Applications*, 1991, R. F. Devellis, Sage, Beverly Hills.]

Guy, William Augustus (1810–1885): Born in Chichester, Guy studied medicine at both Christ's Hospital and Guy's Hospital, London. In 1831 he was awarded the Fothergillian medal of the Medical Society of London for the best paper on asthma. In 1838 Guy was appointed to the Chair of Forensic Medicine at King's College, London. Guy, like *Farr*, was strongly of the opinion that statistics was seriously needed for the study of medical problems, and his contribution to statistics rests primarily on the compilation of bodies of material relating to public health. Guy was a very active member of the Statistical Society of London and because of his work on behalf of the Society, the Royal Statistical Society voted in 1891 to establish in his honour the Guy medal. He was President of the Royal Statistical Society from 1873 to 1875. Guy died on 10 September 1885, in London.

H_0: Symbol for **null hypothesis**.

H_1: Symbol for **alternative hypothesis**.

Haavelmo, Trygve Magnus (1911–1999): Haavelmo was born close to Oslo, Norway in 1911. He went to the United States in 1939 as a Fulbright scholar where he remained until 1947 and was then appointed Professor of Economics and Statistics at the University of Oslo in 1948. During his stay in the United States Haavelmo completed his highly influential work on 'The Probability Approach to Econometrics' where he argued that economic data should be envisioned as being samples selected by Nature, randomly drawn from hypothetical distributions which governed reality but which were unobservable. Haavelmo was awarded the Nobel Memorial Prize in Economics in 1989. He died in Oslo in 1999.

Haar wavelet: See **wavelet analysis**.

Hadamard matrix: An $n \times n$ matrix $\mathbf{H}_n$ consisting entirely of ± 1s and such that $\mathbf{H}_n' \mathbf{H}_n = \text{diag}(n, n, \ldots, n)$. Important in *response surface methodology*. [*Annals of Statistics*, 1978, **6**, 1184–238.]

Haenszel, William (1910–1998): Born in Rochester, New York, Haenszel received a B.A. degree in sociology and mathematics from the University of Buffalo in 1931, and an M.A. degree in statistics in 1932. For the next 20 years he worked first as a statistician at the New York State Department of Health and then as Director of the Bureau of Vital Statistics at the Connecticut State Department of Health. In 1952 Haenszel became Head of the Biometry Section of the National Cancer Institute where he stayed until his retirement in 1976. Haenszel made important contributions to epidemiology including the *Mantel–Haenszel test*. He died on 13 March 1998, in Wheaton, Illinois.

Hajek, Jaroslav (1926–1974): Born in Podebrady, Czechoslovakia, Hajek studied statistical engineering at the College of Special Sciences at the Czech Technical University and graduated in 1950. In 1963 he was awarded a doctorate for his thesis on statistical problems in *stochastic processes*. In 1966 Hajek became professor in the Faculty of Mathematics and Physics of Charles University. His major contributions to statistics were in the fields of stochastic processes and rank tests. Hajek died on June 10th, 1974 in Prague, Czechoslovakia.

Hajnal, John (1924–2008): Born in Darmstadt, Garmany, Hajnal was educated at a Quaker school in Holland after his family left Germany in 1936. Later they moved to London where Hajnal attended University College School, Hampsted and at 16 went on to Oxford to read politics, philosophy and economics gaining a first in 1943. In 1948 he was recruited to work on demography for the UN in New York. In the early 1950s Hajnal left full-time employment to study academic mathematics and in 1953 returned to England to become a medical statistician in Manchester. In 1956 he obtained a lectureship at the London School of Economics eventually becoming professor in 1975. His research interests were a mixture of the practical in demography and theoretical on *Markov chains* and *stochastic processes*. Hajnal died on November 30th, 2008.

Hald, Anders (1913–2007): Born in Jutland, Denmark, Hald studied actuarial science and statistics at the University of Copenhagen from where he graduated in 1939. In 1948 he was awarded a D.Phil, his dissertation involving identifying a trend, seasonal variation and noise in a *time series*. Also in 1948 Hald became professor of statistics in the Social Science Faculty of the University of Copenhagen and in 1960 took up a new chair of mathematical statistics at the university from where he developed mathematical statistics in Denmark. His main contributions to statistics were in the areas of sampling plans and the history of the discipline. Hald died on November 11th, 2007.

Haldane's estimator: An estimator of the *odds ratio* given by

$$\hat{\psi} = \frac{(a + \frac{1}{2})(d + \frac{1}{2})}{(b + \frac{1}{2})(c + \frac{1}{2})}$$

where a, b, c and d are the cell frequencies in the *two-by-two contingency table* of interest. See also **Jewell's estimator**. [*Journal of the Royal Statistical Society, Series D*, 2001, **50**, 309–19.]

Half-mode: A term sometimes used for the mode of a probability distribution or a frequency distribution if this occurs at an extreme value of the variable, for example, in the case of a *J-shaped distribution*.

Half-normal distribution: Synonym for **folded normal distribution**.

Half-normal plot: A plot for diagnosing model inadequacy or revealing the presence of *outliers*, in which the absolute values of, for example, the *residuals* from a *multiple regression* are plotted against the quantiles of the standard normal distribution. Outliers will appear at the top right of the plot as points that are separated from the others, while systematic departures from a straight line could indicate that the model is unsatisfactory. [*Principles of Multivariate Analysis*, 2nd edition, 2000, W. J. Krzanowski, Oxford Science Publications, Oxford.]

Halo effect: The tendency of a subject's performance on some task to be overrated because of the observer's perception of the subject 'doing well' gained in an earlier exercise or when assessed in a different area. [*Journal of Applied Psychology*, 1920, **IV**, 25–9.]

Halperin, Max (1917–1988): Born in Omaha, Nebraska, Halperin graduated from the University of Omaha in 1940 with a degree in mathematics. In 1950 he obtained a Ph.D. degree from the University of North Carolina. His career began as a research mathematician at the RAND corporation and posts at the National Institutes of Health and the Division of Biologic Standards followed. Later he became Research Professor of Statistics and Director of the Biostatistics Center of the Department of Statistics at the George Washington University. Halperin made important contributions in many areas of statistics including *multivariate analysis*, regression analysis, *multiple comparisons* and the detection of *outliers*. He died on 1 February 1988 in Fairfax, Virginia.

Hanging rootogram: A diagram comparing an observed *rootogram* with a fitted curve, in which differences between the two are displayed in relation to the horizontal axis, rather than to the curve itself. This makes it easier to spot large differences and to look for patterns. An example is given in Fig. 69. [*Exploratory Data Analysis*, 1977, J. W. Tukey, Addison-Wesley, Reading, MA.]

Hankel matrix: A *variance-covariance matrix* between past and present values of a *time series* with some future values of the series at some time t. [*Journal of the American Statistical Association*, 1998, **93**, 770–82.]

Hannan, Edward James (1921–1994): Having obtained a commerce degree at the University of Melbourne, Hannan began his career as an economist in the Reserve Bank of Australia. When spending a year at the Australian National University in 1953 he was 'spotted' by *P. A. P. Moran* and began his research on various aspects of the analysis of time series which eventually

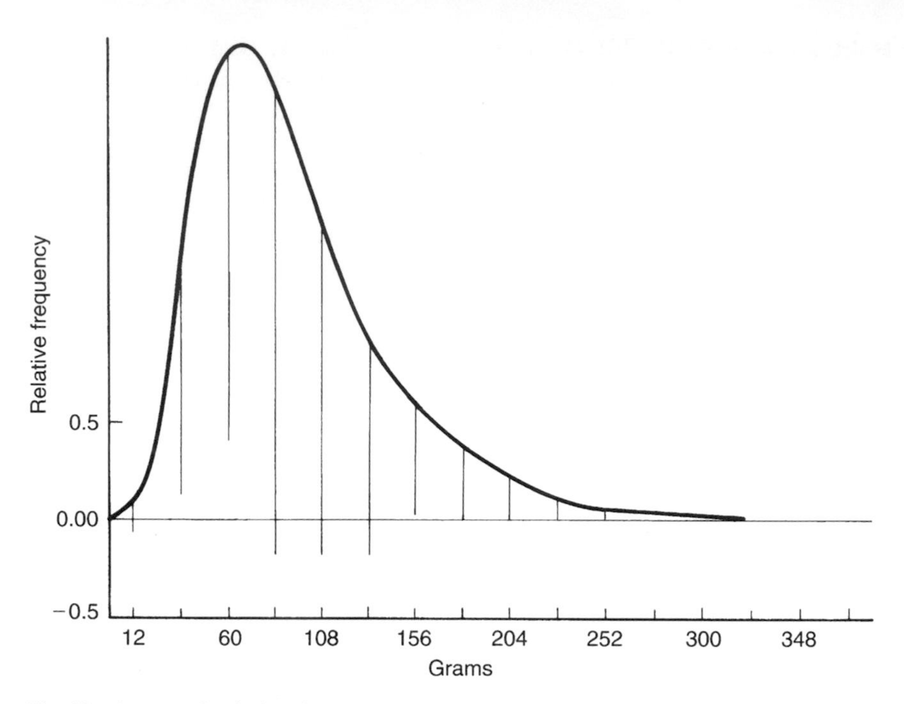

Fig. 69 An example of a hanging rootogram.

brought him international recognition. Hannan received many honours including the Lyle medal for research in mathematics and physics from the Australian Academy of Science and the Pitman medal from the Statistical Society of Australia. He died on 7 January 1994.

Hansen–Hurwitz estimator: An unbiased estimator of the total size of a population given by

$$\hat{\tau} = \frac{1}{n}\sum_{i=1}^{n}\frac{y_i}{p_i}$$

where n is the number of *sampling units* (often regions or areas in this context), y_i is the number of individuals, animals or species, observed in the ith sampling unit, and p_i is the probability of selecting the ith unit for $i = 1, 2, \ldots, N$ where N is the number of population units. (When sampling is with replacement $n = N$.) An unbiased estimator of the variance of $\hat{\tau}$ is

$$\hat{\mathrm{var}}(\hat{\tau}) = \frac{1}{n(n-1)}\sum_{i=1}^{n}\left(\frac{y_i}{p_i} - \hat{\tau}\right)^2$$

See also **Horvitz–Thompson estimator**. [*Experimental and Ecological Statistics*, 2003, **10**, 115–127.]

Hansen, Morris Howard (1910–1990): Born in Thermopolis, Wyoming, Hansen studied accounting at the University of Wyoming obtaining a B.S. degree in 1934. His formal training in statistics consisted of after-hours classes taken at the Graduate School of the U.S. Department of Agriculture. Hansen received a master's degree in statistics in 1940. In 1941 he joined the Census Bureau where he developed the mathematical theory underlying sampling methods. This resulted in the publication in 1953 of the standard reference work, *Sample Survey Methods and Theory*. Elected a member of the National Academy of Sciences in 1976, Hansen died on 9 October 1990.

Haplotype: A combination of two or more *alleles* that are present in the same gamete. [*Statistics in Human Genetics*, 1998, P. Sham, Arnold, London.]

Haplotype analysis: The analysis of *haplotype* frequencies in one or more populations, with the aim of establishing associations between two or more alleles, between a haplotype and a phenotypic trait, or the genetic relationship between populations. [*Statistics in Human Genetics*, 1998, P. Sham, Arnold, London.]

Hardy–Weinberg law: The law stating that both gene frequencies and genotype frequencies will remain constant from generation to generation in an infinitely large interbreeding population in which mating is at random and there is no selection, migration or mutation. In a situation where a single pair of alleles (A and a) is considered, the frequencies of germ cells carrying A and a are defined as p and q, respectively. At equilibrium the frequencies of the genotype classes are $p^2(AA)$, $2pq(Aa)$ and $q^2(aa)$. [*Statistics in Medicine*, 1986, **5**, 281–8.]

Harmonic analysis: A method of determining the period of the cyclical term S_t in a *time series* of the form

$$X_t = S_t + \epsilon_t$$

where ϵ_t represents the random fluctuations of the series about S_t. The cyclical term is represented as a sum of sine and cosine terms so that X_t becomes

$$X_t = \sum_i (A_i \cos \omega_i t + B_i \sin \omega_i t) + \epsilon_t$$

For certain series the periodicity of the cyclical component can be specified very accurately, as, for example, in the case of economic or geophysical series which contain a strict 12-month cycle. In such cases, the coefficients $\{A_i\}$, $\{B_i\}$ can be estimated by regression techniques. For many series, however, there may be several periodic terms present with unknown periods and so not only the coefficients $\{A_i\}$, $\{B_i\}$ have to be estimated but also the unknown frequencies $\{\omega_i\}$. The so-called *hidden periodicities* can often be determined by examination of the *periodogram* which is a plot of $I(\omega)$ against ω where

$$I(\omega) = \frac{2}{N}\left\{\left[\sum_{i=1}^{N} X_t \cos(\omega t)\right]^2 + \left[\sum_{i=1}^{N} X_t \sin(\omega t)\right]^2\right\}$$

and $\omega = 2\pi p/N$, $p = 1, 2, \ldots, [N/2]$; N is the length of the series. Large ordinates on this plot indicate the presence of a cyclical component at a particular frequency. As an example of the application of this procedure Fig. 70 shows the sunspot series and the

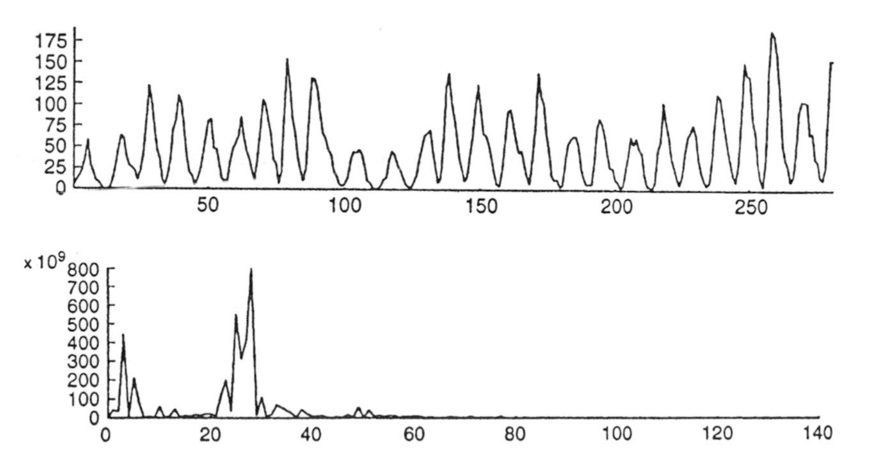

Fig. 70 Sunspot activity and its periodogram (Reproduced from *Applications of Time Series Analysis in Astronomy and Meteorology* by permission of the publisher, Chapman and Hall/CRC Press London.)

periodogram of this series based on 281 observations. It is clear that the ordinate at $\omega = 2\pi \times 28/281$, corresponding to a period of approximately 10 years, is appreciably larger than the other ordinates. If several peaks are observed in the periodogram it cannot be concluded that each of these corresponds to a genuine periodic component in X_t since it is possible that peaks may occur due simply to random fluctuations in the noise term ϵ_t. Various procedures for formally assessing periodic ordinates are available of which the most commonly used are *Schuster's test*, *Walker's test* and *Fisher's g statistic*. See also **spectral analysis, fast Fourier transform** and **window estimates**. [TMS Chapter 7.]

Harmonic mean: The reciprocal of the arithmetic mean of the reciprocals of a set of observations $x_1, x_2, \ldots, x_n$. Specifically obtained from the equation

$$\frac{1}{H} = \frac{1}{n}\sum_{i=1}^{n}\frac{1}{x_i}$$

Used in some methods of analysing *non-orthogonal designs*. The harmonic mean is either smaller than or equal to the arithmetic mean and the *geometric mean*. [ARA Chapter 16.]

Harrington and Fleming G^p tests: A class of linear rank test for comparing two *interval censored samples*. Useful in testing for a difference between two or more *survival curves* or for a single curve against a known alternative. [*Biometrika*, 1982, **69**, 553–66.]

Harris and Stevens forecasting: A method of making short term forecasts in a *time series* that is subject to abrupt changes in pattern and transient effects. Examples of such series are those arising from measuring the concentration of certain biochemicals in biological organisms, or the concentration of plasma growth hormone. The changes are modelled by adding a random perturbation vector having zero mean to a linearly updated parameter vector. [*Bayesian Forecasting and Dynamic Models*, 2nd edition, 1999, M. West and J. Harrison, Springer Verlag, New York.]

Harris walk: A *random walk* on the set of nonnegative integers, for which the matrix of *transition probabilities* consists of zeros except for the elements;

$$\begin{aligned} p_{0,1} &= 1 \\ p_{j,j+1} &= a_j \qquad 0 < a_j < 1, \qquad 1 \leq j < \infty \\ p_{j,j-1} &= 1 - a_j \end{aligned}$$

Hartley, Herman Otto (1912–1980): Born in Berlin, Hartley obtained a Ph.D. degree in mathematics from the University of Berlin in 1933. In 1934 he emigrated to England where he worked with *Egon Pearson* in producing the *Biometrika Tables for Statisticians*. In 1953 Hartley moved to Iowa State College in the USA and from 1963 to 1977 established and ran the Institute of Statistics at Texas A&M University. Contributed to data processing, *analysis of variance*, sampling theory and *sample surveys*. Hartley was awarded the S. S. Wilks medal in 1973. He died on 30 December 1980 in Durham, USA.

Hartley's test: A simple test of the equality of variances of the populations corresponding to the groups in a *one way design*. The test statistic (if each group has the same number of observations) is the ratio of the largest (s^2 largest) to the smallest (s^2 smallest) within group variance, i.e.

$$F = \frac{s^2 \text{ largest}}{s^2 \text{ smallest}}$$

Critical values are available in many statistical tables. The test is sensitive to departures from normality. See also **Bartlett's test** and **Box's test**. [*Biometrika*, 1950, **37**, 308–12.]

Hat matrix: A matrix, $\mathbf{H}$, arising in *multiple regression*, which is used to obtain the predicted values of the response variable corresponding to each observed value via the equation

$$\hat{\mathbf{y}} = \mathbf{H}\mathbf{y}$$

The matrix $\mathbf{H}$ puts the 'hats' on the elements of $\mathbf{y}$; it is a symmetric matrix and is also *idempotent*. Given explicitly in terms of *design matrix*, $\mathbf{X}$ as

$$\mathbf{H} = \mathbf{X}(\mathbf{X}'\mathbf{X})^{-1}\mathbf{X}$$

The diagonal elements of $\mathbf{H}$ are often useful diagnostically in assessing the results from the analysis. See also **Cook's distance**. [*Regression Analysis*, Volume 2, 1993, Ed. M. S. Lewis-Beck, Sage Publications, London.]

Haugh's test: A test for the independence of two *time series* which is based on the sum of finitely many squares of residual cross-correlations. [*Canadian Journal of Statistics*, 1997, **25**, 233–56.]

Hausdorff dimension: Synonym for **fractal dimension**.

Hausman misspecification test: A test for model misspecification that in its standard version considers one estimator $\hat{\boldsymbol{\beta}}$ which is consistent under both correct and incorrect model specification and another estimator $\tilde{\boldsymbol{\beta}}$ which is consistent and efficient under correct specification but inconsistent under misspecification. The null hypothesis is that both $\hat{\boldsymbol{\beta}}$ and $\tilde{\boldsymbol{\beta}}$ are consistent. The test statistic used is

$$(\hat{\boldsymbol{\beta}} - \tilde{\boldsymbol{\beta}})'[\text{cov}(\hat{\boldsymbol{\beta}}) - \text{cov}(\tilde{\boldsymbol{\beta}})]^{-1}(\hat{\boldsymbol{\beta}} - \tilde{\boldsymbol{\beta}})'$$

which asymptotically has a chi-square null distribution with degrees of freedom equal to the dimension of $\boldsymbol{\beta}$. A common application is in testing multilevel models with random effects versus fixed-effects models for clustered data. [*Econometrica*, 1978, **46**, 1251–1271.]

Hausman-Taylor estimator: An *instrumental variable estimator* for linear models for *panel studies* where some *time-dependent covariates* and some time-invariant covariates are endogenous in the sense of being correlated with the random intercept. [*Econometrica*, 1981, **49**, 1377–1398.]

Hawthorne effect: A term used for the effect that might be produced in an experiment simply from the awareness by the subjects that they are participating in some form of scientific investigation. The name comes from a study of industrial efficiency at the Hawthorne Plant in Chicago in the 1920s. [*Annals of Emergency Medicine*, 1995, **26**, 590–4.]

Hazard function: The risk that an individual experiences an event (death, improvement etc.) in a small time interval, given that the individual has survived up to the beginning of the interval. It is a measure of how likely an individual is to experience an event as a function of time. Usually denoted $h(t)$, the function can be expressed in terms of the probability distribution of the *survival times* $f(t)$ and the *survival function* $S(t)$, as $h(t) = f(t)/S(t)$. The hazard function may remain constant, increase, decrease or take on some more complex shape. The function can be estimated as the proportion of individuals experiencing an event in an interval per unit time, given that they have survived to the beginning of the interval, that is

$$h(\hat{t}) = \frac{\text{number of individuals experiencing an event in interval beginning at } t}{(\text{number of individuals surviving at } t)(\text{interval width})}$$

Care is needed in the interpretation of the hazard function because of both selection effects due to variation between individuals and variation within each individual over time. For example, individuals with a high risk are more prone to experience an event early, and those remaining at risk will tend to be a selected group with a lower risk. This will result in the hazard rate being 'pulled down' to an increasing extent as time passes. See also **survival function**, **bathtub curve** and **frailty models**. [SMR Chapter 13.]

Hazard plotting: Based on the *hazard function* of a distribution, this procedure provides estimates of distribution parameters, the proportion of units failing by a given time, percentiles of the distribution, the behaviour of the failure rate of the units as a function of time, and conditional failure probabilities for units at any time. [*Technometrics*, 2000, **42**, 12–25.]

Hazard regression: A procedure for modelling the *hazard function* that does not depend on the assumptions made in *Cox's proportional hazards model*, namely that the log-hazard function is an additive function of both time and the vector of covariates. In this approach, *spline functions* are often used to model the log-hazard function. [*Statistics in Medicine*, 1996, **15**, 1757–70.]

Head-banging smoother: A procedure for smoothing *spatial data*. The basic algorithm proceeds as follows:

- for each point or area whose value y_i is to be smoothed, determine the N nearest neighbours to location x_i
- from among these N nearest neighbours, define a set of points around the point area, such that the 'triple' (pair plus target point at x_i) are roughly collinear. Let NTRIP be the maximum number of such triplets
- let (a_k,b_k) denote the (higher, lower) of the two values in the kth pair and let $A = \text{median}\{a_k\}$, $B = \text{median}\{b_k\}$
- the smoothed value corresponding to y_i is $\tilde{y}_i$ median$\{A, y_i, B\}$. [*IEEE Transactions on Geosciences and Remote Sensing*, 1991, **29**, 369–78.]

Heady, James Austin (1917–2004): Born in Kuling, China, Heady studied mathematics at Merton College, Oxford, United Kingdom from where he graduated in 1939. In 1946 he was appointed to start a new Department of Statistics at St. Bartholomew's Hospital in London, and then in 1949 became a statistician in the Social Medicine Unit of the Medical Research Council where he remained until 1975. During this period Heady produced a long series of publications concerned with the social and biological factors in infant mortality. Later he became interested in the methodology of dietary surveys. Heady died in London on November 4th, 2004.

Healthy life expectancy (HLAE): The average number of years that a newborn child can expect to live in good health. The measure is useful in assessing a health system's effectiveness in reducing the burden of disease. See also **disability adjusted life years (DALYs)**. [*Demographic Research*, 2009, **20**, 467–494.]

Healthy worker effect: The phenomenon whereby employed individuals tend to have lower mortality rates than those unemployed. The effect, which can pose a serious problem in the interpretation of industrial *cohort studies*, has two main components:

- selection at recruitment to exclude the chronically sick resulting in low *standardized mortality rates* among recent recruits to an industry,
- a secondary selection process by which workers who become unfit during employment tend to leave, again leading to lower standardized mortality ratios among long-serving employees.

[*Statistics in Medicine*, 1986, **5**, 61–72.]

Heckman selection model: A model designed to obtain correct inference for regression models based on nonrandom or "selected" samples. The Heckman selection model contains two equations, a linear regression model of primary interest

$$Y_i = \mathbf{x}_i'\boldsymbol{\beta} + \varepsilon_i$$

and a probit selection equation

$$S_i^* = \mathbf{z}_i'\boldsymbol{\gamma} + \delta_i$$

determining that unit i is sampled if $S_i^* > 0$ and not sampled if $S_i^* \leq 0$. The error terms ε_i and δ_i are assumed to be bivariate normal with zero means, $Var(\varepsilon_i) = \sigma^2$ and $Var(\delta_i) = 1$, and correlation ρ. If $\rho = 0$, valid inference regarding the parameters of interest β can be based on the substantive equation alone. However, there is "informative selection" if $\rho \neq 0$. Valid inference must in this case be based on joint estimation of both equations and estimators solely based on the substantive equation would be inconsistent. See also **Selection model**. [*Econometrica*, 1979, **47**, 153–161.]

Hellinger distance: A measure of distance between two probability distributions, $f(\mathbf{x})$ and $g(\mathbf{x})$ given by $\sqrt{2(1-\rho)}$ where

$$\rho = \int \sqrt{f(\mathbf{x})g(\mathbf{x})}\mathbf{dx}$$

See also **Bhattacharya's distance**. [MV2 Chapter 14.]

Hello-goodbye effect: A phenomenon originally described in psychotherapy research, but one which may arise whenever a subject is assessed on two occasions, with some intervention between the visits. Before an intervention a person may present himself/herself in as bad a light as possible, thereby hoping to qualify for treatment, and impressing staff with the seriousness of his/her problems. At the end of the study the person may want to 'please' the staff with his/her improvement, and so may minimize any problems. The result is to make it appear that there has been some improvement when none has occurred, or to magnify the effects that did occur. [*Journal of Clinical Psychology*, 2000, **56**, 853–9.]

Helmert contrast: A *contrast* often used in the *analysis of variance*, in which each level of a factor is tested against the average of the remaining levels. So, for example, if three groups are involved of which the first is a control, and the other two treatment groups, the first contrast tests the control group against the average of the two treatments and the second tests whether the two treatments differ. [KA1 Chapter 11.]

Helsinki declaration: A set of principles to guide clinicians on the ethics of clinical trials and other clinical research. See also **Nuremberg code**. [*Biostatistics: A Methodology for the Health Sciences*, 2nd edition, 2004, E. Van Belle, L. D. Fisher, P. J. Heagerty and T. S. Lumley, Wiley, New York.]

Herbicide bioassay: A procedure for establishing a *dose-response curve* in the development of new herbicides. As no objective death criteria can be given for plants, a graded response such as biomass or height reduction must be considered. [*Herbicide Bioassays*, 1993, ed. J. C. Streibig and P. Kudsk, CRC, Boca Raton, Florida.]

Hess, Irene (1910–2009): Born in Muhlenberg County, Kansas, Hess received a bachelor's degree in mathematics from Indiana University in 1931. After then studying statistics at the University of Michigan, she joined the US. Bureau of Labour Statistics in 1941 and then joined the US Census Bureau where she implemented the new ideas of *probability* *sampling*. Later Hess became director of the Institute of Social Research at the University of Michigan and was responsible for teaching several generations of sampling statisticians how to translate theory into practice. She died on the 5th July, 2009, in Ann Arbor, USA.

Herfindahl index: An index of industry concentration given by

$$H = \sum_{i=1}^{n} [s_i/S]^2$$

where S is the combined size of all firms in an industry (scaled in terms of employees, sales, etc.) and s_i is the size of the ith firm and there are n firms ranked from largest to smallest. Concentration increases with the value of the index. [*Statistical Science*, 1994, **9**, 94–108.]

Hermite functions: Functions $\phi_j(x)$ given by

$$\phi_j(x) = (2^j j! \pi^{\frac{1}{2}})^{-\frac{1}{2}} \exp(-\frac{1}{2}x^2) H_j(x)$$

where $H_j(x)$ is the *Hermite polynomial* defined by

$$H_j(x) = (-1)^j \exp(x^2) \mathrm{d}^j \{\exp(-x^2)\}/\mathrm{d}x^j$$

[KA1 Chapter 6.]

Hermite polynomials: See **Hermite functions**.

Herzberg's formula: See **shrinkage formulae**.

Hessian matrix: The matrix of second-order partial derivatives of a function f of n independent variables $x_1, x_2, \ldots, x_n$ with respect to those variables. The element in the ith row and the jth column of the matrix is therefore $\partial^2 y/\partial x_i \partial x_j$. In statistics the matrix most often occurs for the *log-likelihood* as a function of a set of parameters. The inverse of the matrix then gives standard errors and covariances of the *maximum likelihood estimates* of the parameters. The Hessian is essentially the observed information matrix in contrast to the *Fisher information matrix* where expectations are taken. [MVA Chapter 6.]

Heterogeneous: A term used in statistics to indicate the inequality of some quantity of interest (usually a variance) in a number of different groups, populations, etc. See also **homogeneous**. [MV1 Chapter 3.]

Heteroscedasticity: A random variable is heteroscedastic if its variance depends on the values of another variable. The opposite case is called homoscedastic.

Hettmansperger–McKean test: A *distribution free method* for assessing whether specific subsets of the regression parameters in a *multiple regression* are zero. [NSM Chapter 9.]

Heuristic computer program: A computer program which attempts to use the same sort of selectivity in searching for solutions that human beings use.

Heywood cases: Solutions obtained when using *factor analysis* in which one or more of the variances of the *specific variates* become negative. [MV2 Chapter 12.]

Hidden Markov models: An extension of *finite mixture models* which provides a flexible class of models exhibiting dependence and a possibly large degree of variability. [*Hidden Markov and Other Models for Discrete Valued Time Series*, 1997, I. L. MacDonald and W. Zucchini, Chapman and Hall/CRC Press, London.]

Hidden periodicities: See **harmonic analysis**.

Hidden time effects: Effects that arise in data sets that may simply be a result of collecting the observations over a period of time. See also **cusum**. [SMR Chapter 7.]

Hierarchical clustering: See **agglomerative hierarchical clustering**.

Hierarchical design: Synonym for **nested design**.

Hierarchical likelihood: A somewhat controversial approach to estimating the parameters of *generalized linear mixed models* that treats random effects as parameters. [*Generalized Linear Models with Random Effects: Unified Analysis via H-Likelihood*, 2006, Y. Lee, J. A. Nelder and Y. Pawitan, Chapman and Hall/CRC, Boca Raton.]

Hierarchical models: A series of models for a set of observations where each model results from adding, deleting or constraining parameters from other models in the series. Alternatively, synonym for *multilevel model.*

Higgins's law: A 'law' that states that the prevalance of any condition is inversely proportional to the number of experts whose agreement is required to establish its presence.

High breakdown methods: Methods that are designed to be resistant to even multiple severe *outliers.* Such methods are an extreme example of *robust statistics.* [*Computational Statistics*, 1996, **11**, 137–46.]

High-dimensional data: A term used for data sets which are characterised by a very large number of variables and a much more modest number of observations. In the 21st century such data sets are collected in many areas, for example, text/web *data mining* and *bioinformatics.* The task of extracting meaningful statistical and biological information from such data sets present many challenges for which a number of recent methodological developments, for example, *sure screening methods, lasso,* and *Dantzig selector,* may be helpful. [*Nature Reviews Cancer*, 2008, **8**, 37–49.]

Higher criticism: A *multiple-comparison test* concept arising from the situation where there are many independent tests of significance and interest lies in rejecting the joint null hypothesis. A fraction of the observed significances at a given α-level is compared to the expected fraction under the joint null and the difference between the two quantities is standardized and this standardized value then maximised over a range of significance levels to give the higher criticism statistic. This statistic is effective in assessing whether n normal means are all zero against the alternative that a small fraction is nonzero. [*Annals of Statistics*, 2004, **32**, 962–994.]

Highest posterior density region: See **Bayesian confidence interval**.

High-throughput data: Data produced by modern biological assays and experiments which supply thousands of measurements per sample and where the sheer amount of data increases the need for better models to enhance inference. See also **high-dimensional data**. [*Bioinformatics*, 2007, **23**, 906–909.]

Hill, Austin Bradford (1897–1991): Born in Hampstead, London, Hill served as a pilot in World War I. Contracting tuberculosis prevented him taking a medical qualification, as he would have liked, so instead he studied for a London University degree in economics by correspondence. His interest in medicine drew him to work with the Industrial Fatigue Research Board, a body associated with the Medical Research Council. He improved his knowledge of statistics at the same time by attending *Karl Pearson*'s lectures at University College. In

1933 Hill became Reader in Epidemiology and Vital Statistics at the recently formed London School of Hygiene and Tropical Medicine (LSHTM). He was an extremely successful lecturer and a series of papers on Principles of Medical Statistics published in the *Lancet* in 1937 were almost immediately reprinted in book form. The resulting text remained in print until its ninth edition in 1971. In 1947 Hill became Professor of Medical Statistics at the LSHTM and Honorary Director of the MRC Statistical Research Unit. He had strong influence in the MRC in particular in their setting up of a series of randomized controlled clinical trials, the first involving the use of streptomycin in pulmonory tuberculosis. Hill's other main achievment was his work with Sir Richard Doll on a case-control study of smoking and lung cancer. Hill received the CBE in 1951, was elected Fellow of the Royal Society in 1954 and was knighted in 1961. He received the Royal Statistical Society's Guy medal in gold in 1953. Hill died on 18 April, 1991, in Cumbria, UK.

Hill-climbing algorithm: An algorithm used in those techniques of *cluster analysis* which seek to find a partition of n individuals into g clusters by optimizing some numerical index of clustering. Since it is impossible to consider every partition of the n individuals into g groups (because of the enormous number of partitions), the algorithm begins with some given initial partition and considers individuals in turn for moving into other clusters, making the move if it causes an improvement in the value of the clustering index. The process is continued until no move of a single individual causes an improvement. See also ***K*-means cluster analysis**. [*Cluster Analysis*, 4th edition, 2001, B. S. Everitt, S. Landau and M. Leese, Arnold, London.]

Hinge: A more exotic (but less desirable) term for *quartile*.

Histogram: A graphical representation of a set of observations in which class frequencies are represented by the areas of rectangles centred on the class interval. If the latter are all equal, the heights of the rectangles are also proportional to the observed frequencies. A histogram of heights of elderly women is shown in Fig. 71. [SMR Chapter 2.]

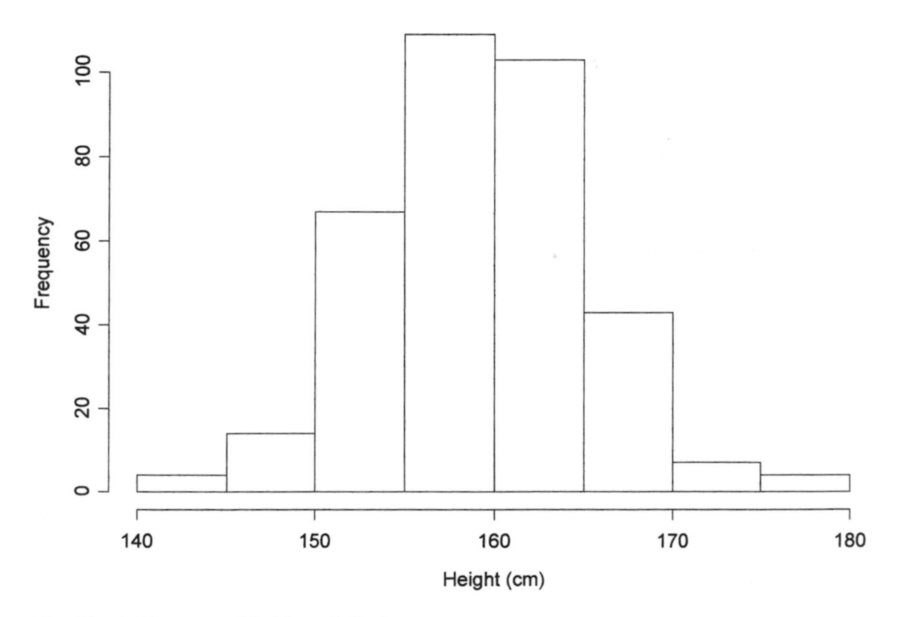

Fig. 71 A histogram of heights of elderly women.

Historical controls: A group of patients treated in the past with a standard therapy, used as the control group for evaluating a new treatment on current patients. Although used fairly frequently in medical investigations, the approach is not to be recommended since possible biases, due to other factors that may have changed over time, can never be satisfactorily eliminated. See also **literature controls**. [SMR Chapter 15.]

Historical prospective studies: A 'prospective study' in which the cohort to be investigated and its subsequent disease history are identified from past records, for example, from information of an individual's work history. [*American Journal of Individual Medicine*, 1983, **4**, 651–7.]

Hit rate: A term occasionally used for the proportion of correct classifications in a *discriminant analysis*.

H-likelihood: See **hierarchical likelihood.**

HLM: Software for the analysis of *multilevel models*. See also **MLwiN**. [Scientific Software International Inc., 7383 N. Lincoln Ave., Suite 100, Lincolnwood, IL 60712, USA.]

Hodges–Lehmann estimator: An estimator for the location difference of two uncensored data samples, $y_1, y_2, \ldots, y_{n_1}$ and $y_{n_1+1}, \ldots, y_n$, $n = n_1 + n_2$, given by

$$\hat{\Delta}_{HL} = \text{median}\{(y_i - y_j) : i = 1, \ldots, n_1;\ j = n_1 + 1, \ldots, n\}$$

[*Nonparametric Statistical Methods*, 2nd edition, 1999, M. Hollander and D. A. Wolfe, Wiley, New York.]

Hoeffding test: A *distribution free method* for testing for the independence of two random variables X and Y, that is able to detect a broader class of alternatives to independence than is possibly by using sample correlation coefficients. [NSM Chapter 8.]

Hoeffding's inequality: A result from probability theory that gives an upper bound on the probability for the sum of random variables to deviate from its expected value. [*Journal of the American Statistical Association*, 1963, **58**, 13–30.]

Hoeffding, Wassily (1914–1991): Born in Mustamaki, Finland, Hoeffding began his university education studying economics but quickly switched to mathematics eventually earning a Ph.D. degree from Berlin University in 1940 with a dissertation on nonparametric measures of association and correlation. He emigrated to the USA in 1946 settling in Chapel Hill, North Carolina. Hoeffding made significant contributions to *sequential analysis*, statistical *decision theory* and *central limit theorems*. He died on 28 February 1991 in Chapel Hill.

Hogben, Lancelot (1895–1975): Born in Southsea, Hampshire, Hogben studied at Cambridge. He was a man of remarkable intellect and a great communicator who made important and original contributions in both theoretical and applied science. Hogben held academic appointments in zoology in England, Scotland, Canada and South Africa before becoming Professor of Zoology at Birmingham from 1941 to 1947, and then Professor of Medical Statistics at the same university from 1947 to 1961. During his career he held five university chairs. Best remembered for his popular book, *Mathematics for the Millions*. Hogben died on 22 August 1975.

Holdout estimate: A method of estimating the *misclassification rate* in a *discriminant analysis*. The data is split into two mutually exclusive sets and the classification rule derived from one and its performance evaluated on the other. The method makes inefficient use of the data (using only part of them to construct the classification rule) and gives a pessimistically biased estimate of the derived rule's misclassification rate. [*Pattern Recognition, A Statistical Approach*, 1982, P. A. Devijver and J. Kittler, Prentice Hall Inc., London.]

Holdover effect: Synonym for **carryover effect**.

Holgate, Philip (1934–1993): Born in Chesterfield, UK, Holgate attended Newton Abbot Grammar School and then entered Exeter University in 1952 to read mathematics. He began his career as a school teacher, teaching mathematics and physics at Burgess Hill School, Hampstead. Holgate's career as a statistician began in 1961 when he joined the Scientific Civil Service at Rothamsted. In 1967 he took up a lecturer post at Birkbeck College, London, progressing to professor in 1970. Holgate's statistical work was primarily in the area of stochastic processes in biology, and he also made seminal contributions to non-associative algebras.

Hollander test: A *distribution free method* for testing for bivariate symmetry. The null hypothesis tested is the two random variables X and Y are *exchangeable*. [NSM Chapter 3.]

Homogeneous: A term that is used in statistics to indicate the equality of some quantity of interest (most often a variance), in a number of different groups, populations, etc. See also **heterogeneous**. [SMR Chapter 7.]

Horseshoe shape: See **seriation**.

Horvitz–Thompson estimator: An unbiased estimator of the total size of a population when sampling is with or without replacement from a finite population and *sampling unit* i has probability p_i of being included in the sample. The estimator does not depend on the number of times a unit may be selected, since each unit is utilized only once in the formula

$$\hat{\tau} = \sum_{i=1}^{\nu} \frac{y_i}{p_i}$$

where ν is the effective sample size (the number of distinct units in the sample) and y_i is the number of individuals, animals or species observed in the ith sampling unit. If the probability that both unit i and unit j are included in the sample is p_{ij} and all these joint *inclusion probabilities* are greater than zero then an unbiased estimator of the variance of $\hat{\tau}$ is

$$var(\hat{\tau}) = \sum_{i=1}^{\nu} \left(\frac{1}{p_i^2} - \frac{1}{p_i} \right) y_i^2 + 2 \sum_{i=1}^{\nu} \sum_{j>i} \left(\frac{1}{p_i p_j} - \frac{1}{p_{ij}} \right) y_i y_j$$

See also **Hansen–Hurwitz estimator**. [*Experimental and Ecological Statistics*, 2003, **10**, 115–127.]

Hosmer-Lemeshow test: A goodness-of-fit test used in *logistic regression*, particularly when there are continuous covariates. Units are divided into *deciles* based on predicted probabilities and a 2 × 10 contingency table of expected and observed frequencies for the deciles obtained. The Hosmer-Lemeshow test statistic is the Pearson chi-square statistic for this contingency table and has an approximate *chi-squared distribution* with 8 degrees of freedom if the model is correct. The test has limited power. [*Applied Logistic Regression*, 2nd edn, 2000, D. W. Hosmer and S. Lemeshow, Wiley, New York.].

Hospital controls: See **control group**.

Hot deck: A method widely used in surveys for imputing missing values. In its simplest form the method involves sampling with replacement m values from the sample respondents A_r to an item y, where m is the number of non-respondents to the item and r is the number of respondents. The sampled values are used in place of the missing values. In practice, the accuracy of imputation is improved by first forming two or more imputation classes using control variables observed in all sample units, and then applying the procedure separately within each imputation class for each item with missing values. [*Statistics in Medicine*, 1997, **16**, 5–19.]

Hotelling, Harold (1895–1973): Born in Fulda, Minnesota, Hotelling first studied journalism at the University of Washington but eventually turned to mathematics gaining a Ph.D. in 1924 for his dissertation in the field of topology. Hotelling worked first at Stanford University before, in 1931, being appointed Professor of Economics at Columbia University. His major contributions to statistical theory were in multivariate analysis and probably his most important paper was 'The generalization of Student's ratio' now known as Hotelling's T^2. He also played a major role in the development of *principal components analysis* and of *canonical correlations*. Elected to the National Academy of Sciences in 1972 and in 1973 to a membership of The Academia Nazionale dei Lincei in Rome. Hotelling died on 26 December 1973 in Chapel Hill, North Carolina.

Hotelling–Lawley trace: See **multivariate analysis of variance**.

Hotelling's T^2 test: A generalization of *Student's t-test* for *multivariate data*. Can be used to test either whether the population mean vector of a set of q variables is the null vector or whether the mean vectors of two populations are equal. In the latter case the relevant test statistic is calculated as

$$T^2 = \frac{n_1 n_2 (\bar{\mathbf{x}}_1 - \bar{\mathbf{x}}_2)' \mathbf{S}^{-1} (\bar{\mathbf{x}}_1 - \bar{\mathbf{x}}_2)}{n_1 + n_2}$$

where n_1 and n_2 are sample sizes, $\bar{\mathbf{x}}_1$ and $\bar{\mathbf{x}}_2$ are sample mean vectors, and $\mathbf{S}$ is a weighted average of the separate sample *variance–covariance matrices*. Under the hypothesis that the population mean vectors are the same,

$$\frac{n_1 + n_2 - q - 1}{(n_1 + n_2 - 2)q} T^2$$

has an *F-distribution* with q and $(n_1 + n_2 - q - 1)$ degrees of freedom. See also **Mahalanobis D^2**. [*Principles of Multivariate Analysis*, 2nd edition, 2000, W. J. Krzanowski, Oxford Science Publications, Oxford.]

Hot hand hypothesis: Synonymous with **streaky hypothesis**.

Household interview surveys: Surveys in which the primary *sampling units* are typically geographic areas such as counties or cities. For each such unit sampled, there are additional levels of subsampling involving successively smaller geographic areas, for example, census districts, neighbourhoods within census districts and households within neighbourhoods. Individuals within sampled households may also be subsampled. The main purpose of the multistage cluster sampling is to lessen the number of areas to which interviewers must travel. [*Nonresponse in Household Interview Surveys*, 1998, R. M. Groves, M. P. Couper, Jossey Bass.]

Hsu, Pao-Lu (1910–1970): Born in Beijing, China, Hsu first studied chemistry at what was later to become Beijing University, but transferred to the Department of Mathematics in Qin Huo University in 1930. In 1938 he received a Ph.D. degree from University College, London. Hsu worked in a number of areas of probability theory and mathematical statistics particularly on the distribution of sample variances from nonnormal populations. In 1956 he was made Director of the first research institute for probability and statistics to be established in China. Hsu died on 18 December 1970 in Beijing.

Huberized estimator: Synonym for **sandwich estimator**.

Huber's condition: A necessary and sufficient design condition for the estimates from using *least squares estimation* in linear models to have an asymptotic normal distribution provided the error terms are independently and identically distributed with finite variance. Given explicitly by

$$\lim_{n\to\infty} \max_{1\le i\le n} h_{ii_n} = 0$$

where h_{ii_n} are the diagonal elements of the *hat matrix*. [*Robust Statistics*, 2003, P. J. Huber, Wiley, New York.]

Human capital model: A model for evaluating the economic implication of disease in terms of the economic loss of a person succumbing to morbidity or mortality at some specified age. Often such a model has two components, the direct cost of disease, for example, medical management and treatment, and the indirect cost of disease, namely the loss of economic productivity due to a person being removed from the labour force. [*Berichte über Landwirtschaft*, 1996, **74**, 165–85.]

Human height growth curves: The growth of human height is, in general, remarkably regular, apart from the pubertal growth spurt. A satisfactory longitudinal *growth curve* is extremely useful as it enables long series of measurements to be replaced by a few parameters, and might permit early detection and treatment of growth abnormalities. Several such curves have been proposed, of which perhaps the most successful is the following five-parameter curve

$$X = A - \frac{2(A-B)}{\exp[C(t-E)] + \exp[D(t-E)]}$$

where t=time (prenatal age measured from the day of birth), X=height reached at age t, A=adult height, B=height reached by child at age E, C=a first time-scale factor in units of inverse time, D=a second time-scale factor in units of inverse time, E=approximate time at which the pubertal growth spurt occurs. [*Biometrics*, 1988, **44**, 995–1003.]

Hurdle model: A model for count data that postulates two processes, one generating the zeros in the data and one generating the positive values. A binomial model governs the binary outcome of whether the count variable has a zero or a positive value. If the value is positive the 'hurdle is crossed' and the *conditional distribution* of the positive values is a suitable *zero-truncated probability distribution*. [*Regression Analysis of Count Data*, 1998, C. A. Cameron and P. K. Trivedi, Cambridge University Press, Cambridge.]

Huynh–Feldt correction: A correction term applied in the analysis of data from *longitudinal studies* by simple *analysis of variance* procedures, to ensure that the within subject *F-tests* are approximately valid even if the assumption of *sphericity* is invalid. See also **Greenhouse–Geisser correction** and **Mauchly test**. [MV2 Chapter 13.]

Hybrid log-normal distribution: The distribution of a random variable, X, when $Y = \log(\rho X) + \rho X, \rho > 0$ is normally distributed with mean μ and variance σ^2. As ρ increases the shape of the distribution changes from log-normal to normal. For fixed ρ the distribution is of log-normal type for small values of X but for large X it is of normal shape. The distribution has been used as a model for worker's periods of exposure to radiation. [*Japanese Journal of Applied Statistics*, 1986, **15**, 1–14.]

Hyperbolic distributions: Probability distributions, $f(x)$, for which the graph of $\log f(x)$ is a hyperbola. [*Statistical Distributions in Scientific Work*, Volume 4, 1981, edited by C. Taillie, G. P. Patil and B. Baldessari, Reidel, Dordrecht.]

Hyperexponential distribution: A term sometimes used for a mixture of two *exponential distributions* with different means, λ_1 and λ_2, and mixing proportion p, i.e. the probability distribution given by

$$f(x) = p\lambda_1 e^{-\lambda_1 x} + (1-p)\lambda_2 e^{-\lambda_2 x}$$

Hypergeometric distribution: A probability distribution associated with *sampling without replacement* from a population of finite size. If the population consists of r elements of one kind and N - r of another, then the probability of finding x elements of the first kind when a random sample of size n is drawn is given by

$$P(x) = \frac{\binom{r}{x}\binom{N-r}{n-x}}{\binom{N}{n}}$$

The mean of x is nr/N and its variance is

$$\left(\frac{nr}{N}\right)\left(1-\frac{r}{n}\right)\left(\frac{N-n}{N-1}\right)$$

When N is large and n is small compared to N, the hypergeometric distribution can be approximated by the *binomial distribution.* [STD Chapter 20.]

Hyperparameter: A parameter (or vector of parameters) θ_2 that indexes a family of possible *prior distributions* for a parameter θ_1 in *Bayesian inference*, i.e. θ_2 is a parameter of a distribution on parameters. An investigator needs to specify the value of θ_2 in order to specify the chosen prior distribution. [*Kendall's Advanced Theory of Statistics*, Volume 2B, 2nd edition, 2004, A. O'Hagan and J. Forster, Arnold, London.]

Hyperprior: A *prior distribution* on a hyperparameter. [*Bayesian Theory,* 2000, J. M. Bernardo and A. M. F. Smith, Wiley, New York.]

Hypothesis testing: A general term for the procedure of assessing whether sample data is consistent or otherwise with statements made about the population. See also **null hypothesis, alternative hypothesis, composite hypothesis, significance test, significance level, type I** and **type II error**. [SMR Chapter 8.]

I

ICM: Abbreviation for **iterated conditional modes algorithm**.

IDA: Abbreviation for **initial data analysis**.

Idempotent matrix: A matrix, $\mathbf{A}$, with the property that $\mathbf{A} = \mathbf{A}^2$. [*Linear Models*, 1971, S. R. Searle, Wiley, New York.]

Identification: A parametric model is said to be identified if there is a unique value $\boldsymbol{\theta}_0$ of the parameter vector $\boldsymbol{\theta}$ that can produce a given probability distribution for the data. If the model is not identified consistent estimation is generally precluded. More formally, two values of $\boldsymbol{\theta}$, $\boldsymbol{\theta}_1$ and $\boldsymbol{\theta}_2$ are called *observationally equivalent* if they produce the same distribution of the data $\boldsymbol{Y}$ (i.e. $f(\boldsymbol{Y};\boldsymbol{\theta}_1) = f(\boldsymbol{Y};\boldsymbol{\theta}_2)$ for all $\boldsymbol{Y}$). $\boldsymbol{\theta}$ is *locally identified* at a parameter point $\boldsymbol{\theta}_0$ if there does not exist a value $\boldsymbol{\theta}'$ such that $\boldsymbol{\theta}'$ and $\boldsymbol{\theta}_0$ are observationally equivalent. $\boldsymbol{\theta}$ is *globally identified* if for any parameter point there is no other observationally equivalent point. [*Econometrica*, 1971, **39**, 577–591.]

Identification keys: Devices for identifying samples from a set of known taxa, which have a tree-structure where each node corresponds to a diagnostic question of the kind 'which one of a named set of attributes does the specimen to be identified possess?' The outcome determines the branch of the tree to follow and hence the next diagnostic question leading ultimately to the correct identification. Often there are only two attributes, concerning the presence and absence of a particular character or the response to a binary test, but multi-state characters or tests are permissible. See also **decision tree**. [*Journal of the Royal Statistical Society, Series A*, 1980, **143**, 253–92.]

Identity matrix: A diagonal matrix in which all the elements on the leading diagonal are unity and all other elements are zero.

Idle column method: See **Taguchi's idle column method**.

Ignorability: A missing data mechanism is said to be ignorable for *likelihood* inference if (1) the joint likelihood for the responses of interest and missing data indicators can be decomposed into two separate components (containing the parameters of main interest and the parameters of the missingness mechanism, respectively) and (2) the parameters for each component are distinct in the sense that there are no parameter restrictions across the components. The component for the missingness mechanism can then be ignored in statistical inference for the parameters of interest. Ignorability follows if *missing values* are missing completely at random or missing at random and the parameters are distinct. [*Biometrika*, 1976, **63**, 581–592.]

Ill conditioned matrix: A matrix $\mathbf{X}$ for which $\mathbf{XX}'$ has at least one *eigenvalue* near zero so that numerical problems arise in computing $(\mathbf{XX}')^{-1}$. *Linear Regression Analysis*, 2nd edition, 2003, G. A. F. Seber and A. J. Lee, Wiley, New York.]

Image restoration: Synonym for **segmentation**.

Immigration-emigration models.: Models for the development of a population that is augmented by the arrival of individuals who found families independently of each other. [*Branching Processes with Biological Applications*, 1975, P. Jagers, Wiley, New York.]

Immune proportion: The proportion of individuals who are not subject to death, failure, relapse, etc., in a sample of *censored survival times*. The presence of such individuals may be indicated by a relatively high number of individuals with large censored survival times. *Finite mixture models* allowing for immunes can be fitted to such data and data analysis similar to that usually carried out on survival times performed. An important aspect of such analysis is to consider whether or not an immune proportion does in fact exist in the population. See also **cure models**. [*Biometrics*, 1995, **51**, 181–201.]

Imperfect detectability: A problem characteristic of many surveys of natural and human populations, that arises because even when a unit (such as a spatial plot) is included in the sample, not all individuals in the selected unit may be detected by the observer. For example, in a survey of homeless people, some individuals in the selected units may be missed. To estimate the population total in a survey in which this problem occurs, both the sampling design and the detection probabilities must be taken into account. [*Biometrics*, 1994, **50**, 712–24.]

Importance sampling: The estimation of expectations of random variables with respect to a distribution of interest (the target distribution) using samples from another distribution (the trial distribution). One motivation is that sampling from the trial distribution can serve as a variance reduction technique, another that it may be easier to simulate from the trial distribution than from the target distribution. More formally, let $g(x)$ be the target density we wish to integrate and $g(x)/h(x)$ the trial distribution. Then

$$\int g(x)dx = \int g(x)\frac{h(x)}{h(x)}dx = \int \frac{g(x)}{h(x)}h(x)dx = E_h\left(\frac{g(x)}{h(x)}\right),$$

as long as $h(x) \neq 0$, where E_h denotes expectation with respect to h. This gives a Monte Carlo estimator

$$\bar{g}(x) = \frac{1}{n}\sum_{i=1}^{n}\frac{g(X_i)}{h(X_i)}$$

where $X_i \sim h(x)$. The Monte Carlo variance of $\bar{g}(x)$ is minimized if the importance sampling function $h(x)$ is proportional to the target distribution $g(x)$. Hence, a good importance sampling function is almost proportional to the target distribution. [*Statistical Science*, 1998, **13**, 163–185.]

Imprecise probabilities: An approach used by *soft methods* in which uncertainty is represented by closed, convex sets of probability distributions and the probability of an event is specified as an interval of possible values rather than only as a precise one. The amount of imprecision is the difference between the upper and lower probabilities defining an interval. See also **belief functions**. [*Journal of the Royal Statistical Society, Series B*, 1996, **58**, 3–57.]

Improper prior distribution: In Bayesian statistics a *prior distribution* is called improper if it depends on the data or does not integrate to 1. [*Bayesian Data Analysis*, 2004, A. Gelman, J. B. Carlin, H. S. Stern and D. B. Rubin, Chapman and Hall/CRC, Boca Raton.]

Imputation: A process for estimating *missing values* using the non-missing information available for a subject. Many methods have been developed most of which can be put into the following *multiple regression* framework.

$$\hat{y}_{mi} = b_{r0} + \sum b_{rj}x_{mij} + \hat{\epsilon}_{mi}$$

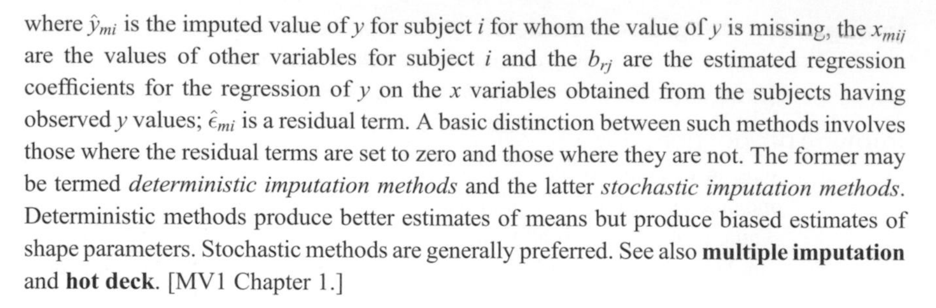

where $\hat{y}_{mi}$ is the imputed value of y for subject i for whom the value of y is missing, the x_{mij} are the values of other variables for subject i and the b_{rj} are the estimated regression coefficients for the regression of y on the x variables obtained from the subjects having observed y values; $\hat{\epsilon}_{mi}$ is a residual term. A basic distinction between such methods involves those where the residual terms are set to zero and those where they are not. The former may be termed *deterministic imputation methods* and the latter *stochastic imputation methods*. Deterministic methods produce better estimates of means but produce biased estimates of shape parameters. Stochastic methods are generally preferred. See also **multiple imputation** and **hot deck**. [MV1 Chapter 1.]

IMSL STAT/LIBRARY: A Fortran subprogram library of useful statistical and mathematical procedures. [Visual Numerics Inc. Corporate Headquarters, 12657 Alcosta Boulevard, Suite 450, San Ramon, CA 94583, USA.]

Incidence: A measure of the rate at which people without a disease develop the disease during a specific period of time. Calculated as

$$\text{incidence} = \frac{\text{number of new cases of a disease over a period of time}}{\text{population at risk of the disease in the time period}}$$

it measures the appearance of disease. See also **prevalence**. [SMR Chapter 14.]

Incidental parameter problem: A problem that sometimes arises when the number of parameters increases in tandem with the number of observations. For example, models for *panel studies* may include individual-specific fixed intercepts, called “incidental parameters”, in addition to the “structural parameters” of interest, making the number of parameters tend to infinity as the number of individuals tends to infinity. In some models, such as logistic models with individual-specific fixed intercepts, this leads to inconsistent *maximum likelihood estimates* of the structural parameters. [*Econometrica*, 1948, **16**, 1–32.]

Inclusion and exclusion criteria: Criteria that define the subjects that can be accepted into a study, particularly a *clinical trial*. Inclusion criteria define the population of interest, exclusion criteria remove people for whom the study treatment is contraindicated. [*Statistics in Medicine*, 1994, **9**, 73–86.]

Inclusion probability: See **simple random sampling**.

Incomplete beta function: See **beta function**.

Incomplete block design: An experimental design in which not all treatments are represented in each block. See also **balanced incomplete block design**. [*Planning of Experiments,* 1992, D. R. Cox, Wiley, New York.]

Incomplete contingency tables: *Contingency tables* containing *structural zeros.*

Incomplete gamma function: See **gamma function**.

Incubation period: The time elapsing between the receipt of infection and the appearance of symptoms. The length of the incubation period depends on the disease, ranging from days in, for example, malaria, to years in something like AIDS. Estimation of the incubation period is important in investigations of how a disease may spread and in the projection of the evolution of an epidemic. [*Statistics in Medicine,* 1990, **9**, 1387–416.]

Independence: Essentially, two events are said to be independent if knowing the outcome of one tells us nothing about the other. More formally the concept is defined in terms of the probabilities of the two events. In particular two events A and B are said to be independent if

$$\Pr(A \text{ and } B) = \Pr(A) \times \Pr(B)$$

where Pr(A) and Pr(B) represent the probabilities of A and B. See also **conditional probability** and **Bayes' theorem**. [KA1 Chapter 1.]

Independent component analysis (ICA): A method for analyzing complex measured quantities thought to be mixtures of other more fundamental quantities, into their underlying components. Typical examples of the data to which ICA might be applied are:

- electroencephalogram (EEG) signal, which contains contributions from many different brain regions,
- person's height, which is determined by contributions from many different genetic and environmental factors.

[*Independent Component Analysis*, 2001, A. Hyvarinen, J. Karhunen and E. Oja, Wiley, New York.]

Independent samples *t*-test: See **Student's *t*-test**.

Independent variables: See **explanatory variables**.

Index of clumping: An index used primarily in the analysis of *spatial data*, to investigate the pattern of the population under study. The index is calculated from the counts, $x_1, x_2, \ldots, x_n$, obtained from applying *quadrant sampling* as

$$\text{ICS} = s^2/\bar{x} - 1$$

where $\bar{x}$ and S^2 are the mean and variance of the observed counts. If the population is 'clustered', the index will be large, whereas if the individuals are regularly spaced the index will be negative. The *sampling distribution* of ICS is unknown even for simple models of the underlying mechanism generating the population pattern. [*Spatial Data Analysis by Example*, 1987, G. Upton and B. Fingleton, Wiley, Chichester.]

Index of dispersion: A statistic most commonly used in assessing whether or not a random variable has a *Poisson distribution*. For a set of observations $x_1, x_2, \ldots, x_n$ the index is given by

$$D = \sum_{i=1}^{n} (x_i - \bar{x})^2/\bar{x}$$

If the population distribution is Poisson, then D has approximately a *chi-squared distribution* with $n - 1$ degrees of freedom. See also **binomial index of dispersion**. [*Biometrika*, 1966, **53**, 167–82.]

Index number: A measure of the magnitude of a variable at one point (in time, for example), to its value at another. An example is the *consumer price index*. The main application of such numbers is in economics. [*The Making of Index Numbers*, 1922, I. Fisher, Houghton, Mifflin, Boston.]

Index plot: A plot of some diagnostic quantity obtained after the fitting of some model, for example, *Cook's distances*, against the corresponding observation number. Particularly suited to the detection of *outliers*. [ARA Chapter 10.]

Indian buffet process: A *stochastic process* used in *unsupervised pattern recognition* when deriving a particular probability distribution. The process is defined by a culinary metaphor based on Indian restaurants in London that offer lunchtime buffets with an apparently infinite number of dishes; N customers enter such a restaurant one after another. Each customer encounters a buffet consisting of infinitely many dishes arranged in a line. The first customer starts at the left of the buffet and takes a serving from each dish, stopping after a Poisson(α) number of dishes as his plate becomes overburdened. The ith customer moves along the buffet, sampling dishes in proportion to their popularity, serving himself with probability $\frac{m_k}{i}$, where m_k is the number of previous customers who have sampled a dish. Having reached the end of all previous sampled dishes, the ith customer then tries a Poisson $\left(\frac{\alpha}{i}\right)$ number of new dishes. [Technical Report 2005–001, Gatsby Computational Neuroscience Unit, University College, London.]

Indicator variable: Synonym for *binary variable.* Also used for a *manifest variable* that is thought to be related to an underlying *latent variable* in the context of *structural equation models.*

Indirect least squares: An estimation method used in the fitting *of structural equation models. Ordinary least squares* is first used to estimate the *reduced form* parameters. Using the relations between structural and reduced form parameters it is subsequently solved for the structural parameters in terms of the estimated reduced form parameters. Indirect least squares estimates are consistent but not unbiased. [*Econometric Methods*, 4th edn, 1997, J. Johnston and J. DiNardo, McGraw-Hill, New York.]

Indirect standardization: The process of adjusting a crude mortality or morbidity rate for one or more variables by using a known *reference population.* It might, for example, be required to compare cancer mortality rates of single and married women with adjustment being made for the likely different age distributions in the two groups. *Age-specific mortality rates* in the reference population are applied separately to the age distributions of the two groups to obtain the expected number of deaths in each. These can then be combined with the observed number of deaths in the two groups to obtain comparable mortality rates. See **standardized mortality ratio (SMR)** and **direct standardization**. [*Statistics in Medicine*, 1987, **6**, 61–70.]

Individual differences scaling: A form of *multidimensional scaling* applicable to data consisting of a number of *proximity matrices* from different sources, i.e. different subjects. The method allows for individual differences in the perception of the stimuli by deriving weights for each subject that can be used to stretch or shrink the dimensions of the recovered geometrical solution. [MV1 Chapter 5.]

INDSCAL: Acronym for **individual differences scaling**.

Infant mortality rate: The ratio of the number of deaths during a calendar year among infants under one year of age to the total number of live births during that year. Often considered as a particularly responsive and sensitive index of the health status of a country or geographical area. The table below gives the rates per 1000 births in England, Wales, Scotland and Northern Ireland in both 1971 and 1992.

	1971	**1992**
England	17.5	6.5
Wales	18.4	5.9
Scotland	19.9	6.8
NI	22.7	6.0

Infectious period: A term used in describing the progress of an *epidemic* for the time following the latent period and during which a patient infected with the disease is able to discharge infectious matter in some way and possibly communicate the disease to other susceptibles. [*Analysis of Infectious Disease,* 1989, N. G. Becker, Chapman and Hall/CRC Press, London.]

Inference: The process of drawing conclusions about a population on the basis of measurements or observations made on a sample of units from the population. See also **frequentist inference** and **Bayesian inference**. [KA1 Chapter 8.]

Infertile worker effect: The observation that working women may be relatively infertile since having children may keep women away from work. See also **healthy worker effect**. [*Journal of Occupational and Environmental Medicine,* 1996, **38**, 352–8.]

Infinitely divisible distribution: A probability distribution $f(x)$ with corresponding *characteristic function* $\phi(t)$, which has the property that for every positive integer n there exists a characteristic function $\phi_n(t)$ such that

$$\phi_n(t) = [\phi(t)]^n$$

This implies that for each n the distribution can be represented as the distribution of the convolution (sum) of n independent random variables with a common distribution. The *chi-squared distribution* is an example. [KA1 Chapter 4.]

Influence: A term used primarily in regression analysis to denote the effect of each observation on the estimated regression parameters. One useful index of the influence of each observation is provided by the diagonal elements of the *hat matrix*. [*Technometrics*, 1977, **19**, 15–18.]

Influence statistics: A range of statistics designed to assess the effect or influence of an observation in determining the results of a regression analysis. The general approach adopted is to examine the changes that occur in the regression coefficients when the observation is omitted. The statistics that have been suggested differ in the particular regression results on which the effect of deletion is measured and the standardization used to make them comparable over observations. All such statistics can be computed from the results of the single regression using all data. See also **Cook's distance, DFFITS, DFBETAS, COVRATIO** and **hat matrix**. [ARA Chapter 10.]

Influential observation: An observation that has a disproportionate influence on one or more aspects of the estimate of a parameter, in particular, regression coefficients. This influence may be due to differences from other subjects on the explanatory variables, an extreme value for the response variable, or a combination of these. *Outliers*, for example, are often also influential observations. [ARA Chapter 10.]

Information matrix: See **Fisher's information**.

Information theory: A branch of applied probability theory applicable to many communication and signal processing problems in engineering and biology. Information theorists devote their efforts to quantitative examination of the following three questions;:

- What is information?
- What are the fundamental limitations on the accuracy with which information can be transmitted?
- What design methodologies and computational algorithms yield practical systems for communication and storing information that perform close to the fundamental limits mentioned previously?

See also **entropy**. [*Information Theory and Coding*, N. Abrahamson, 1968, McGraw-Hill, New York.]

Informative censoring: *Censored observations* that occur for reasons related to treatment, for example, when treatment is withdrawn as a result of a deterioration in the physical condition of a patient. This form of censoring makes most of the techniques for the analysis of survival times, for example, strictly invalid. [*Statistics in Medicine*, 1992, **11**, 1861–70.]

Informative missing values: See **missing values**.

Informative prior: A term used in the context of *Bayesian inference* to indicate a *prior distribution* that reflects empirical or theoretical information regarding the value of an unknown parameter. [KA1 Chapter 8.]

Informed consent: The consent required from each potential participant prior to random assignment in a *clinical trial* as specified in the 1996 version of the *Helsinki declaration* intended to guide physicians conducting therapeutic trials, namely: in any research on human beings, each potential subject must be adequately informed of the aims, methods, anticipated benefits and potential hazards of the study and the discomfort it may entail. He or she should be informed that he or she is at liberty to abstain from participation in the study and he or she is free to withdraw his or her consent to participation at any time. The physician should then obtain the subject's freely-given informed consent, preferably in writing. [*Journal of the American Medical Association*, 1997, **277**, 925–6.]

Initial data analysis (IDA): The first phase in the examination of a data set which consists of a number of informal steps including

- checking the quality of the data,
- calculating simple summary statistics and constructing appropriate graphs.

The general aim is to clarify the structure of the data, obtain a simple descriptive summary, and perhaps get ideas for a more sophisticated analysis. [*Problem Solving, A Statistician's Guide*, 1988, C. Chatfield, Chapman and Hall/CRC Press, London.]

Inliars: A facetious term for *inliers*.

Inliers: A term used for the observations most likely to be subject to error in situations where a dichotomy is formed by making a 'cut' on an ordered scale, and where errors of classification can be expected to occur with greatest frequency in the neighbourhood of the cut. Suppose, for example, that individuals are classified say on a hundred point scale that indicates degree of illness. A cutting point is chosen on the scale to dichtomize individuals into well and ill categories. Errors of classification are certainly more likely to occur in the neighbourhood of the cutting point. [*Transformation and Weighing in Regression*, 1988, R. J. Carroll and D. Ruppert, Chapman and Hall/CRC Press, London.]

Innovational outlier: See **additive outlier**.

Instantaneous count procedure: A sampling method used in biological research for estimating population numbers.

Instantaneous death rate: Synonym for **hazard rate**.

Institutional surveys: Surveys in which the primary *sampling units* are institutions, for example, hospitals. Within each sampled institution, a sample of patient records is selected. The main purpose of the two-stage design (compared to a simple random sample) are to lessen the number of institutions that need to be subsampled, and to avoid constructing a sampling frame of patient records for the entire population. Stratified sampling involving selecting institutions with differing probabilities based on some institutional characteristic (e.g. size) is typically used to lessen the variability of estimators. See also **cluster sampling**. [*A Concept in Corporate Planning*, 1970, R. L. Ackoff, Wiley, New York.]

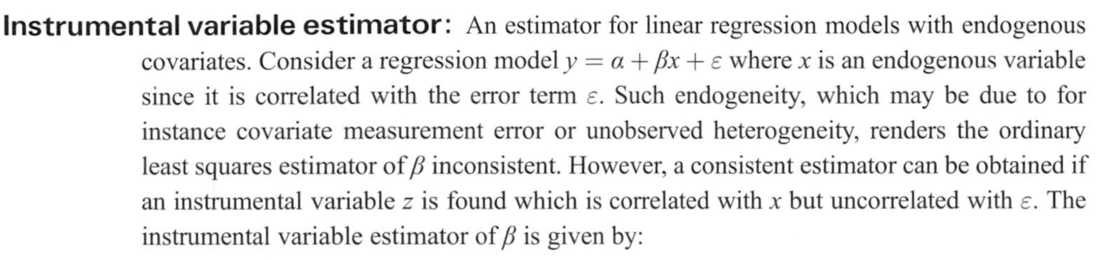

Instrumental variable estimator: An estimator for linear regression models with endogenous covariates. Consider a regression model $y = \alpha + \beta x + \varepsilon$ where x is an endogenous variable since it is correlated with the error term ε. Such endogeneity, which may be due to for instance covariate measurement error or unobserved heterogeneity, renders the ordinary least squares estimator of β inconsistent. However, a consistent estimator can be obtained if an instrumental variable z is found which is correlated with x but uncorrelated with ε. The instrumental variable estimator of β is given by:

$$\hat{\beta}_{IV} = (\mathbf{z}'\mathbf{x})^{-1}\mathbf{z}'\mathbf{y}$$

where **z**, **x** and **y** are n x 1 vectors and n is the sample size. [*Microeconometrics,* 2005, A. C. Cameron and P. K. Trivedi, Cambridge University Press, Cambridge.]

Integrated hazard function: Synonym for **cumulative hazard function**.

Intention-to-treat analysis: A procedure in which all patients randomly allocated to a treatment in a *clinical trial* are analysed together as representing that treatment, whether or not they completed, or even received it. Here the initial random allocation not only decides the allocated treatment, it decides there and then how the patient's data will be analysed, whether or not the patient actually receives the prescribed treatment. This method is adopted to prevent disturbances to the prognostic balance achieved by randomization and to prevent possible bias from allowing compliance, a factor often related to outcome, to determine the groups for comparison. [*Controlled Clinical Trials*, 2000, **21**, 167–189.]

Interaction: A term applied when two (or more) explanatory variables do not act independently on a response variable. Figure 72 shows an example from a 2×2 factorial design. See also **additive effect**. [SMR Chapter 12.]

Intercept: The parameter in an equation derived from a regression analysis corresponding to the expected value of the response variable when all the explanatory variables are zero.

Intercropping experiments: Experiments involving growing two or more crops at the same time on the same piece of land. The crops need not be planted nor harvested at exactly the same time, but they are usually grown together for a significant part of the growing season. Used extensively in the tropics and subtropics, particularly in developing countries where people are rapidly depleting scarce resources but not producing enough food. [*Statistical Design and Analysis for Intercropping Experiments*, 1999, T. Walter, Springer, New York.].

Interim analyses: Analyses made prior to the planned end of a *clinical trial*, usually with the aim of detecting treatment differences at an early stage and thus preventing as many patients as possible receiving an 'inferior' treatment. Such analyses are often problematical particularly if carried out in a haphazard and unplanned fashion. See also **alpha spending function**. [*Statistics in Medicine*, 1994, **13**, 1401–10.]

Interior analysis: A term sometimes applied to analysis carried out on the fitted model in a regression problem. The basic aim of such analyses is the identification of problem areas with respect to the original *least squares* fit of the fitted model. Of particular interest is the disposition of individual data points and their relative *influence* on global measures used as guides in subsequent stages of analysis or as estimates of parameters in subsequent models. *Outliers* from the fitted or predicted response, multivariate outliers among the explanatory variables, and points in the space of explanatory variables with great *leverage* on the fitted model should be identified, and their influence evaluated before further analysis is undertaken. [*Quality and Quantity*, 1994, **28**, 21–53.]

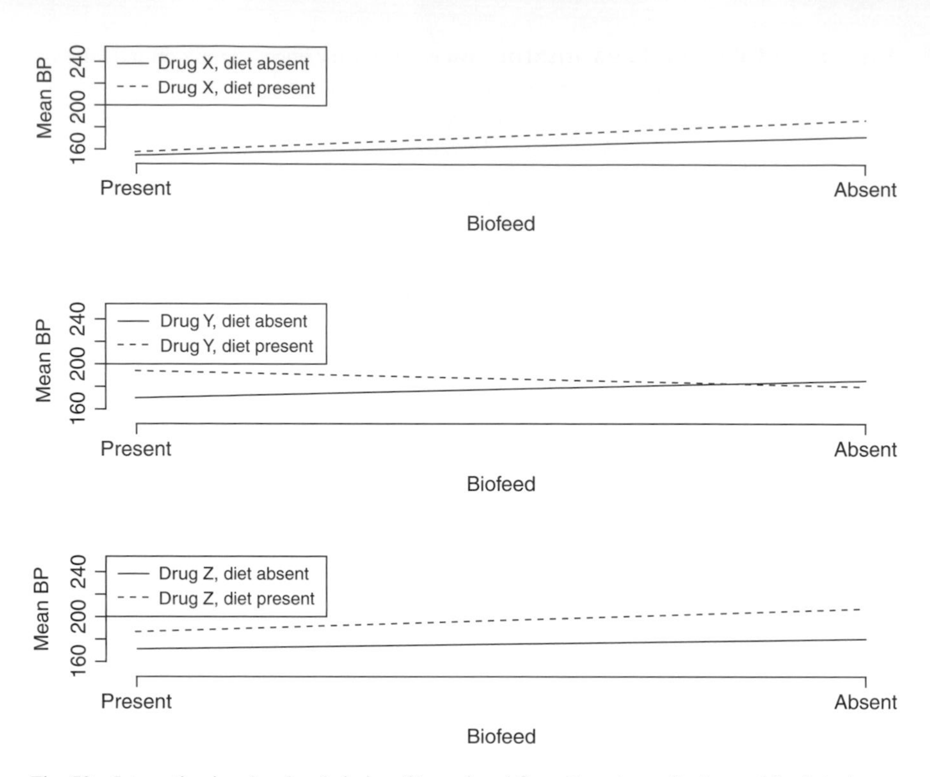

Fig. 72 Interaction in a 2 × 2 × 2 design. (Reproduced from *Experiment Design and Statistical Methods*, D. R. Boniface, by permission of the publishers Chapman and Hall, London.)

Interpolation: The process of determining a value of a function between two known values without using the equation of the function itself. [SMR Chapter 3.]

Interquartile range: A measure of spread given by the difference between the first and third quartiles of a sample. [SMR Chapter 3.]

Interrupted time series design: A study in which a single group of subjects is measured several times before and after some event or manipulation. Often also used to describe investigations of a single subject. See also **longitudinal data** and **N of 1 clinical trial**. [*Experimental and Quasi-Experimental Designs for Generalized Casual Inference*, 2002, W. R. Shadish, T. D. Cook and D. T. Campbell, Houghton Mifflin, Boston.]

Interruptible designs: Experimental designs that attempt to limit the information lost if an experiment is prematurely ended. [*Technometrics*, 1982, **24**, 55–8.]

Interval-censored observations: Observations that often arise in the context of studies of time elapsed to a particular event when subjects are not monitored continuously. Instead the prior occurrence of the event of interest is detectable only at specific times of observation, for example, at the time of medical examination. [*Statistics in Medicine*, 1988, **7**, 1139–46.]

Interval estimate: See **estimate**.

Interval estimation: See **estimation**.

Interval variable: Synonym for **continuous variable**.

Intervened Poisson distribution: A probability distribution that can be used as a model for a disease in situations where the incidence is altered in the middle of a data collection period due to preventative treatments taken by health service agencies. The mathematical form of the distribution is

$$P(\text{number of cases} = x) = [e^{\theta\rho}(e^{\theta} - 1)]^{-1}[(1+\rho)^x - \rho^x]\theta^x/x!$$

where $x = 1, 2, \ldots$. The parameters $\theta(>0)$ and $\rho(0 \le \rho \le \infty)$ measure incidence and intervention, respectively. A zero value of ρ is indicative of completely successful preventive treatments, whereas $\rho = 1$ is interpreted as a status quo in the incidence rate even after the preventive treatments are applied. [*Communications in Statistics – Theory and Methods*, 1995, **24**, 735–54.]

Intervention analysis in time series: An extension of *autoregressive integrated moving average models* applied to *time series* allowing for the study of the magnitude and structure of changes in the series produced by some form of intervention. An example is assessing how efficient is a preventive programme to decrease monthly number of accidents. [*Journal of the American Statistical Association*, 1975, **70**, 70–9.]

Intervention study: Synonym for **clinical trial**.

Interviewer bias: The *bias* that occurs in surveys of human populations because of the direct result of the action of the interviewer. This bias can arise for a variety of reasons including failure to contact the right persons and systematic errors in recording the answers received from the respondent. [*Journal of Occupational Medicine*, 1992, **34**, 265–71.]

Intraclass contingency table: A table obtained from a *square contingency table* by pooling the frequencies of cells corresponding to the same pair of categories. Such tables arise frequently in genetics when the genotypic distribution at a single locus with r alleles, $A_1, A_2, \ldots, A_r$, is observed. Since A_iA_j is indistinguishable from A_jA_i, $i \neq j$, only the total frequency of the unordered pair A_iA_j is observed. Thus the data consist of the frequencies of homozygotes and the combined frequencies of heterozygotes. [*Statistics in Medicine*, 1988, **7**, 591–600.]

Intraclass correlation: Although originally introduced in genetics to judge sibling correlations, the term is now most often used for the proportion of variance of an observation due to between-subject variability in the 'true' scores of a measuring instrument. Specifically if an observed value, x, is considered to be true score (t) plus measurement error (e), i.e.

$$x = t + e$$

the intraclass correlation is

$$\frac{\sigma_t^2}{(\sigma_t^2 + \sigma_e^2)}$$

where σ_t^2 is the variance of t and σ_e^2 the variance of e. The correlation can be estimated from a study involving a number of raters giving scores to a number of patients. Also used to quantity the dependence of responses for units nested in clusters according to a *random intercept model*. [KA2 Chapter 26.]

Intrinsic aliasing: See **aliasing**.

Intrinsically non-linear models: See **non-linear models**.

Intrinsic error: A term most often used in a clinical laboratory to describe the variability in results caused by the inate imprecision of each analytical step.

Intrinsic rate of natural increase: Synonym for **Malthusian parameter**.

Invariance: A property of a set of variables or a statistic that is left unchanged by a transformation. The variance of a set of observations is, for example, invariant under addition or subtraction of a constant from the data.

Inverse Bernoulli sampling: A series of *Bernoulli trials* that is continued until a preassigned number, r, of successes have been obtained; the total number of trials necessary to achieve this, n, is the observed value of a random variable, n, having a *negative binomial distribution*. [*Statistica Neerlandica*, 2002, **56**, 387–399.]

Inverse distribution function: A function $G(\alpha)$ such that the probability that the random variable X takes a value less than or equal to it is α, i.e.

$$\Pr[X \leq G(\alpha)] = \alpha$$

Inverse Gaussian distribution: Synonym for **inverse normal distribution**.

Inverse normal distribution: The probability distribution, $f(x)$, given by

$$f(x) = \left(\frac{\lambda}{2\pi x^3}\right)^{\frac{1}{2}} \exp -\left\{\frac{\lambda(x-\mu)^2}{2\mu^2 x}\right\}, \quad x>0$$

where μ and λ are both positive. The mean, variance, *skewness* and *kurtosis* of the distribution are as follows:

$$\text{means} = \mu$$
$$\text{variance} = \mu^3/\lambda$$
$$\text{skewness} = 3(\mu/\lambda)^{\frac{1}{2}}$$
$$\text{kurtosis} = 3 + \frac{15\mu}{\lambda}$$

A member of the *exponential family* which is skewed to the right. Examples of the disitribution are shown in Fig. 73. [STD Chapter 21.]

Inverse polynomial functions: Functions useful for modelling many *dose–response* relationships in biology. For a particular dose or stimulus x, the expected value of the response variable, y, is defined by

$$E(y) = \frac{x+\alpha}{\sum_{i=0}^{d} \beta_i (x+\alpha)^i} \quad x \geq 0$$

The parameters, $\beta_1, \beta_2, \ldots, \beta_d$, define the shape of the dose–response curve and α defines its position on the x axis. A particularly useful form of the function is obtained by setting $\alpha = 0$ and $d = 1$. The resulting curve is

$$E(y) = \frac{x}{\beta_0 + \beta_1 x} \quad x \geq 0$$

which can be rewritten as

$$E(y) = \frac{k_1 x}{k_2 + x}$$

where $k_1 = 1/\beta_1$ and $k_2 = \beta_0/\beta_1$. This final equation is equivalent to the *Michaelis–Menten* equation. [*Biometrics*, 1966, **22**, 128–41.]

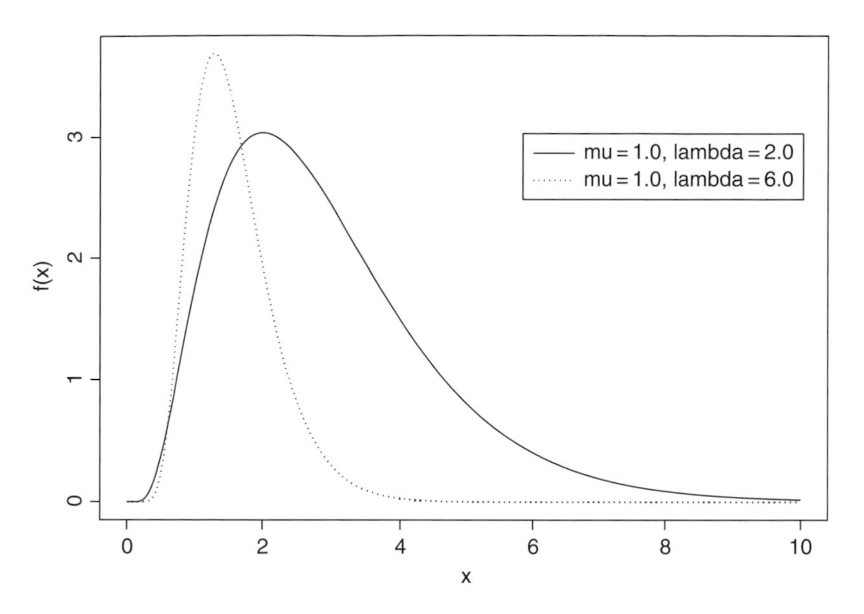

Fig. 73 Examples of inverse normal distributions: $\mu = 1.0$; $\lambda = 2.0, 6.0$.

Inverse sine transformation: Synonymous with **arc sine transformation**.

Inverse survival function: The quantile, $Z(\alpha)$ that is exceeded by the random variable x with probability α, i.e.

$$P[X > Z(\alpha)] = \alpha$$

$Z(\alpha) = G(1 - \alpha)$ where G is the *inverse distribution function*.

Inversion theorem: A theorem that proves that a probability distribution, $f(x)$, is uniquely determined by its *characteristic function*, $\phi(t)$. The theorem states that

$$f(x) = \frac{1}{2\pi}\int_{-\infty}^{\infty} e^{-itx}\phi(t)\mathrm{d}t$$

[KA1 Chapter 4.]

Inverted Wishart distribution: The distribution of the inverse of a positive definite matrix, **A**, if and only if $\mathbf{A}^{-1}$ has a Wishart distribution. [*Aspects of Multivariate Statistical Theory*, 1982, R. J. Muirhead, Wiley, New York.]

IRLS: Abbreviation for **iteratively reweighted least squares**.

Irreducible chain: A *Markov chain* in which all states intercommunicate. [*Introduction to Probability Theory and its Applications,* Volume 1, 2nd edition, 1967, W. Feller, Wiley, New York.]

Irwin–Hall distribution: The probability distribution of the sum, S, of n independent random variables each with a *uniform distribution* in (0,1). The distribution is given by

$$\Pr(S = s) = \frac{1}{(n-1)!}\sum_{j=0}^{k}(-1)^j\binom{n}{j}(s-j)^{n-1}, \quad k \le s \le k+1,\ 0 \le k \le n-1$$
$$= 0, \quad \text{elsewhere}$$

[*Continuous Univariate Distributions*, Volume 2, 2nd edition, 1995, N. L. Johnson, S. Kotz and N. Balakrishnan, Wiley, New York.]

Irwin, Joseph Oscar (1898–1982): Born in London, Irwin was awarded a mathematics scholarship to Christ's College, Cambridge in December 1917, but because of the war did not graduate until 1921, when he immediately joined *Karl Pearson*'s staff at University College, London. In 1928 he joined *Fisher* at Rothamsted Experimental Station, working there until 1931, when he joined the Medical Research Council, where he stayed until 1965. He worked on a variety of statistical problems arising from areas such as animal carcinogenicity, accident proneness, vaccines and hot environments for soldiers in the tropics. Received the Royal Statistical Society's Guy Medal in silver in 1953 and acted as President of the British Region of the Biometric Society in 1958 and 1959. Irwin died on 27 July 1982 in Schaffhausen, Switzerland.

Ising–Stevens distribution: The probability distribution of the number of runs, X, of either of two types of objects (n_1 of one kind and n_2 of the other) arranged at random in $n = n_1 + n_2$ positions around a circle. Given by

$$\Pr(X = x) = \frac{\binom{n_1}{x}\binom{n_2 - 1}{x - 1}}{\binom{n_1 + n_2 - 1}{n_2}}$$

[*Annals of Eugenics*, 1939, **9**, 10–17.]

Isobole: See **isobologram**.

Isobologram: A diagram used to characterize the interactions among jointly administered drugs or chemicals. The contour of constant response (i.e. the *isobole*), is compared to the 'line of additivity', i.e. the line connecting the single drug doses that yield the level of response associated with that contour. The interaction is described as *synergistic*, additive, or *antagonistic* according to whether the isobole is below, coincident with, or above the line of additivity. See Fig. 74 for an example. [*Statistics in Medicine*, 1994, **13**, 2289–310.]

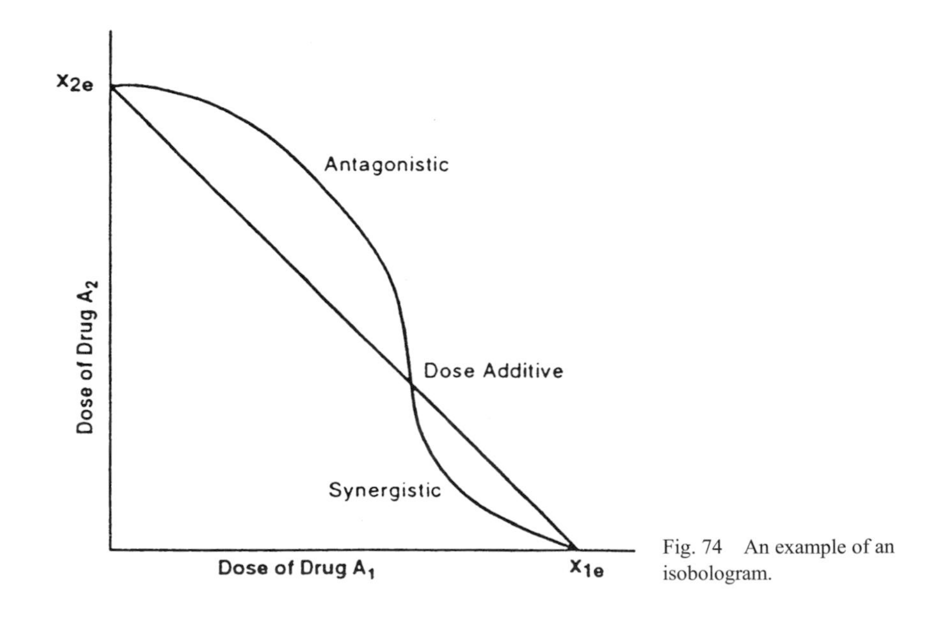

Fig. 74 An example of an isobologram.

Isotonic regression: A form of regression analysis that minimizes a weighted sum of squares subject to the condition that the regression function is order preserving. [*Statistical Inference under Order Restrictions*, 1972, R. E. Barlow, D. J. Bartholomew, J.M Bremmer and H. D. Brunk, Wiley, New York.]

Item difficulty: See **Rasch model**.

Item non-response: A term used about data collected in a survey to indicate that particular questions in the survey attract refusals, or responses that cannot be coded. Often this type of missing data makes reporting of the overall response rate for the survey less relevant. See also **non-response**. [*Journal of the American Statistical Association*, 1994, **89**, 693–6.]

Item-response theory: A set of statistical models and methods designed to measure abilities from items answered in ability tests. See also **Rasch model**. [*Psychometrika*, 1981, 443–59.]

Item-total correlation: A widely used method for checking the homogeneity of a scale made up of several items. It is simply the *Pearson's product moment correlation coefficient* of an individual item with the scale total calculated from the remaining items. The usual rule of thumb is that an item should correlate with the total above 0.20. Items with lower correlation should be discarded. [*Health Measurement Scales*, 3rd edition, 2003, D. L. Streiner and G. R. Norman, Oxford Medical Publications, Oxford.]

Iterated bootstrap: A two-stage procedure in which the samples from the original *bootstrap* population are themselves bootstrapped. The technique can give *confidence intervals* of more accurate coverage than simple bootstrapping. [*Biometrika*, 1994, **81**, 331–40.]

Iterated conditional modes algorithm (ICM): A procedure analogous to *Gibbs sampling*, with the exception that the mode of each conditional *posterior distribution* is determined at each update, rather than sampling a value from these conditional distributions. [*Ordinal Data Modelling*, 1999, V. E. Johnson and J. H. Abbers, Springer, New York.]

Iteration: The successive repetition of a mathematical process, using the result of one stage as the input for the next. Examples of procedures which involve iteration are *iterative proportional fitting*, the *Newton–Raphson method* and the *EM algorithm*.

Iteratively reweighted least squares (IRLS): A *weighted least squares* procedure in which the weights are revised or re-estimated at each iteration. In many cases the result is equivalent to *maximum likelihood estimation*. Widely used when fitting *generalized linear models*. [GLM Chapter 2.]

Iterative proportional fitting: An algorithm that allows the construction of two-way contingency tables with specified *marginal totals* and a specified degree of association. The algorithm can be used to find *maximum likelihood estimates* of the frequencies to be expected assuming a particular *log-linear model* when such estimates cannot be found directly from simple calculations using relevant marginal totals. [*The Analysis of Contingency Tables*, 2nd edition, 1992, B. S. Everitt, Chapman and Hall/CRC Press, London.]

J

Jaccard coefficient: A *similarity coefficient* for use with data consisting of a series of *binary variables* that is often used in *cluster analysis*. The coefficient is given by

$$s_{ij} = \frac{a}{a + b + c}$$

where a, b and c are three of the frequencies in the 2×2 cross-classification of the variable values for subjects i and j. The critical feature of this coefficient is that 'negative matches' are excluded. See also **matching coefficient**. [MV1 Chapter 3.]

Jackknife: A procedure for reducing bias in estimation and providing approximate *confidence intervals* in cases where these are difficult to obtain in the usual way. The principle behind the method is to omit each sample member in turn from the data thereby generating n separate samples each of size $n - 1$. The parameter of interest, θ, can now be estimated from each of these subsamples giving a series of estimates, $\hat{\theta}_1, \hat{\theta}_2, \ldots, \hat{\theta}_n$. The jackknife estimator of the parameter is now

$$\tilde{\theta} = n\hat{\theta} - (n - 1)\frac{\sum_{i=1}^{n} \hat{\theta}_i}{n}$$

where $\hat{\theta}$ is the usual estimator using the complete set of n observations. The jackknife estimator of the standard error of $\hat{\theta}$ is

$$\hat{\sigma}_J = \left[\frac{(n-1)}{n}\sum_{i=1}^{n}(\hat{\theta}_i - \bar{\theta})^2\right]^{\frac{1}{2}}$$

where

$$\bar{\theta} = \frac{1}{n}\sum_{i=1}^{n} \hat{\theta}_i$$

A frequently used application is in *discriminant analysis*, for the estimation of *misclassification rates*. Calculated on the sample from which the *classification rule* is derived, these are known to be optimistic. A jackknifed estimate obtained from calculating the discriminant function n times on the original observations, each time with one of the values removed, is usually a far more realistic measure of the performance of the derived classification rule. [KA1 Chapter 10.]

Jacobian leverage: An attempt to extend the concept of *leverage* in regression to *non-linear models*, of the form

$$y_i = f_i(\boldsymbol{\theta}) + \epsilon_i \qquad (i = 1, \ldots, n)$$

where f_i is a known function of the vector of parameters $\boldsymbol{\theta}$. Formally the Jacobian leverage, $\hat{j}_{ik}$ is the limit as $b \to 0$ of

$$\{f_i(\hat{\boldsymbol{\theta}}) - f_i[\hat{\boldsymbol{\theta}}(k,b)]\}/b$$

where $\hat{\boldsymbol{\theta}}$ is the *least squares* estimate of $\boldsymbol{\theta}$ and $\hat{\boldsymbol{\theta}}(k,b)$ is the corresponding estimate when y_k is replaced by $y_k + b$. Collecting the $\hat{j}_{ik}$s in an $n \times n$ matrix $\mathbf{J}$ give

$$\hat{\mathbf{J}} = \hat{\mathbf{V}}(\hat{\mathbf{V}}'\hat{\mathbf{V}} - [\hat{\boldsymbol{\epsilon}}'][\hat{\mathbf{W}}])^{-1}\hat{\mathbf{V}}'$$

where $\hat{\boldsymbol{\epsilon}}$ is the $n \times 1$ vector of residuals with elements $\epsilon_i = y_i - f_i(\hat{\boldsymbol{\theta}})$ and $[\hat{\boldsymbol{\epsilon}}'][\hat{\mathbf{W}}] = \sum_{i=1}^{n} \hat{\epsilon}_i \hat{\mathbf{W}}_i$: $\mathbf{W}_i$ denotes the *Hessian matrix* of $f_i(\boldsymbol{\theta})$ and $\hat{\mathbf{V}}$ is the matrix whose ith row is $\partial f_i(\boldsymbol{\theta})/\partial \boldsymbol{\theta}'$ evaluated at $\boldsymbol{\theta} = \hat{\boldsymbol{\theta}}$. The elements of $\mathbf{J}$ can exhibit *superleverage*. [*Biometrika*, 1993, **80**, 99–106.]

James–Stein estimators: A class of estimators that arise from discovery of the remarkable fact that in a *multivariate normal distribution* with dimension at least three, the vector of sample means, $\bar{\mathbf{x}}$, may be an inadmissable estimator of the vector of population means, i.e. there are other estimators whose risk functions are everywhere smaller than the risk of $\bar{\mathbf{x}}$, where the risk function of an estimator is defined in terms of variance plus bias squared. If $\mathbf{X}' = [X_1, X_2, \ldots, X_p]$ then the estimators in this class are given by

$$\delta^{a,b}(\mathbf{X}) = \left(1 - \frac{a}{b + \mathbf{X}'\mathbf{X}}\right)\mathbf{X}$$

If a is sufficiently small and b is sufficiently large, then $\delta^{a,b}$ has everywhere smaller risk than $\mathbf{X}$. [*Journal of Multivariate Analysis*, 2006, **97**, 1984–1986.]

Jeffreys's distance: See **Kullback–Leibler information**.

Jeffreys, Sir Harold (1891–1989): Born in Fatfield, Tyne and Wear, Jeffreys studied at St John's College, Cambridge where he was elected to a Fellowship in 1914, an appointment he held without a break until his death. From 1946 to 1958 he was Plumian Professor of Astronomy. Made significant contributions to both geophysics and statistics. His major work *Theory of Probability* is still in print nearly 70 years after its first publication. Jeffreys was made a Fellow of the Royal Society in 1925 and knighted in 1953. He died in Cambridge on 18 March 1989.

Jeffreys's prior: A probability distribution proportional to the square root of the *Fisher information*. [*Kendall's Advanced Theory of Statistics*, Volume 2B, 2nd edition, 2004, A. O'Hagan and J. Forster, Arnold, London.]

Jelinski–Moranda model: A model of software reliability that assumes that failures occur according to a Poisson process with a rate decreasing as more faults are discovered. In particular if ϕ is the true failure rate per fault and N is the number of faults initially present, then the failure rate for the ith fault (after $i - 1$ faults have already been detected, not introducing new faults in doing so) is

$$\lambda_i = (N - i + 1)\,\phi$$

[*Statistical Computer Performance Evaluation*, 1972, ed. P. Freberger, Academic Press, New York.]

Jenkins, Gwilym Meirion (1933–1982): Jenkins obtained a first class honours degree in mathematics from University College, London in 1953 and a Ph.D. in 1956. He worked for two years as a junior fellow at the Royal Aircraft Establishment gaining invaluable experience of the spectral analysis of time series. Later he become lecturer and then reader at

Imperial College London where he continued to make important contributions to the analysis of *time series*. Jenkins died on 10 July 1982.

Jensen's inequality: For a random variable y, the inequality

$$g(\mu) \geq E[g(y)]$$

where $\mu = E(y)$ and g is a concave function. Used in information theory to find bounds or performance indicators. [*Bulletin of the Australian Mathematical Society*, 1997, **55**, 185–9.]

Jewell's estimator: An estimator of the *odds ratio* given by

$$\hat{\psi} = \frac{ad}{(b+1)(c+1)}$$

where a, b, c and d are the cell frequencies in the *two-by-two contingency table* of interest. See also **Haldane's estimator**. [*Statistics in Medicine*, 1989, **8**, 987–96.]

Jittered sampling: A term sometimes used in the analysis of *time series* to denote the sampling of a continuous series where the intervals between points of observation are the values of a random variable. [*Annals of the Institute of Statistical Mathematics*, 1996, **48**, 29–48.]

Jittering: A procedure for clarifying *scatterplots* when there is a multiplicity of points at many of the plotting locations, by adding a small amount of random variation to the data before graphing. An example is shown in Fig. 75.

Johnson-Neyman technique: A technique that can be used in situations where *analysis of covariance* is not valid because of heterogeneity of slopes. With this technique the expected value of the response variable is assumed to be a linear function of the one or two covariates of interest, but not the same linear function in the different subject groups. [*Psychometrika*, 1964, **29**, 241–256.]

Johnson's system of distributions: A flexible family of probability distributions that can be used to summarize a set of data by means of a mathematical function which will fit the data and allow the estimation of percentiles. Based on the following transformation

$$z = \gamma + \eta k_i(x, \lambda, \epsilon)$$

where γ is a standard normal variable and the $k_i(x, \lambda, \epsilon)$ are chosen to cover a wide range of possible shapes. The following are most commonly used:

$$k_1(x, \lambda, \epsilon) = \sin\text{h}^{-1}\left(\frac{x-\epsilon}{\lambda}\right)$$

the S_U *distribution*;

$$k_2(x, \lambda, \epsilon) = \ln\left(\frac{x-\epsilon}{\lambda+\epsilon-x}\right)$$

the S_B *distribution*;

$$k_3(x, \lambda, \epsilon) = \ln\left(\frac{x-\epsilon}{\lambda}\right)$$

the S_L *distribution*. The S_L is in essence a three-parameter *lognormal distribution* since the parameter λ can be eliminated by setting $\gamma^* = \gamma - \eta \ln \lambda$ so that $z = \gamma^* + \eta \ln(x - \epsilon)$. The S_B is a distribution bounded on $(\epsilon, \epsilon + \lambda)$ and the S_U is an unbounded distribution. [KA1 Chapter 6.]

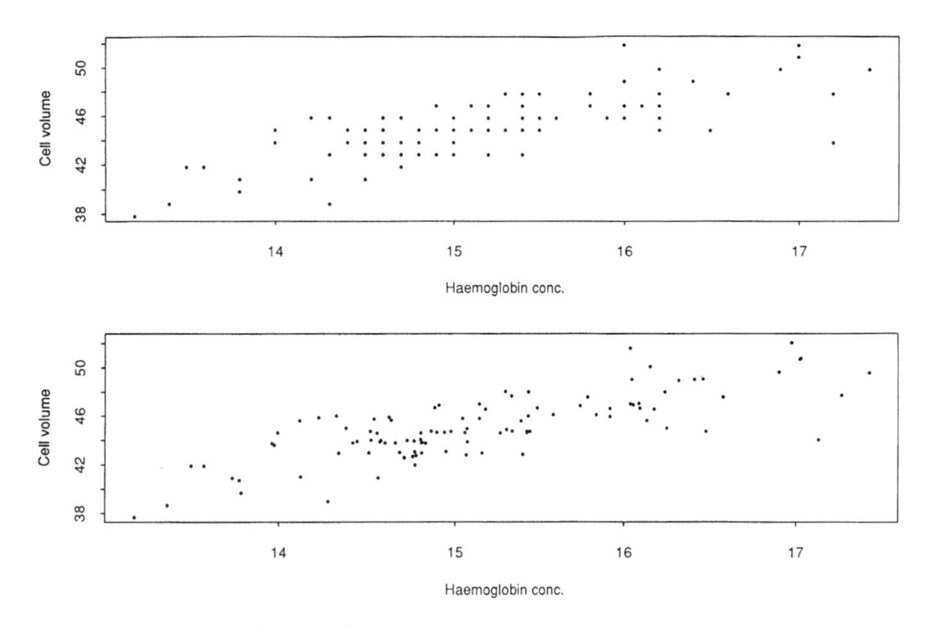

Fig. 75 An example of 'jittering': first scatterplot shows raw data; second scatterplot shows data after being jittered.

Joint distribution: Essentially synonymous with *multivariate distribution*, although used particularly as an alternative to *bivariate distribution* when two variables are involved. [KA1 Chapter 6.]

Jolly–Seber model: A model used in *capture-recapture* sampling which allows for capture probabilities and survival probabilities to vary among sampling occasions, but assumes these probabilities are homogeneous among individuals within a sampling occasion. [*Biometrics*, 2003, **59**, 786–794.]

Jonckheere, Aimable Robert (1920–2005): Born in Lille in the north of France, Jonckheere was educated in England from the age of seven. On leaving Edmonton County School he became apprenticed to a firm of actuaries. As a pacifist, Jonckheere was forced to spend the years of World War II as a farmworker in Jersey, and it was not until the end of hostilities that he was able to enter University College London, where he studied psychology. But Jonckheere also gained a deep understanding of statistics and acted as statistical collaborator to both Hans Eysenck and Cyril Burt. He developed a new statistical test for detecting trends in categorical data and collaborated with Jean Piaget and Benoit Mandelbrot on how children acquire concepts of probability. Jonckheere died in London on 24 September 2005.

Jonckheere's k-sample test: A *distribution free method* for testing the equality of a set of location parameters against an *ordered alternative hypothesis*. [*Nonparametrics: Statistical Methods Based on Ranks*, 1975, E. L. Lehmann, Holden-Day, San Francisco.]

Jonckheere–Terpstra test: A test for detecting specific types of departures from independence in a *contingency table* in which both the row and column categories have a natural order (a *doubly ordered contingency table*). For example, suppose the r rows represent r distinct drug therapies at progressively increasing drug doses and the c columns represent c ordered responses. Interest in this case might centre on detecting a departure from independence in which drugs administered at larger doses are more responsive than drugs administered at

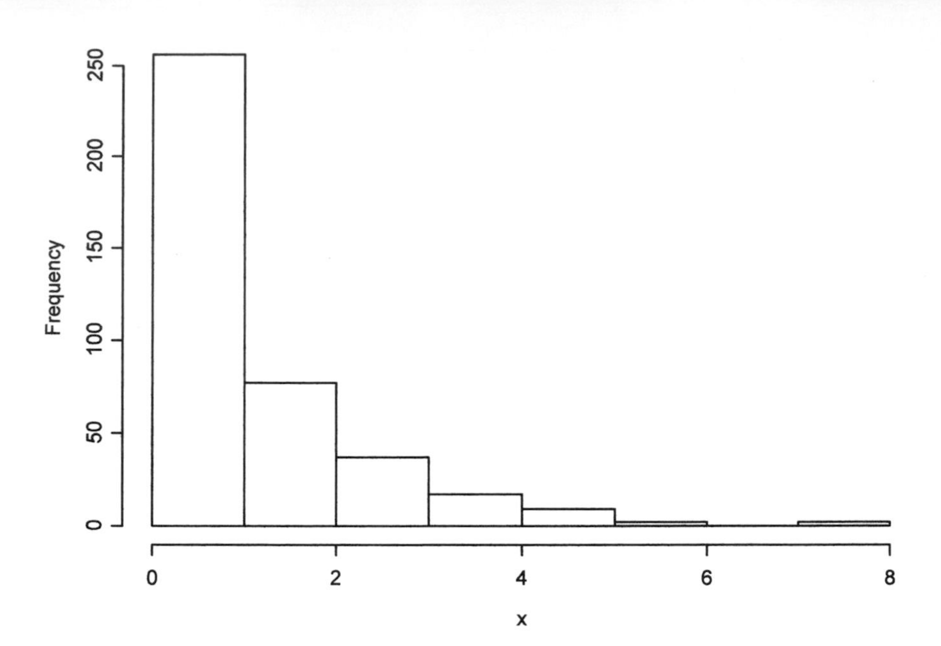

Fig. 76 An example of a J-shaped frequency distribution.

smaller ones. See also **linear-by-linear association test**. [*Biostatics: A Methodology for the Health Sciences*, 2nd edn, 2004, G. Van Belle, L. D. Fisher, P. J. Heagesty and T. S. Lumley, Wiley, New York.]

J-shaped distribution: An extremely *asymmetrical distribution* with its maximum frequency in the initial (or final) class and a declining or increasing frequency elsewhere. An example is given in Fig. 76.

Just identified model: See **identification**.

K

(k_1, k_2)-design: An unbalanced design in which two measurements are made on a sample of k_2 individuals and only a single measurement made on a further k_1 individuals.

Kalman filter: A recursive procedure that provides an estimate of a signal when only the 'noisy signal' can be observed. The estimate is effectively constructed by putting exponentially declining weights on the past observations with the rate of decline being calculated from various variance terms. Used as an estimation technique in the analysis of *time series* data. [TMS Chapter 10.]

Kaiser's rule: A rule often used in *principal components analysis* for selecting the appropriate number of components. When the components are derived from the *correlation matrix* of the observed variables, the rule suggests retaining only those components with eigenvalues (variances) greater than one. See also **scree plot**. [*Educational and Psychological Measurement*, 1960, **20**, 141–51.]

Kaplan–Meier estimator: Synonym for **product limit estimator**.

Kappa coefficient: A chance corrected index of the *agreement* between, for example, judgements and diagnoses made by two raters. Calculated as the ratio of the observed excess over chance agreement to the maximum possible excess over chance, the coefficient takes the value one when there is perfect agreement and zero when observed agreement is equal to chance agreement. See also **Aickin's measure of agreement**. [SMR Chapter 14.]

Karber method: A nonparametric procedure for estimating the *median effective dose* in an quantal assay. [*Encyclopedia of Statistical Sciences*, 2006, eds. S. Kotz, C. B. Read, N. Balakrishnan and B. Vidakovic, Wiley, New York.]

KDD: Abbreviation for **knowledge discovery in data bases**.

Kellerer, Hans (1902–1976): Born in a small Bavarian village, Kellerer graduated from the University of Munich in 1927. In 1956 he became Professor of Statistics at the University of Munich. Kellerer played an important role in getting modern statistical methods accepted in Germany.

Kempthorne, Oscar (1919–2000): Born in Cornwall, England, Kempthorne studied at Cambridge University where he received both B.A. and M.A. degress. In 1947 he joined the Iowa State College statistics faculty where he remained an active member until his retirement in 1989. In 1960 Kempthorne was awarded an honorary Doctor of Science degree from Cambridge in recognition of his contributions to statistics in particular to *experimental design* and the *analysis of variance*. He died on 15 November 2000 in Annapolis, Maryland.

Kempton, Rob (1946–2003): Born in Isleworth, Middlesex, Kempton read mathematics at Wadham College, Oxford. His first job was as a statistician at Rothamsted Experimental Station, and in 1976 he was appointed Head of Statistics at the Plant Breeding Institute in

Cambridge where he made contributions to the design and analysis of experiments with spatial trends and treatment *carryover effects.* In 1986 Kempton became the founding director of the Scottish Agricultural Statistics Service where he carried out work on the statistical analysis of health risks from food and from genetically modified organisms. He died on 11 May 2003.

Kendall's coefficient of concordance: Synonym for **coefficient of concordance**.

Kendall, Sir Maurice George (1907–1983): Born in Kettering, Northamptonshire, Kendall studied mathematics at St John's College, Cambridge, before joining the Administrative Class of the Civil Service in 1930. In 1940 he left the Civil Service to become Statistician to the British Chamber of Shipping. Despite the obvious pressures of such a post in war time, Kendall managed to continue work on *The Advanced Theory of Statistics* which appeared in two volumes in 1943 and 1946. In 1949 he became Professor of Statistics at the London School of Economics where he remained until 1961. Kendall's main work in statistics involved the theory of *k-statistics, time series* and *rank correlation methods.* He also helped organize a number of large sample survey projects in collaboration with governmental and commercial agencies. Later in his career, Kendall became Managing Director and Chairman of the computer consultancy, SCICON. In the 1960s he completed the rewriting of his major book into three volumes which were published in 1966. In 1972 he became Director of the World Fertility Survey. Kendall was awarded the Royal Statistical Society's Guy Medal in gold and in 1974 a knighthood for his services to the theory of statistics. He died on 29 March 1983 in Redhill, UK.

Kendall's tau statistics: Measures of the correlation between two sets of rankings. Kendall's tau itself (τ) is a rank correlation coefficient based on the number of inversions in one ranking as compared with another, i.e. on S given by

$$S = P - Q$$

where P is the number of concordant pairs of observations, that is pairs of observations such that their rankings on the two variables are in the same direction, and Q is the number of discordant pairs for which rankings on the two variables are in the reverse direction. The coefficient τ is calculated as

$$\tau = \frac{2S}{n(n-1)}$$

A number of other versions of τ have been developed that are suitable for measuring association in an $r \times c$ *contingency table* with both row and column variables having ordered categories. (Tau itself is not suitable since it assumes no tied observations.) One example is the coefficient, τ_C given by

$$\tau_C = \frac{2mS}{n^2(m-1)}$$

where $m = \min(r, c)$. See also **phi-coefficient, Cramer's V** and **contingency coefficient**. [SMR Chapter 11.]

Kernel density estimators: Methods of estimating a probability distribution using estimators of the form

$$\hat{f}(x) = \frac{1}{nh}\sum_{i=1}^{n} K\left(\frac{x - X_i}{h}\right)$$

where h is known as *window width* or *bandwidth* and K is the *kernel function* which is such that

$$\int_{-\infty}^{\infty} K(u)\mathrm{d}u = 1$$

Essentially such *kernel estimators* sum a series of 'bumps' placed at each of the observations. The kernel function determines the shape of the bumps while h determines their width.

Three commonly used kernel functions are

Rectangular: $K(x)=0.5$ if $|x|<1$ and $K(x)=0$ otherwise
Triangular: $K(x)=1-|x|$ if $|x|<1$ and $K(x)=0$ otherwise
Gaussian: $K(x)=\frac{1}{\sqrt{2\pi}}e^{-\frac{1}{2}x^2}$

The three kernel functions are shown in Figure 77 The kernel estimator is a series of 'bumps' placed at the observations, with the kernel function determining the shape of the bumps and the window width h determining their width. Figure 78 shows the individual bumps and the estimate, $\hat{f}$, obtained from adding them up for an artificial set of data points using a Gaussian kernel. The extension of kernel density estimation from univariate situations to multivariate data is relatively straightforward. An example of the application of the method to univariate data on waiting times between two eruptions of the Old Faithful geyser in Yellowstone National Park is shown in Figure 79, and a bivariate example using data on body measurements is given in Figure 80. [*Density Estimation in Statistics and Data Analysis*, 1986, B. Silverman, Chapman and Hall/CRC Press, London.]

Kernel estimators: See **kernel density estimators**.

Kernel function: See **kernel density estimators**.

Kernel regression smoothing: A *distribution free method* for smoothing data. In a single dimension, the method consists of the estimation of $f(x_i)$ in the relation

$$y_i = f(x_i) + e_i$$

where $e_i, i = 1, \ldots, n$ are assumed to be symmetric errors with zero means. There are several methods for estimating the regression function, f, for example, averaging the y_i

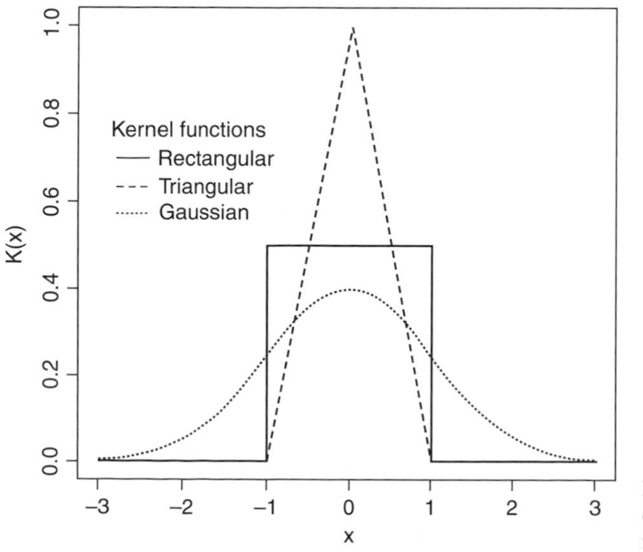

Fig. 77 Three commonly used kernel functions.

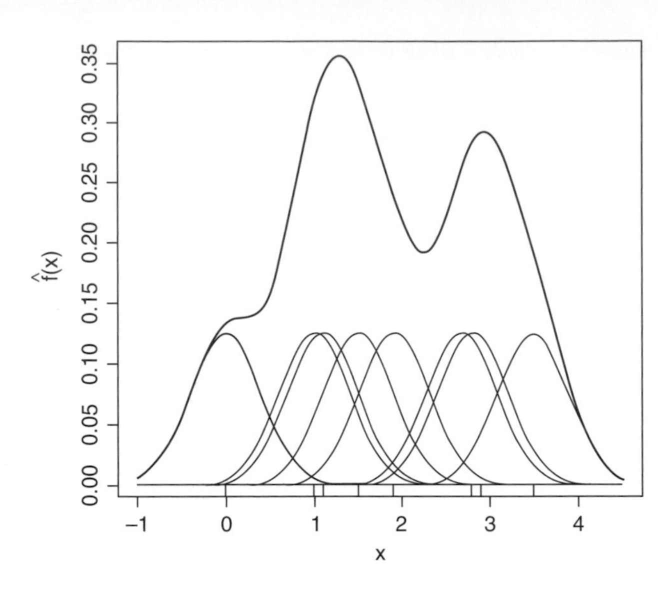

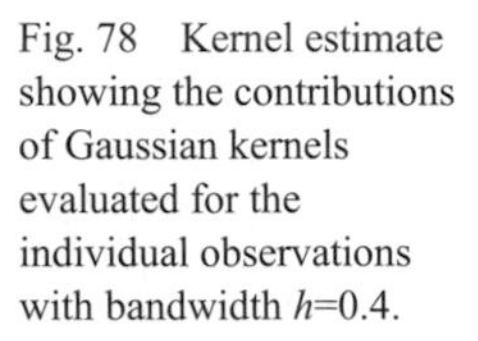

Fig. 78 Kernel estimate showing the contributions of Gaussian kernels evaluated for the individual observations with bandwidth h=0.4.

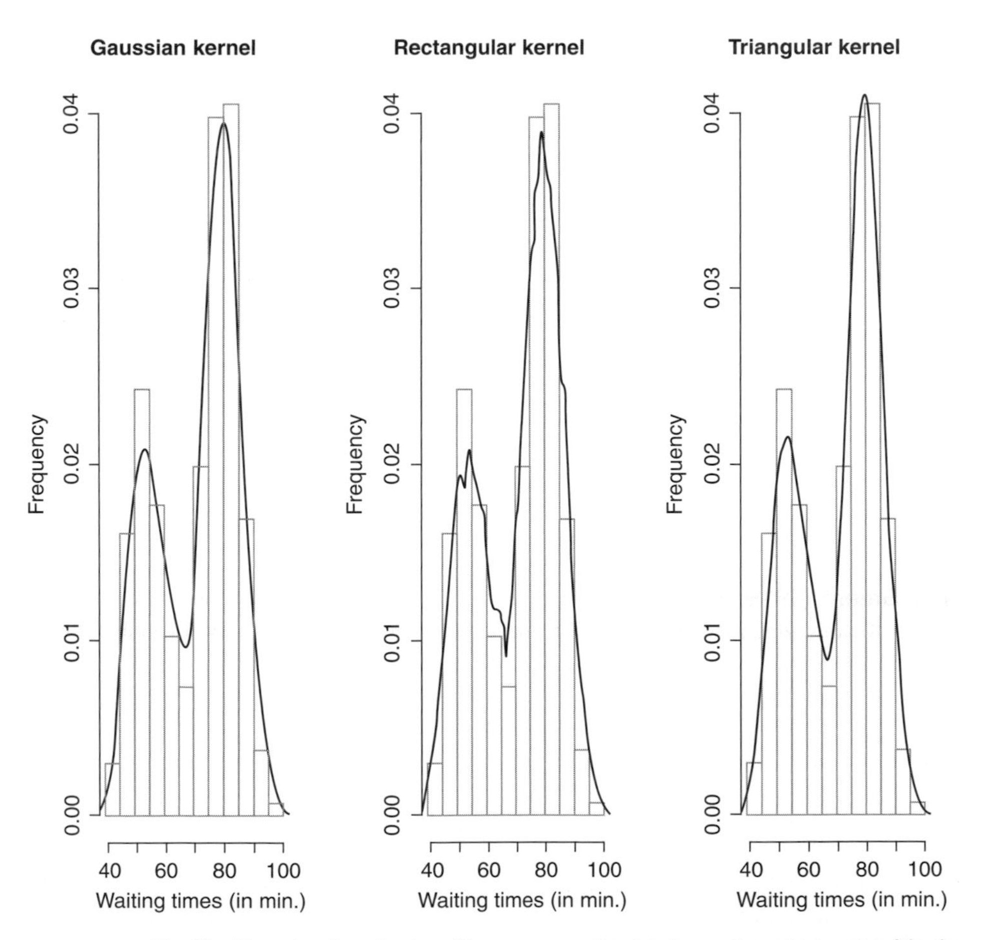

Fig. 79 Three density estimates of the geyser eruption data imposed on a histogram of the data.

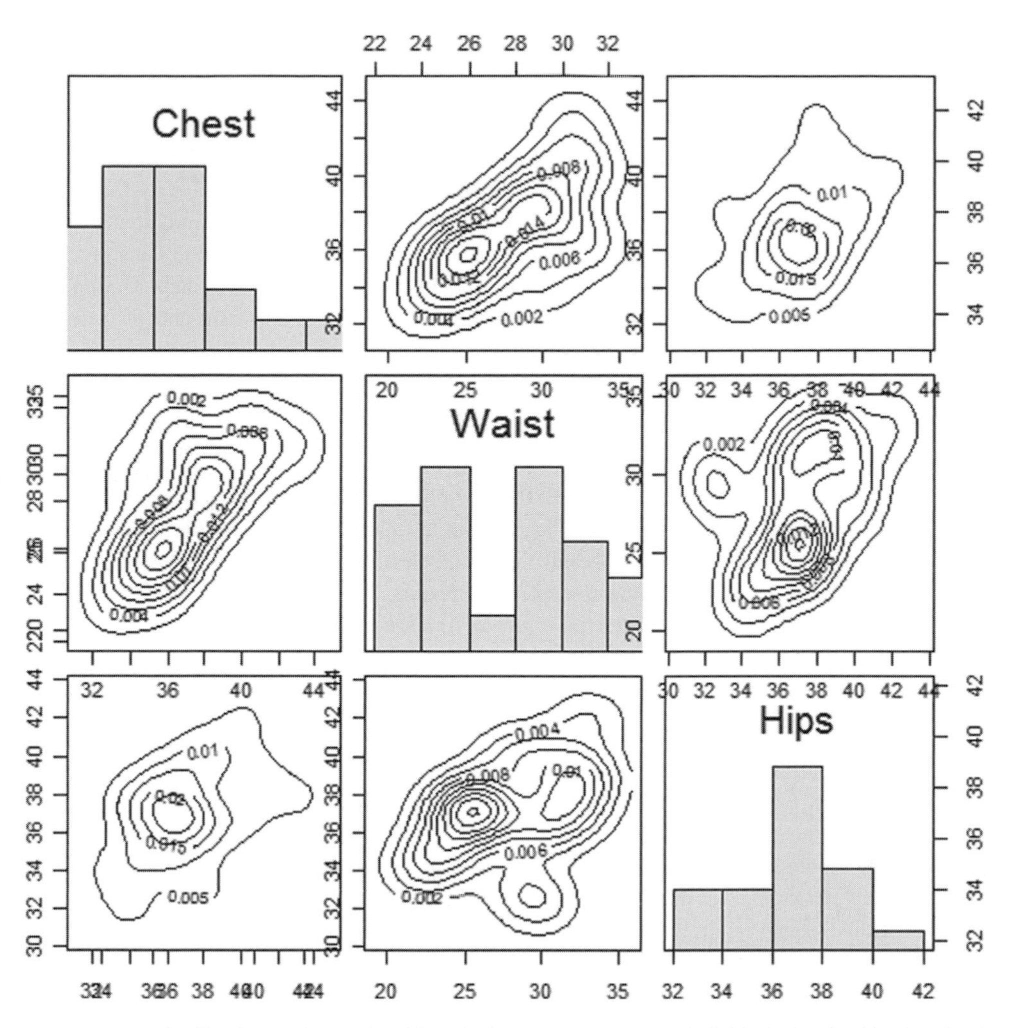

Fig. 80 Scatterplot matrix of three body measurements on 20 individuals showing bivariate density estimates on each panel and histograms on main diagonal.

values that have x_i close to x. See also **regression modelling**. [*Transformation and Weighting in Regression*, 1988, R. J. Carroll and D. Ruppert, Chapman and Hall/CRC Press, London.]

Khamis, Salem (1919–2005): Born in a small village in the Nazareth district of Palestine, Khamis graduated with a BA in mathematics and an MA in physics both from the American University of Beirut. On receiving a British Council Fellowship he travelled to England to undertake doctoral studies in statistics at University College, London. Khamis spent much of his working life as an international civil servant for the United Nations but he made significant contributions to statistics in the areas of sampling theory and, in particular, the computation of purchasing power parities (PPPs) of currencies for the conversion of national currency-dominated economic aggregates, like gross domestic product, into a common comparable currency unit; the most well known PPP is the *Big Mac Index*. Khamis was vice-president of the International Statistical Institute from 1979–1981. He died in Hemel Hempsted, England on the 16th June 2005.

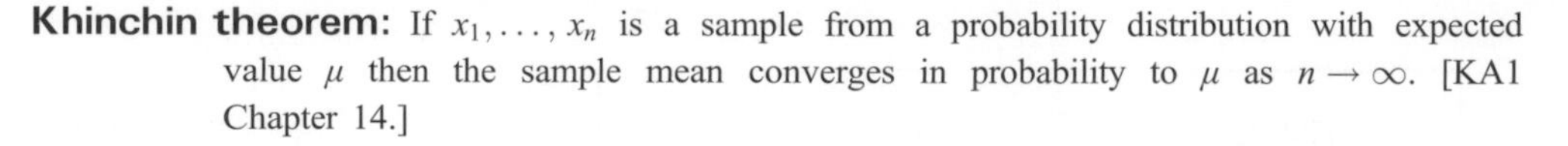

Khinchin theorem: If $x_1, \ldots, x_n$ is a sample from a probability distribution with expected value μ then the sample mean converges in probability to μ as $n \to \infty$. [KA1 Chapter 14.]

Kish, Leslie (1910–2000): Born in Poprad at the time it was part of the Austro-Hungarian Empire (it is now in Slovakia), Kish received a degree in mathematics from the City College of New York in 1939. He first worked at the Bureau of the Census in Washington and later at the Department of Agriculture. After serving in the war, Kish moved to the University of Michigan in 1947, and helped found the Institute for Social Research. During this period he received both MA and Ph.D degrees. Kish made important and far reaching contributions to the theory of sampling much of it published in his pioneering book *Survey Sampling*. Kish received many awards for his contributions to statistics including the Samuel Wilks Medal and an Honorary Fellowship of the ISI.

Kitagawa,Tosio (1909–1993): Born in Otaru City, Japan, Kitagawa studied in the Department of Mathematics at the University of Tokyo for where he graduated in 1934. And in 1934 he was awarded a Doctorate of Science from the same university for his work on functional equations. In 1943 he became the first professor of mathematical statistics in Japan and in 1948–1949 was acting director of the Institute of Statistical Mathematics in Tokyo. Kitagawa made contributions in *statistical quality control, sample survey theory* and *design of experiments*. He died on March 13th, 1993 in Tokyo.

Kleiner–Hartigan trees: A method for displaying *multivariate data* graphically as 'trees' in which the values of the variables are coded into the length of the terminal branches and the terminal branches have lengths that depend on the sums of the terminal branches they support. One important feature of these trees that distinguishes them from most other compound characters, for example, *Chernoff's faces*, is that an attempt is made to remove the arbitrariness of assignment of the variables by performing an *agglomerative hierarchical cluster analysis* of the variables and using the resultant *dendrogram* as the basis tree. Some examples are shown in Fig. 81. [*Journal of the American Statistical Association*, 1981, **76**, 260–9.]

Klotz test: A *distribution free method* for testing the equality of variance of two populations having the same median. More efficient than the *Ansari–Bradley test*. See also **Conover test**. [*Encyclopedia of Statistical Sciences*, 2006, eds. S. Kotz, C. B. Read, N. Balakrishnan and B. Vidakovic, Wiley, New York.]

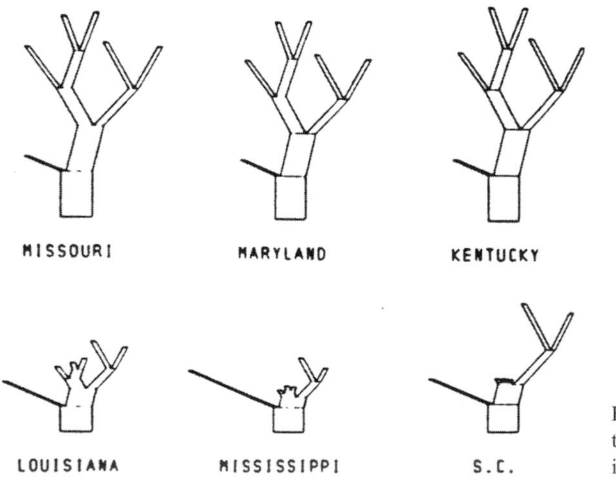

Fig. 81 Kleiner–Hartigan trees for Republican vote data in six Southern states.

K-means cluster analysis: A method of *cluster analysis* in which from an initial partition of the observations into K clusters, each observation in turn is examined and reassigned, if appropriate, to a different cluster in an attempt to optimize some predefined numerical criterion that measures in some sense the 'quality' of the cluster solution. Many such clustering criteria have been suggested, but the most commonly used arise from considering features of the *within groups, between groups* and *total matrices of sums of squares and cross products* (**W, B, T**) that can be defined for every partition of the observations into a particular number of groups. The two most common of the clustering criteria arising from these matrices are

$$\text{minimization of trace}(\mathbf{W})$$
$$\text{minimization of determinant}(\mathbf{W})$$

The first of these has the tendency to produce 'spherical' clusters, the second to produce clusters that all have the same shape, although this will not necessarily be spherical. See also **agglomerative hierarchical clustering methods, divisive methods** and **hill-climbing algorithm**. [MV2 Chapter 10.]

K-means inverse regression: An extension of *sliced inverse regression* to *multivariate regression* with any number of response variables. The method may be particularly useful at the 'exploration' part of an analysis, before suggesting any specific multivariate model. [*Technometrics*, 2004, **46**, 421–429.]

Knots: See **spline functions**.

Knowledge discovery in data bases (KDD): A form of *data mining* which is interactive and iterative requiring many decisions by the researcher. [*Communication of the ACM*, 1996, **39 (II)**, 27–34.]

Knox's tests: Tests designed to detect any tendency for patients with a particular disease to form a *disease cluster* in time and space. The tests are based on a *two-by-two contingency table* formed from considering every pair of patients and classifying them as to whether the members of the pair were or were not closer than a critical distance apart in space and as to whether the times at which they contracted the disease were closer than a chosen critical period. [*Statistics in Medicine*, 1996, **15**, 873–86.]

Kolmogorov, Andrei Nikolaevich (1903–1987): Born in Tambov, Russia, Kolmogorov first studied Russian history at Moscow University, but turned to mathematics in 1922. During his career he held important administrative posts in the Moscow State University and the USSR Academy of Sciences. He made major contributions to probability theory and mathematical statistics including laying the foundations of the modern theory of Markov processes. Kolmogorov died on 20 October 1987 in Moscow.

Kolmogorov–Smirnov two-sample method: A *distribution free method* that tests for any difference between two population probability distributions. The test is based on the maximum absolute difference between the *cumulative distribution functions* of the samples from each population. Critical values are available in many statistical tables. [*Biostatistics: A Methodology for the Health Sciences*, 2nd edn, 2004, G. Van Belle, L. D. Fisher, P. J. Heagerty and T. S. Lumley, Wiley, New York.]

Korozy, Jozsef (1844–1906): Born in Pest, Korozy worked first as an insurance clerk and then a journalist, writing a column on economics. Largely self-taught he was appointed director of a municipal statistical office in Pest in 1869. Korozy made enormous contributions to the statistical and demographic literature of his age, in particular developing the first fertility tables. Joined the University of Budapest in 1883 and received many honours and awards both at home and abroad.

Kovacsics, Jozsef (1919–2003): Jozsef Kovacsics was one of the most respected statisticians in Hungary. Starting his career in the Hungarian Statistical Office, he was appointed Director of the Public Library of the Statistical Office in 1954 and in 1965 became a full professor of statistics at the Eotvos Lorand University where he stayed until his retirement in 1989. Kovacsics carried out important work in several areas of statistics and *demography*. He died on 26 December 2003.

KPSS test: Abbreviation for the **Kwiatkowski-Phillips-Schmidt-Shin test**.

Kriging: A method for providing the value of a random process at a specified location in a region, given a dataset consisting of measurements of the process at a variety of locations in the region. For example, from the observed values $y_1, y_2, \ldots, y_n$ of the concentration of a pollutant at n sites, $t_1, t_2, \ldots, t_n$, it may be required to estimate the concentration at a new nearby site, t_0. Named after D. G. Krige who first introduced the technique to estimate the amount of gold in rock. [*Interpolation for Spatial Data: Some Theory for Kriging*, 1999, M. L. Stein, Springer, New York.]

Kronecker product of matrices: The result of multiplying the elements of an $m \times m$ matrix **A** term by term by those of an $n \times n$ matrix **B**. The result is an $mn \times mn$ matrix. For example, if

$$\mathbf{A} = \begin{pmatrix} 1 & 2 \\ 3 & 4 \end{pmatrix} \quad \text{and } \mathbf{B} = \begin{pmatrix} 1 & 4 & 7 \\ 2 & 5 & 8 \\ 3 & 6 & 9 \end{pmatrix}$$

then

$$\mathbf{A} \bigotimes \mathbf{B} = \begin{pmatrix} 1 & 4 & 7 & 2 & 8 & 14 \\ 2 & 5 & 8 & 4 & 10 & 16 \\ 3 & 6 & 9 & 6 & 12 & 18 \\ 3 & 12 & 21 & 4 & 16 & 28 \\ 6 & 10 & 16 & 8 & 20 & 32 \\ 9 & 18 & 27 & 12 & 24 & 36 \end{pmatrix}$$

$\mathbf{A} \bigotimes \mathbf{B}$ is not in general equal to $\mathbf{B} \bigotimes \mathbf{A}$. [MV1 Chapter 2.]

Kruskal–Wallis test: A *distribution free method* that is the analogue of the *analysis of variance* of a *one-way design*. It tests whether the groups to be compared have the same population median. The test statistic is derived by ranking all the N observations from 1 to N regardless of which group they are in, and then calculating

$$H = \frac{12 \sum_{i=1}^{k} n_i (\bar{R}_i - \bar{R})^2}{N(N-1)}$$

where n_i is the number of observations in group i, $\bar{R}_i$ is the mean of their ranks, $\bar{R}$ is the average of all the ranks, given explicitly by $(N+1)/2$. When the null hypothesis is true the test statistic has a *chi-squared distribution* with $k - 1$ degrees of freedom. [SMR Chapter 9.]

***k*-statistics:** A set of symmetric functions of sample values proposed by *Fisher*, defined by requiring that the pth k-statistic, k_p, has expected value equal to the pth *cumulant*, κ_p, i.e.

$$E(k_p) = \kappa_p$$

Originally devised to simplify the task of finding the *moments* of sample statistics. [KA1 Chapter 12.]

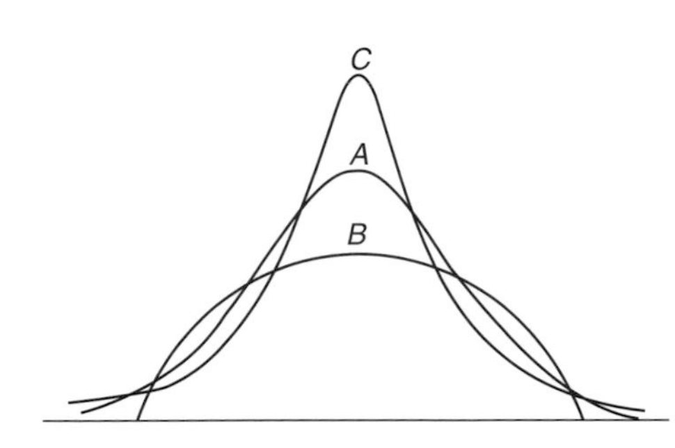

Fig. 82 Curves with differing degrees of kurtosis: *A*, mesokurtic; *B*, platykurtic; *C*, leptokurtic.

Kuiper's test: A test that a circular random variable has an *angular uniform distribution*. Given a set of observed values, $\theta_1, \theta_2, \ldots, \theta_n$ the test statistic is

$$V_n = D_n^+ + D_n^-$$

where

$$D_n^+ = \max(j/n - x_j), \quad j = 1, \ldots, n$$
$$D_n^- = \max\left(x_j - \frac{j-1}{n}\right), \quad j = 1, \ldots, n$$

and $x_j = \theta_j/2\pi$. $V_n\sqrt{n}$ has an approximate standard normal distribution under the hypothesis of an angular uniform distribution. See also **Watson's test**. [MV2 Chapter 14.]

Kullback–Leibler information: A function, *I*, defined for two probability distributions, *f(x)* and *g(x)* and given by

$$I(f:g) = \int_{-\infty}^{\infty} f(x)\log\left\{\frac{f(x)}{g(x)}\right\}\mathrm{dx}$$

Essentially an asymmetric distance function for the two distributions. *Jeffreys's distance* measure is a symmetric combination of the Kullback–Leibler information given by

$$I(f;g) + I(g;f)$$

See also **Bhattacharya's distance** and **Hellinger distance**. [MV2 Chapter 11.]

Kurtosis: The extent to which the peak of a unimodal probability distribution or frequency distribution departs from the shape of a normal distribution, by either being more pointed (*leptokurtic*) or flatter (*platykurtic*). Usually measured for a probability distribution as

$$\mu_4/\mu_2^2$$

where μ_4 is the fourth central *moment* of the distribution, and μ_2 is its variance. (Corresponding functions of the sample moments are used for frequency distributions.) For a normal distribution this index takes the value three and often the index is redefined as the value above minus three so that the normal distribution would have a value zero. (Other distributions with zero kurtosis are called *mesokurtic*.) For a distribution which is leptokurtic the index is positive and for a platykurtic curve it is negative. See Fig. 82. See also **skewness**. [KA1 Chapter 3.]

Kurtosis procedure: See **many-outlier detection procedures**.

Kwiatkowski-Phillips-Schmidt-Shin test (KPSS test): A test of the hypothesis that a *time series* is *stationary* around a deterministic trend. [*Journal of Economics*, 1992, **54**, 159–178.]

L

L'Abbé plot: A plot often used in the *meta-analysis* of *clinical trials* where the outcome is a binary response. The event risk (number of events/number of patients in a group) in the treatment groups are plotted against the risk for the controls of each trial. If the trials are fairly homogeneous the points will form a 'cloud' close to a line, the gradient of which will correspond to the pooled treatment effect. Large deviations or scatter would indicate possible heterogeneity. [*Annals of Internal Medicine*, 1987, **107**, 224–33.]

Labour force survey: A survey carried out in the UK on a quarterly basis since the spring of 1992. It covers 60 000 households and provides labour force and other details for about 120 000 people aged 16 and over. The survey covers not only unemployment, but employment, self-employment, hours of work, redundancy, education and training.

Lack of memory property: A property possessed by a random variable Y, namely that

$$\Pr\{Y \geq x + y \,|Y \geq y\} = \Pr\{Y \geq x\}$$

Variables having an *exponential distribution* have this property, as do those following a *geometric distribution*. [*A Primer on Statistical Distributions*, 2003, N. Balakrishnan and V. B. Nevzorov, Wiley, New York.]

Ladder of re-expression: See **one-bend transformation**.

Lagging indicators: Part of a collection of economic *time series* designed to provide information about broad swings in measures of aggregate economic activity known as business cycles. Used primarily to confirm recent turning points in such cycles. Such indicators change after the overall economy has changed and examples include labour costs, business spending and the unemployment rate.

Lagrange multipliers: A method of evaluating maxima or minima of a function of possibly several variables, subject to one or more linear constraints. [*Optimization, Theory and Applications*, 1979, S. S. Rao, Wiley Eastern, New Delhi.]

Lagrange multiplier test: Synonym for **score test**.

Lancaster, Henry Oliver (1913–2001): Born in Kempsey, New South Wales, Lancaster intended to train as an actuary, but in 1931 enrolled as a medical student at the University of Sydney, qualifying in 1937. Encountering Yule's *Introduction to the Theory of Statistics*, he became increasingly interested in statistical problems, particularly the analysis of 2 × 2 × 2 tables. After World War II this interest led him to obtain a post as Lecturer in Statistics at the Sydney School of Public Health and Tropical Medicine. In 1948 Lancaster left Australia to study at the London School of Hygiene and Tropical Medicine under *Bradford Hill*, returning to Sydney in 1950 to study trends in Australian mortality. Later he undertook more theoretical work that led to his 1969 book, *The Chi-squared Distribution*. In 1966 he was made President of the Statistical Society of Australia. Lancaster died in Sydney on 2 December 2001.

Lancaster models: A means of representing the *joint distribution* of a set of variables in terms of the *marginal distributions*, assuming all interactions higher than a certain order vanish. Such models provide a way to capture dependencies between variables without making the sometimes unrealisitic assumptions of total independence on the one hand, yet having a model that does not require an unrealistic number of observations to provide precise parameter estimates. [*Statistical Pattern Recognition*, 1999, A. Webb, Arnold, London.]

Landmark analysis: A term applied to a form of analysis occasionally applied to data consisting of survival times in which a test is used to assess whether 'treatment' predicts subsequent survival time among subjects who survive to a 'landmark' time (for example, 6 months post-randomization) and who have, at this time, a common prophylaxis status and history of all other covariates. [*Statistics in Medicine*, 1996, **15**, 2797–812.]

Landmark-based shape analysis: An approach to quantifying the shape and shape differences of widely-spaced landmarks (coordinate points) that allows visualization of these differences between groups. [*Statistical Shape Analysis*, 1998, I. Dryden and K. V. Mardia, Wiley, New York.]

Landmark registration: A method for aligning the profiles in a set of *repeated measures data*, for example *human growth curves*, by identifying the timing of salient features such as peaks, troughs, or inflection points. Curves are then aligned by transforming individual time so that landmark events become synchronized. [*The Annals of Statistics*, 1992, **16**, 82–112.]

Laplace approximation: An approach to approximating the integral of a function $\int f(x)dx$ by fitting a multivariate normal density at the maximum $\hat{x}$ of $f(x)$ and computing the volume under the density. The covariance matrix of the fitted *multivariate normal distribution* is determined by the *Hessian matrix* of log $f(x)$ at the maximum $\hat{x}$. The term is also used for approximating the *posterior distribution* with a multivariate normal centered at the maximum a posteriori estimate. [*Bayesian Data Analysis*, 2004, A.Gelman, J. B. Carlin, H. S. Stern and D. B. Rubin, Chapman and Hall/CRC, Boca Raton.]

Laplace distribution: The probability distribution, $f(x)$, given by

$$f(x) = \frac{1}{2}\phi^{-1}\exp\left[-\left(\frac{|x-\theta|}{\phi}\right)\right] \qquad \phi > 0, -\infty < x < \infty$$

Can be derived as the distribution of the difference of two independent random variables each having an identical *exponential distribution*. Examples of the distribution are shown in Fig. 83. The mean, variance, *skewness* and *kurtosis* of the distribution are as follows:

$$\text{mean} = \theta$$
$$\text{variance} = 2\phi^2$$
$$\text{skewness} = 0$$
$$\text{kurtosis} = 6$$

Also known as the *double exponential distribution*. [STD Chapter 22.]

Laplace, Pierre-Simon, Marquis de (1749–1827): Born in Beaumont-en-Auge near to Caen where he first studied, Laplace made immense contributions to physics, particularly astronomy, and mathematics. Became a professor at the École Militaire in 1771 and entered the Académie des Sciences in 1773. Most noted for his five-volume work *Traité de Mécanique Céleste* he also did considerable work in probability including independently discovering *Bayes' theorem*, ten years after Bayes' essay on the topic. Much of this work was published in *Théorie Analytique des Probabilités*. Laplace died on 5 March 1827, in Paris.

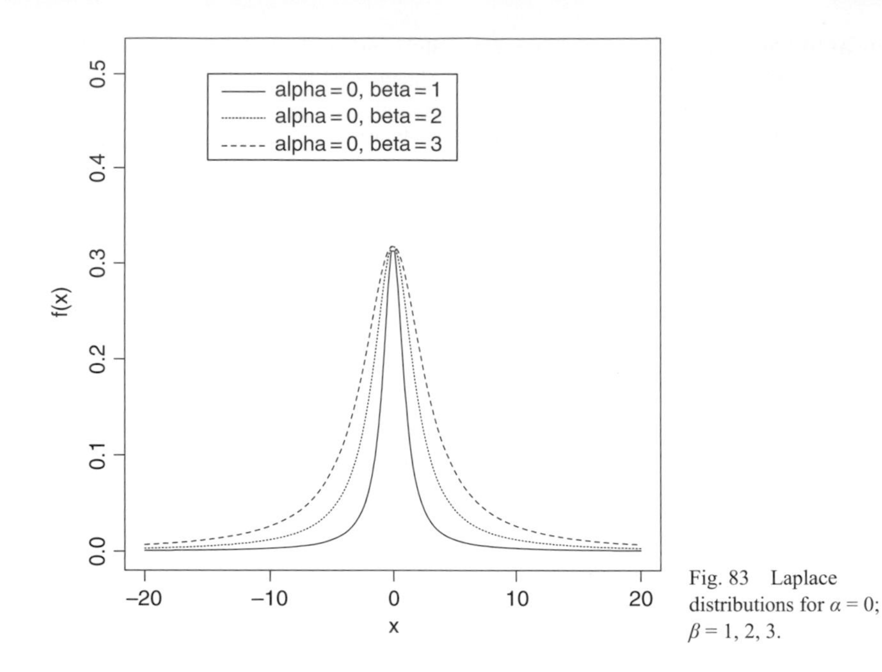

Fig. 83 Laplace distributions for $\alpha = 0$; $\beta = 1, 2, 3$.

Large sample methods: Methods for evaluating the performance of estimators and statistics as the sample size tends to infinity. See also **asymptotic distribution**.

Lasagna's law: Once a *clinical trial* has started the number of suitable patients dwindles to a tenth of what was calculated before the trial began. [*British Medical Journal*, 2001, **322**, 1457–62.]

Lasso: A penalized *least squares regression* method that can be used for variable selection on high-dimensional data. For a *multiple linear regression* model of the form

$$y_i = \mathbf{x}_i'\boldsymbol{\beta} + \epsilon_i$$

The lasso estimator of $\boldsymbol{\beta}$ is the value that minimizes

$$\sum_{i=1}^{n}(y_i - \mathbf{x}_i'\boldsymbol{\beta})^2 + \lambda \sum_{j=1}^{q} |\beta_j|$$

where q is the number of variables and λ is the penalty parameter. The estimator minimizes the usual sum of squared errors but has a bound on the sum of the absolute values of these coefficents. When the bound is large enough the constraint has no effect and the solution is simply that of multiple regression. When however the bound is smaller the parameter estimates are 'shrunken' versions of the least squares estimates. [*Journal of the Royal Statistical Society, Series B*, 1996, **58**, 267–288.]

Last observation carried forward: A method for replacing the observations of patients who drop out of a *clinical trial* carried out over a period of time. It consists of substituting for each *missing value* the subject's last available assessment of the same type. Although widely applied, particularly in the pharmaceutical industry, its usefulness is very limited since it makes very unlikely assumptions about the data, for example, that the (unobserved) post drop-out response remains frozen at the last observed value. See also **imputation**. [*Statistics in Medicine*, 1992, **11**, 2043–62.]

Latent class analysis: A method of assessing whether a set of observations involving q categorical variables, in particular, binary variables, consists of a number of different groups or classes within which the variables are independent. Essentially a *finite mixture model* in which the component distributions are the product of q Bernoulli distributions, one for each of the binary variables in the data. Parameters in such models can be estimated by *maximum likelihood estimation* via the *EM algorithm*. Can be considered as either an analogue of *factor analysis* for categorical variables, or a model of *cluster analysis* for such data. See also **grade of membership model**. [*An Introduction to Latent Variable Models*, 1984, B. S. Everitt, Chapman and Hall, London.]

Latent class identifiability display: A graphical diagnostic for recognizing weakly identified models in applications of *latent class analysis*. [*Biometrics*, 2000, **56**, 1055–67.]

Latent period: The time interval between the initiation time of a disease process and the time of the first occurrence of a specifically defined manifestation of the disease. An example is the period between exposure to a tumorigenic dose of radioactivity and the appearance of tumours. [*The Mathematical Theory of Infectious Diseases and its Applications*, 1975, N. T. J. Bailey, Arnold, London.]

Latent root distributions: Probability distributions for the latent roots of a square matrix whose elements are random variables having a *joint distribution*. Those of primary importance arise in multivariate analyses based on the assumption of *multivariate normal distributions*. See also **Bartlett's test for eigenvalues**. [*An Introduction to Multivariate Statistics*, 1979, M. S. Srivastava and C. G. Khatri, North Holland, New York.]

Latent roots: Synonym for **eigenvalues**.

Latent variable: A variable that cannot be measured directly, but is assumed to be related to a number of observable or *manifest variables*. Examples include racial prejudice and social class. See also **indicator variable**. [*An Introduction to Latent Variable Models*, 1984, B. S. Everitt, Chapman and Hall/CRC Press, London.]

Latent vectors: Synonym for **eigenvectors**.

Latin hypercube sampling (LHS): A stratified random sampling technique in which a sample of size N from multiple (continuous) variables is drawn such that for each individual variable the sample is (marginally) maximally stratified, where a sample is maximally stratified when the number of strata equals the sample size N and when the probability of falling in each of the strata equals N^{-1}. An example is shown in Fig. 84, involving two independent uniform [0,1] variables, the number of categories per variable equals the sample size (6), each row or each column contains one element and the width of rows and columns is 1/6. [*Reliability Engineering and System Safety*, 2003, **81**, 23–69.]

Latin square: An experimental design aimed at removing from the experimental error the variation from two extraneous sources so that a more sensitive test of the treatment effect can be achieved. The rows and columns of the square represent the levels of the two extraneous factors. The treatments are represented by Roman letters arranged so that no letter appears more than once in each row and column. The following is an example of a 4 × 4 Latin square

A	B	C	D
B	C	D	A
C	D	A	B
D	A	B	C

See also **Graeco-Latin square**. [*Handbook of Experimental Methods for Process Improvement*, 1997, D. Drain, Chapman and Hall, London.]

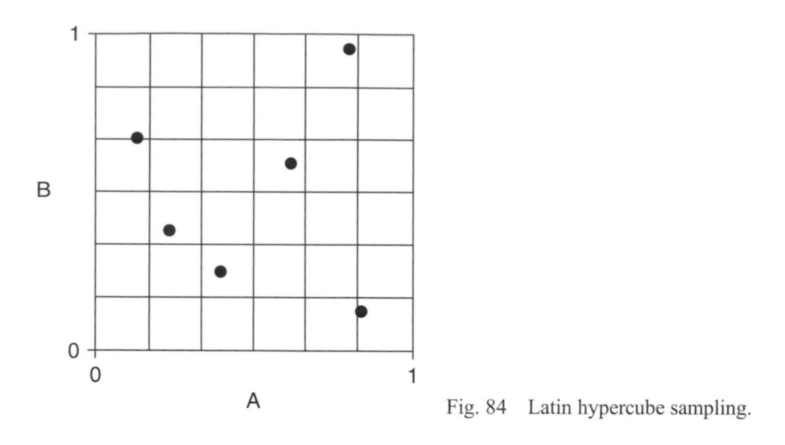

Fig. 84 Latin hypercube sampling.

Lattice designs: A class of *incomplete block designs* introduced to increase the precision of treatment comparisons particularly in agricultural crop trials. [*The Design of Experiments*, 1988, R. Mead, Cambridge University Press, Cambridge.]

Lattice distribution: A class of probability distributions to which most distributions for discrete random variables used in statistics belong. In such distributions the intervals between values of any one random variable for which there are non-zero probabilities are all integral multiples of one quantity. Points with these coordinates thus form a lattice. By an approximate linear transformation it can be arranged that all variables take values which are integers. [*Univariate Discrete Distributions*, 2005, N. I. Johnson, A. W. Kemp and S. Kotz, Wiley, New York.]

Lattice square: An *incomplete block design* for $v = (s - 1)s$ treatments in $b = rs$ *blocks*, each containing $s - 1$ units. In addition the blocks can be arranged into r complete replications. [*Biometrics*, 1977, **33**, 410–13.]

Law of large numbers: A 'law' that attempts to formalize the intuitive notion of probability which assumes that if in n identical trials an event A occurs n_A times, and if n is very large, then n_A/n should be near the probability of A. The formalization involves translating 'identical trials' as *Bernoulli trials* with probability p of a success. The law then states that as n increases, the probability that the average number of successes deviates from p by more then any preassigned value ϵ where $\epsilon > 0$ is arbitrarily small but fixed, tends to zero. [*An Introduction to Probability Theory*, Volume 1, 3rd edition, 1968, W. Feller, Wiley, New York.]

Law of likelihood: Within the framework of a statistical model, a particular set of data supports one statistical hypothesis better than another if the *likelihood* of the first hypothesis, on the data, exceeds the likelihood of the second hypothesis. [*Likelihood*, 1992, A. W. F. Edwards, Cambridge University Press, Cambridge.]

Law of primacy: A 'law' relevant to work in market research which says that an individual for whom, at the moment of choice, n brands are tied for first place in brand strength chooses each of these n brands with probability $1/n$. [*Marketing Research: State of the Art Perspectives*, edited by C. Chakrapani, 2000, American Marketing Association, Chicago.]

Law of truly large numbers: The law that says that, with a large enough sample, any outrageous thing is likely to happen. See also **coincidences**.

LD50: Abbreviation for **lethal dose 50**.

LDU test: A test for the rank of a matrix **A** using an estimate of **A** based on a sample of observations. [*Journal of the American Statistical Association*, 1996, **91**, 1301–9.]

Lead time: An indicator of the effectiveness of a screening test for chronic diseases given by the length of time the diagnosis is advanced by screening. [*International Journal of Epidemiology*, 1982, **11**, 261–7.]

Lead time bias: A term used particularly with respect to cancer studies for the bias that arises when the time from early detection to the time when the cancer would have been symptomatic is added to the *survival time* of each case. [*Journal of the National Cancer Institute,* 1968, **41**, 665–81.]

Leaps-and-bounds algorithm: An algorithm used to find the optimal solution in problems that may have a very large number of possible solutions. Begins by splitting the possible solutions into a number of exclusive subsets and limits the number of subsets that need to be examined in searching for the optimal solution by a number of different strategies. Often used in *all subsets regression* to restrict the number of models that have to be examined. [ARA Chapter 7.]

Least absolute deviation regression: An alternative to *least squares estimation* for determining the parameters in *multiple regression*. The criterion minimized to find the estimators of the regression coefficients is S, given by;

$$S = \sum_{i=1}^{n} | y_i - \beta_0 - \sum_{j=1}^{q} \beta_j x_{ij} |$$

where y_i and $x_{ij}, j = 1, \ldots, q$ are the response variable and explanatory variable values for individual i. The estimators, $b_0, b_1, \ldots, b_q$ of the regression coefficients, $\beta_0, \beta_1, \ldots, \beta_q$ are such that the median$(b_i) = \beta_i$ (*median unbiased*) and are *maximum likelihood estimators* when the errors have a *Laplace distribution*. This type of regression has greater power than that based on least squares for asymmetric error distributions and heavy tailed, symmetric error distributions; it also has greater resistance to the influence of a few outlying values of the dependent variable. [*Transformation and Weighting in Regression*, 1988, R. J. Carroll and D. Ruppert, Chapman and Hall/CRC Press, London.]

Least significant difference test: An approach to comparing a set of means that controls the *familywise error rate* at some particular level, say α. The hypothesis of the equality of the means is tested first by an α-level *F-test*. If this test is not significant, then the procedure terminates without making detailed inferences on pairwise differences; otherwise each pairwise difference is tested by an α-level, *Student's t-test*.

Least squares cross-validation: A method of cross validation in which models are assessed by calculating the sum of squares of differences between the observed values of a sub-set of the data and the relevant predicted values calculated from fitting a model to the remainder of the data. [*Biometrika*, 1984, **71**, 353–60.]

Least squares estimation: A method used for estimating parameters, particularly in regression analysis, by minimizing the difference between the observed response and the value predicted by the model. For example, if the expected value of a response variable y is of the form

$$E(y) = \alpha + \beta x$$

where x is an explanatory variable, then least squares estimators of the parameters α and β may be obtained from n pairs of sample values $(x_1, y_1), (x_2, y_2), \ldots, (x_n, y_n)$ by minimizing S given by

$$S = \sum_{i=1}^{n}(y_i - \alpha - \beta x_i)^2$$

to give

$$\hat{\alpha} = \bar{y} - \hat{\beta}\bar{x}$$

$$\hat{\beta} = \frac{\sum_{i=1}^{n}(x_i - \bar{x})(y_i - \bar{y})}{\sum_{i=1}^{n}(x_i - \bar{x})^2}$$

Often referred to as *ordinary least squares* to differentiate this simple version of the technique from more involved versions such as, *weighted least squares* and *iteratively reweighted least squares.* [ARA Chapter 1.]

Leave-one-out estimator: See **error-rate estimation**.

LeCam, Lucien (1924–2000): Born in Felletin, France, LeCam studied at the University of Paris and received a Licence en Sciences (university diploma) in 1945. His career began with work on power and hydraulic systems at the Electricité de France, but in 1950 he went to Berkeley at the invitation of *Jerzy Neyman.* LeCam received a Ph.D. from the university in 1952. He was made Assistant Professor of Mathematics in 1953 and in 1955 joined the new Department of Statistics at Berkeley. He was departmental chairman from 1961 to 1965. LeCam was a brilliant mathematician and made important contributions to the asymptotic theory of statistics, much of the work being published in his book *Asymptotic Methods in Statistical Decision Theory*, published in 1986.

Left-censored: See **censored observations**.

Lehmann, Erich Leo (1917–2009): Lehmann obtained his MA and PhD from the University of California, Berkeley where he worked from 1942. During the Second World War he worked as an analyst for the US Air Force. In the early 1950s Lehmann worked at Columbia, Princeton and Stanford universities, before returning to Berkeley. Erich Lehmann made numerous important contributions to mathematical statistics, particularly to estimation theory and the theory of hypothesis testing. He served as editor of *The Annals of Mathematical Statistics*, was president of the Institute of Mathematical Statistics, and an elected member of the American Academy of Arts and Sciences and the US National Academy of Science. Erich Lehmann died on September 12th 2009.

Length-biased data: Data that arise when the probability that an item is sampled is proportional to its length. A prime example of this situation occurs in *renewal theory* where inter-event time data are of this type if they are obtained by sampling lifetimes in progress at a randomly chosen point in time. [*Biometrika*, 1996, **83**, 343–54.]

Length-biased sampling: The bias that arises in a sampling scheme based on patient visits, when some individuals are more likely to be selected than others simply because they make more frequent visits. In a *screening study* for cancer, for example, the sample of cases detected is likely to contain an excess of slow-growing cancers compared to the sample diagnosed positive because of their symptoms. [*Australian Journal of Statistics*, 1981, **23**, 91–4.]

Lenth's method: An objective method for deciding which effects are active in the analysis of unreplicated experiments, when the model is saturated and hence there are no degrees of freedom for estimating the error variance. [*Technometrics*, 1989, **31**, 469–473.]

Lepage test: A *distribution free test* for either location or dispersion. The test statistic is related to that used in the *Mann–Whitney test* and the *Ansari–Bradley test.* [NSM Chapter 5.]

Leptokurtic curve: See **kurtosis**.

Leslie matrix model: A model often applied in demographic and animal population studies in which the vector of the number of individuals of each age at time t, $\mathbf{N}_t$, is related to the initial number of individuals of each age, $\mathbf{N}_0$, by the equation

$$\mathbf{N}_t = \mathbf{M}^t \mathbf{N}_0$$

where $\mathbf{M}$ is what is known as the *population projection matrix* given by

$$\mathbf{M} = \begin{pmatrix} B_0 & B_1 & B_2 & \cdots & B_{v-1} & B_v \\ P_0 & 0 & 0 & \cdots & 0 & 0 \\ 0 & P_1 & 0 & \cdots & 0 & 0 \\ \vdots & \vdots & \vdots & \cdots & \vdots & \vdots \\ 0 & 0 & 0 & \cdots & P_{v-1} & P_v \end{pmatrix}$$

where B_x equals the number of females born to females of age x in one unit of time that survive to the next unit of time, P_x equals the proportion of females of age x at time t that survive to time $t+1$ and v is the greatest age attained. See also **population growth model**. [*Bulletin of Mathematical Biology*, 1995, **57**, 381–99.]

Lethal dose 50: The administered dose of a compound that causes death to 50% of the animals during a specified period, in an experiment involving toxic material. [*Modelling Binary Data*, 2nd edition, 2003, D. Collett, Chapman and Hall/CRC Press, London.]

Levene test: A test used for detecting heterogeneity of variance, which consists of an *analysis of variance* applied to *residuals* calculated as the differences between observations and group means. [*Journal of the American Statistical Association*, 1974, **69**, 364–7.]

Leverage points: A term used in regression analysis for those observations that have an extreme value on one or more explanatory variables. The effect of such points is to force the fitted model close to the observed value of the response leading to a small residual. See also **hat matrix, influence statistics** and **Cook's distance**. [ARA Chapter 10.]

Lévy concentration function: A function $Q(x; \lambda)$ of a random variable X defined by the equality

$$Q(X; \lambda) = \sup_x \Pr(x \leq X \leq x + \lambda\eta)$$

for every $\lambda \geq 0$; $Q(X; \lambda)$ is a non-decreasing function of λ satisfying the inequalities $0 \leq Q(X; \lambda) \leq 1$ for every $\lambda \geq 0$. A measure of the variability of the random variable that is used to investigate convergence problems for sums of independent random variables. [*Concentration Functions*, 1973, W. Hengarner and R. Theodorescu, Academic Press, New York.]

Lévy distribution: A probability distribution, $f(x)$, given by

$$f(x) = (2\pi)^{-\frac{1}{2}} \exp[-\frac{1}{2x}] x^{-\frac{3}{2}}, \quad 0 \leq x < \infty$$

None of the moments of the distribution exist. An example of such a distribution is given in Fig. 85. [KA1 Chapter 4.]

Lévy, Paul-Pierre (1886–1971): Born in Paris, Lévy was educated at the Ecole Polytechnique between 1904 and 1906. His main work was in the calculus of probability where he introduced the *characteristic function*, and in *Markov chains, martingales* and *game theory*. Lévy died on 15 December 1971 in Paris.

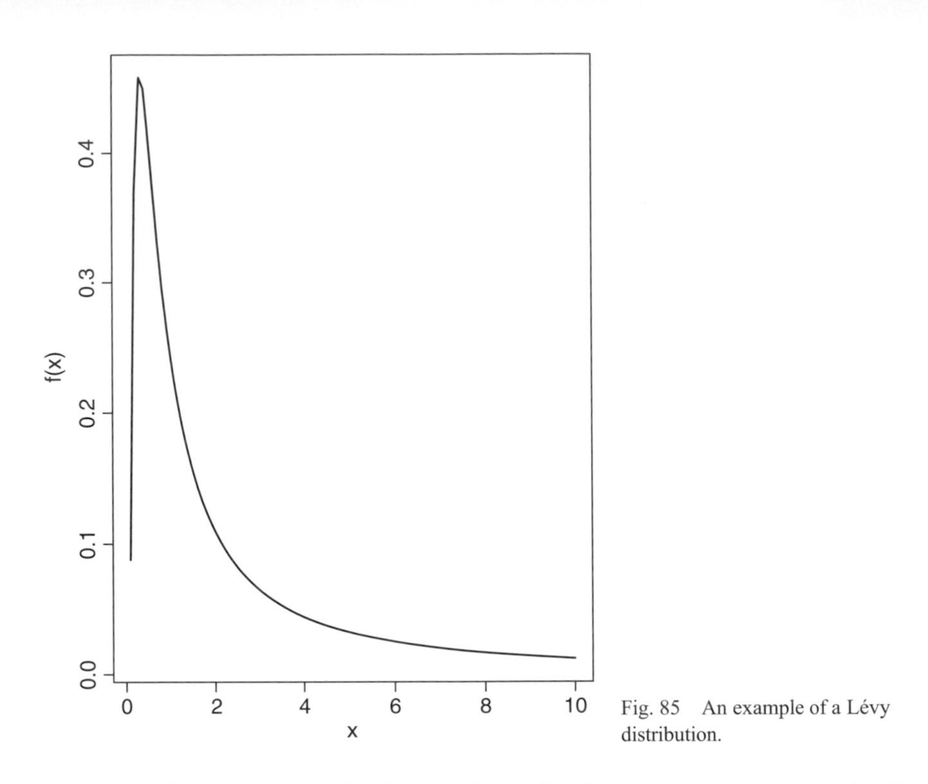

Fig. 85 An example of a Lévy distribution.

Lévy process: A *stochastic process* that has independent and stationary increments and a stochastically continuous sample path. Such processes provide a natural generalization of the sum of independent and identically distributed random variables. [*Journal of Financial Economics*, 2004, **71**, 113–141.]

Lexian distributions: *Finite mixture distributions* having component *binomial distributions* with common *n*. [*Univariate Discrete Distributions*, 2005, N. I. Johnson, A. W. Kemp and S. Kotz, Wiley, New York.]

Lexicostatistics: A term sometimes used for investigations of the evolution times of languages, involving counting the number of cognate words shared by each pair of present-day languages and using these data to reconstruct the ancestry of the family. [*Statistique Textuelle*, 1994, L. Lebart and A. Salem, Dunod, Paris.]

Lexis diagram: A diagram for displaying the simultaneous effects of two time scales (usually age and calendar time) on a rate. For example, mortality rates from cancer of the cervix depend upon age, as a result of the age-dependence of the *incidence*, and upon calendar time as a result of changes in treatment, population screening and so on. The main feature of such a diagram is a series of rectangular regions corresponding to a combination of two time bands, one from each scale. Rates for these combinations of bands can be estimated by allocating failures to the rectangles in which they occur and dividing the total observation time for each subject between rectangles according to how long the subjects spend in each. An example of such a diagram is given in Fig. 86. [*Statistics in Medicine*, 1987, **6**, 449–68.]

Lexis, Wilhelm (1837–1914): Lexis's early studies were in science and mathematics. He graduated from the University of Bonn in 1859 with a thesis on analytic mechanics and a degree in mathematics. In 1861 he went to Paris to study social science and his most important statistical work consisted of several articles published between 1876 and 1880 on population and *vital statistics*.

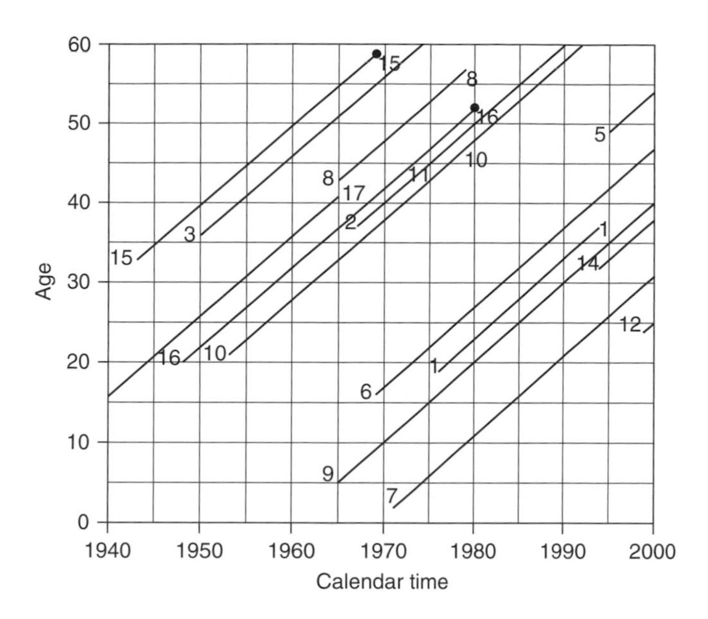

Fig. 86 A Lexis diagram showing age and calendar period.

Liapunov, Alexander Mikhailovich (1857–1918): Born in Yaroslavl, Russia, Liapunov attended Saint Petersburg University from 1876 where he worked under *Chebyshev.* In 1885 he was appointed to Kharkov University to teach mechanics and then in 1902 he left Kharkov for St. Petersburg on his election to the Academy of Sciences. Liapunov made major contributions to probability theory. He died on November 3rd, 1918 in Odessa, USSR.

Liddell, Francis Douglas Kelly (1924–2003): Liddell was educated at Manchester Grammar School from where he won a scholarship to Trinity College, Cambridge, to study mathematics. He graduated in 1945 and was drafted into the Admiralty to work on the design and testing of naval mines. In 1947 he joined the National Coal Board where in his 21 years he progressed from Scientific Officer to Head of the Mathematical Statistics Branch. In 1969 Liddell moved to the Department of Epidemiology at McGill University, where he remained until his retirement in 1992. Liddell contributed to the statistical aspects of investigations of occupational health, particularly exposure to coal, silica and asbestos. He died in Wimbledon, London, on 5 June 2003.

Lie factor: A quantity suggested by Tufte for judging the honesty of a graphical presentation of data. Calculated as

$$\frac{\text{apparent size of effect shown in graph}}{\text{actual size of effect in data}}$$

Values close to one are desirable but it is not uncommon to find values close to zero and greater than five. The example shown in Fig. 87 has a lie factor of about 2.8. [*The Visual Display of Quantitative Information*, 1983, E. R. Tufte, Graphics Press, Cheshire, Connecticut.]

Life expectancy: The expected number of years remaining to be lived by persons of a particular age. Estimated life expectancies at birth for a number of countries in 2008 are as follows:

Country	Life expectancy at birth (years)
Japan	82.67
Iceland	80.43
United Kingdom	78.70
Afghanistan	43.77

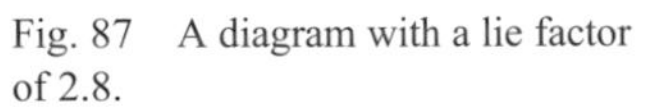
Fig. 87 A diagram with a lie factor of 2.8.

Life table: A procedure used to compute chances of survival and death and remaining years of life, for specific years of age. An example of part of such a table is as follows:

Life table for white females, United States, 1949–1951

1	2	3	4	5	6	7
0	23.55	100000	2355	97965	7203179	72.03
1	1.89	97465	185	97552	7105214	72.77
2	1.12	97460	109	97406	7007662	71.90
3	0.87	97351	85	97308	6910256	70.98
4	0.69	92266	67	97233	6812948	70.04
⋮	⋮	⋮	⋮	⋮	⋮	⋮
100	388.39	294	114	237	566	1.92

1 = Year of age
2 = Death rate per 1000
3 = Number surviving of 100 000 born alive
4 = Number dying of 100 000 born alive
5 = Number of years lived by cohort
6 = Total number of years lived by cohort until all have died
7 = Average future years of life
[SMR Chapter 13.]

Life table analysis: A procedure often applied in *prospective studies* to examine the distribution of mortality and/or morbidity in one or more diseases in a *cohort study* of patients over a fixed period of time. For each specific increment in the follow-up period, the number entering the period, the number leaving during the period, and the number either dying from the disease (mortality) or developing the disease (morbidity), are all calculated. It is assumed that an individual not completing the follow-up period is exposed for half this period, thus enabling the data for those 'leaving' and those 'staying' to be combined into an appropriate denominator for the estimation of the percentage dying from or developing the disease. The advantage of this approach is that all patients, not only those who have been involved for an extended period, can be be included in the estimation process. See also **actuarial estimator** [SMR Chapter 13.]

Lifting scheme: A method for constructing new wavelets with prescribed properties for use in *wavelet analysis*. [*SIAM Journal of Mathematical Analysis,* 1998, **29**, 511–546.]

Likelihood: The probability of a set of observations given the value of some parameter or set of parameters. For example, the likelihood of a random sample of n observations, $x_1, x_2, \ldots, x_n$ with probability distribution, $f(x,\theta)$ is given by

$$L = \prod_{i=1}^{n} f(x_i, \theta)$$

This function is the basis of *maximum likelihood estimation*. In many applications the likelihood involves several parameters, only a few of which are of interest to the investigator. The remaining *nuisance parameters* are necessary in order that the model make sense physically, but their values are largely irrelevant to the investigation and the conclusions to be drawn. Since there are difficulties in dealing with likelihoods that depend on a large number of incidental parameters (for example, maximizing the likelihood will be more difficult) some form of modified likelihood is sought which contains as few of the uninteresting parameters as possible. A number of possibilities are available. For example, the *marginal likelihood*, eliminates the nuisance parameters by integrating them out of the likelihood. The *profile likelihood* with respect to the parameters of interest, is the original likelihood, partially maximized with respect to the nuisance parameters. See also **quasi-likelihood, pseudo-likelihood, partial likelihood, hierarchical likelihood, conditional likelihood, law of likelihood** and **likelihood ratio**. [KA2 Chapter 17.]

Likelihood distance test: A procedure for the detection of *outliers* that uses the difference between the *log-likelihood* of the complete data set and the log-likelihood when a particular observation is removed. If the difference is large then the observation involved is considered an outlier. [*Statistical Inference Based on the Likelihood*, 1996, A. Azzalini, CRC/Chapman and Hall, London.]

Likelihood principle: Within the framework of a statistical model, all the information which the data provide concerning the relative merits of two hypotheses is contained in the *likelihood ratio* of these hypotheses on the data. [*Likelihood*, 1992, A. W. F. Edwards, Cambridge University Press, Cambridge.]

Likelihood ratio test: The ratio of the *likelihoods* of the data under two hypotheses, H_0 and H_1, can be used to assess H_0 against H_1 since under H_0, the statistic, λ, given by

$$\lambda = -2 \ln \frac{L_{H_0}}{L_{H_1}}$$

has approximately a *chi-squared distribution* with degrees of freedom equal to the difference in the number of parameters in the two hypotheses. See also **Wilks' theorem G^2, deviance, goodness-of-fit** and **Bartlett's adjustment factor**. [KA2 Chapter 23.]

Likert, Rensis (1903–1981): Likert was born in Cheyenne, Wyoming and studied civil engineering and sociology at the University of Michigan. In 1932 he obtained a Ph.D. at Columbia

University's Department of Psychology. His doctoral research dealt with the measurement of attitudes. Likert's work made a tremendous impact on social statistics and he received many honours, including being made President of the American Statistical Association in 1959.

Likert scales: Scales often used in studies of attitudes in which the raw scores are based on graded alternative responses to each of a series of questions. For example, the subject may be asked to indicate his/her degree of agreement with each of a series of statements relevant to the attitude. A number is attached to each possible response, e.g. 1:strongly approve; 2:approve; 3:undecided; 4:disapprove; 5:strongly disapprove; and the sum of these used as the composite score. A commonly used Likert-type scale in medicine is the Apgar score used to appraise the status of newborn infants. This is the sum of the points (0,1 or 2) allotted for each of five items:

- heart rate (over 100 beats per minute 2 points, slower 1 point, no beat 0);
- respiratory effort;
- muscle tone;
- response to simulation by a catheter in the nostril;
- skin colour.

[*Scale Development: Theory and Applications*, 1991, R. F. DeVellis, Sage, Newbury Park.]

Lilliefors test: A test used to assess whether data arise from a normally distributed population where the mean and the variance of the population are not specified. See also **Kolmogorov-Smirnov test**. [*Journal of the American Statistical Association*, 1967, **62**, 399–402.]

Lim–Wolfe test: A rank based multiple test procedure for identifying the dose levels that are more effective than the zero-dose control in *randomized block designs*, when it can be assumed that the efficacy of the increasing dose levels in monotonically increasing up to a point, followed by a monotonic decrease. [*Biometrics*, 1997, **53**, 410–18.]

Lindley's paradox: A name used for situations where using *Bayesian inference* suggests very large odds in favour of some null hypothesis when a standard sampling-theory test of significance indicates very strong evidence against it. [*Biometrika*, 1957, **44**, 187–92.]

Linear birth process: Synonym for **Yule–Furry process**.

Linear-by-linear association test: A test for detecting specific types of departure from independence in a *contingency table* in which both the row and column categories have a natural order. See also **Jonckheere–Terpstra test**. [*The Analysis of Contingency Tables*, 2nd edition, 1992, B. S. Everitt, Chapman and Hall/CRC Press, London.]

Linear-circular correlation: A measure of correlation between an interval scaled random variable X and a *circular random variable*, Θ, lying in the interval $(0,2\pi)$. For example, X may refer to temperature and Θ to wind direction. The measure is given by

$$\rho^2_{X\Theta} = (\rho^2_{12} + \rho^2_{13} - 2\rho_{12}\rho_{13}\rho_{23})/(1 - \rho^2_{23})$$

where $\rho_{12} = \text{corr}(X, \cos\ \Theta)$, $\rho_{13} = \text{corr}(X, \sin\Theta)$ and $\rho_{23} = \text{corr}(\cos\Theta, \sin\Theta)$. The sample quantity $R^2_{x\theta}$ is obtained by replacing the ρ_{ij} by the sample coefficients r_{ij}. [*Biometrika*, 1976, **63**, 403–5.]

Linear estimator: An estimator which is a linear function of the observations, or of sample statistics calculated from the observations.

Linear filters: Suppose a series $\{x_t\}$ is passed into a 'black box' which produces $\{y_t\}$ as output. If the following two restrictions are introduced: (1) The relationship is linear; (2) The relationship is invariant over time, then for any t, y_t is a weighted linear combination of past and future values of the input

$$y_t = \sum_{j=-\infty}^{\infty} a_j x_{t-j} \quad \text{with} \quad \sum_{j=-\infty}^{\infty} a_j^2 < \infty$$

It is this relationship which is known as a linear filter. If the input series has power spectrum $h_x(\omega)$ and the output a corresponding spectrum $h_y(\omega)$, they are related by

$$h_y(\omega) = \left| \sum_{j=\infty}^{\infty} a_j e^{-i\omega j} \right|^2 h_x(\omega)$$

if we write $h_y(\omega) = |\Gamma(\omega)|^2 h_x(\omega)$ where $\Gamma(\omega) = \sum_{j=-\infty}^{\infty} a_j e^{-ij\omega}$, then $\Gamma(\omega)$ is called the *transfer function* while $|\Gamma(\omega)$ is called the amplitude gain. The squared value is known as the *gain* or the *power transfer function* of the filter. [TMS Chapter 2.]

Linear function: A function of a set of variables, parameters, etc., that does not contain powers or cross-products of the quantities. For example, the following are all such functions of three variables, x_1, x_2 and x_3,

$$y = x_1 + 2x_2 + x_3$$
$$z = 6x_1 - x_3$$
$$w = 0.34x_1 - 2.4x_2 + 12x_3$$

Linearizing: The conversion of a *non-linear model* into one that is linear, for the purpose of simplifying the estimation of parameters. A common example of the use of the procedure is in association with the *Michaelis–Menten equation*

$$B = \frac{B_{\max} F}{K_D + F}$$

where B and F are the concentrations of bound and free ligand at equilibrium, and the two parameters, $B_{\max}$ and K_D are known as capacity and affinity. This equation can be reduced to a linear form in a number of ways, for example

$$\frac{1}{B} = \frac{1}{B_{\max}} + \frac{K_D}{B_{\max}} \left(\frac{1}{F} \right)$$

is linear in terms of the variables $1/B$ and $1/F$. The resulting linear equation is known as the *Lineweaver–Burk equation*.

Linear logistic regression: Synonym for **logistic regression**.

Linearly separable: A term applied to two groups of observations when there is a linear function of a set of variables $\mathbf{x}' = [x_1, x_2, \ldots, x_q]$, say, $\mathbf{x}'\mathbf{a} + b$ which is positive for one group and negative for the other. See also **discriminant function analysis** and **Fisher's linear discriminant function**. [*Pattern Recognition,* 1979, **11**, 109–114]

Linear model: A model in which the expected value of a random variable is expressed as a linear function of the parameters in the model. Examples of linear models are

$$E(y|x) = \alpha + \beta x$$
$$E(y|x) = \alpha + \beta x + \gamma x^2$$

where x and y represent variable values and α, β and γ parameters. Note that the linearity applies to the parameters not to the variables. See also **linear regression** and **generalized linear models**. [ARA Chapter 6.]

Linear parameter: See **non-linear model**.

Linear regression: A term usually reserved for the simple linear model involving a response, y, that is a continuous variable and a single explanatory variable, x, related by the equation

$$E(y|x) = \alpha + \beta x$$

where E denotes the expected value. See also **multiple regression** and **least squares estimation**. [ARA Chapter 1.]

Linear transformation: A transformation of q variables $x_1, x_2, \ldots, x_q$ given by the q equations

$$\begin{aligned} y_1 &= a_{11}x_1 + a_{12}x_2 + \cdots + a_{1q}x_q \\ y_2 &= a_{21}x_1 + a_{22}x_2 + \cdots + a_{2q}x_q \\ \vdots &= \\ y_p &= a_{p1}x_1 + a_{p2}x_2 + \cdots + a_{qq}x_q \end{aligned}$$

Such a transformation is the basis of *principal components analysis.* [MV1 Chapter 2.]

Linear trend: A relationship between two variables in which the values of one change at a constant rate as the other increases.

Line-intersect sampling: A method of *unequal probability sampling* for selecting *sampling units* in a geographical area. A sample of lines is drawn in a study region and, whenever a sampling unit of interest is intersected by one or more lines, the characteristic of the unit under investigation is recorded. The procedure is illustrated in Fig. 88. As an example, consider an ecological habitat study in which the aim is to estimate the total quantity of berries of a certain plant species in a specified region. A random sample of lines each of the same length is selected and drawn on a map of the region. Field workers walk each of the lines and whenever the line intersects a bush of the correct species, the number of berries on the bush are recorded. [*Sampling Biological Populations*, 1979, edited by R. M. Cormack, G. P. Patil and D. S. Robson, International Co-operative Publishing House, Fairland.]

Lineweaver–Burk equation: See **linearizing**.

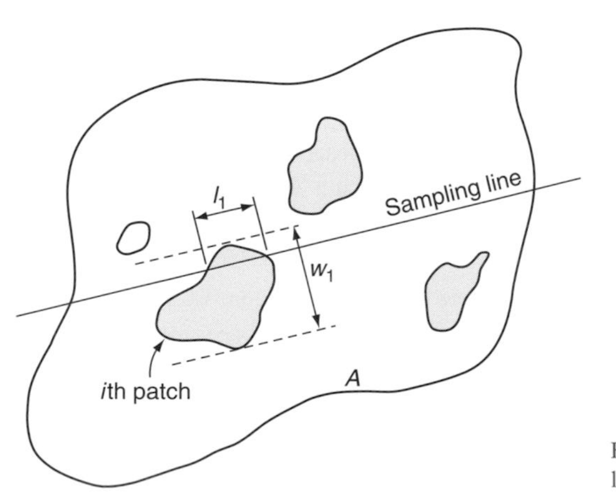

Fig. 88 An illustration of line-intersect sampling.

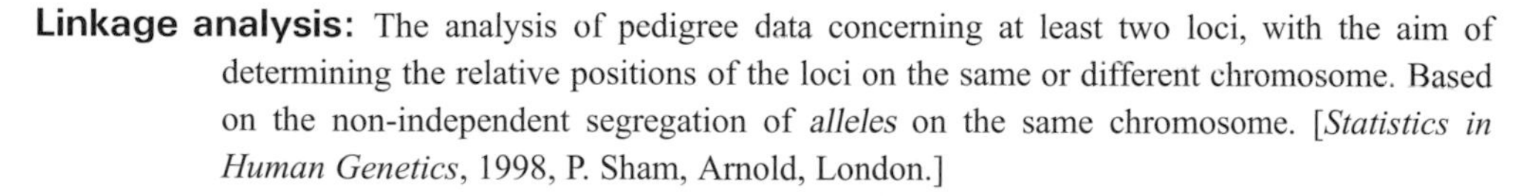
Linkage analysis: The analysis of pedigree data concerning at least two loci, with the aim of determining the relative positions of the loci on the same or different chromosome. Based on the non-independent segregation of *alleles* on the same chromosome. [*Statistics in Human Genetics*, 1998, P. Sham, Arnold, London.]

Linked micromap plot: A plot that provides a graphical overview and details for spatially indexed statistical summaries. The plot shows spatial patterns and statistical patterns while linking regional names to their locations on a map and to estimates represented by statistical panels. Such plots allow the display of *confidence intervals* for estimates and inclusion of more than one variable. [*Statistical Computing and Graphics Newsletter*, 1996, **7**, 16–23.]

Link function: See **generalized linear model**.

Linnik distribution: The probability distribution with *characteristic function*

$$\phi(t) = \frac{1}{1 + |t|^{\alpha}}$$

where $0 < \alpha \leq 2$. [*Ukrainian Mathematical Journal*, 1953, **5**, 207–243.]

Linnik, Yuri Vladimirovich (1915–1972): Born in Belaya Tserkov', Ukraine, Linnik entered the University of Leningrad in 1932 to study physics but eventually switched to mathematics. From 1940 until his death he worked at the Leningrad Division of the Steklov Mathematical Institute and was also made a professor of mathematics at the University of Leningrad. His first publications were on the rate of convergence in the *central limit theorem* for independent symmetric random variables and he continued to make major contributions to the arithmetic of probability distributions. Linnik died on June 30th, 1972 in Leningrad.

LISREL: A computer program for fitting *structural equation models* involving *latent variables*. See also **EQS** and **Mplus**. [Scientific Software Inc., 1369, Neitzel Road, Mooresville, IN 46158-9312, USA.]

Literature controls: Patients with the disease of interest who have received, in the past, one of two treatments under investigation, and for whom results have been published in the literature, now used as a control group for patients currently receiving the alternative treatment. Such a control group clearly requires careful checking for comparability. See also **historical controls**.

Loading matrix: See **factor analysis**.

Lobachevsky distribution: The probability distribution of the sum (X) of n independent random variables, each having a *uniform distribution* over the interval $[-\frac{1}{2}, \frac{1}{2}]$, and given by

$$f(x) = \frac{1}{(n-1)!} \sum_{j=0}^{[x+n/2]} (-1)^j \binom{n}{j} \left[x + \frac{1}{2}n - j\right]^{n-1} \quad -\frac{1}{2}n \leq x \leq \frac{1}{2}n$$

where $[x + n/2]$ denotes the integer part of $x + n/2$. [*Probability Theory: A Historical Sketch*, L. E. Maistrov, 1974, Academic Press, New York.]

Local dependence fuction: An approach to measuring the dependence of two variables when both the degree and direction of the dependence is different in different regions of the plane. [*Biometrika*, 1996, **83**, 899–904.]

Locally weighted regression: A method of regression analysis in which polynomials of degree one (linear) or two (quadratic) are used to approximate the regression function in particular 'neighbourhoods' of the space of the explanatory variables. Often useful for smoothing scatter diagrams to allow any structure to be seen more clearly and for identifying possible non-linear relationships between the response and explanatory variables. A *robust*

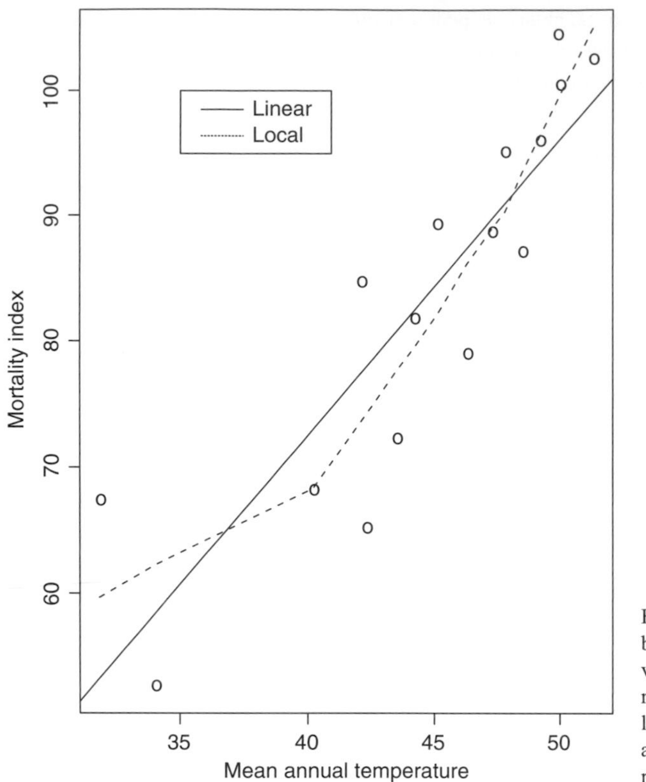

Fig. 89 Scatterplot of breast cancer mortality rate versus temperature of region with lines fitted by least squares calculation and by locally weighted regression.

estimation procedure (usually known as *loess*) is used to guard against deviant points distorting the smoothed points. Essentially the process involves an adaptation of *iteratively reweighted least squares*. The example shown in Fig. 89 illustrates a situation in which the locally weighted regression differs considerably from the linear regression of y on x as fitted by *least squares estimation*. See also **kernel regression smoothing**. [*Journal of the American Statistical Association*, 1979, **74**, 829–36.]

Local odds ratio: The *odds ratio* of the *two-by-two contingency tables* formed from adjacent rows and columns in a larger *contingency table*.

Location: The notion of central or 'typical value' in a sample distribution. See also **mean, median** and **mode**.

LOCF: Abbreviation for **last observation carried forward**.

Lods: A term sometimes used in *epidemiology* for the logarithm of an *odds ratio*. Also used in genetics for the logarithm of a *likelihood ratio*.

Lod score: The common logarithm (base 10) of the ratio of the *likelihood* of pedigree data evaluated at a certain value of the recombination fraction to that evaluated at a recombination fraction of a half (that is no linkage). [*Statistics in Human Gentics*, 1998, P. Sham, Arnold, London.]

Loess: See **locally weighted regression**.

Loewe additivity model: A model for studying and understanding the joint effect of combined treatments. Used in pharmacology and in the development of combination therapies. [*Journal of Pharmaceutical Statistics*, 2007, **17**, 461–480.]

Logarithmic series distribution: The probability distribution of a discrete random variable, X, given by

$$\Pr(X = r) = k\gamma^r/r \qquad x \geq 0$$

where γ is a *shape parameter* lying between zero and one, and $k = -1/\log(1-\gamma)$. The distribution is the limiting form of the *negative binomial distribution* with the zero class missing. The mean of the distribution is $k\gamma/(1-\gamma)$ and its variance is $k\gamma(1-k\gamma)/(1-\gamma)^2$. The distribution has been widely used by entomologists to describe species abundance. [STD Chapter 23.]

Logarithmic transformation: The transformation of a variable, x, obtained by taking $y = \ln(x)$. Often used when the frequency distribution of the variable, x, shows a moderate to large degree of *skewness* in order to achieve normality.

Log-cumulative hazard plot: A plot used in *survival analysis* to assess whether particular parametric models for the survival times are tenable. Values of $\ln(-\ln \hat{S}(t))$ are plotted against ln t, where $\hat{S}(t)$ is the estimated *survival function.* For example, an approximately linear plot suggests that the survival times have a *Weibull distribution* and the plot can be used to provide rough estimates of its two parameters. When the slope of the line is close to one, then an *exponential distribution* is implied. [*Modelling Survival Data in Medical Research*, 2nd edition, 2003, D. Collett, Chapman and Hall/CRC Press, London.]

Log-F accelerated failure time model: An *accelerated failure time model* with a generalized F-distribution for *survival time.* [*Statistics in Medicine*, 1988, **5**, 85–96.]

Logistic distribution: The limiting probability distribution as n tends to infinity, of the average of the largest to smallest sample values, of random samples of size n from an *exponential distribution.* The distribution is given by

$$f(x) = \frac{\exp[-(x-\alpha)/\beta]}{\beta\{1+\exp[-(x-\alpha)/\beta]\}^2} \qquad -\infty < x < \infty, \beta > 0$$

The *location parameter,* α is the mean. The variance of the distribution is $\pi^2\beta^2/3$, its *skewness* is zero and its *kurtosis,* 4.2. The *standard logistic distribution* with $\alpha = 0, \beta = 1$ with *cumulative probability distribution function,* $F(x)$, and probability distribution, $f(x)$, has the property

$$f(x) = F(x)[1 - F(x)]$$

[STD Chapter 24.]

Logistic growth model: The model appropriate for a *growth curve* when the rate of growth is proportional to the product of the size at the time and the amount of growth remaining. Specifically the model is defined by the equation

$$y = \frac{\alpha}{1 + \gamma e^{-\beta t}}$$

where α, β and γ are parameters. [*Journal of Zoology*, 1997, **242**, 193–207.]

Logistic normal distributions: A class of distributions that can model dependence more flexibly that the *Dirichlet distribution.* [*Biometrika*, 1980, **67**, 261–72.]

Logistic regression: A form of regression analysis used when the response variable is a binary variable. The method is based on the *logistic transformation* or *logit* of a proportion, namely

$$\text{logit}(p) = \ln \frac{p}{1-p}$$

As p tends to 0, logit(p) tends to $-\infty$ and as p tends to 1, logit(p) tends to ∞. The function logit(p) is a *sigmoid curve* that is symmetric about $p = 0.5$. Applying this transformation, this form of regression is written as;

$$\ln \frac{p}{1-p} = \beta_0 + \beta_1 x_1 + \cdots + \beta_q x_q$$

where p = Pr(dependent variable = 1) and $x_1, x_2, \ldots, x_q$ are the explanatory variables. Using the logistic transformation in this way overcomes problems that might arise if p was modelled directly as a linear function of the explanatory variables, in particular it avoids fitted probabilities outside the range (0,1). The parameters in the model can be estimated by *maximum likelihood estimation.* See also **generalized linear models**, **mixed effects logistic regression, multinomial regression**, and **ordered logistic regression.** [SMR Chapter 12.]

Logistic transformation: See **logistic regression**.

Logit: See **logistic regression**.

Logit confidence limits: The upper and lower ends of the *confidence interval* for the logarithm of the *odds ratio*, given by

$$\ln \hat{\psi} \pm z_{\alpha/2} \sqrt{\text{var}(\ln \hat{\psi})}$$

where $\hat{\psi}$ is the estimated odds ratio, $z_{\alpha/2}$ the normal equivalent deviate corresponding to a value of $\alpha/2$, $(1-\alpha)$ being the chosen size of the confidence interval. The variance term may be estimated by

$$\hat{\text{var}}(\ln \hat{\psi}) = \frac{1}{a} + \frac{1}{b} + \frac{1}{c} + \frac{1}{d}$$

where a,b,c and d are the frequencies in the *two-by-two contingency table* from which $\hat{\psi}$ is calculated. The two limits may be exponentiated to yield a corresponding confidence interval for the odds ratio itself. [*The Analysis of Contingency Tables*, 2nd edition, 1992, B. S. Everitt, Chapman and Hall/CRC Press, London.]

Logit rank plot: A plot of logit{Pr(E|S)} against logit(r) where Pr(E|S) is the *conditional probability* of an event E given a risk score S and is an increasing function of S, and r is the proportional rank of S in a sample of population. The slope of the plot gives an overall measure of effectiveness and provides a common basis on which different risk scores can be compared. [*Applied Statistics*, 199, **48**, 165–83.]

Log-likelihood: The logarithm of the *likelihood.* Generally easier to work with than the likelihood itself when using *maximum likelihood estimation.*

Log-linear models: Models for *count data* in which the logarithm of the expected value of a count variable is modelled as a linear function of parameters; the latter represent associations between pairs of variables and higher order interactions between more than two variables. Estimated expected frequencies under particular models can be found from *iterative proportional fitting.* Such models are, essentially, the equivalent for frequency data, of the models for continuous data used in *analysis of variance*, except that interest usually now centres on parameters representing interactions rather than those for main effects. See also **generalized linear model**. [*The Analysis of Contingency Tables*, 2nd edition, 1992, B. S. Everitt, Chapman and Hall/CRC Press, London.]

Lognormal distribution: The probability distribution of a random variable, X, for which ln (X) has a normal distribution with mean μ and variance σ^2. The distribution is given by

$$f(x) = \frac{1}{x\sigma(2\pi)^{\frac{1}{2}}} \exp\left[-\frac{1}{2\sigma^2}(\ln x - \mu)^2\right] \qquad 0 \leq x < \infty$$

The mean, variance, *skewness* and *kurtosis* of the distribution are

$$\begin{aligned}
\text{mean} &= \exp(\mu + \frac{1}{2}\sigma^2) \\
\text{variance} &= \exp(2\mu + \sigma^2)(\exp(\sigma^2) - 1) \\
\text{skewness} &= (\exp(\sigma^2) + 2)(\exp(\sigma^2) - 1)^{\frac{1}{2}} \\
\text{kurtosis} &= \exp(4\sigma^2) + 2\exp(3\sigma^2) + 3\exp(2\sigma^2) - 3
\end{aligned}$$

For small σ the distribution is approximated by the normal distribution. Some examples of the distribution are given in Fig. 90. [STD Chapter 25.]

Logrank test: A test for comparing two or more sets of *survival times*, to assess the null hypothesis that there is no difference in the survival experience of the individuals in the different groups. For the two-group situation the *test statistic* is

$$U = \sum_{j=1}^{r}(d_{1j} - e_{1j})$$

where d_{1j} is the number of deaths in the first group at $t_{(j)}$, the jth ordered death time, $j = 1, 2, \ldots, r$, and e_{1j} is the corresponding expected number of deaths given by

$$e_{1j} = n_{1j}d_j/n_j$$

where d_j is the total number of deaths at time $t_{(j)}$, n_j is the total number of individuals at risk at this time, and n_{1j} the number of individuals at risk in the first group. The expected value of U is zero and its variance is given by

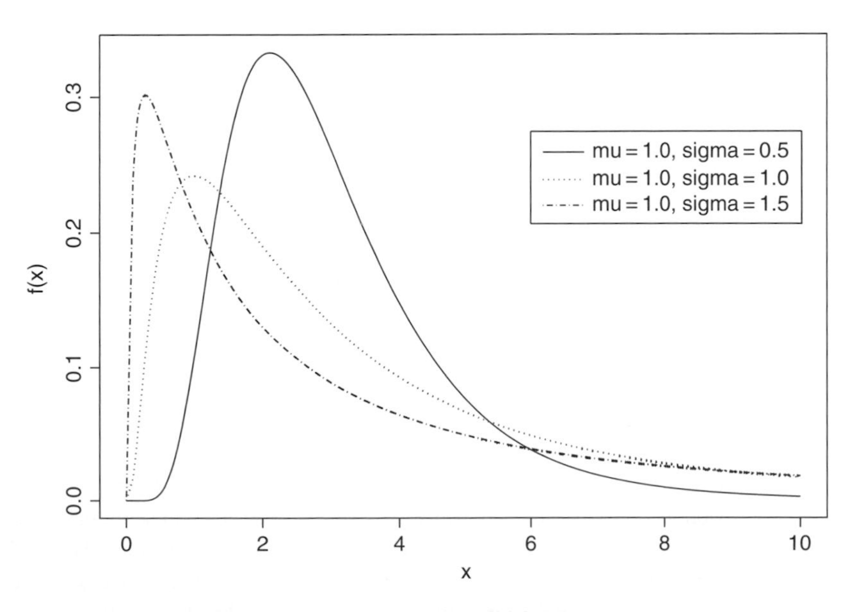

Fig. 90 Lognormal distributions for $\mu == 1.0$; $\sigma = 0.5, 1.0, 1.5$.

$$V = \sum_{j=1}^{r} n_j v_{1j}$$

where

$$v_{1j} = \frac{n_{1j} n_{2j} d_j (n_j - d_j)}{n_j^2 (n_j - 1)}$$

Consequently $U/\sqrt{V}$ can be referred to a standard normal distribution to assess the hypothesis of interest. Other tests use the same test statistic with different values for the weights. The *Tarone–Ware test*, for example, uses $w_j = \sqrt{n_j}$ and the *Peto-Prentice test* uses

$$w_j = \prod_{i=1}^{j} \frac{n_i - d_i + 1}{n_i + 1}$$

[SMR Chapter 13.]

LOGXACT: A specialized statistical package that provides exact inference capabilities for *logistic regression.* [Cytel Software Corp., 675 Massachusetts Avenue, Cambridge, MA 02139 USA.]

Lomb periodogram: A generalization of the *periodogram* for unequally spaced *time series.* [*Biological Rhythm Research*, 2001, **32**, 341–5.]

Longini–Koopman model: In *epidemiology* a model for primary and secondary infection, based on the characterization of the *extra-binomial variation* in an infection rate that might arise due to the 'clustering' of the infected individual within households. The assumptions underlying the model are:

- a person may become infected at most once during the course of the epidemic;
- all persons are members of a closed 'community'. In addition each person belongs to a single 'household'. A household may consist of one or several individuals;
- the sources of infection from the community are distributed homogeneously throughout the community. Household members mix at random within the household;
- each person can be infected either from within the household or from the community. The probability that a person is infected from the community is independent of the number of infected members in his or her household;

The probability that exactly k additional individuals will become infected for a household with s initial susceptibles and j initial infections is

$$P(k|s,j) = \binom{s}{k} P(k|k,j) B^{(s-k)} Q^{(j+k)(s-k)} \qquad k = 0, 1, \ldots, s-1$$

$$P(s|s,j) = 1 - \sum_{k=0}^{s-1} P(k|s,k)$$

where B is the probability that a susceptible individual is not infected from the community during the course of the infection, and Q is the probability that a susceptible person escapes infection from a single infected household member. [*Statistics in Medicine*, 1994, **13**, 1563–74.]

Longitudinal data: Data arising when each of a number of subjects or patients give rise to a vector of measurements representing the same variable observed at a number of different time points. Such data combine elements of multivariate data and *time series* data. They differ from the former, however, in that only a single variable is involved, and from the latter in consisting of a (possibly) large number of short series, one from each subject, rather than a single long series. Such data can be collected either prospectively, following subjects forward in time, or retrospectively, by extracting measurements on each person from historical records. This type of data is also often known as *repeated measures data*, particularly in the social and behavioural sciences, although in these disciplines such data are more likely to arise from observing individuals repeatedly under different experimental conditions rather than from a simple time sequence. Special statistical methods are often needed for the analysis of this type of data because the set of measurements on one subject tend to be intercorrelated. This correlation must be taken into account to draw valid scientific inferences. The design of most such studies specifies that all subjects are to have the same number of repeated measurements made at equivalent time intervals. Such data is generally referred to as *balanced longitudinal data*. But although balanced data is generally the aim, *unbalanced longitudinal data* in which subjects may have different numbers of repeated measurements made at differing time intervals, do arise for a variety of reasons. Occasionally the data are unbalanced or incomplete by design; an investigator may, for example, choose in advance to take measurements every hour on one half of the subjects and every two hours on the other half. In general, however, the main reason for unbalanced data in a longitudinal study is the occurrence of *missing values* in the sense that intended measurements are not taken, are lost or are otherwise unavailable. See also **Greenhouse and Geisser correction, Huynh–Feldt correction, compound symmetry, generalized estimating equations, Mauchly test, response feature analysis, time-by-time ANOVA** and **split-plot design**. [*Analysis of Longitudinal Data*, 2nd edition, 2002, P. J. Diggle, P. J. Heagerty, K.-Y. Liang and S. Zeger, Oxford Scientific Publications, Oxford.]

Longitudinal studies: Studies that give rise to *longitudinal data*. The defining characteristic of such a study is that subjects are measured repeatedly through time.

Long memory processes: A *stationary stochastic process* with slowly decaying or long-range correlations. See also **long-range dependence**. [*Statistics for Long Memory Processes*, 1995, J. Beran, Chapman and Hall/CRC Press, London.]

Long-range dependence: Small but slowly decaying correlations in a stochastic process. Such correlations are often not detected by standard tests, but their effect can be quite strong. [*Journal of the American Statistical Association*, 1997, **92**, 881–93.]

Lord, Frederic Mather (1912–2000): Born in Hanover, New Hampshire, Lord graduated from Dartmouth College in 1936 and received a Ph.D. from Princeton in 1952. In 1944 he joined the Educational Testing Service and is recognized as the principal developer of the statistical machinery underlying modern mental testing. Lord died on 5 February 2000 in Naples, Florida.

Lord's paradox: The fact that estimates of treatment effects using *change scores*, where (posttest–pretest) is regressed on treatment, differ from *analysis of covariance*, where the posttest is regressed on both treatment and pretest. [*Sociological Methodology*, 1990, **20**, 93–114.]

Lorenz curve: Essentially a graphical representation of the cumulative distribution of a variable, most often used for income or wealth. If the risks of disease are not monotonically increasing as the exposure becomes heavier, the data have to be rearranged from the lowest to the highest risk before the calculation of the cumulative percentages. Associated with such a curve is the *Gini index* defined as twice the area between the curve and the diagonal line. This index is between zero and one, with larger values indicating greater variability while smaller ones indicate greater uniformity. The further the Lorenz curve lies below the line of equality, the more unequal is the distribution of, in Figure 91 for example, income. [KA1 Chapter 2.]

Loss function: See **decision theory**.

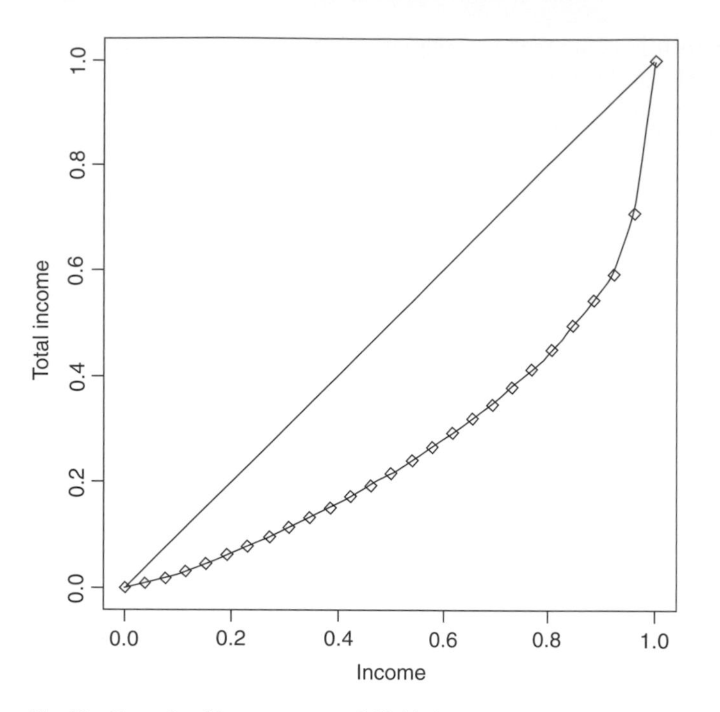

Fig. 91 Example of Lorenz curve and Gini index.

Lotteries: See **national lotteries**.

Low-dose extrapolation: The estimation of the potential risks associated with a low dose of, for example, a suspected carcinogen, for observed data on individuals exposed to moderately high levels of the agent of interest. One approach is to rely on an assumed mathematical function relating risk to exposure, for example the *probit* or *logit.* In many cases however it is possible to postulate different models that fit the observed data equally well, but which provide point estimates of risk at low exposures that differ by several orders of magnitude. See also **detection limits** and **dose-response curve**. [*Communications in Statistics: Simulation and Computation*, 1988, **11**, 27–45.]

Lower triangular matrix: A matrix in which all the elements above the main diagonal are zero. An example is the following,

$$\mathbf{L} = \begin{pmatrix} 1 & 0 & 0 & 0 \\ 2 & 3 & 0 & 0 \\ 1 & 1 & 3 & 0 \\ 1 & 5 & 6 & 7 \end{pmatrix}$$

LSD: Abbreviation for **least significant difference**.

LST: Abbreviation for **large simple trial**.

L-statistics: Linear functions of order statistics often used in estimation problems because they are typically computationally simple. [*Statistics in Civil Engineering*, 1997, A. V. Metcalfe, Edward Arnold, London.]

Luganni and Rice formula: A *saddlepoint method* approximation to a probability distribution from a corresponding *cumulant generating function.* [KA1 Chapter 11.]

Lyapunov exponent: A measure of the experimental divergence of trajectories in chaotic systems [*Fractal Geometry*, 1990, K. Falconer, Wiley, Chichester.]

Lynden–Bell method: A method for estimating the *hazard rate*, or probability distribution of a random variable observed subject to data truncation. It is closely related to the *product-limit estimator* for censored data. [*Monthly Notices of the Royal Astronomical Society*, 1987, **226**, 273–80.]

Machine learning: A term that literally means the ability of a machine to recognize patterns that have occurred repeatedly and to improve its performance based on past experience. In essence this reduces to the study of computer algorithms that improve automatically through experience. A computer program is said to learn from experience E with respect to some class of tasks T and performance measure P, if its performance at tasks in T, as measured by P, improves with experience E. Machine learning is inherently a multi-disciplinary field using results and techniques from probability and statistics, computational complexity theory, *information theory* etc; it is closely related to *pattern recognition* and *artificial intelligence* and is widely used in modern *data mining*. [*Introduction to Machine Learning*, 2004, E. Alpaydin, MIT Press, Boston.]

Mack–Wolfe test: A *distribution free test* for one way designs for testing a null hypothesis of equality against an alternative specifying an *umbrella ordering*. [NSM Chapter 6.]

MAD: Abbreviation for **median absolute deviation**.

MADAM: Abbreviation for **mean and dispersion additive model**.

Mahalanobis D^2: A measure of the distance between two groups of individuals given by

$$D^2 = (\bar{\mathbf{x}}_1 - \bar{\mathbf{x}}_2)'\mathbf{S}^{-1}(\bar{\mathbf{x}}_1 - \bar{\mathbf{x}}_2)$$

where $\bar{\mathbf{x}}_1$ and $\bar{\mathbf{x}}_2$ are the mean vectors of the two groups and $\mathbf{S}$ is a weighted average of the *variance–covariance matrices* of the two groups, $\mathbf{S_1}$ and $\mathbf{S_2}$, i.e.

$$\mathbf{S} = \frac{n_1\mathbf{S}_1 + n_2\mathbf{S}_2}{n_1 + n_2}$$

where n_1 and n_2 are the sample sizes in the two groups. See also **Hotelling's T^2 test**. [MV1 Chapter 4.]

Mahalanobis, Pransanta Chandra (1893–1972): Born in Calcutta, India, Mahalanobis first studied physics in Calcutta and then in Cambridge where he was elected a senior research scholar in 1915. He eventually returned to India to teach at the Presidency College in Calcutta. Mahalanobis made major contributions to many areas of statistics. His D^2 statistic arose from work on anthropometric problems. Mahalanobis' work on field experiments led to a close friendship with *Fisher*. In 1931 he founded the Indian Statistical Institute. Mahalanobis was made a Fellow of the Royal Society in 1945 and given one of his country's highest awards the *Padma Vibhushan*. He died on 28 June 1972 in Calcutta.

Mahalonobis-Taguchi system: A diagnostic and predictive method for analyzing patterns in *multivariate data*. The system is reported to make accurate forecasts using small, correlated data sets. [*Concurrent Engineering*, 2006, **14**, 343–354.]

Main effect: An estimate of the independent effect of usually a factor variable on a response variable in *analysis of variance*. [SMR Chapter 12.]

Mainframes: High speed, large and expensive computers with a very large storage capacity and capable of supporting thousands of users simultaneously. Mainframes support more simultaneous programs than *supercomputers*, but the latter can execute a single program faster.

Mainland, Donald (1902–1985): Mainland graduated in medicine at Edinburgh University and in 1930 he became Professor and Chairman of the Department of Anatomy at Dalhousie University. He later developed an interest in measurement issues and statistics, publishing a book on statistics in medicine in 1937. In 1950 Mainland became Professor of Medical Statistics at New York University. A prolific writer on statistical issues, Mainland died in July 1985 in Kent, Connecticut.

Majority rule: A requirement that the majority of a series of *diagnostic tests* are positive before declaring that a patient has a particular complaint. See also **unanimity rule**. [*Statistics in Medicine*, 1987, **7**, 549–557.]

Mallows' C_p statistic: An index used in *multiple regression analysis* as an aid in choosing the 'best' subset of explanatory variables. The index is defined as

$$C_p = \sum_{i=1}^{n} (y_i - \hat{y}_i^{(k)})^2 / s^2 - n + 2p$$

where n is the number of observations, y_i is the observed value of the response variable for individual i, $\hat{y}_i^{(k)}$ is the corresponding predicted value from a model with a particular set of k explanatory variables and s^2 is the residual mean square after regression on the complete set of p explanatory variables. The model chosen is the one with the minimum value of C_p. See also **Akaike's information criterion** and **all subsets regression**. [ARA Chapter 7.]

Malthus, Thomas Robert (1766–1834): Malthus studied history, poetry, modern languages, classics and mathematics at Jesus College, Cambridge. After being elected to a Fellowship at Jesus in 1793 he became a curate in a small town, Albury in 1798, the year in which he published the first version of his famous work, *Essay on the Principle of Population as it affects the Future Improvement of Society.* Malthus was a strong advocate of statistical investigation and was a founder member of the Statistical Society of London. He died in December 1834 in Bath, UK.

Malthusian parameter: The rate of increase that a population would ultimately attain if its age-specific birth and death rates were to continue indefinitely. Explicitly the parameter λ in the exponential equation of population growth, $N(t) = N_0\, e^{\lambda t}$, where N_0 is the initial population size ($t = 0$) and t is the elapsed time. See also **population growth model**. [*Proceedings of the National Academy of Science, USA*, 1996, **93**, 276–8.]

Mandel, John (1914–2007): Born in Antwerp, Belgium, Mandel received a master's degree in chemistry from the University Libre de Bruxelles in 1937 and worked as a chemist in Belgium up to 1940 when he was forced to flee to France and then Portugal before travelling to the United States. In New York he again worked as a chemist, studying mathematics at night at Brooklyn College. In 1965 Mandel received a doctorate in mathematical statistics from the Eindhoven University of Technology in the Netherlands. Mandel then spent 40 years as a statistician at the National Institute of Standards and Technology in Washington and became an internationally recognized authority on the application of mathematical and statistical methods in the physical sciences. He received many honours including the Shewhart Medal and the American Statistical Association's W. J. Youden Award. Mandel died on the 1st October 2007 in Silver Springs, USA.

Manhattan distance: Synonym for **city block distance**.

Manifest variable: A variable that can be measured directly, in contrast to a *latent variable*.

Mann–Whitney test: A *distribution free test* used as an alternative to the *Student's t-test* for assessing whether two populations have the same median. The test statistic U is calculated from comparing each pair of values, one from each group, scoring these pairs 1 or 0 depending on whether the first group observation is higher or lower than that from the second group and summing the resulting scores over all pairs. In the case of no ties then

$$U = W - n(n+1)/2$$

where n is the number of observations in the smaller group and W is the rank sum statistic associated with *Wilcoxon's rank sum test*. The two tests are thus equivalent. Tables giving critical values of the test statistic are available, and for moderate and large sample sizes, a normal approximation can be used. [SMR Chapter 6.]

MANOVA: Acronym for **multivariate analysis of variance**.

Mantel, Nathan (1919–2002): Born in New York, Mantel graduated from City College in 1939 with a major in statistics. His career as a statistician began in 1940 in the War Production Board. After the war Mantel joined the National Cancer Institute where he remained until his retirement. Mantel made substantial contributions to biostatistics and epidemiology, including the *Mantel–Haenszel estimator*. He died on 26 May 2002, in Potomac, Maryland.

Mantel–Haenszel estimator: An estimator of the assumed common *odds ratio* in a series of *two-by-two contingency* tables arising from different populations, for example, occupation, country of origin, etc. Specifically the estimator is defined as

$$\omega = \sum_{i=1}^{k} a_i d_i / \sum_{i=1}^{k} c_i b_i$$

where k is the number of two-by-two tables involved and a_i, b_i, c_i d_i are the four counts in the ith table. [*Case-Control Studies: Design, Conduct, Analysis*, 1982, J. J. Schlesselman, Oxford University Press, Oxford.]

Many-outlier detection procedures: Procedures for detecting multiple *outliers* in a sample of n observations, $x_1, x_2, \ldots, x_n$. Generally based on assuming that k of the observations $x_{i_1}, x_{i_2}, \ldots, x_{i_k}$ are outliers distributed normally with mean $\mu + \lambda_{i_j}, j = 1, \ldots, k$ and variance σ^2, while the remaining n-k observations are normally distributed with mean μ and again variance σ^2. Three such procedures are:

1. *The extreme standardized deviate* (ESD), given by

$$\text{ESD}_i = |x^{(S_i-1)} - \bar{x}(S_{i-1})|/s(S_{i-1}), \quad i = 1, \ldots, k$$

where $x^{(S_i-1)}$ is the observation farthest away from the mean $\bar{x}(S_{i-1})$ in the subset S_{i-1}, and

$$s^2(S_{i-1}) = \sum_{j=1}^{n-i+1} [x_j(S_{i-1}) - \bar{x}(S_{i-1})]^2/(n-i)$$

$S_0 = \{x_1, x_2, \ldots, x_n\}$ and subsets $S_i, i = 1, \ldots, k-1$ are formed by deleting an observation farthest from the mean of the subset S_{i-1}

2. *The standardized range procedure* (STR):

$$\text{STR}_i = [x_l(S_{i-1}) - x_s(S_{i-1})]/s(S_{i-1}), \quad i = 1, \ldots, k$$

where $x_l(S_{i-1})$ is the largest observation in the subset S_{i-1}, and $x_s(S_{i-1})$ is the smallest observation in the subset S_{i-1}.

3. *The kurtosis procedure* (KUR):

$$\text{KUR}_i = (n-i+1)\sum_{j=1}^{n-i+1}[x_j(S_{i-1}) - \bar{x}(S_{i-1})]^4/[\sum_{j=1}^{n-i+1}\{x_j(S_{i-1}) - \bar{x}(S_{i-1})]^2\}^2,$$
$$i = 1, \ldots, k$$

Sample values of any of these statistics are compared with the corresponding critical values for $i=k$. If a sample value is found to be greater than the corresponding critical value, k outliers are declared to be present in the sample; otherwise, similar comparisons are made for $i = k-1, \ldots, 1$ until for a particular value of $i=l$, the sample value is greater than the corresponding ordeal value in which case l outliers are declared to be present in the sample. If for $i=1$ the sample value of the statistics is less than or equal to the corresponding critical value, no outliers are detected. [*Technometrics*, 1975, **17**, 221–7.]

MAP estimate: Abbreviation for **maximum a posteriori estimate**.

Maple: A computer system for both mathematical computation and computer algebra. See also **Mathematica** [http://www.maplesoft.com/]

Mardia's multivariate normality test: A test that a set of *multivariate data* arise from a *multivariate normal distribution* against departures due to *kurtosis*. The test is based on the following *multivariate kurtosis measure*

$$b_{2,q} = \frac{1}{n}\sum_{i=1}^{n}\{(\mathbf{x}_i - \bar{\mathbf{x}})'\mathbf{S}^{-1}(\mathbf{x}_i - \bar{\mathbf{x}})\}^2$$

where q is the number of variables, n is the sample size, $\mathbf{x_i}$ is the vector of observations for subject i, $\bar{\mathbf{x}}$ is the mean vector of the observations, and $\mathbf{S}$ is the sample *variance–covariance matrix*. For large samples under the hypothesis that the data arise from a multivariate normal distribution, $b_{2,q}$, has a normal distribution with mean $q(q+2)$ and variance $8q(q+2)/n$. [*Multivariate Analysis*, 1979, K. V. Mardia, J. T. Kent and J. M. Bibby, Academic Press, London.]

MAREG: A software package for the analysis of marginal regression models. The package allows the application of *generalized estimating equations* and *maximum likelihood methods*, and includes techniques for handling *missing values*. [*Statistical Software Newsletter in Computational Statistics and Data Analysis*, 1997, **24**, 237–41 and http://www.stat.uni-muenchen.de/sfb386/software/mareg/winmareg.html]

Marginal distribution: The probability distribution of a single variable, or combinations of variables, in a multivariate distribution. Obtained from the multivariate distribution by integrating over the other variables. [KA1 Chapter 1.]

Marginal homogeneity: A term applied to *square contingency tables* when the probabilities of falling in each category of the row classification equal the corresponding probabilities for the column classification. See also **Stuart–Maxwell test**.

Marginal likelihood: See **likelihood**.

Marginal matching: The matching of treatment groups in terms of means or other summary characteristics of the matching variables. Has been shown to be almost as efficient as the matching of individual subjects in some circumstances. [*Work, Employment and Society*, 1994, **8**, 421–31.]

Marginal models: See **population averaged model.**

Marginal structural model (MSM): A model for estimation of the causal effect of a time-dependent exposure in the presence of *time-dependent covariates* that may simultaneously be confounders and intermediate variables. Inverse-probability-of-treatment weighting estimators (extensions of the *Horvitz-Thompson estimator*) are used to estimate causal effects based on the concept of *potential outcomes* (see *Neyman-Rubin causal framework*). A simpler but more limited approach than using a *structural nested model.* [*Epidemiology*, 2000, **11**, 550–560.]

Marginal totals: A term often used for the total number of observations in each row and each column of a *contingency table.*

Markers of disease progression: Quantities that form a general monotonic series throughout the course of a disease and assist with its modelling. In general such quantities are highly prognostic in predicting the future course. An example is CD4 cell count (cells per μl), which is generally accepted as the best marker of HIV disease progression. [*Chemical and Biological Interactions*, 1994, **91**, 181–6.]

Market model: A *linear regression* model of the form

$$r_t = \alpha + \beta m_t + \epsilon_t$$

where r_t, m_t represent the returns of the stock and of the market in period t; ϵ_t is the error term which is uncorrelated to the market return in period t. This model is widely used in modern portfolio management. [*Mathematical Finance*, 1997, **7**, 127–55.]

Markov, Andrey Andreyevich (1856–1922): Born in Ryazan, Russia, Markov studied at St Petersburg where he became professor in 1893. In 1905 he left for self-imposed exile in the town of Zaraisk. A student of *Chebyshev*, he worked on number theory and probability theory. Best remembered for the concept of a *Markov chain* which has found many applications in physics, biology and linguistics.

Markov chain: A *stochastic process*, $\{X_t\}, t = 0, 1, 2, \ldots$ where X_t takes values in the finite set $S = \{1, 2, \ldots, N\}$, and is such that

$$\Pr(X_n = i_n | X_0 = i_0, \ldots, X_{n-1} = i_{n-1}) = \Pr(X_n = i_n | X_{n-1} = i_{n-1})$$

(Strictly this defines a *discrete time Markov chain*; if the set S contains an uncountable number of elements then a *continuous time Markov chain* results.) The equation given implies that to make predictions about the future behaviour of the system it suffices to consider only its present state and not its past history. The probability $\Pr(X_n = i_n | X_{n-1} = i_{n-1})$ is known as a *one step transition probability.* The more general *transition probability*, $\Pr(X_n = i | X_m = j)$ satisfies the famous *Chapman–Kolmogorov equation* given by

$$\Pr(X_n = i | X_m = j) = \sum_k \Pr(X_r = k | X_m = j) \Pr(X_n = i | X_r = k) \quad m < r < n$$

A *time-homogeneous Markov chain* is one for which the transition probabilities depend only on the difference $n - m$ rather than on n or m. In particular, the one-step transition probability, $\Pr(X_n = i | X_{n-1} = j)$ can be written as simply p_{ij}. The $N \times N$ matrix, **P**, with ijth element p_{ij} is a *stochastic matrix*, i.e.

$$0 \leq p_{ij} \leq 1, \quad 1 \leq i, j \leq N$$

$$\sum_{j=1}^{N} p_{ij} = 1, \quad 1 \leq i \leq N$$

P is known as a *transition matrix*. The *ij*th element of $\mathbf{P}^n$ gives the *n*-step transition probability. If the limit as $n \to \infty$ is π_j then the π_js constitute the *stationary distribution* of the chain. [*Statistics in Civil Engineering*, 1997, A. V. Metcalfe, Edward Arnold, London.]

Markov chain Monte Carlo methods (MCMC): Powerful methods for indirectly simulating random observations from complex, and often high dimensional probability distributions. Originally used in image processing for the restoration of blurred or distorted pictures, the technique has now become of considerable importance in many branches of statistics, in particular in applications of *Bayesian inference* where the integrations needed to evaluate the *posterior distribution* itself or particular features of interest such as moments, quantiles, etc., have, until recently, been a source of considerable practical difficulties. But although most applications of MCMC have been in Bayesian inference, the methods are often also extremely useful in classical *likelihood* calculations. In general terms the problem attacked by MCMC methodology is that of obtaining characteristics of interest (for example, mean, variance, etc.) of the *marginal distribution*, *f(x)* arising from a given *joint distribution*, $g(x, y_1, \ldots, y_q)$ as

$$f(x) = \int \cdots \int g(x, y_1, \ldots, y_q) \mathrm{d}y_1 \ldots \mathrm{d}y_q$$

The most natural and straightforward approach would be to calculate $f(x)$ and then use it to obtain the required characteristics. In general, however, the necessary integrations are extremely difficult to perform, either analytically or numerically. MCMC methods effectively allow generation of samples from $f(x)$ without requiring $f(x)$ explicitly. By simulating a large enough sample, the mean, variance or any other characteristic of $f(x)$ (or $f(x)$ itself) can be calculated to any desired degree of accuracy. The essential feature of the MCMC approach is a cleverly constructed *Markov chain*, the *stationary distribution* of which is precisely the distribution of interest f. A number of different methods are available for creating this Markov chain consisting of a sequence of random variables $\{X_0, X_1, X_2, \ldots\}$. The *Metropolis–Hastings algorithm*, for example, samples at each stage a candidate point Y from a so-called proposal distribution that may depend on the current point X_t. The candidate point is then accepted with probability $\alpha(X_t, Y)$ where

$$\alpha(X, Y) = \min(1, \frac{f(Y)q(X|Y)}{f(X)q(Y|X)})$$

and q is the proposal distribution. If the candidate point is accepted, the next state becomes $X_{t+1} = Y$ otherwise $X_{t+1} = X_t$ and the chain does not move. Remarkably, the proposal distribution q can take any form and the stationary distribution of the chain will still be f. An example of the application of this algorithm in a simple case is shown in Fig. 92. The most widely used form of MCMC in statistical applications is *Gibbs sampling* where, for two variables for example, a sample from $f(x)$ is generated by sampling instead from the two conditional distributions, $h(x|y)$ and $r(y|x)$. So now, beginning with an initial value Y_0 a sequence $\{Y_0, X_0, Y_1, X_1, Y_2, X_2, \ldots, Y_k, X_k\}$ is generated by sampling X_j from $h(.|Y_j)$ and Y_{j+1} from $r(.|X_j)$. Under reasonably general conditions the distribution of X_k converges to $f(x)$ as $k \to \infty$. So for k large enough the final observation in the sequence is effectively a sample point from $f(x)$. See also **data augmentation algorithm**. [*Markov Chain Monte Carlo in Practice*, 1996, edited by W. R. Gilks, S. Richardson and D. J. Spiegelhalter, Chapman and Hall/CRC Press, London.]

Markov illness–death model: A model in which live individuals are classified as either having, or not having, a disease A, and then move between these possibilities and death as indicated in Fig. 93. [*Medical Decision Making*, 1994, **14**, 266–72.]

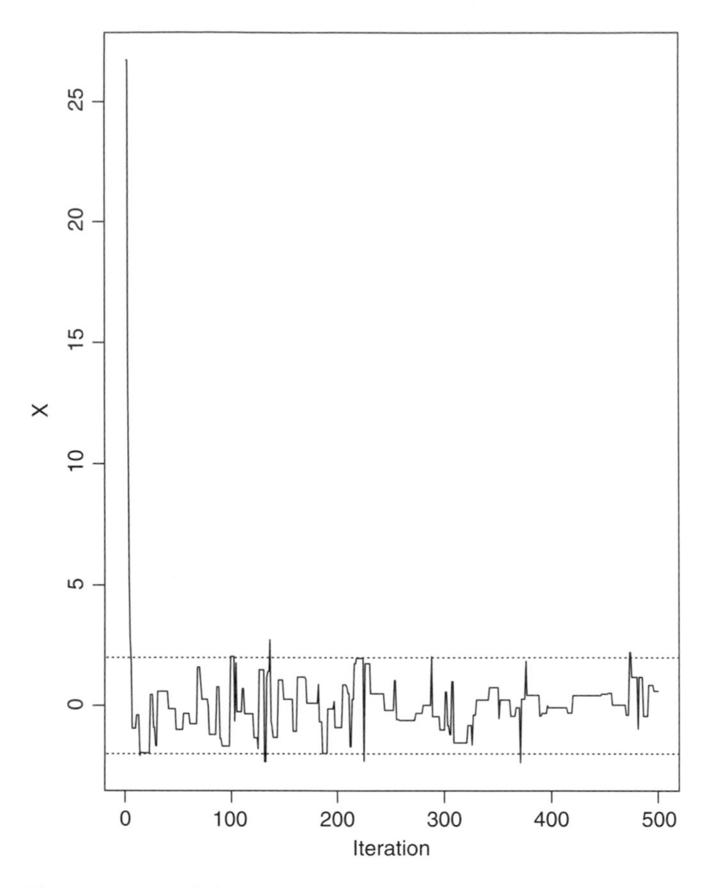

Fig. 92 Markov chain Monte Carlo methods illustrated by 500 iterations from the Metropolis algorithm with stationary distribution $N(0, 1)$ and proposed distribution $N(x, 5.5)$.

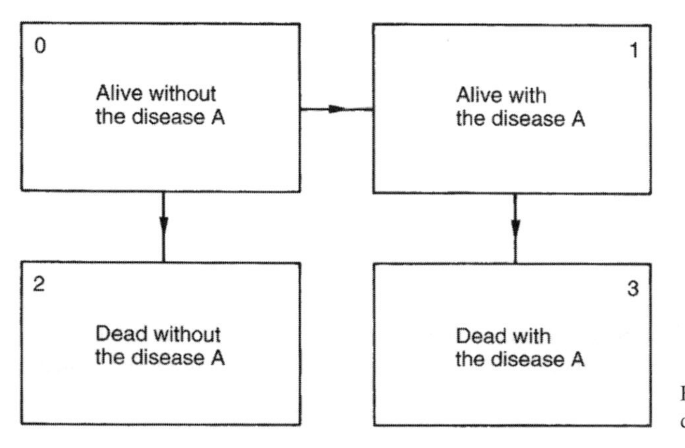

Fig. 93 Markov illness–death model diagram.

Markov inequality: See **Chebyshev's inequality**.

Markov process: Synonym for **continuous time Maskov chain**.

Marquis of Lansdowne (1780–1863): William Petty Fitzmaurice Lansdowne was a British Whig politician who became the first President of the Royal Statistical Society. He was a champion of Catholic emancipation, the abolition of the slave-trade and the cause of popular education.

MARS: Acronym for **multivariate adaptive regression splines**.

Martingale: In a gambling context the term originally referred to a system for recouping losses by doubling the stake after each loss. The modern mathematical concept can also be viewed in such terms as a mathematical formalization of the concept of a 'fair' game defined as one in which the expected size of the gambler's bank after the nth game given the results of all previous games is unchanged from the size after the $(n-1)$th game, i.e. the gambler's average gain or loss after each game, given the previous history of the games is zero. More formally the term refers to a *stochastic process*, or a sequence of random variables $S_1, S_2, \ldots$, which is such that the expected value of the $(n+1)$th random variable conditional on the values of the first n, equals the value of the nth random variable, i.e.

$$E(S_{n+1}|S_1, S_2, \ldots, S_n) = S_n \qquad n \geq 1$$

so that $E(S_n)$ does not depend on n. [*Discrete Parameter Martingales*, 1975, J. Neveu, North Holland, Amsterdam.]

Masking: A term usually applied in the context of a sample containing multiple *outliers* whose presence prevents many methods from detecting them. May also be used as a synonym for *blinding*. [*Clinical Trials in Psychiatry*, 2008, B. S. Everitt and S. Wessely, Wiley, Chichester.]

Matched case-control study: See **retrospective study**.

Matched pairs: A term used for observations arising from either two individuals who are individually matched on a number of variables, for example, age, sex, etc., or where two observations are taken on the same individual on two separate occasions. Essentially synonymous with **paired samples**.

Matched pairs *t*-test: A *Student's t-test* for the equality of the means of two populations, when the observations arise as *paired samples*. The test is based on the differences between the observations of the matched pairs. The *test statistic* is given by

$$t = \frac{\bar{d}}{s_d/\sqrt{n}}$$

where n is the sample size, $\bar{d}$ is the mean of the differences, and s_d their standard deviation. If the null hypothesis of the equality of the population means is true then t has a *Student's t-distribution* with $n-1$ degrees of freedom. [SMR Chapter 9.]

Matched set: See **one:m matching**.

Matching: The process of making a study group and a comparison group comparable with respect to extraneous factors. Often used in *retrospective studies* when selecting cases and controls to control variation in a response variable due to sources other than those immediately under investigation. Several kinds of matching can be identified, the most common of which is when each case is individually matched with a control subject on the matching variables, such as age, sex, occupation, etc. When the variable on which the matching takes place is continous it is usually transformed into a series of categories (e.g. age), but a second method is to say that two values of the variable match if their difference lies between defined limits. This method is known as *caliper matching*. Also important is *group matching* in which the distributions of the extraneous factors are made similar in the groups to be compared. See also **paired samples**. [SMR Chapter 5.]

Matching coefficient: A *similarity coefficient* for data consisting of a number of binary variables that is often used in *cluster analysis*. Given by

$$s_{ij} = \frac{a}{a+b+c+d}$$

where *a, b, c* and *d* are the four frequencies in the two-by-two cross-classification of the variable values for subjects *i* and *j*. See also **Jaccard coefficient**. [*Cluster Analysis*, 4th edition, 2001, B. S. Everitt, S. Landau and M. Leese, Arnold, London.]

Matching distribution: A probability distribution that arises in the following way. Suppose that a set of *n* subjects, numbered $1, \ldots, n$ respectively, are arranged in random order, then it is the distribution of the number of subjects for which the position in the random order equals the number originally assigned to them. Found from application of *Boole's formula* to be

$$\Pr(X = x) = [x!]^{-1}\{1 - (1!)^{-1} + (2!)^{-1} + \cdots + (-1)^{n-x}(n-x)!^{-1}\} \quad x = 0, 1, \ldots, n$$

Note that $\Pr(X = n-1) = 0$. [*Univariate Discrete Distributions*, 2005, N. L. Johnson, A. W. Kemp and S. Kotz, Wiley, New York.]

Maternal mortality: A maternal death is the death of a woman while pregnant, or within 42 days of the termination of pregnancy, from any cause related to or aggravated by the pregnancy or its management, but not from accidental or incidental causes. Some 585 000 women a year die as a result of maternal mortality, 99% of them in the developing world. [*World Health Statistics Quarterly*, 1996, **49**, 77–87.]

Mathematica: A fully integrated software environment for technical and scientific computing that combines numerical and symbolic computation. See also **Maple**. [http://www.wolfram.com/]

Mathisen's test: *A distribution free test* for the *Behrens–Fisher problem*. [*Robust Non-parametric Statistical Methods*, 1998, T. P. Hettmansperger and J. W. McKean, Arnold, London.]

Matrix exponential transformation: A transformation for any $p \times p$ matrix $\mathbf{A}$, defined as follows:

$$\mathbf{C} = \exp(\mathbf{A}) = \sum_{s=0}^{\infty} \frac{\mathbf{A}^s}{s!}$$

where $\mathbf{A}^0$ is the $p \times p$ identity matrix and $\mathbf{A}^s$ denotes ordinary matrix multiplication of $\mathbf{A}$, s times. Used in the modelling of a *variance–covariance matrix* in particular in studying the dependence of the covariances on explanatory variables. [*International Journal of Applied Mathematics*, 2001, **7**, 289–308.]

Mauchly test: A test that a *variance–covariance matrix* of pairwise differences of responses in a set of *longitudinal data* is a scalar multiple of the identity matrix, a property known as *sphericity*. Of most importance in the analysis of *longitudinal data* where this property must hold for the *F*-tests in the *analysis of variance* of such data to be valid. See also **compound symmetry, Greenhouse–Geisser correction** and **Huynh–Feldt correction**. [MV2 Chapter 13.]

MAVE method: Abbreviation for **minimum average variance estimation method**.

Maximum a posteriori estimate (MAP): Essentially the mode of a *posterior distribution*. Often used as a measure of location in image analysis applications. [*Markov Chain Monte Carlo in Practice*, 1996, W. R. Gilks, S. Richardson, and D. J. Spiegelhalter, Chapman and Hall/CRC Press, London.]

Maximum entropy principle: A principle which says that to assign probabilities using limited information, the *entropy* of the distribution, *S*, should be maximized where

$$S = \sum_i p_i \log(p_i)$$

subject to the constraints of known expectation values. [*The Statistician*, 1998, **4**, 629–41.]

Maximum F-ratio: Equivalent to *Hartley's test.* See also **Bartlett's test** and **Box's test**.

Maximum likelihood estimation: An estimation procedure involving maximization of the *likelihood* or the *log-likelihood* with respect to the parameters. Such estimators are particularly important because of their many desirable statistical properties such as *consistency,* and *asymptotic efficiency.* As an example consider the number of successes, X, in a series of *random variables* from a *Bernoulli distribution* with success probability p. The likelihood is given by

$$P(X = x|p) = \binom{n}{x} p^x (1-p)^{n-x}$$

Differentiating the log-likelihood, L, with respect to p gives

$$\frac{\partial L}{\partial p} = \frac{x}{p} - \frac{n-x}{1-p}$$

Setting this equal to zero gives the estimator $\hat{p} = x/n$. See also **EM algorithm**. [KA2 Chapter 18.]

Maximum likelihood estimator: The estimator of a parameter obtained from applying *maximum likelihood estimation.*

Maximum normed residual: A statistic that may be used to test for a single *outlier,* when fitting a linear model. Defined as the maximum of the absolute values of the normed residuals, that is

$$|z|^{(1)} = \max\{|z_1|, |z_2|, \ldots, |z_n|\}$$

where $z_i = e_i/(\sum_{i=1}^{n} e_i^2)^{\frac{1}{2}}$ and e_i is the usual residual, namely the observed minus the predicted value of the response variable. [*Annals of Mathematical Statistics*, 1971, **42**, 35–45.]

Maxwell, Albert Ernest (1916–1996): Maxwell was educated at the Royal School, Cavan and at Trinity College, Dublin where he developed interests in psychology and mathematics. His first career was as school teacher and then headmaster at St. Patrick's Cathedral School, Dublin. In 1952 Maxwell left school teaching to take up a post of lecturer in statistics at the Institute of Psychiatry, a post graduate school of the University of London. He remained at the Institute until his retirement in 1978. Maxwell's most important contributions were in *multivariate analysis,* particularly *factor analysis.* He died in 1996 in Leeds, UK.

Mayo-Smith, Richard (1854–1901): Born in Troy, Ohio, Mayo-Smith graduated from Amherst college eventually obtaining a Ph.D. Joined the staff of Columbia College as an instructor in history and political science. One of the first to teach statistics at an American university, he was also the first statistics professor to open a statistical laboratory. Mayo-Smith was made an Honorary Fellow of the Royal Statistical Society in 1890, the same year in which he was elected a member of the National Academy of Sciences.

MAZ experiments: *Mixture–amount experiments* which include control tests for which the total amount of the mixture is set to zero. Examples include drugs (some patients do not receive any of the formulations being tested) fertilizers (none of the fertilizer mixtures are applied to certain plots) and paints/coatings (some specimens are not painted/coated). [*Experiments with Mixtures; Design Models and The Analysis of Mixture Data*, 1981, J. A. Cornell, Wiley, New York.]

MCAR: Abbreviation for **missing completely at random**.

McCabe–Tremayne test: A test for the *stationarity* of particular types of *time series.* [*Annals of Statistics*, 1995, **23**, 1015–28.]

MCMC: Abbreviation for **Markov chain Monte Carlo method**.

MCML: Abbreviation for **Monte Carlo maximum likelihood**.

McNemar's test: A test for comparing proportions in data involving *paired samples*. The test statistic is given by

$$X^2 = \frac{(b-c)^2}{b+c}$$

where b is the number of pairs for which the individual receiving treatment A has a positive response and the individual receiving treatment B does not, and c is the number of pairs for which the reverse is the case. If the probability of a positive response is the same in each group, then X^2 has a *chi-squared distribution* with a single degree of freedom. [SMR Chapter 10.]

MCP: Abbreviation for **minimum convex polygon**.

MDL: Abbreviation for **minimum description length**.

MDP: Abbreviation for **minimum distance probability**.

MDS: Abbreviation for **multidimensional scaling**.

Mean: A measure of location or central value for a continuous variable. For a definition of the population value see *expected value*. For a sample of observations $x_1, x_2, \ldots, x_n$ the measure is calculated as

$$\bar{x} = \frac{\sum_{i=1}^{n} x_i}{n}$$

Most useful when the data have a *symmetric distribution* and do not contain *outliers*. See also **median** and **mode**. [SMR Chapter 2.]

Mean and dispersion additive model (MADAM): A flexible model for variance heterogeneity in a normal error model, in which both the mean and variance are modelled using semi-parametric additive models. [*Statistics and Computing*, 1996, **6**, 52–65.]

Mean deviation: For a sample of observations, $x_1, x_2, \ldots, x_n$ this term is defined as $\frac{1}{n}\sum_{i=1}^{n} |x_i - \bar{x}|$ where $\bar{x}$ is the arithmetic mean of the observations. Used in the calculation of **Geary's ratio**.

Mean-range plot: A graphical tool useful in selecting a transformation in *time series* analysis. The range is plotted against the mean for each seasonal period, and a suitable transformation chosen according to the appearance of the plot. If the range appears to be independent of the mean, for example, no transformation is needed. If the plot displays random scatter about a straight line then a logarithmic transformation is appropriate.

Mean square contingency coefficient: The square of the *phi-coefficient*.

Mean squared error: The expected value of the square of the difference between an estimator and the true value of a parameter. If the estimator is unbiased then the mean squared error is simply the variance of the estimator. For a biased estimator the mean squared error is equal to the sum of the variance and the square of the bias. [KA2 Chapter 17.]

Mean square error estimation: Estimation of the parameters in a model so as to minimize the *mean square error*. This type of estimation is the basis of the *empirical Bayes method*.

Mean square ratio: The ratio of two mean squares in an *analysis of variance*.

Mean squares: The name used in the context of *analysis of variance* for estimators of particular variances of interest. For example, in the analysis of a *one way design*, the *within groups* *mean square* estimates the assumed common variance in the k groups (this is often also referred to as the *error mean square*). The *between groups mean square* estimates the variance of the group means.

Mean vector: A vector containing the mean values of each variable in a set of multivariate data.

Measurement error: Errors in reading, calculating or recording a numerical value. The difference between observed values of a variable recorded under similar conditions and some fixed true value. [*Statistical Evaluation of Measurement Errors*, 2004, G. Dunn, Arnold, London.]

Measures of association: Numerical indices quantifying the strength of the statistical dependence of two or more qualitative variables. See also **phi-coefficient** and **Goodman–Kruskal measures of association**. [*The Analysis of Contingency Tables*, 2nd edition, 1992, B. S. Everitt, Chapman and Hall/CRC Press, London.]

Median: The value in a set of ranked observations that divides the data into two parts of equal size. When there is an odd number of observations the median is the middle value. When there is an even number of observations the measure is calculated as the average of the two central values. Provides a measure of location of a sample that is suitable for *aysmmetric distributions* and is also relatively insensitive to the presence of *outliers*. See also **mean**, **mode**, **spatial median** and **bivariate Oja median**. [SMR Chapter 2.]

Median absolute deviation (MAD): A very robust estimator of scale given by

$$\text{MAD} = \text{median}_i(|x_i - \text{median}_j(x_j)|)$$

or, in other words, the median of the absolute deviations from the median of the data. In order to use MAD as a consistent estimator of the standard deviation it is multiplied by a scale factor that depends on the distribution of the data. For normally distributed data the constant is 1.4826 and the expected value of 1.4826 MAD is approximately equal to the population standard deviation. [*Annals of Statistics*, 1994, **22**, 867–85.]

Median centre: Synonym for **spatial median**.

Median effective dose: A quantity used to characterize the potency of a stimulus. Given by the amount of the stimulus that produces a response in 50% of the cases to which it is applied. [*Modelling Binary Data*, 2nd edition, 2002, D. Collett, Chapman and Hall, London.]

Median lethal dose: Synonym for **lethal dose 50**.

Median survival time: A useful summary of a *survival curve* defined formally as $F^{-1}(\frac{1}{2})$ where $F(t)$, the cumulative distribution function of the *survival times*, is the probability of surviving less than or equal to time t. [*Biometrics*, 1982, **38**, 29–41.]

Median unbiased: See **least absolute deviation regression**.

Medical audit: The examination of data collected from routine medical practice with the aim of identifying areas where improvements in efficiency and/or quality might be possible.

MEDLINE: Abbreviation for Medical Literature Analysis Retrieval System Online.

Mega-trial: A large-scale *clinical trial*, generally involving thousands of subjects, that is designed to assess the effects of one or more treatments on endpoints such as death or disability. Such trials are needed because the majority of treatments have only moderate effects on such endpoints. [*Lancet*, 1994, **343**, 311–22.]

Meiotic mapping: A procedure for identifying complex disease *genes*, in which the trait of interest is related to a map of marker loci. [*Statistics in Medicine*, 2000, **19**, 3337–43.]

Mendelian randomization: A term applied to the random assortment of alleles at the time of gamete formation, a process that results in population distributions of genetic variants that are generally independent of behavioural and environmental factors that typically confound epidemiologically derived associations between putative risk factors and disease. In some circumstances taking advantage of Mendelian randomization may provide an epidemiological study with the properties of a randomized comparison, for example, a *randomized clinical trial.* In particular, the use of Mendelian randomization may allow causal inferences to be drawn from epidemiological studies, typically using an *instrumental variable estimator.* [*Statistical Methods in Medical Research*, 2007, **16**, 309–330.]

Merrell, Margaret (1900–1995): After graduating from Wellesley College in 1922, Merrell taught mathematics at the Bryn Mawr School in Baltimore until 1925. She then became one of the first graduate students in biostatistics at the Johns Hopkins University School of Hygiene and Public Health. Merrell obtained her doctorate in 1930. She made significant methodological contributions and was also a much admired teacher of biostatistics. Merrell died in December 1995 in Shelburne, New Hampshire.

Mesokurtic curve: See **kurtosis**.

M-estimators: *Robust estimators* of the parameters in statistical models. (The name derives from 'maximum likelihood-like' estimators.) Such estimators for a location parameter μ might be obtained from a set of observations $y_1, \ldots, y_n$ by solving

$$\sum_{i=1}^{n} \psi(y_1 - \hat{\mu}) = 0$$

where ψ is some suitable function. When $\psi(x) = x^2$ the estimator is the mean of the observations and when $\psi(x) = |x|$ it is their median. The function

$$\begin{aligned} \psi(x) &= x, \ |x|<c \\ &= 0 \text{ otherwise} \end{aligned}$$

corresponds to what is know as *metric trimming* and large outliers have no influence at all. The function

$$\begin{aligned} \psi(x) &= -c, \ x < -c \\ &= x, \ |x|<c \\ &= c, \ x > c \end{aligned}$$

is known as *metric Winsorizing* and brings in extreme observations to $\mu \pm c$. See also **bisquare regression estimation**. [*Transformation and Weighting in Regression*, 1988, R. J. Carroll and D. Ruppert, Chapman and Hall/CRC Press, London.]

Meta-analysis: A collection of techniques whereby the results of two or more independent studies are statistically combined to yield an overall answer to a question of interest. The rationale behind this approach is to provide a test with more *power* than is provided by the separate studies themselves. The procedure has become increasingly popular in the last decade or so but it is not without its critics particularly because of the difficulties of knowing which studies should be included and to which population final results actually apply. See also **funnel plot**. [SMR Chapter 10.]

Meta regression: An extension of *meta-analysis* in which the relationship between the treatment effect and covariates characterizing the studies is modeled using some form of regression, for example, weighted regression. In this way insight can be gained into how outcome is related to the design and population studied. [*Biometrics*, 1999, **55**, 603–29.]

Method of moments: A procedure for estimating the parameters in a model by equating sample *moments* to their population values. A famous early example of the use of the procedure is in *Karl Pearson*'s description of estimating the five parameters in a *finite mixture distribution* with two univariate normal components. Little used in modern statistics since the estimates are known to be less efficient than those given by alternative procedures such as *maximum likelihood estimation.* [KA1 Chapter 3.]

Method of statistical differentials: Synonymous with the **delta method**.

Metric inequality: A property of some *dissimilarity coefficients* (δ_{ij}) such that the dissimilarity between two points i and j is less than or equal to the sum of their dissimilarities from a third point k. Specifically

$$\delta_{ij} \leq \delta_{ik} + \delta_{jk}$$

Indices satisfying this inequality are referred to as *distance measures*. [MV1 Chapter 5.]

Metric trimming: See **M-estimators**.

Metric Winsorizing: See **M-estimators**.

Metropolis–Hastings algorithm: See **Markov chain Monte Carlo methods**.

Michael's test: A test that a set of data arise from a normal distribution. If the ordered sample values are $x_{(1)}, x_{(2)}, \ldots, x_{(n)}$, the test statistic is

$$D_{SP} = \max_i |g(f_i) - g(p_i)|$$

where $f_i = \Phi\{(x_{(i)} - \bar{x})/s\}, g(y) = \frac{2}{\pi}\sin^{-1}(\sqrt{(y)}), p_i = (i - 1/2)/n, \bar{x}$ is the sample mean,

$$s^2 = \frac{1}{n}\sum_{i=1}^{n}(x_{(i)} - \bar{x})^2$$

and

$$\Phi(w) = \int_{-\infty}^{w} \frac{1}{\sqrt{2\pi}} e^{-\frac{1}{2}u^2} \mathrm{d}u$$

Critical values of D_{SP} are available in some statistical tables.

Michaelis–Menten equation: See **linearizing** and **inverse polynomial functions**.

Microarrays: A technology that facilitates the simultaneous measurement of thousands of *gene* expression levels. A typical microarray experiment can produce millions of data points, and the statistical task is to efficiently reduce these numbers to simple summaries of genes' structures. [*Journal of American Statistical Association*, 2001, **96**, 1151–60.]

Microdata: Survey data recording the economic and social behaviour of individuals, firms etc. and the environment in which they exist. Such data is an essential tool for the understanding of economic and social interactions and the design of government policy.

Mid P-value: An alternative to the conventional p-value that is used, in particular, in some analyses of discrete data, for example, *Fisher's exact test* on *two-by-two contingency tables.* In the latter if $x = a$ is the observed value of the frequency of interest, and this is larger than the value expected, then the mid P-value is defined as

$$\text{mid P-value} = \tfrac{1}{2}\text{Prob}(x = a) + \text{Prob}(x > a)$$

In this situation the usual p-value would be defined as $\text{Prob}(x \geq a)$. [*Statistics in Medicine*, 1993, **12**, 777–88.]

Mid-range: The mean of the smallest and largest values in a sample. Sometimes used as a rough estimate of the mean of a *symmetrical distribution*.

Midvariance: A robust estimator of the variation in a set of observations. Can be viewed as giving the variance of the middle of the observation's distribution. [*Annals of Mathematical Statistics*, 1972, **43**, 1041–67.]

Migration process: A process involving both immigration and emigration, but different from a *birth–death process* since in an immigration process the immigration rate is independent of the population size, whereas in a birth–death process the birth rate is a function of the population size at the time of birth. [*Reversibility and Stochastic Networks*, 1979, F. P. Kelly, Wiley, Chichester.]

Mills ratio: The ratio of the *survival function* of a random variable to the probability distribution of the variable, i.e.

$$\text{Mills ratio} = \frac{(1 - F(x))}{f(x)}$$

where $F(x)$ and $f(x)$ are the *cumulative probability distribution* and the probability distribution function of the variable, respectively. Can also be written as the reciprocal of the *hazard function*. [KA1 Chapter 5.]

MIMIC model: Abbreviation for **multiple indicator multiple cause model**.

Minimal sufficient statistic: See **sufficient statistic**.

Minimax rule: A term most often encountered when deriving *classification rules* in *discriminant analysis*. It arises from attempting to find a rule that safeguards against doing very badly on one population, and so uses the criterion of minimizing the maximum probability of misclassification in deriving the rule. [*Discrimination and Classification*, 1981, D. J. Hand, Wiley, Chichester.]

Minimization: A method for allocating patients to treatments in *clinical trials* which is usually an acceptable alternative to random allocation. The procedure ensures balance between the groups to be compared on prognostic variables, by allocating with high probability the next patient to enter the trial to whatever treatment would minimize the overall imbalance between the groups on the prognostic variables, at that stage of the trial. See also **biased coin method** and **block randomization**. [SMR Chapter 15.]

Minimum aberration criterion: A criterion for finding *fractional factorial designs* which minimizes the bias incurred by nonnegligible interactions. [*Technometrics*, 1980, **22**, 601–608.]

Minimum average variance estimation (MAVE) method: An approach to dimension reduction when applying regression models to *high-dimensional data*. [*Journal of the Royal Statistical Society, Series B*, 2002, **64**, 363–410.]

Minimum chi-squared estimator: A statistic used to estimate some parameter of interest which is found by minimizing with respect to the parameter, a *chi-square statistic* comparing observed frequencies and expected frequencies which are functions of the parameter. [KA2 Chapter 19.]

Minimum convex polygon: Synonym for **convex hull**.

Minimum description length (MDL): An approach to choosing statistical models for data when the purpose is simply to describe the given data rather than to estimate the parameters of some hypothetical population. [*Stochastic-Complexity in Statistical Inquiry*, 1989, J. Rissanen, World Scientific Publishing Co., Singapore.]

Minimum distance probability (MDP): A method of *discriminant analysis* based on a distance which can be used for continuous, discrete or mixed variables with known or unknown distributions. The method does not depend on one specific distance, so it is the investigator who has to decide the distance to be used according to the nature of the data. The method also makes use of the knowledge of prior probabilities and provides a numerical value of the confidence in the goodness of allocation of every individual studied. [*Biometrics*, 2003, **59**, 248–53.]

Minimum spanning tree: See **tree**.

Minimum variance bound: Synonym for **Cramér–Rao lower bound**.

Minimum volume ellipsoid: A term for the ellipsoid of minimum volume that covers some specified proportion of a set of *multivariate data*. Used to construct *robust estimators* of mean vectors and *variance–covariance matrices*. Such estimators have a high *breakdown point* but are computationally expensive. For example, for an $n \times q$ data matrix **X**, if h is the integer part of $n(n+1)/2$ then the volumes of $n!/h!(n-h)!$ ellipsoids need to be considered to find the one with minimum volume. So for $n = 20$ there are 184 756 ellipsoids and for $n = 30$ more than 155 million ellipsoids. [*Journal of the American Statistical Association*, 1990, **85**, 633–51.]

MINITAB: A general purpose statistical software package, specifically designed to be useful for teaching purposes. [MINITAB, Brandon Court, Unit E1, Progress Way, Coventry CV3 2TE, UK; MINITAB Inc., Quality Plaza, 1829 Pine Hall Road, State College, PA 16801-3008, USA; www.minitab.com]

Minkowski distance: A *distance measure*, $d_{\mathbf{xy}}$, for two observations $\mathbf{x}' = [x_1, x_2, \ldots, x_q]$ and $\mathbf{y}' = [y_1, y_2, \ldots, y_q]$ from a set of multivariate data, given by

$$d_{\mathbf{xy}} = \left[\sum_{i=1}^{q} (x_i - y_i)^r \right]^{\frac{1}{r}}$$

When $r = 2$ this reduces to *Euclidean distance* and when $r = 1$ to *city block distance*. [MV1 Chapter 3.]

Mirror-match bootstrapping: A specific form of the *bootstrap* in which the sample is subsampled according to the design that was used to select the original sample from the population. [*Journal of the American Statistical Association,* 1992, **87**, 755–65.]

Misclassification error: The misclassification of categorical variables that can happen in epidemiological studies and which may induce problems of analysis and interpretation, in particular it may affect the assessment of exposure-disease associations. Also used for the misclassification when using a *discriminant function*. [*Statistics in Medicine*, 1989, **8**, 1095–1106.]

Mis-interpretation of P-values: A p-value is commonly interpreted in a variety of ways that are incorrect. Most common are that it is the probability of the null hypothesis, and that it is the probability of the data having arisen by chance. For the correct interpretation see the entry for **P-value**. [SMR Chapter 8.]

Missing at random (MAR): See **missing values**.

Missing completely at random (MCAR): See **missing values**.

Missing information principle: This principle states that the missing information in data with *missing values* is equal to the difference between the *information matrix* for complete data and the observed information matrix. [*Proceedings of the 6th Berkeley Symposium on Mathematical Statistics and Probability*, 1972, **1**, 697–715.]

Missing values: Observations missing from a set of data for some reason. In *longitudinal studies*, for example, they may occur because subjects drop out of the study completely or do not appear for one or other of the scheduled visits or because of equipment failure. Common causes of subjects prematurely ceasing to participate include recovery, lack of improvement, unwanted signs or symptoms that may be related to the investigational treatment, unpleasant study procedures and intercurrent health problems. Such values greatly complicate many methods of analysis and simply using those individuals for whom the data are complete can be unsatisfactory in many situations. A distinction can be made between values *missing completely at random* (MCAR), *missing at random* (MAR) and *non-ignorable* (or *informative*). The MCAR variety arise when individuals drop out of the study in a process which is independent of both the observed measurements and those that would have been available had they not been missing; here the observed values effectively constitute a *simple random sample* of the values for all study subjects. Random drop-out (MAR) occurs when the drop-out process depends on the outcomes that have been observed in the past, but given this information is conditionally independent of all future (unrecorded) values of the outcome variable following drop-out. Finally, in the case of informative drop-out, the drop-out process depends on the unobserved values of the outcome variable. It is the latter which cause most problems for the analysis of data containing missing values. See also **last observation carried forward, attrition, imputation, multiple imputation** and **Diggle–Kenward model for drop-outs**. [*Analysis of Longitudinal Data*, 2nd edition, 2002, P. J. Diggle, P. J. Heagerty, K.-Y. Liang and S. Zeger, Oxford Science Publications, Oxford.]

Misspecification: A term applied to describe assumed statistical models which are incorrect for one of a variety of reasons, for example, using the wrong probability distribution, omitting important covariates, or using the wrong *link function*. Such errors can produce inconsistent or inefficient estimates of parameters. See also **White's information matrix test** and **Hausman misspecification test**. [*Biometrika*, 1986, **73**, 363–9.]

Mitofsky–Waksberg scheme: See **telephone interview surveys**.

Mitscherlich curve: A curve which may be used to model a *hazard function* that increases or decreases with time in the short term and then becomes constant. Its formula is

$$h(t) = \theta - \beta e^{-\gamma t}$$

where all three parameters, θ, β and γ, are greater than zero. [*Australian Journal of Experimental Agriculture*, 2001, **41**, 655–61.]

Mixed data: Data containing a mixture of continuous variables, ordinal variables and categorical variables.

Mixed-effects logistic regression: A generalization of standard *logistic regression* in which the intercept terms, α_i are allowed to vary between subjects according to some probability distribution, $f(\alpha)$. In essence these terms are used to model the possible different *frailties* of the subjects. For a single covariate x, the model often called a random intercept model, is

$$\text{logit}[P(y_{ij}|\alpha_i, x_{ij})] = \alpha_i + \beta x_{ij}$$

where y_{ij} is the binary response variable for the jth measurement on subject i, and x_{ij} is the corresponding covariate value. Here β measures the change in the conditional logit of the probability of a response of unity with the covariate x, for individuals in each of the underlying risk groups described by α_i. The *population averaged model* for y_{ij} derived from this model is

$$P(y_{ij} = 1|x_{ij}) = \int (1 + e^{-\alpha-\beta x_{ij}})^{-1} f(\alpha)\mathrm{d}\alpha$$

Can be used to analyse *clustered binary data* or *longitudinal studies* in which the outcome variable is binary. In general interest centres on making inferences about the regression coefficient, β (in practice a vector of coefficients) with the αs being regarded as nuisance parameters. Parameter estimation typically proceeds via the *marginal likelihood* where the random effects are integrated out of the likelihood. Alternatively, estimation can be based on the *conditional likelihood*, with conditioning on the *sufficient statistics* for the αs which are consequently eliminated from the likelihood function. [*Statistics in Medicine*, 1996, **15**, 2573–88.]

Mixed effects models: See **multilevel models**.

Mixture–amount experiment: One in which a *mixture experiment* is performed at two or more levels of total amount and the response is assumed to depend on the total amount of the mixture as well as on the component proportions. [*Experiments with Mixtures: Designs, Models and the Analysis of Mixture Data*, 2nd edition, 1990, J. A. Cornell, Wiley, New York.]

Mixture distribution: See **finite mixture distribution**.

Mixture experiment: An experiment in which two or more ingredients are mixed or blended together to form an end product. Measurements are taken on several blends of the ingredients in an attempt to find the blend that produces the 'best' response. The measured response is assumed to be a function only of the components in the mixture and not a function of the total amount of the mixture. [*Experiments with Mixtures*, 2nd edition, 1990, J. A. Cornell, Wiley, New York.]

Mixture transition distribution model: Models for *time series* in which the conditional distribution of the current observation given the past is a mixture of conditional distributions given each one of the last r observations. Such models can capture features such as flat stretches, bursts of activity, *outliers* and changepoints. [*Statistical Science*, 2002, **17**, 328–56.]

MLE: Abbreviation for **maximum likelihood estimation**.

MLwiN: A software package for fitting *multilevel models*. [Centre for Multilevel Modelling, Graduate School of Education, University of Bristol, 35 Berkeley Square, Bristol BS8 1JA, UK.]

Mode: The most frequently occurring value in a set of observations. Occasionally used as a measure of location. See also **mean** and **median**. [SMR Chapter 2.]

Model: A description of the assumed structure of a set of observations that can range from a fairly imprecise verbal account to, more usually, a formalized mathematical expression of the process assumed to have generated the observed data. The purpose of such a description is to aid in understanding the data. See also **deterministic model, logistic regression, multiple regression, random model** and **generalized linear models**. [SMR Chapter 8.]

Model-based inference: Statistical inference for parameters of a statistical model, sometimes called a *data generating mechanism* or infinite *superpopulation*, where variability is interpreted as due to hypothetical replicated samples from the model. Often contrasted to *design-based inference*. [*Canadian Journal of Forest Research*, 1998, **88**, 1429–1447.]

Model building: A procedure which attempts to find the simplest model for a sample of observations that provides an adequate fit to the data. See also **parsimony principle** and **Occam's razor**.

Mojena's test: A test for the number of groups when applying *agglomerative hierarchical clustering methods*. In detail the procedure is to select the number of groups corresponding to the first stage in the *dendrogram* satisfying

$$\alpha_{j+1} > \bar{\alpha} + ks_{\alpha}$$

where $\alpha_0, \alpha_1, \ldots, \alpha_{n-1}$ are the fusion levels corresponding to stages with $n, n-1, \ldots, 1$ clusters, and n is the sample size. The terms $\bar{\alpha}$ and s_{α} are, respectively, the mean and unbiased standard deviation of the α values and k is a constant, with values in the range 2.75–3.50 usually being recommended. [*Cluster Analysis*, 4th edition, 2001, B. S. Everitt, S. Landau and M. Leese, Arnold, London.]

Moment generating function: A function, $M(t)$, derived from a probability distribution, $f(x)$, as

$$M(t) = \int_{-\infty}^{\infty} e^{tx} f(x) \mathrm{d}x$$

When $M(t)$ is expanded in powers of t the coefficient of t^r gives the rth central *moment* of the distribution, μ'_r. If the *probability generating function* is $P(t)$, the moment generating function is simply $P(e^t)$. See also **characteristic function**. [KA1 Chapter 3.]

Moments: Values used to characterize the probability distributions of random variables. The kth moment about the origin for a variable x is defined as

$$\mu'_k = E(x^k)$$

so that μ'_1 is simply the mean and generally denoted by μ. The kth moment about the mean, μ_k, is defined as

$$\mu_k = E(x - \mu)^k$$

so that μ_2 is the variance. Moments of samples can be defined in an analogous way, for example,

$$m'_k = \frac{\sum_{i=1}^{n} x_i^k}{n}$$

where $x_1, x_2, \ldots, x_n$ are the observed values. See also **method of moments**. [KA1 Chapter 3.]

Moments of the correlation matrix determinant: See **correlation matrix distribution**.

Moments of the generalized variance: The moments of the *generalized variance*, i.e., the determinant of the sample variance-covariance matrix, **S**, are given by

$$E(|\mathbf{S}|^t) = \frac{2^{pt}}{n^{pt}} \prod_{j=1}^{p} \frac{\Gamma\{\frac{1}{2}(n-j) + t\}}{\Gamma\{\frac{1}{2}(n-j)\}} |\mathbf{\Sigma}|^t$$

where n is the sample size, p is the number of variables and $\mathbf{\Sigma}$ is the population variance-covariance matrix of the underlying *multivariate normal distribution*. See also **Wishart distribution**. [KA1]

Monotonic decreasing: See **monotonic sequence**.

Monotonic increasing: See **monotonic sequence**.

Monotonic regression: A procedure for obtaining the curve that best fits a set of points, subject to the constraint that it never decreases. A central component of *non-metric scaling* where quantities known as disparities are fitted to Euclidean distances subject to being monotonic with the corresponding dissimilarities. [MV2 Chapter 12.]

Monotonic sequence: A series of numerical values is said to be *monotonic increasing* if each value is greater than or equal to the previous one, and *monotonic decreasing* if each value is less than or equal to the previous one. See also **ranking**.

Monte Carlo maximum likelihood (MCML): A procedure which provides a highly effective computational solution to problems involving dependent data when the use of the *likelihood function* may be intractable. [*Journal of the Royal Statistical Society, Series B*, 1992, **54**, 657–60.]

Monte Carlo methods: Methods for finding solutions to mathematical and statistical problems by *simulation.* Used when the analytic solution of the problem is either intractable or time consuming. [*Simulation and the Monte Carlo Method*, 1981, R. Y. Rubenstein, Wiley, New York.]

Monty Hall problem: A seemingly counter-intuitive problem in probability that gets its name from the TV game show, 'Let's Make a Deal' hosted by Monty Hall. On the show a participant is shown three doors behind one of which is a valuable prize and behind the other two booby prizes. The participant selects a door and then, before the chosen door is opened, the host opens one the two remaining doors to reveal one of the booby prizes. The participant is then asked if he/she would like to stay with the originally selected door or switch to the other, as yet, unopened door. Many people think that switching doors makes no difference to the probability of winning the valuable prize but many people are wrong because switching doubles this probability from a third to two thirds. [*Chance Rules*, 2nd edn, 2008, B. S. Everitt, Springer, New York.]

Mood's test: A *distribution free* test for the equality of variability in two poulations assumed to be symmetric with a common median. If the samples from the two populations are denoted $x_1, x_2, \ldots, x_{n_1}$ and $y_1, y_2, \ldots, y_{n_2}$, then the test statistic is

$$M = \sum_{i=1}^{n_1} \left[R_i - \frac{n_1 + n_2 + 1}{2} \right]^2$$

where R_i is the rank of x_i in the combined sample arranged in increasing order. For moderately large values of n_1 and n_2 the test statistic can be transformed into Z which under the null hypothesis of equal variability has a standard normal distribution, where Z is given by

$$Z = \frac{M - \frac{n_1(N^2 - 1)}{12}}{\sqrt{\frac{n_1 n_2 (N+1)(N^2 - 4)}{180}}}$$

where $N = n_1 + n_2$. [*Annals of Mathematical Statistics*, 1954, **25**, 514–22.]

Moore–Penrose inverse: A matrix **X** which satisfies the following conditions for a $n \times m$ matrix A:

$$\mathbf{AXA} = \mathbf{A}$$
$$\mathbf{XAX} = \mathbf{X}$$
$$(\mathbf{AX})' = \mathbf{AX}$$
$$(\mathbf{XA})' = \mathbf{XA}$$

[*Linear Regression Analysis*, 1977, G. A. F. Seber, Wiley, New York.]

Moral graph: Synonym for **conditional independence graph**.

Moran, Patrick Alfred Pierce (1917–1988): Born in Sydney, Australia, Moran entered Sydney University to study mathematics and physics at the age of only 16. He graduated with a First Class Honours in mathematics in 1937 and continued his studies at St. John's College, Cambridge. In 1946 he took up a position of Senior Research Officer at the Institute of Statistics in Oxford and in 1952 became the first holder of the Chair in Statistics at the Australian National University. Here he founded a Research Department which became the beginning of modern Australian statistics. Moran worked on dam theory, genetics and geometrical probability. He was elected a Fellow of the Royal Society in 1975. Moran died on 19 September 1988 in Canberra, Australia.

Moran's *I*: A statistic used in the analysis of *spatial data* to detect *spatial autocorrelation.* If $x_1, x_2, \ldots, x_n$ represent data values at n locations in space and $\mathbf{W}$ is a matrix with elements w_{ij} equal to one if i and j are neighbours and zero otherwise ($w_{ii}=0$) then the statistic is defined as

$$I = \frac{n}{\mathbf{1}'\mathbf{W}\mathbf{1}} \frac{\mathbf{z}'\mathbf{W}\mathbf{z}}{\mathbf{z}'\mathbf{z}}$$

where $z_i = x_i - \bar{x}$ and $\mathbf{1}$ is an n-dimensional vector of ones. The statistic is like an ordinary Pearson's product moment correlation coefficient but the cross-product terms are calculated only between neighbours. If there is no spatial autocorrelation then I will be close to zero. Clustering in the data will lead to positive values of I, which has a maximum value of approximately one. See also **rank adjacency statistic**. [*Statistics in Medicine*, 1993, **12**, 1883–94.]

Morbidity: A term used in epidemiological studies to describe sickness in human populations. The WHO Expert Committee on Health Statistics noted in its sixth report that morbidity could be measured in terms of three units:

- persons who were ill,
- the illnesses (periods or spells of illness) that those persons experienced,
- the duration of these illnesses.

Morgenstern's hypothesis: The statement that the more rudimentary the theory of the user, the less precision is required of the data. Consequently the maximum precision of measurement needed is dependent upon the power and fine structure of the theory.

Morgenstern's uniform distribution: A *bivariate probability distribution*, $f(x,y)$, of the form

$$f(x,y) = 1 + \alpha(2x-1)(2y-1) \qquad 0 \leq x,y \leq 1 \qquad -1 \leq \alpha \leq 1$$

A perspective plot of the distribution with $\alpha = 0.5$ is shown in Fig. 94. [KA1 Chapter 7.]

Morphometrics: A branch of *multivariate analysis* in which the aim is to isolate measures of 'size' from those of 'shape'. [MV2 Chapter 14.]

Mortality: A term used in studies in *epidemiology* to describe death in human populations. Statistics on mortality are compiled from the information contained in death certificates. [*Mortality Pattern in National Populations*, 1976, S. N. Preston, Academic Press, New York.]

Mortality odds ratio: A ratio equivalent to the *odds ratio* used in case-control studies where the equivalent of cases are deaths from the cause of interest and the equivalent of controls are deaths from all other causes. Finally the equivalent of exposure is membership of a study group of interest, for example, a particular occupation group. See also **proportionate mortality ratio** and **standardized mortality ratio**. [*American Journal of Epidemiology*, 1981, **114**, 144–148.]

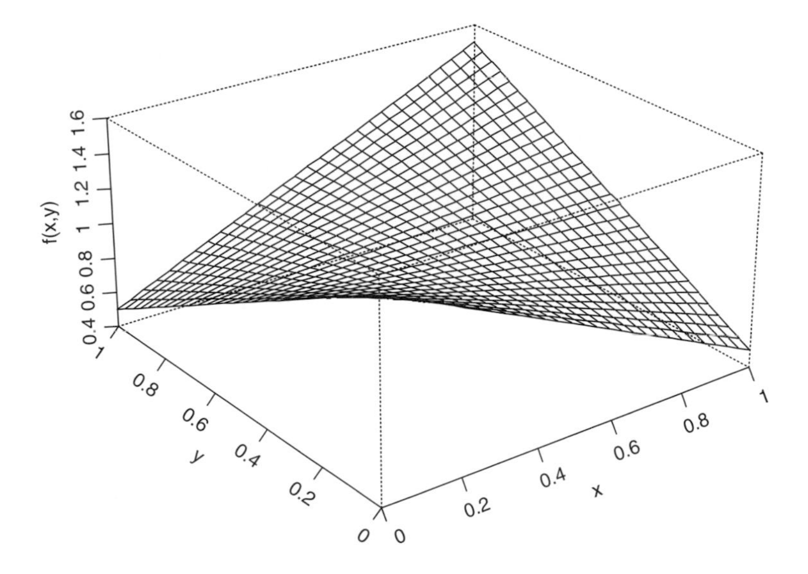

Fig. 94 Perspective plot of Morgenstern's uniform distribution.

Mortality rate: Synonym for **death rate**.

Mosaic displays: A graphical display of the *standardized residuals* from fitting a *log-linear model* to a *contingency table* in which colour and outline of the mosaic's 'tiles' are used to indicate the size and the direction of the residuals with the area of the 'tiles' being proportional to the corresponding number of cases. An example is shown in Figure 95; here the display represents the residuals from fitting an independence model to counts of hair and eye colour from men and women. The plot indicates that there are more blue-eyed blond females than expected under independence and too few brown-eyed blond females. [*Journal of the American Statistical Association*, 1994, **89**, 190–200.]

Most powerful test: A test of a null hypothesis which has greater power than any other test for a given alternative hypothesis. [*Testing Statistical Hypotheses*, 2nd edition, 1986, E. Lehmann, Wiley, New York.]

Most probable number: See **serial dilution assay**.

Mosteller, Frederick (1916–2008): Born in Clarksburg, West Virgina, USA Mosteller obtained and M.Sc degree at Carnegie Tech in 1939 before moving to Princeton to work on a Ph.D under the supervision of *Wilks* and *Tukey*. He obtained his Ph.D in mathematics in 1946 Mosteller founded the Department of Statistics at Harvard and served as its first chairman from 1957–1969. He contributed to many areas of statistics and the application of statistics and statistics education. Mosteller was an avid fan of the Boston Red Sox baseball team and one of his most well-known papers published in 1952 in the *Journal of the American Statistical Association* concerns the baseball World Series inspired by the Boston Red Sox's loss to the St. Louis Cardinals in 1946. In the paper Mosteller showed that the 'stronger' team, i.e. the one with a higher winning percentage, would still often loose to a 'weaker' team, simply because of chance. Mosteller died on July 23rd, 2008 in Falls Church, USA.

Mother wavelet: See **wavelet functions**.

Mover–stayer model: A model defined by a *Markov chain* with *state space* $1, \ldots, m$ in which the *transition probabilities* are of the form

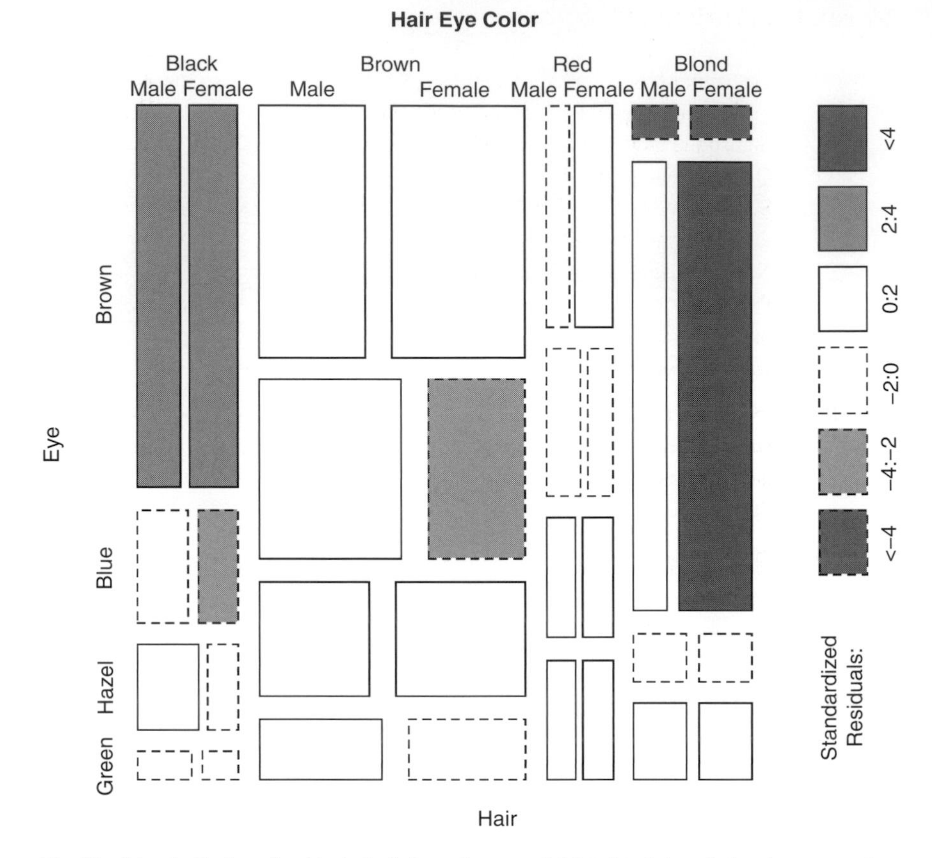

Fig. 95 Mosaic display of residuals for independence model fitted to data on hair colour, eye colour and gender.

$$\pi_{ij} = (1 - s_i)p_j, \quad i \neq j = 1, \ldots, m$$
$$\pi_{ii} = (1 - s_i)p_i + s_i, \qquad i = 1, \ldots, m$$

where $\{p_k\}$ represents a probability distribution and

$$1 - s_i \geq 0, \ (1 - s_i)p_i + s_i \geq 0, \ \text{for all } i = 1, \ldots, m$$

The conditional probabilities of state change are given by

$$P_{ij} = \pi_{ij}/(1 - \pi_{ii}) = p_j/(1 - p_i)$$

The model has been widely applied in medicine, for example, in models for the HIV/AIDS epidemic. [*Journal of Applied Statistics*, 1997, **24**, 265–78.]

Moving average: A method used primarily for the smoothing of *time series*, in which each observation is replaced by a weighted average of the observation and its near neighbours. The aim is usually to smooth the series enough to distinguish particular features of interest. [TMS Chapter 3.]

Moving average process: A *time series* having the form

$$x_t = a_t - \theta_1 a_{t-1} - \theta_2 a_{t-2} - \cdots - \theta_p a_{t-p}$$

where $a_t, a_{t-1}, \ldots, a_{t-p}$ are a *white noise sequence* and $\theta_1, \theta_2, \ldots, \theta_p$ are the parameters of the model. [TMS Chapter 3.]

Moyal, Jose Enriqué (1910–1998): Born in Jerusalem, Moyal was educated at high school in Tel Aviv and later studied mathematics at Cambridge University and electrical engineering at the Institut d'Electrotéchnique in Grenoble. He first worked as an engineer, but after 1945 he moved into statistics obtaining a diploma in mathematical statistics from the Institut de Statistique at the University of Paris. After the war Moyal began his career in statistics at Queen's University, Belfast and then in 1948 was appointed Lecturer in Mathematical Statistics at the University of Manchester. Later positions held by Moyal included those in the Department of Statistics at the Australian National University and at the Argonne National Laboratory of the US Atomic Energy Commission in Chicago. In 1972 he became Professor of Mathematics at Macquarie University, Sydney. Moyal made important contributions in engineering and mathematical physics as well as statistics where his major work was on *stochastic processes*. He died in Canberra on 22 May 1998.

Mplus: Software for fitting an extensive range of statistical models, including *factor analysis* models, *structural equation models* and *confirmatory factor analysis* models. Particularly useful for data sets having both continuous and categorical variables. [Muthén and Muthén, 3463 Stoner Avenue, Los Angeles CA 90066, USA.]

MTMM: Abbreviation for **multitrait–multimethod model**.

Muller-Griffiths procedure: A procedure for estimating the population mean in situations where the sample elements can be ordered by inspection but where exact measurements are costly. [*Journal of Statistical Planning and Inference*, 1980, **4**, 33–44.]

Multicentre study: A *clinical trial* conducted simultaneously in a number of participating hospitals or clinics, with all centres following an agreed-upon study protocol and with independent random allocation within each centre.The benefits of such a study include the ability to generalize results to a wider variety of patients and treatment settings than would be possible with a study conducted in a single centre, and the ability to enrol into the study more patients than a single centre could provide. [SMR Chapter 15.]

Multicollinearity: A term used in regression analysis to indicate situations where the explanatory variables are related by a linear function, making the estimation of regression coefficients impossible. Including the sum of the explanatory variables in the regression analysis would, for example, lead to this problem. Approximate multicollinearity can also cause problems when estimating regression coefficients. In particular if the *multiple correlation* for the regression of a particular explanatory variable on the others is high, then the variance of the corresponding estimated regression coefficient will also be high. See also **ridge regression**. [ARA Chapter 12.]

Multidimensional scaling (MDS): A generic term for a class of techniques that attempt to construct a low-dimensional geometrical representation of a *proximity matrix* for a set of stimuli, with the aim of making any structure in the data as transparent as possible. The aim of all such techniques is to find a low-dimensional space in which points in the space represent the stimuli, one point representing one stimulus, such that the distances between the points in the space match as well as possible in some sense the original dissimilarities or similarities. In a very general sense this simply means that the larger the observed dissimilarity value (or the smaller the similarity value) between two stimuli, the further apart should be the points representing them in the derived spatial solution. A general approach to finding the required coordinate values is to select them so as to minimize some least squares type fit criterion such as

$$\sum_{i<j}[d_{ij}(\mathbf{x}_i, \mathbf{x}_j) - \delta_{ij}]^2$$

where δ_{ij} represent the observed dissimilarities and $d_{ij}(\mathbf{x}_i, \mathbf{x}_j)$ represent the distance between the points with q-dimensional coordinates $\mathbf{x}_i$ and $\mathbf{x}_j$ representing stimuli i and j. In most applications d_{ij} is chosen to be *Euclidean distance* and the fit criterion is minimized by some optimization procedure such as steepest descent. The value of q is usually determined by one or other of a variety of generally *ad hoc* procedures. See also **classical scaling, individual differences scaling** and **nonmetric scaling**. [MV1 Chapter 5.]

Multidimensional unfolding: A form of *multidimensional scaling* applicable to both rectangular proximity matrices where the rows and columns refer to different sets of stimuli, for example, judges and soft drinks, and asymmetric proximity matrices, for example, citations of journal A by journal B and vice versa. Unfolding was introduced as a way of representing judges and stimuli on a single straight line so that the rank-order of the stimuli as determined by each judge is reflected by the rank order of the distance of the stimuli to that judge. See also **unfolding**. [*Psychological Review*, 1950, **57**, 148–158.]

Multiepisode models: Models for *event history data* in which each individual may undergo more than a single transition, for example, lengths of spells of unemployment, or time period before moving to another region. [*Regression with Social Data*, 2004, A. De Maris, Wiley.]

Multi-hit model: A model for a toxic response that results from the random occurrence of one or more fundamental biological events. A response is assumed to be induced once the target tissue has been 'hit' by a number, k, of biologically effective units of dose within a specified time period. Assuming that the number of hits during this period follows a *Poisson process*, the probability of a response is given by

$$P(\text{response}) = P(\text{at least } k \text{ hits}) = 1 - \sum_{j=0}^{k-1} \exp(-\lambda)\frac{\lambda^j}{j!}$$

where λ is the expected number of hits during this period. When $k=1$ the multi-hit model reduces to the *one-hit model* given by

$$P(\text{response}) = 1 - e^{-\lambda}$$

[*Communication in Statistics – Theory and Methods*, 1995, **24**, 2621–33.]

Multilevel models: Regression models for multilevel or *clustered data* where units i are nested in clusters j, for instance a *cross-sectional study* where students are nested in schools or *longitudinal studies* where measurement occasions are nested in subjects. In multilevel data the responses are expected to be dependent or correlated even after conditioning on observed covariates. Such dependence must be taken into account to ensure valid statistical inference.

Multilevel regression models include random effects with normal distributions to induce dependence among units belonging in a cluster. The simplest multilevel model is a linear random intercept model

$$\mathrm{Y_{ij}} = \beta_0 + \beta_1 x_{1\mathrm{ij}} + \ldots + \beta_{\mathrm{q}} x_{\mathrm{qij}} + \zeta_{0\mathrm{j}} + \epsilon_{\mathrm{ij}} = (\beta_0 + \zeta_{0\mathrm{j}}) + \beta_1 x_{1\mathrm{ij}} + \ldots + \beta_{\mathrm{q}} x_{\mathrm{qij}} + \epsilon_{\mathrm{ij}}$$

where the normally distributed random intercept $\zeta_{0\mathrm{j}}$ vary between clusters. The terms ϵ_{ij} are often called level-1 residuals and represent the residual variability within clusters. The level-1 residuals are assumed to be normally distributed mutually independent and independent from the random intercepts. In a linear random intercept model the residual correlation between the units in a cluster, given the covariates, is the *intra-class correlation*. An important assumption in multilevel models is that the observed convariates $x_{1\mathrm{ij}}, \ldots, x_{\mathrm{qij}}$ are independent from the random effects (and the level-1 residuals). If a *Hausman test* suggests that this so-called exogeneity assumption is violated, *fixed-effects models* with fixed cluster-specific effects α_j are sometimes preferred. Multilevel models are sometimes called mixed models because they contain both fixed effects $\beta_0, \beta_1, \ldots, \beta_q$ and a random effects ζ_{0j}.

More complex linear multilevel models also include at least one random coefficient ζ_{1j}

$$
\begin{aligned}
Y_{ij} &= \beta_0 + \beta_1 x_{1ij} + \beta_2 x_{2ij} + \ldots + \beta_q x_{qij} + \zeta_{0j} + \zeta_{1j} x_{1ij} + \epsilon_{ij} \\
&= (\beta_0 + \zeta_{0j}) + (\beta_1 + \zeta_{1j}) x_{1ij} + \beta_2 x_{2ij} + \ldots + \beta_q x_{qij} + \epsilon_{ij}
\end{aligned}
$$

where the effect of the covariate X_{1ij} (which varies within clusters) vary between clusters with random slope $\beta + \zeta_{1j}$. The random effects ζ_{0j} and ζ_{1j} have a bivariate normal distribution with zero means.

Multilevel generalized linear models or generalized linear mixed models are multilevel models where random effects are introduced in the linear predictor of *generalized linear models*. In addition to linear models for continuous responses, such models include, for instance, logistic random effects models for dichotomous, ordinal and nominal responses and log-linear random effects models for counts.

Multilevel models can also be specified for higher-level data where units are nested in clusters which are nested in superclusters. An example of such a design would be measurement occasions nested in subjects who are nested in communities. Other terms sometimes used for multilevel models include *mixed models. hierarchical models, random effects models* and *random coeffiencnt models*. [*Multilevel Analysis*, 1999, T. A. B. Snijders, Sage, London]

Multimodal distribution: A probability distribution or frequency distribution with several modes. Multimodality is often taken as an indication that the observed distribution results from the mixing of the distributions of relatively distinct groups of observations. An example of a distribution with four modes is shown in Fig. 96.

Multinomial coefficient: The number of ways that k distinguishable items can be distributed into n containers so that there are k_i items in the ith container. Given by

$$
\frac{k!}{k_1!k_2!\cdots k_n!}
$$

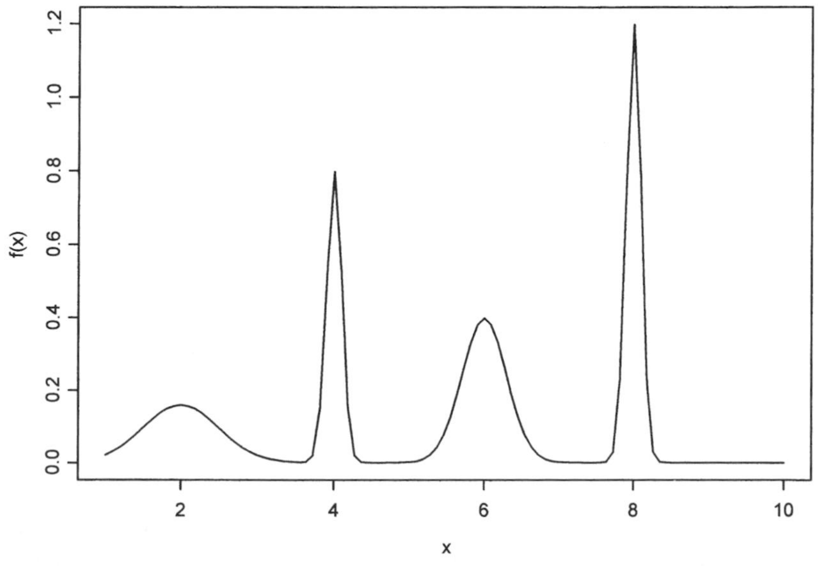

Fig. 96 A frequency distribution with four modes.

Multinomial distribution: A generalization of the *binomial distribution* to situations in which r outcomes can occur on each of n trials, where $r > 2$. Specifically the distribution is given by

$$P(n_1, n_2, \ldots, n_r) = \frac{n!}{n_1!n_2! \cdots n_r!} p_1^{n_1} p_2^{n_2} \cdots p_r^{n_r}$$

where n_i is the number of trials with outcome i, and p_i is the probability of outcome i occurring on a particular trial. The expected value of n_i is np_i and its variance is $np_i(1 - p_i)$. The covariance of n_i and n_j is $-np_ip_j$. [STD Chapter 26.]

Multinomial logit model: See **multinomial logistic regression**.

Multinomial logistic regression: A form of *logistic regression* for use when the categorical response variable has more than two unordered categories. If we let k be the number of categories of the response variable, Y, then the model used (sometimes called the *multinomial logit model*) is the following;

$$\pi_r(\mathbf{x}) = \frac{\exp(\mathbf{x}'\boldsymbol{\beta}_r)}{\sum_{s=1}^{k} \exp(\mathbf{x}'\boldsymbol{\beta}_r)}, r = 1, \ldots, k$$

where $\pi_r(\mathbf{x}) = \Pr(Y = r|\mathbf{x}), r = 1, \ldots, k$, $\mathbf{x}$ is a vector of explanatory variables and $\boldsymbol{\beta}_r$ is a vector of regression coefficients for category r. Because it is only possible to investigate the effect of **x** upon the 'preference' of a response category compared to other categories, not all the parameter vectors are identifiable so it is necessary to choose a reference category, commonly category k and to take $\boldsymbol{\beta}_k = \mathbf{0}$. See also **proportional odds model**. [*Applied Logistic Regression*, 2nd edn, 2000, D. W. Hosmer and S. Lemeshow, Wiley, New York.]

Multinormal distribution: Synonym for **multivariate normal distribution**.

Multiphasic screening: A process in which tests in *screening studies* may be performed in combination. For example, in cancer screening, two or more anatomic sites may be screened for cancer by tests applied to an individual during a single screening session. [*American Journal of Public Health*, 1964, **54**, 741–50.]

Multiple comparison tests: Procedures for detailed examination of the differences between a set of means, usually after a general hypothesis that they are all equal has been rejected. No single technique is best in all situations and a major distinction between techniques is how they control the possible inflation of the type I error. See also **Bonferroni correction, Duncan's multiple range test, Scheffé's test** and **Dunnett's test**. [*Biostatistics: A Methodology for the Health Sciences*, 2nd edition, 2004, G. Van Belle, L. D. Fisher, P. J. Heagerty and T. S. Lumley, Wiley, New York.]

Multiple correlation coefficient: The correlation between the observed values of the dependent variable in a *multiple regression*, and the values predicted by the estimated regression equation. Often used as an indicator of how useful the explanatory variables are in predicting the response. The square of the multiple correlation coefficient gives the proportion of variance of the response variable that is accounted for by the explanatory variables. [SMR Chapter 12.]

Multiple endpoints: A term used to describe the variety of outcome measures used in many *clinical trials*. Typically there are multiple ways to measure treatment success, for example, length of patient survival, percentage of patients surviving for two years, or percentage of patients experiencing tumour regression. The aim in using a variety of such measures is to gain better overall knowledge of the differences between the treatments being compared. The danger with such an approach is that the performance of multiple significance tests incurs an increased risk of a *false positive result.* See also **Bonferroni correction**. [*Statistics in Medicine*, 14, 1995, 1163–76.]

Multiple-frame surveys: Surveys that refer to two or more *sampling frames* that can cover a target population. For example, in a study of AIDS a general population frame might be used in addition to drug treatment centres and hospitals. Information from the samples is then combined to estimate population quantities. Such surveys are very useful for sampling rare or hard-to-reach populations and can result in considerable cost savings over a single frame design with comparable precision. See also **respondent-driven sampling** and **snowball sampling**. [*Journal of the American Statistical Association*, 2006, **10**, 1019–1030.]

Multiple imputation: A *Monte Carlo method* in which *missing values* in a data set are replaced by $m > 1$ simulated versions, where m is typically small (say 3–10). Each of the simulated complete datasets is analysed by the method appropriate to the investigation at hand, and the results are later combined to produce estimates, confidence intervals etc. The imputations are created by a Bayesian approach which requires specification of a parametric model for the complete data and, if necessary, a model for the mechanism by which data become missing. Also required is a *prior distribution* for the unknown model parameters. *Bayes' theorem* is used to simulate m independent samples form the conditional distribution of the missing values given the observed values. In most cases special computation techniques such as *Markov chain Monte Carlo methods* will be needed. See also **SOLAS**. [*The Analysis of Incomplete Multivariate Data*, 1997, J. Schafer, CRC/Chapman and Hall.]

Multiple indicator multiple cause model (MIMIC): A *structural equation model* in which there are multiple indicators and multiple causes of each *latent variable.* [*Modelling Covariance and Latent Variables using EQS*, 1993, G. Dunn, B. Everitt and A. Pickles, Chapman and Hall/CRC Press, London.]

Multiple linear regression: Synonym for **multiple regression**.

Multiple regression: A term usually applied to models in which a continuous response variable, y, is regressed on a number of explanatory variables, $x_1, x_2, \ldots, x_q$. Explicitly the model fitted is

$$E(y|x_1, \ldots, x_q) = \beta_0 + \beta_1 x_1 + \cdots + \beta_q x_q$$

where E denotes expected value. By introducing a vector $\mathbf{y}' = [y_1, y_2, \ldots, y_n]$ and an $n \times (q+1)$ matrix $\mathbf{X}$ given by

$$\mathbf{X} = \begin{pmatrix} 1 & x_{11} & x_{12} & \cdots & x_{1q} \\ 1 & x_{21} & x_{22} & \cdots & x_{2q} \\ \vdots & \vdots & \vdots & \vdots & \vdots \\ 1 & x_{n1} & x_{n2} & \cdots & x_{nq} \end{pmatrix}$$

the model for n observations can be written as

$$\mathbf{y} = \mathbf{X}\beta + \epsilon$$

where $\epsilon' = [\epsilon_1, \epsilon_2, \ldots, \epsilon_n]$ contains the residual error terms and $\beta' = [\beta_0, \beta_1, \beta_2, \ldots, \beta_q]$. *Least squares estimation* of the parameters involves the following set of equations

$$\hat{\beta} = (\mathbf{X}'\mathbf{X})^{-1}\mathbf{X}'\mathbf{y}$$

The regression coefficients, $\beta_1, \beta_2, \ldots, \beta_q$ give the change in the response variable corresponding to a unit change in the appropriate explanatory variable, conditional on the other variables remaining constant. Significance tests of whether the coefficients take the value zero can be derived on the assumption that for a given set of values of the explanatory variables, y has a normal distribution with constant variance. See also **regression diagnostics, selection methods in regression** and **beta coefficient**. [ARA Chapter 3.]

Multiple time response data: Data arising in studies of episodic illness, such as bladder cancer and epileptic seizures. In the former, for example, individual patients may suffer multiple bladder tumours at observed times, $t_1 < t_2 < \cdots < t_k$.

Multiple time series: Observations taken simultaneously on two or more *time series*. Also known as *multivariate time series* and *vector time series*. Environmental examples include readings of lead concentrations at several sites at 5 minute intervals, air temperature readings taken at hourly intervals at a fixed height above sea level at several locations and monthly ozone levels at several recording stations. If the m series are represented as $\mathbf{X}'_t = [x_{1,t}, x_{2,t}, \ldots, x_{m,t}]$ for $t = 1, 2, \ldots, n$ then the *cross-covariance function*, $\gamma_{ij}(k)$ is $\text{cov}(x_{i,t}, x_{j,\,t-k})$ and the *cross-correlation function*, $\rho_{ij}(k)$ is defined as

$$\rho_{ij}(k) = \frac{\gamma_{ij}(k)}{[\gamma_{ii}(0)\gamma_{jj}(0)]^{\frac{1}{2}}}$$

The matrix $\boldsymbol{\rho}(k)$ with ijth entry equal to $\rho_{ij}(k)$ can be used to define the *spectral density matrix*, $\mathbf{f}(\omega)$, defined as

$$\mathbf{f}(\omega) = \sum_{l=-\infty}^{\infty} \exp(-2\pi i \omega l)\boldsymbol{\rho}(k) \qquad -0.5 \leq \omega \leq 0.5$$

where $i^2 = -1$. The ijth element of this matrix is the *cross-spectral density*. The real and imaginary parts of this matrix define the *co-spectrum* and *quadrature spectrum*. Multivariate generalizations of *autoregressive moving average models* are available for modelling such a collection of time series. See also **coherence**. [TMS Chapter 8.]

Multiplication rule for probabilities: For events A and B that are independent, the probability that both occur is the product of the separate probabilities, i.e.

$$\Pr(A \text{ and } B) = \Pr(A)\Pr(B)$$

where Pr denotes probability. Can be extended to k independent events, $B_1, B_2, \ldots, B_k$ as

$$\Pr(\cap_{i=1}^{k} B_i) = \prod_{i=1}^{k} \Pr(B_i)$$

[KA1 Chapter 8.]

Multiplicative intensity model: A generalization of the *Cox's proportional hazards model*. [*The Annals of Statistics*, 1978, **6**, 701–26.]

Multiplicative model: A model in which the combined effect of a number of factors when applied together is the product of their separate effects. See also **additive model**.

MULTISCALE: Software for *multidimensional scaling* set in an inferential framework. [J. O. Ramsay, Department of Psychology, McGill University, Steward Biological Sciences Building, 1205 Dr. Penfield Avenue, Montreal, QC, Canada H34 1B1.]

Multistage sampling: Synonym for **cluster sampling**.

Multistate models: Models that arise in the context of the study of *survival times*. The experience of a patient in such a study can be represented as a process that involves two (or more) states. In the simplest situation at the point of entry to the study, the patient is in a state that corresponds to being alive. Patients then transfer from this 'live' state to the 'dead' state at some rate measured by the *hazard function* at a given time. More complex models will involve more states. For example, a three-state model might have patients alive and tumour free, patients alive and tumour present and the 'dead' state. See also **Markov illness–death model**. [*Statistics in Medicine*, 1988, **7**, 819–42.]

Multitaper spectral estimators: A very powerful class of procedures for estimating the *spectral density* of a *time series*, which use the average of several direct spectral estimators. [*Signal Processing*, 1997, **58**, 327–32.]

Multitrait–multimethod model (MTMM): A form of *confirmatory factor analysis* model in which different methods of measurement are used to measure each latent variable. Allows the variance of each measure to be decomposed into trait variance, method variance and measurement error variance. [*Modelling Covariances and Latent Variables using EQS*, 1993, G. Dunn, B. Everitt and A. Pickles, Chapman and Hall/CRC Press, London.]

Multivariate adaptive regression splines (MARS): A method of flexible *nonparametric regression modelling* that works well with moderate sample sizes and with more than two explanatory variables. [*Journal of Computational and Graphical Statistics*, 1997, **6**, 74–91.]

Multivariate analysis: A generic term for the many methods of analysis important in investigating multivariate data. Examples include *cluster analysis, principal components analysis* and *factor analysis*. [MV1 and MV2.]

Multivariate analysis of variance: A procedure for testing the equality of the mean vectors of more than two populations for a multivariate response variable. The technique is directly analogous to the *analysis of variance* of univariate data except that the groups are compared on q response variables simultaneously. In the univariate case, F-tests are used to assess the hypotheses of interest. In the multivariate case, however, no single test statistic can be constructed that is optimal in all situations. The most widely used of the available test statistics is *Wilk's lambda* (Λ) which is based on three matrices $\mathbf{W}$ (the *within groups matrix of sums of squares and products*), $\mathbf{T}$ (the *total matrix of sums of squares and cross-products*) and $\mathbf{B}$ (the *between groups matrix of sums of squares and cross-products*), defined as follows:

$$\mathbf{T} = \sum_{i=1}^{g}\sum_{j=1}^{n_i}(\mathbf{x}_{ij} - \bar{\mathbf{x}})(\mathbf{x}_{ij} - \bar{\mathbf{x}})'$$

$$\mathbf{W} = \sum_{i=1}^{g}\sum_{j=1}^{n_i}(\mathbf{x}_{ij} - \bar{\mathbf{x}}_i)(\mathbf{x}_{ij} - \bar{\mathbf{x}}_i)'$$

$$\mathbf{B} = \sum_{i=1}^{g} n_i(\bar{\mathbf{x}}_i - \bar{\mathbf{x}})(\bar{\mathbf{x}}_i - \bar{\mathbf{x}})'$$

where $\mathbf{x}_{ij}, i = 1, \ldots, g, j = 1, \ldots, n_i$ represent the jth multivariate observation in the ith group, g is the number of groups and n_i is the number of observations in the ith group. The mean vector of the ith group is represented by $\bar{\mathbf{x}}_i$ and the mean vector of all the observations by $\bar{\mathbf{x}}$. These matrices satisfy the equation

$$\mathbf{T} = \mathbf{W} + \mathbf{B}$$

Wilk's lambda is given by the ratio of the determinants of $\mathbf{W}$ and $\mathbf{T}$, i.e.

$$\Lambda = \frac{|\mathbf{W}|}{|\mathbf{T}|} = \frac{|\mathbf{W}|}{|\mathbf{W} + \mathbf{B}|}$$

The statistic, Λ, can be transformed to give an F-test to assess the null hypothesis of the equality of the population mean vectors. In addition to Λ a number of other test statistics are available

- *Roy's largest root criterion*: the largest *eigenvalue* of $\mathbf{BW}^{-1}$,
- The *Hotelling–Lawley trace*: the sum of the eigenvalues of $\mathbf{BW}^{-1}$,
- The *Pillai–Bartlett trace*: the sum of the eigenvalues of $\mathbf{BT}^{-1}$.

It has been found that the differences in power between the various test statistics are generally quite small and so in most situations which one is chosen will not greatly affect conclusions. [MV1 Chapter 6.]

Multivariate Bartlett test: A test for the equality of a number *(k)* of *variance-covariance matrices*. The test statistic used is M given by

$$M = \frac{|\mathbf{S}_1|^{\nu_1/2}|\mathbf{S}_2|^{\nu_2/2}\ldots|\mathbf{S}_k|^{\nu_k/2}}{|\mathbf{S}|^{\nu}}$$

where $\nu_i = n_i - 1$ where n_i is the sample size of the ith sample, $\mathbf{S}_i$ is the covariance matrix of the ith sample, $\nu = \sum_{i=1}^{k} \nu_i$ and $\mathbf{S}$ is the pooled sample covariance matrix given by

$$\mathbf{S} = \frac{\sum_{i=1}^{k} \nu_i \mathbf{S}_i}{\sum_{i=1}^{k} \nu_i}$$

The statistic $u = -2(1 - c_1)\log M$ is under the null hypothesis of the equality of the population covariance matrices, approximately distributed as chi-squared with $\frac{1}{2}(k-1)p(p+1)$ degrees of freedom where p is the number of variables and

$$c_1 = \left[\sum_{i=1}^{k}\frac{1}{\nu_i} - (1/\sum_{i=1}^{k}\nu_i)\right]\left[\frac{2p^2+3p-1}{6(p+1)(k-1)}\right]$$

[*Encyclopedia of Statistical Sciences*, 2006, eds. S. Kotz, C. B. Read, N. Balakrishnan and B. Vidakovic, Wiley, New York.]

Multivariate Cauchy distribution: See **Student's *t*-distribution**.

Multivariate counting process: A *stochastic process* with k components counting the occurrences, as time passes, of k different types of events, none of which can occur simultaneously. [*Statistical Analysis of Counting Processes*, 1982, M. Jacobsen, Springer-Verlag, New York.]

Multivariate data: Data for which each observation consists of values for more than one random variable. For example, measurements on blood pressure, temperature and heart rate for a number of subjects. Such data are usually displayed in the form of a *data matrix*, i.e.

$$\mathbf{X} = \begin{pmatrix} x_{11} & x_{12} & \cdots & x_{1q} \\ x_{21} & x_{22} & \cdots & x_{2q} \\ \vdots & \vdots & \cdots & \vdots \\ x_{n1} & x_{n2} & \cdots & x_{nq} \end{pmatrix}$$

where n is the number of subjects, q the number of variables and x_{ij} the observation on variable j for subject i. [MV1 Chapter 1.]

Multivariate distribution: The simultaneous probability distribution of a set of random variables. See also **multivariate normal distribution** and **Dirichlet distribution**. [KA1 Chapter 15.]

Multivariate growth data: Data arising in studies investigating the relationships in the growth of several organs of an organism and how these relationships evolve. Such data enable biologists to examine growth gradients within an organism and use these as an aid to understanding its form, function and biological niche as well as the role of evolution in bringing it to its present form. [*Psychometrika*, 2004, **69**, 65–79.]

Multivariate hypergeometric distribution: A probability distribution associated with *sampling without replacement* from a population of finite size. The population consists of r_i elements of type i, $i = 1, 2, \ldots, k$ with $\sum_{i=1}^{k} r_i = N$. Then the probability of finding x_I elements of the ith type when a random sample of size n is drawn $(n = \sum x_i)$ is

$$Pr(x_1, x_2, \ldots, x_k) = \binom{r_1}{x_1}\binom{r_2}{x_2}\cdots\binom{r_k}{x_k}/\binom{N}{n}$$

When $k=2$ this reduces to the *hypergeometric distribution*. The distribution is important in conditional inference for $2 \times k$ *contingency tables*. [*Statistical Analysis of Categorial Data*, 1999, C. J. Lloyd, Wiley, New York.]

Multivariate kurtosis measure: See **Mardia's multivariate normality test**.

Multivariate normal distribution: The probability distribution of a set of variables $\mathbf{x}' = [x_1, x_2, \ldots, x_q]$ given by

$$f(x_1, x_2, \ldots, x_q) = (2\pi)^{-q/2}|\boldsymbol{\Sigma}|^{-\frac{1}{2}}\exp\left\{-\frac{1}{2}(\mathbf{x}-\boldsymbol{\mu})'\boldsymbol{\Sigma}^{-1}(\mathbf{x}-\boldsymbol{\mu})\right\}$$

where $\boldsymbol{\mu}$ is the mean vector of the variables and $\boldsymbol{\Sigma}$ is their *variance–covariance matrix*. This distribution is assumed by multivariate analysis procedures such as *multivariate analysis of variance*. See also **bivariate normal distribution**. [KA1 Chapter 15.]

Multivariate power exponential distribution: A multivariate probability distribution, *f(x)* given by

$$f(\mathbf{x}) = \frac{n\Gamma\left(\frac{n}{2}\right)}{\pi^{\frac{n}{2}}\sqrt{|\boldsymbol{\Sigma}|}\Gamma\left(1+\frac{n}{2\beta}\right)2^{1+\frac{n}{2\beta}}}\exp\left\{-\frac{1}{2}[(\mathbf{x}-\mu)\prime\boldsymbol{\Sigma}^{-1}(\mathbf{x}-\mu)]^{\beta}\right\}$$

The distribution has mean vector μ and *variance-covariance matrix* given by

$$\frac{2^{\frac{1}{\beta}}\Gamma\left(\frac{n+2}{2\beta}\right)}{n\Gamma\left(\frac{n}{2\beta}\right)}\boldsymbol{\Sigma}$$

When $\beta = 1, f(\mathbf{x})$ becomes the *multivariate normal distribution*. In general $f(\mathbf{x})$ is an elliptically contoured distribution in which β determines the *kurtosis*. For $\beta < 1$ the distribution has heavier tails than the normal distribution. [*Nonlinear Models in Medical Statistics*, 2001, J. K. Lindsey, Oxford University Press, Oxford.]

Multivariate probit analysis: A method for assessing the effect of explanatory variables on a set of two or more correlated binary response variables. See also **probit analysis**. [*Biometrics*, 1981, **37**, 541–51.]

Multivariate skew-normal distribution: See **skew-normal distribution**.

Multivariate Student's *t*-distribution: See **Student's *t*-distribution**.

Multivariate time series: Synonym for **multiple time series**.

Multivariate ZIP model (MZIP): A multivariate version of *zero-inflated Poisson regression* useful in monitoring manufacturing processes when several types of defect are possible. [*Technometrics*, 1999, **41**, 29–38.]

Mutation distance: A distance measure for two amino acid sequences, defined as the minimal number of nucleotides that would need to be altered in order for the gene of one sequence to code for the other. [*Branching Processes with Biological Applications*, 1975, P. Jagers, Wiley, New York.]

Mutually exclusive events: Events that cannot occur jointly.

MYCIN: An *expert system* developed at Stanford University to assist physicians in the diagnosis and treatment of infectious diseases. The system was never used in practice, not because of any weakness in performance, but because of ethical and legal problems related to the use of computers in medicine – if it gives the wrong diagnosis who would you sue? [*Rule-Based Expert Systems*, 1985, B. G. Buchanan and E. H. Shortliffe, Addison-Wesley, Reading, MA.]

MZIP: Abbreviation for **multivariate ZIP model**.

N

Nadaraya-Watson estimator: An estimator used to estimate regression functions based on data $\{x_i, y_i\}$. The estimator produces an estimate of Y at any requested value of X (not only the ones in the data). A *kernel function* is used to assign weights to be put on values in the data near X in estimating Y. [*Nonparametric Functional Data Analysis: Theory and Practice*, 2006, F. F. Erraty and P. V. Ieu, Springer, New York.]

NAG: Numerical Algorithms Group producing many useful subroutines relevant to statistics. [NAG Ltd., Wilkinson House, Jordan Hill Road, Oxford, OX2 8DR; NAG Inc., 1431 Opus Place, Suite 220, Downers Grove, Illinois 60515-1362, USA; www.nag.co.uk]

Naor's distribution: A discrete probability distribution that arises from the following model; Suppose an urn contains n balls of which one is red and the remainder are white. Sampling with replacement of a white ball (if drawn) by a red ball continues until a red ball is drawn. Then the probability distribution of the required number of draws, Y, is

$$\Pr(Y = y) = \frac{(n-1)!y}{(n-y)!n^y}$$

After $n - 1$ draws the urn contains only red balls and so no more than n draws are required. [*Univariate Discrete Distributions*, 2005, N. L. Johnson, A. W. Kemp and S. Kotz, Wiley, New York.]

National lotteries: Games of chance held to raise money for particular causes. The first held in the UK took place in 1569 principally to raise money for the repair of the Cinque Ports. There were 400 000 tickets or lots, with prizes in the form of plate, tapestries and money. Nowadays lotteries are held in many countries with proceeds either used to augment the exchequer or to fund good causes. The current UK version began in November 1994 and consists of selecting six numbers from 49 for a one pound stake. The winning numbers are drawn 'at random' using one of a number of 'balls-in-drum' type of machine. [*Chance Rules*, 2nd edn, 2008, B. S. Everitt, Springer-Verlag, New York.]

Natural-pairing: See **paired samples**.

Nearest-neighbour clustering: Synonym for **single linkage clustering**.

Nearest-neighbour methods I: Methods of *discriminant analysis* based on studying the *training set* subjects most similar to the subject to be classified. Classification might then be decided according to a simple majority verdict among those most similar or 'nearest' training set subjects, i.e. a subject would be assigned to the group to which the majority of the 'neighbours' belonged. Simple nearest-neighbour methods just consider the most similar neighbour. More general methods consider the k nearest neighbours, where $k > 1$. [*Discrimination and Classification*, 1981, D. J. Hand, Wiley, Chichester.]

Nearest-neighbour methods II: Methods used in the preliminary investigation of *spatial data* to assess depatures from *complete spatial randomness*. For example, histograms of distances between nearest neighbours are often useful. [*Spatial Processes: Models and Applications*, 1981, A. D. Cliff and K. Ord, Penn, London.]

Necessarily empty cells: Synonym for **structural zeros**.

Negative binomial distribution: The probability distribution of the number of failures, X, before the kth success in a sequence of *Bernoulli trials* where the probability of success at each trial is p and the probability of failure is $q=1-p$. The distribution is given by

$$\Pr(X=x)=\binom{k+x-1}{x}p^k q^x \qquad 0\leq x<\infty$$

The mean, variance, *skewness* and *kurtosis* of the distribution are as follows:

$$\text{mean} = kq/p$$
$$\text{variance} = kq/p^2$$
$$\text{skewness} = (1+q)(kq)^{-\frac{1}{2}}$$
$$\text{kurtosis} = 3+(1+4p+p^2)/(kp)$$

Often used to model *overdispersion* in count data. [STD Chapter 28.]

Negative exponential distribution: Synonym for **exponential distribution**.

Negative hypergeometric distribution: In *sampling without replacement* from a population consisting of r elements of one kind and $N-r$ of another, if two elements corresponding to that selected are replaced each time, then the probability of finding x elements of the first kind in a random sample of n elements is given by

$$\Pr(x)=\frac{\binom{r+x-1}{x}\binom{N-r+n-x-1}{n-x}}{\binom{N+n-1}{n}}$$

The mean of the distribution is Nr/N and the variance is $(nr/N)\ (1-r/N)\ (N+n)/(N+1)$. Corresponds to a *beta binomial distribution* with integral parameter values. [KA1 Chapter 5.]

Negative multinominal distribution: A generalization of the *negative binomial distribution* in which $r>2$ outcomes are possible on each trial and sampling is continued until m outcomes of a particular type are obtained. [*Biometrika*, 1949, **36**, 47–58.]

Negative predictive value: The probability that a person having a negative result on a *diagnostic test* does not have the disease. See also **positive predictive value**. [SMR Chapter 14.]

Negative skewness: See **skewness**.

Negative study: A study that does not yield a statistically significant result.

Negative synergism: See **synergism**.

Neighbourhood controls: Synonym for **community controls**.

Nelder–Mead simplex algorithm: Type of **simplex algorithm**.

Nelson–Aalen estimator: A nonparametric estimator of the *cumulative hazard function* from censored *survival data*. Essentially a *method of moments* estimator. [*Biometrics*, 1996, **52**, 1034–41.]

Nested case-control study: A commonly used design in *epidemiology* in which a cohort is followed to identify cases of some disease of interest and the controls are selected for each case from those at risk when a subject becomes a case for comparison of exposures. See also **retrospective design, nested case-control study** and **case-cohort study**. [*Statistics in Medicine*, 1993, **12**, 1733–46.]

Nested design: A design in which levels of one or more factors are subsampled within one or more other factors so that, for example, each level of a factor B occurs at only one level of another factor A. Factor B is said to be nested within factor A. An example might be where interest centres on assessing the effect of hospital and doctor on a response variable, patient satisfaction. The doctors can only practice at one hospital so they are nested within hospitals. See also **multilevel model**.

Nested model: Synonym for **hierarchical model** and **multilevel model**.

Network: A linked set of computer systems, capable of sharing computer power and/or storage facilities.

Network sampling: A *sampling design* in which a simple random sample or stratified sample of *sampling units* is made and all observational units which are linked to any of the selected sampling units are included. Different observational units may be linked to different numbers of the sampling units. In a survey to estimate the prevalence of a rare disease, for example, a random sample of medical centres might be selected. From the records of each medical centre in the sample, records of the patients treated for the disease of interest could be extracted. A given patient may have been treated at more than one centre and the more centres at which treatment has been received, the higher the *inclusion probability* for the patient's records. [*Aids*, 1996, **10**, 657–66.]

Neural networks: See **artificial neural networks**.

Newman–Keuls test: A *multiple comparison test* used to investigate in more detail the differences existing between a set of means as indicated by a significant *F-test* in an *analysis of variance*. The test proceeds by arranging the means in increasing order and calculating the *test statistic*

$$S = \frac{\bar{x}_A - \bar{x}_B}{\sqrt{\frac{s^2}{2}\left(\frac{1}{n_A} + \frac{1}{n_B}\right)}}$$

where $\bar{x}_A$ and $\bar{x}_B$ are the two means being compared, s^2 is the within groups mean square from the analysis of variance, and n_A and n_B are the number of observations in the two groups. Tables of critical values of S are available, these depending on a parameter r that specifies the interval between the ranks of the two means being tested. For example, when comparing the largest and smallest of four means, $r=4$, and when comparing the second smallest and smallest means, $r=2$. [SMR Chapter 9.]

Newton–Raphson method: A numerical procedure that can be used to optimize (minimize or maximize) a function f with respect to a set of parameters $\boldsymbol{\theta}' = [\theta_1, \ldots, \theta_m]$. The iterative scheme is

$$\boldsymbol{\theta}_{i+1} = \boldsymbol{\theta}_i - \mathbf{G}^{-1}(\boldsymbol{\theta}_i)\mathbf{g}(\boldsymbol{\theta}_i)$$

where $\mathbf{g}(\boldsymbol{\theta}_i)$ is the vector of derivatives of f with respect to $\theta_1, \ldots, \theta_m$ evaluated at $\boldsymbol{\theta}_i$ and $\mathbf{G}(\boldsymbol{\theta}_i)$ is the $m \times m$ matrix of second derivatives of f with respect to the parameters again evaluated at $\boldsymbol{\theta}_i$. The convergence of the method is very fast when $\boldsymbol{\theta}$ is close to the optimum but when this is not so $\mathbf{G}$ may become negative definite and the method may fail to converge. A further disadvantage of the method is the need to invert $\mathbf{G}$ on each iteration. See also **Fisher's scoring method** and **simplex method**. [*Introduction to Optimization Methods and Their Applications in Statistics*, 1987, B.S. Everitt, Chapman and Hall/CRC Press, London.]

Neyman, Jerzy (1894–1981): Born in Bendery, Moldova, Neyman's paternal grandfather was a Polish nobleman and a revolutionary who was burned alive in his house during the 1863 Polish uprising against the Russians. His doctoral thesis at the University of Warsaw was on probabilistic problems in agricultural experiments. Until 1938 when he emigrated to the USA he had worked in Poland though making academic visits to France and England. Between 1928 and 1933 he developed, in collaboration with *Egon Pearson*, a firm basis for the theory of hypothesis testing, supplying the logical foundation and mathematical rigour that were missing in the early methodology. In 1934 Neyman created the theory of survey sampling and also laid the theoretical foundation of modern *quality control procedures*. He moved to Berkeley in 1938. Neyman was one of the founders of modern statistics and received the Royal Statistical Society's Guy medal in gold and in 1968 the US Medal of Science. Neyman died on 5 August 1981 in Oakland, California.

Neyman–Pearson lemma: An important result for establishing most powerful statistical tests. Suppose the set of values taken by random variables $\mathbf{X}' = [X_1, X_2, \ldots, X_n]$ are represented by points in n-dimensional space (the *sample space*) and associated with each point x is the value assigned to x by two possible probability distributions P_0 and P_1 of X. It is desired to select a set S_0 of sample points x in such a way that if $P_0(S_0) = \sum_{x \in S_0} P_0(x) = \alpha$ then for any set S satisfying $P(S) = \sum_{x \in S} P_0(x) \leq \alpha$ one has $P_1(S) \leq P_1(S_0)$. The lemma states that the set $S_0 = \{x : r(x) > C\}$ is a solution of the stated problem and that this is true for every value of C where $r(x)$ is the *likelihood ratio*, $P_1(x)/P_0(x)$. [*Testing Statistical Hypotheses*, 2nd edition, 1986, E. L. Lehmann, Wiley, New York.]

Neyman-Rubin causal framework: A counterfactual framework of causality which is useful for understanding the assumptions required for valid causal inference under different research designs. Consider a simple study where either an active treatment (or intervention) $T_i = 1$ or a control treatment (or intervention) $T_i = 0$ is administered to each unit i. Although each unit is only given one of the treatments, two "potential outcomes" are denoted $Y_i(1)$ if i were given the active treatment and $Y_i(0)$ if it were given the control treatment. The causal effect of interest is $\delta_i = Y_i(1) - Y_i(0)$, but this cannot be estimated since only one of $Y_i(1)$ and $Y_i(0)$ is observed. However, under certain assumptions that are explicated in the framework, an average treatment effect δ can be estimated by $\hat{\delta} = \hat{E}(Y_i = 1|T_i = 1) - \hat{E}(Y_i = 1|T_i = 0)$. [*Journal of the American Statistical Association*, 1986, **81**, 945–970.]

Neyman smooth test: A goodness-of-fit test for testing uniformity. [*Journal of Applied Mathematics and Decision Sciences*, 2001, **5**, 181–191.]

Nightingale, Florence (1820–1910): Born in Florence, Italy, Florence Nightingale trained as a nurse at Kaiserswerth and Paris. In the Crimean War (1854) she led a party of 38 nurses to organize a nursing department as Scutari, where she substantially improved the squalid hospital conditions. She devoted much of her life to campaigns to reform the health and living conditions of the British Army, basing her arguments on massive amounts of data carefully collated and tabulated and often presented in the form of *pie charts* and *bar charts*.

Ahead of her time as an epidemiologist, Florence Nightingale was acutely aware of the need for suitable comparisons when presenting data and of the possible misleading effects of crude death rates. Well known to the general public as the Lady of the Lamp, but equally deserving of the lesser known alternative accolade, the Passionate Statistician, Florence Nightingale died in London on 13 August 1910.

NLM: Abbreviation for **non-linear mapping**.

NOEL: Abbreviation for **no-observed-effect level**.

N of 1 clinical trial: A special case of a *crossover design* aimed at determining the efficacy of a treatment (or the relative merits of alternative treatments) for a specific patient. The patient is repeatedly given a treatment and placebo, or different treatments, in successive time periods. See also **interrupted time series design**. [*Statistical Issues in Drug Development*, 2nd edition, 2008, S. Senn, Wiley-Blackwell, Chichester.]

No free lunch theorem: A theorem concerned with optimization that states (in very general terms) that a general-purpose universal optimization strategy is theoretically impossible, and the only way one strategy can outperform another is if it is specialized to the specific problem under consideration. [*Complexity*, 1996, **1**, 40–46.]

Noise: A *stochastic process* of irregular fluctuations. See also **white noise sequence**.

Nominal significance level: The significance level of a test when its assumptions are valid.

Nominal variable: Synonym for **unordered categorical variable**.

Nomogram: Graphic method that permit the representation of more than two quantities on a plane surface. The example shown in Fig. 97 is of such a chart for calculating sample size or power. [SMR Chapter 15.]

Noncentral chi-squared distribution: The probability distribution, $f(x)$, of the sum

$$\sum_{i=1}^{\nu}(Z_i + \delta_i)^2$$

where $Z_1, \ldots, Z_\nu$ are independent standard normal random variables and $\delta_1, \ldots, \delta_\nu$ are constants. The distribution has degrees of freedom ν and is given explicitly by

$$f(x) = \frac{e^{-(x+\lambda)/2}x^{(\nu-2)/2}}{2^{\nu/2}}\sum_{r=0}^{\infty}\frac{\lambda^r x^r}{2^{2r}r!\Gamma(\frac{1}{2}\nu + r)} \qquad x > 0$$

where $\lambda = \sum_{i=1}^{\nu}\delta_i^2$ is known as the noncentrality parameter. Arises as the distribution of sums of squares in *analysis of variance* when the hypothesis of the equality of group means does not hold. [KA2 Chapter 23.]

Noncentral distributions: A series of probability distributions each of which is an adaptation of one of the standard *sampling distributions* such as the *chi-squared distribution*, the *F-distribution* or *Student's* t-distribution for the distribution of some test statistic under the alternative hypothesis. Such distributions allow the *power* of the corresponding hypothesis tests to be calculated. See also **noncentral chi-squared distribution**, **noncentral *F*-distribution** and **noncentral *t*-distribution**. [KA2 Chapter 23.]

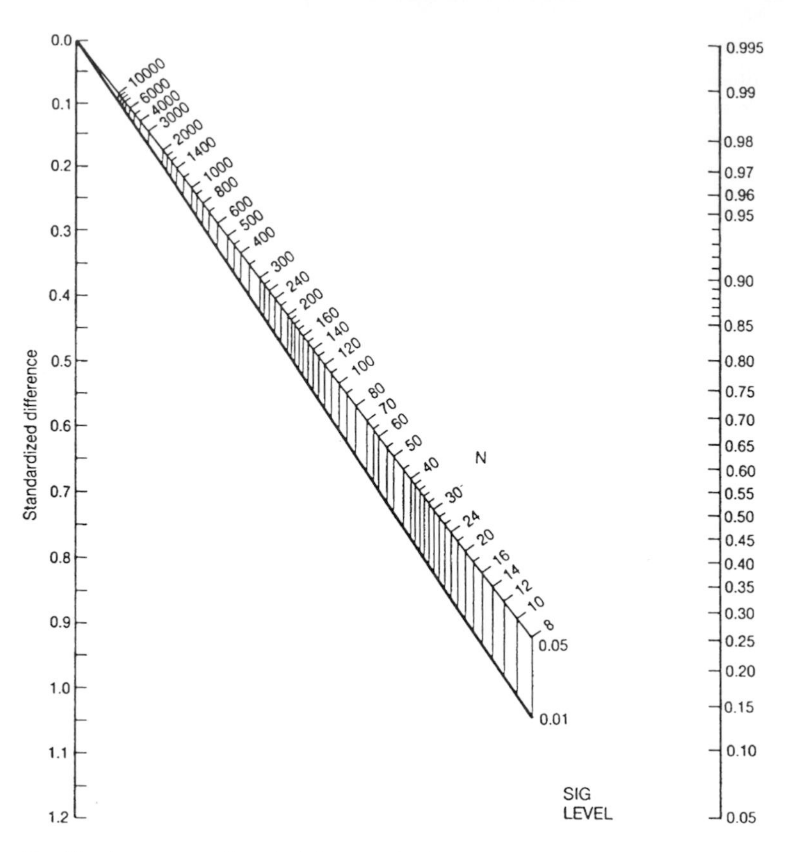

Fig. 97 A nomogram for calculating sample size.

Noncentral *F*-distribution: The probability distribution of the ratio of a random variable having a *noncentral chi-squared distribution* with noncentrality parameter λ divided by its degrees of freedom (ν_1), to a random variable with a *chi-squared distribution* also divided by its degrees of freedom (ν_2). Given explicitly by

$$f(x) = e^{-\lambda/2} \sum_{r=0}^{\infty} \frac{\dfrac{(\frac{1}{2}\lambda)^r}{r!}\left(\dfrac{\nu_1}{\nu_2}\right)^{(\nu_1/2)+r}}{B(\frac{1}{2}\nu_1 + r, \frac{1}{2}\nu_2)} \frac{x^{(\nu_1/2)+r-1}}{\left(1+\frac{\nu_1}{\nu_2}x\right)^{r+(\nu_1+\nu_2)/2}}$$

where B is the *beta function*. The *doubly noncentral F-distribution* arises from considering the ratio of two noncentral chi-squared variables each divided by their respective degrees of freedom. [KA2 Chapter 23.]

Noncentral hypergeometric distribution: A probability distribution constructed by supposing that in *sampling without replacement*, the probability of drawing say a white ball, given that there are X' white and $N' - X'$ black balls is not $\frac{X'}{N'}$ but $X'[X' + \theta(N' - X')]^{-1}$ with $\theta \neq 1$. [*Univariate Discrete Distributions*, 2005, N. L. Johnson, A. W. Kemp and S. Kotz, Wiley, New York.]

Noncentral *t*-distribution: The probability distribution, $f(x)$, of the ratio

$$\frac{Z+\delta}{W^{\frac{1}{2}}}$$

where Z is a random variable having a standard normal distribution and W is independently distributed as χ^2/ν with ν degrees of freedom; δ is a constant. Given explicitly by

$$f(x) = \{\Gamma(\tfrac{1}{2})\Gamma(\tfrac{1}{2}\nu)\nu^{1/2}\exp(\tfrac{1}{2}\delta^2)\}^{-1}\sum_{r=0}^{\infty}\frac{\{(2/\nu)^{1/2}\delta x\}^r}{r!}\frac{\Gamma\{\tfrac{1}{2}(\nu+r+1)\}}{\left(1+\frac{x^2}{\nu}\right)^{(\nu+r+1)/2}}$$

[KA2 Chapter 23.]

Non-compliance: See **protocol violations**.

Nondifferential measurement error: When the measurement error for a fallibly measured covariate in a regression model is conditionally independent of the outcome, given the true covariate. [*American Journal of Epidemiology*, 1991, **134**, 1233–1246.]

Non-Gaussian time series: *Time series*, often not *stationary* with respect to both mean and period, which exhibit a non-Gaussian random component. [*Biometrics*, 1994, **50**, 798–812.]

Non-identified response A term used to denote *censored observations* in *survival data*, that are not independent of the endpoint of interest. Such observations can occur for a variety of reasons:

- Misclassification of the response; e.g. death from cancer, the response of interest, being erroneously misclassified as death from another unrelated cause.
- Response occurrence causing prior censoring; e.g. relapse to heroin use causing a subject to quit a rehabilitation study to avoid chemical detection.

[*Biometrics*, 1994, **50**, 1–10.]

Noninferiority trial: See **superiority trial**.

Non-informative censoring: *Censored observations* that can be considered to have the same probabilities of failure at later times as those individuals remaining under observation. [*Journal of the American Statistical Association*, 1988, **83**, 772–9.]

Non-informative prior distribution: A *prior distribution* which is specified in an attempt to be non-commital about a parameter, for example, a *uniform distribution*. [*Biometrika*, 1987, **74**, 557–562]

Nonlinear mapping (NLM): A method for obtaining a low-dimensional representation of a set of multivariate data, which operates by minimizing a function of the differences between the original inter-individual *Euclidean distances* and those in the reduced dimensional space. The function minimized is essentially a simple sum-of-squares. See also **multidimensional scaling** and **ordination**. [*Pattern Analysis and Applications*, 2000, **3**, 61–68.]

Nonlinear model: A model that is non-linear in the parameters, for example

$$E(y|x_1,x_2) = \beta_1 e^{\beta_2 x_1} + \beta_3 e^{\beta_4 x_2}$$
$$E(y|x) = \beta_1 e^{-\beta_2 x}$$

Some such models can be converted into linear models by *linearization* (the second equation above, for example, by taking logarithms throughout). Those that cannot are often referred to as *intrinsically non-linear*, although these can often be approximated by linear equations in some circumstances. Parameters in such models usually have to be estimated using an optimization procedure such as the *Newton–Raphson method.* In such models *linear parameters* are those for which the second partial derivative of the model function with respect to the parameter is zero (β_1 and β_3 in the first example above); when this is not the case (β_2 and β_4 in the first example above) they are called *non-linear parameters*. [ARA Chapter 14.]

Nonlinear regression: Synonym for **non-linear model**.

Nonlinear time series: See **time series.**

Nonmasked study: Synonym for **open label study**.

Nonmetric scaling: A form of *multidimensional scaling* in which only the ranks of the observed *dissimilarity coefficients* or *similarity coefficients* are used in producing the required low-dimensional representation of the data. See also **monotonic regression**. [MV1 Chapter 5.]

Nonnegative garrotte: An approach to choosing subsets of explanatory variables in regression problems that eliminates some variables, 'shrinks' the regression coefficients of others (similar to what happens in *ridge regression*), and gives relatively stable results unlike many of the usual subset selection procedures. The method operates by finding $\{c_k\}$ to minimize

$$\sum_k \left(y_n - \sum_k c_k \hat{\beta}_k x_k n \right)^2$$

where $\{\hat{\beta}_k\}$ are the results of *least squares estimation*, and y and x represent the response and explanatory variables. The $\{c_k\}$ satisfy the constraints;

$$c_k \geq 0 \qquad \sum c_k \leq s$$

The new regression coefficients are $\tilde{\beta}_k(s) = c_k \hat{\beta}_k$. As the garrotte is drawn tighter by decreasing s, more of the $\{c_k\}$ become zero and the remaining non-zero $\tilde{\beta}_k(s)$ are shrunken. In general the procedure produces regression equations having more non-zero coefficients than other subset selection methods, but the loss of simplicity is offset by substantial gains in accuracy. [*Technometrics*, 1995, **37**, 373–84.]

Nonparametric Bayesian models: Models used in *Bayesian inference* that do not commit a *prior distribution* to a particular number of parameters; instead the data is allowed to dictate how many parameters there are. Possible *overfitting* is prevented by integrating out all parameters and latent variables. [*Bayesian Data Analysis*, 2nd edn, 2003, A.Gelman, J. B. Carlin, H. S. Stern and D. B. Rubin, Chapman and Hall/CRC, Boca Raton.]

Nonparametric maximum likelihood (NPML): A *likelihood* approach which does not require the specification of a full parametric family for the data. Usually, the nonparametric maximum likelihood is a multinomial likelihood on a sample. Simple examples include the empirical cumulative distribution function and the *product-limit estimator.* Also used to relax parametric assumptions regarding random effects in *multilevel models.* Closely related to the *empirical likelihood.* [*Journal of the American Statistical Association*, 1978, **73**, 805–811.]

Non-orthogonal designs: *Analysis of variance* designs with two or more factors in which the number of observations in each cell are not equal.

Nonparametric analysis of covariance: An *analysis of covariance* model in which the covariate effect is assumed only to be 'smooth' rather than of some specific linear or perhaps non-linear form. See also **kernel regression smoothing**. [*Biometrics*, 1995, **51**, 920–31.]

Nonparametric methods: See **distribution free methods**.

Nonparametric regression modelling: See **regression modelling**.

Non-randomized clinical trial: A *clinical trial* in which a series of consecutive patients receive a new treatment and those that respond (according to some pre-defined criterion) continue to receive it. Those patients that fail to respond receive an alternative, usually the conventional, treatment. The two groups are then compared on one or more outcome variables. One of the problems with such a procedure is that patients who respond may be healthier than those who do not respond, possibly resulting in an apparent but not real benefit of the treatment. [*Statistics in Medicine*, 1984, **3**, 341–6.]

Non-response: A term generally used for failure to provide the relevant information being collected in a survey. Poor response can be due to a variety of causes, for example, if the topic of the survey is of an intimate nature, respondents may not care to answer particular questions. Since it is quite possible that respondents in a survey differ in some of their characteristics from those who do not respond, a large number of non-respondents may introduce bias into the final results. See also **item non-response**. [SMR Chapter 5.]

No-observed-effect level (NOEL): The dose level of a compound below which there is no evidence of an effect on the response of interest. [*Food and Chemical Toxicology*, 1997, **35**, 349–55.]

Norm: Most commonly used to refer to 'what is usual', for example, the range into which body temperatures of healthy adults fall, but also occasionally used for 'what is desirable', for example, the range of blood pressures regarded as being indicative of good health.

Normal approximation: Normal distributions that approximate other distributions; for example, a normal distribution with mean np and variance $np(1-p)$ that acts as an approximation to a *binomial distribution* as n, the number of trials, increases. The term, p represents the probability of a 'success' on any trial. See also **DeMoivre–Laplace theorem**. [*Handbook of the Normal Distribution*, 1996, J. K. Patel and C. B. Read, CRC, Boca Raton.]

Normal distribution: A probability distribution, $f(x)$, of a random variable, X, that is assumed by many statistical methods. Specifically given by

$$f(x) = \frac{1}{\sigma\sqrt{2\pi}} \exp\left[-\frac{1}{2}\frac{(x-\mu)^2}{\sigma^2}\right]$$

where μ and σ^2 are, respectively, the mean and variance of x. This distribution is *bell-shaped* as shown in the example given in Fig. 98. [STD Chapter 29.]

Normal equations: The linear equations arising in applying *least squares estimation* to determining the coefficients in a linear model.

Normal equivalent deviate: A value, x_p, corresponding to a proportion, p, that satisfies the following equation

$$\int_{-\infty}^{x_p} \frac{1}{\sqrt{2\pi}} \exp\left(\frac{-u^2}{2}\right) \mathrm{d}u = p$$

Also known as the *normit*. See also **probit analysis**.

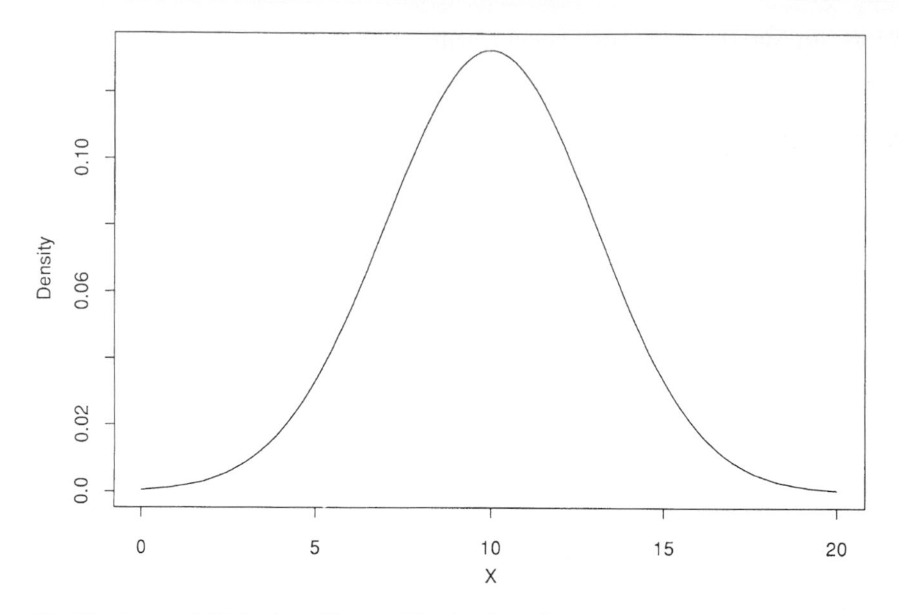

Fig. 98 A normal distribution with mean 10 and variance 9.

Normality: A term used to indicate that some variable of interest has a normal distribution.

Normal probability plot: See **probability plot**.

Normal range: Synonym for **reference interval**.

Normal scores: The expectations of the order statistics of a sample from the standard normal distribution. The basis of *probability plots*. [*Biometrika*, 1961, **48**, 151–65.]

Normal scores test: An alternative to the *Mann–Whitney test* for comparing populations under shift alternatives. [*Practical Nonparametric Statistics*, 1980, W. J. Conover, Wiley, New York.]

Normit: See **normal equivalent deviate**.

NORMIX: A computer program for the *maximum likelihood estimation* of the parameters in a *finite mixture distribution* in which the components are *multivariate normal distributions* with different mean vectors and possibly different *variance–covariance matrices*. [http://alumnus.caltech.edu/~wolfe/normix.htm]

nQuery advisor: A software package useful for determining sample sizes when planning research studies. [Statistical Solutions Ltd., 8 South Bank, Crosse's Green, Cork, Ireland; Stonehill Corporate Center, Suite 104, 999 Broadway, Saugus, MA 01906, USA.]

Nuisance parameter: A parameter of a model in which there is no scientific interest but whose values are usually needed (but in general are unknown) to make inferences about those parameters which are of such interest. For example, the aim may be to draw an inference about the mean of a normal distribution when nothing certain is known about the variance. The *likelihood* for the mean, however, involves the variance, different values of which will lead to different likelihood. To overcome the problem, test statistics or estimators for the parameters that are of interest are sought which do not depend on the unwanted parameter(s). See also **conditional likelihood**. [KA2 Chapter 20.]

Null distribution: The probability distribution of a test statistic when the null hypothesis is true.

Null hypothesis: Typically, the 'no difference' or 'no association' hypothesis to be tested (usually by means of a significance test) against an alternative hypothesis that postulates non-zero difference or association.

Null matrix: A matrix in which all elements are zero.

Null vector: A vector the elements of which are all zero.

Number needed to treat: The reciprocal of the reduction in absolute risk between treated and control groups in a *clinical trial.* It is interpreted as the number of patients who need to be treated to prevent one adverse event. [*British Medical Journal*, 1995, **310**, 452–4.]

Number numbness: The inability to fathom, compare or appreciate very big or very small numbers. [*Chance Rules, 2nd edn*, 2008, B. S. Everitt, Springer, New York.]

Number of partitions: A general expression for the number of partitions, N, of n individuals or objects into g groups is given by

$$N = \frac{1}{g!}\sum_{i=1}^{g}(-1)^{g-i}\binom{g}{i}i^n$$

For example, when $n=25$ and $g=4$, then N is 45 232 115 901. [*Introduction to Combinatorial Mathematics*, 1968, G. L. Liu, McGraw-Hill, New York.]

Numerical integration: The study of how the numerical value of an integral can be found. Also called *quadrature* which refers to finding a square whose area is the same as the area under a curve. See also **Simpson's rule, Gaussian quadrature** and **trapeziodal rule**. [*Methods of Numerical Integration*, 1984, P. J. Davis and P. Rabinowitz, Academic Press, New York.]

Numerical taxonomy: In essence a synonym for **cluster analysis**.

Nuremberg code: A list of ten standards for carrying out clinical research involving human subjects, drafted after the trials of Nazi war criminals at Nuremberg. See also **Helsinki declaration**. [*Trials of War Criminals before the Nuremberg Military Tribunals Under Control Council Law No 10, volume 2*, 1949, 181–182, Washington DC, US Government Printing Office.]

Nyquist frequency: The frequency above which there is no information in a continuous *time series* which has been digitized by taking values at time intervals δt apart. Explicitly the frequency is $1/2\delta t$ cycles per unit time.

O

Oblique factors: A term used in *factor analysis* for *common factors* that are allowed to be correlated. [*Psychometrika*, 2003, **68**, 299–321.]

O'Brien–Fleming method: A method of *interim analysis* in a *clinical trial* in which very small p-values are required for the early stopping of a trial, whereas later values for stopping are closer to conventional levels of significance. [*Statistics in Medicine*, 1994, **13**, 1441–52.]

O'Brien's two-sample tests: Extensions of the conventional tests for assessing differences between treatment groups that take account of the possible heterogeneous nature of the response to treatment and which may be useful in the identification of subgroups of patients for whom the experimental therapy might have most (or least) benefit. [*Statistics in Medicine*, 1990, **9**, 447–56.]

Observational study: A study in which the objective is to uncover cause-and-effect relationships but in which it is not feasible to use controlled experimentation, in the sense of being able to impose the procedure or treatments whose effects it is desired to discover, or to assign subjects at random to different procedures. Surveys and most epidemiological studies fall into this class. Since the investigator does not control the assignment of treatments there is no way to ensure that similar subjects receive different treatments. The classical example of such a study that successfully uncovered evidence of an important causal relationship is the smoking and lung cancer investigation of *Doll* and *Hill*. See also **experimental study, prospective study** and **retrospective study**. [SMR Chapter 5.]

Observation-driven model: A term usually applied to models for *longitudinal data* or *time series* which introduce *within unit correlation* by specifying the *conditional distribution* of an observation at time t as a function of past observations. An example is the *ante-dependence model*. In contrast in a *parameter-driven model*, *correlation* is introduced through a latent process, for example, by introducing a random subject effect. [*Biometrika*, 2003, **90**, 777–90.]

Observed-score equating: The practice of transforming observed scores on different forms of (usually) a cognitive test to a common normal distribution. [*Test Equating*, 1982, edited by P. W. Holland and D. B. Rubin, Academic Press, New York.]

Obuchowski and Rockette method: An alternative to the *Dorfman–Berbaum–Metz method* for analyzing multiple reader *receiver operating curve* data. Instead of modelling the *jack-knife* pseudovalues, in this approach a mixed-effects *analysis of variance* model is used to model the accuracy of the jth reader using the ith diagnostic test. [*Communications in Statistics; Simulation and Computation*, 1995, **24**, 285–308.]

Occam's razor: An early statement of the *parsimony principle*, given by William of Occam (1280–1349) namely 'entia non sunt multiplicanda praeter necessitatem'; i.e. 'A plurality (of reasons) should not be posited without necessity'. In other words one should not make more assumptions than the minimum needed. The concept underlies all scientific modelling and theory building, and helps to eliminate those variables and constructs that are not really needed to explain a particular phenomenon, with the consequence that there is less chance of introducing inconsistencies, ambiguities and redundancies. [*Mind*, 1915, **24**, 287–288.]

Occam's window: A procedure used in *Bayesian inference* for selecting a small set of models over which a model average can be computed. [*Markov Chain Monte Carlo in Practice*, 1996, W. R. Gilks, S. Richardson and D. J. Spiegelhalter, Chapman and Hall/CRC Press, London.]

Occupancy problems: Essentially problems concerned with the probability distribution of arrangements of say r balls in n cells. For example, if the r balls are distinguishable and distributed at random, so that each of the n^r arrangements is equally likely, then the number of balls in a given cell has a *binomial distribution* in which the number of trials is r and the probability of success is $1 / n$; the probability that every cell is occupied is

$$\sum_{i=0}^{n}(-1)^n\binom{n}{i}\left(1-\frac{i}{n}\right)^r$$

[*An Introduction to Probability Theory and Its Applications*, Volume 1, 3rd edition, 1968, W. Feller, Wiley, New York.]

Odds: The ratio of the probabilities of the two possible states of a binary variable. See also **odds ratio** and **logistic regression**. [SMR Chapter 10.]

Odds ratio: The ratio of the *odds* for a binary variable in two groups of subjects, for example, males and females. If the two possible states of the variable are labelled 'success' and 'failure' then the odds ratio is a measure of the odds of a success in one group relative to that in the other. When the odds of a success in each group are identical then the odds ratio is equal to one. Usually estimated as

$$\hat{\psi} = \frac{ad}{bc}$$

where a, b, c and d are the appropriate frequencies in the *two-by-two contingency table* formed from the data. See also **Haldane's estimator, Jewell's estimator** and **logistic regression**. [SMR Chapter 10.]

Offset: A term used in *generalized linear models* to indicate a known regression coefficient that is to be included in a model, i.e. one that does not have to be estimated. For example suppose the number of deaths for district i and age class j is assumed to follow a *Poisson distribution* with mean $N_{ij}\,\theta_{ij}$ where N_{ij} is the total person years for district i and age class j. Further it is postulated that the parameter θ_{ij} is the product of district (θ_i) and age (θ_j) effects. The model for the mean (μ_i) is thus

$$\ln(\mu_i) = \ln(N_{ij}) + \ln\theta_i + \ln\theta_j$$

The first term on the right-hand side is the offset. [GLM Chapter 6.]

Offspring distribution: See **Bienaymé–Galton–Watson process**.

Ogawa, Junjiro (1915–2000): Born in Saitama Prefecture, Japan, Ogawa obtained his first degree from the University of Tokyo in 1938, followed by a D.Sc. from the same university in 1954. After military service he joined the Institute of Statistical Mathematics in Tokyo. From 1955–1964 he was a member of the University of North Carolina's Department of Statistics. During this period he made important contributions to theoretical statistics particularly in the areas of *multivariate analysis, experimental design* and *order statistics*. Ogawa was President of the Japanese Statistical Society in 1981 and 1982. He died in Chiba, Japan on 8 March 2000.

Ogive: A term often applied to the graphs of *cumulative frequency distributions*. Essentially synonymous with *sigmoid*, which is to be preferred.

Oja median: See **bivariate Oja median**.

O. J. Simpson paradox: A term arising from a claim made by the defence lawyer in the murder trial of O. J. Simpson. The lawyer stated that the statistics demonstrate that only one-tenth of one percent of men who abuse their wives go on to murder them, with the implication that one or two instances of alleged abuse provides very little evidence that the wife's murder was committed by the abusive husband. The argument simply reflects that most wives are not murdered and has no relevance once a murder has been committed and there is a body. What needs to be considered here is, given that a wife with an abusive partner has been murdered, what is the probability that the murderer is her abuser? It is this *conditional probability* that provides the relevant evidence for the jury to consider, and estimates of the probability range from 0.5 to 0.8. [*Dicing with Death*, 2003, S. Senn, Cambridge University Press, Cambridge.]

Oliveira, José Tiago da Fonseca (1928–1993): Born in Mozambique, Oliveira first studied mathematics at the University of Oporto, before being awarded a doctorate in algebra from the University of Lisbon in 1957. His interest in statistics began during his employment at the Institute for Marine Biology in Lisbon, and further developed at Columbia University with pioneering work on bivariate extremes. Became an Honorary Fellow of the Royal Statistical Society in 1987. Oliveira died on 23 June 1993.

Olkin and Tate model: Synonymous with **general location model**.

OLS: Abbreviation for **ordinary least squares**.

Omitted covariates: A term usually found in connection with *regression modelling*, where the model has been incompletely specified by not including important covariates. The omission may be due either to an incorrect conceptual understanding of the phenomena under study or to an inability to collect data on all the relevant factors related to the outcome under study. Mis-specifying regression models in this way can result in seriously biased estimates of the effects of the covariates actually included in the model. [*Statistics in Medicine*, 1992, **11**, 1195–208.]

Omori's law: The first empirical law of seismology, namely that the frequency of after shocks from an earthquake at time t, $\lambda(t)$, decays hyperbolically after the main shock. The law has no clear physical explanation. [*Journal of Physics of the Earth*, 1995, **43**, 1–33.]

One-bend transformation: A power family of transformations, $y = x^k$, that provides a useful approach to 'straightening' a single bend in the relationship between two variables. Ordering the transformations according to the exponent k gives a sequence of power transformations, which is sometimes referred to as the *ladder of re-expression*. The common powers considered are:

$$k = -1, -\tfrac{1}{2}, 0, \tfrac{1}{2}, 1, 2$$

where $k = 0$ is interpreted as the *logarithmic transformation* and $k = 1$ implies no transformation. See also **two-bend transformation, Box–Cox transformation** and **Box–Tidwell transformation**. [ARA Chapter 11.]

One-compartment model: See **compartment models.**

One-dimensional random walk: A *Markov chain* on the integers 0, 1, . . ., with the *one-step transition probabilities*

$$p_{i,i-1} = q_i, \quad p_{ii} = r_i, \quad p_{i,i+1} = p_i$$

for $i \geq 0$ with $q_0 = 0$. For $|i - j| \geq 2$, $p_{ij} = 0$ so that the parameters, q_i, r_i and p_i sum to one. [*European Physics Journal B*, 1999, **12**, 569–77.]

One-hit model: See **multi-hit model.**

One:*m* (1:*m*) matching: A form of matching often used when control subjects are more readily obtained than cases. A number, m ($m > 1$), of controls are attached to each case, these being known as the *matched set*. The theoretical efficiency of such matching in estimating, for example, *relative risk*, is $m / (m + 1)$ so one control per case is 50% efficient, while four per case is 80% efficient. Increasing the number of controls beyond 5–10 brings rapidly diminishing returns. [*Biometrics*, 1969, **22**, 339–55.]

One-sided test: A significance test for which the alternative hypothesis is directional; for example, that one population mean is greater than another. The choice between a one-sided and two-sided test must be made before any test statistic is calculated. [SMR Chapter 8.]

One-step method: A procedure for obtaining a pooled estimate of an *odds ratio* from a set of *two-by-two contingency tables*. Not recommended for general use since it can lead to extremely biased results particularly when applied to unbalanced data. The *Mantel–Haenszel estimator* is usually far more satisfactory. [*Statistics in Medicine*, 1990, **9**, 247–52.]

One-step transition probability: See **Markov Chain.**

One-tailed test: Synonym for **one-sided test.**

One-way design: See **analysis of variance.**

Open label study: An investigation in which patient, investigator and peripheral staff are all aware of what treatment the patient is receiving. [SMR Chapter 15.]

Open sequential design: See **sequential analysis.**

Operational research: Research concerned with applying scientific methods to the problems facing executive and administrative authorities. [*Introduction to Operations Research*, 8th edn, 2005, F. S. Hillier and G. J. Luberman, McGraw-Hill, Boston, MA.]

Opinion survey: A procedure that aims to ascertain opinions possessed by members of some population with regard to particular topics. See also **sample survey.**

Optimal scaling: The process of simultaneously transforming proximity data and representing the transformed data by a geometrical (often a *Euclidean distance*) model. See also **multi-dimensional scaling.** [*Psychometrika*, 1981, **46**, 357–88.]

Optimization methods: Procedures for finding the maxima or minima of functions of, generally, several variables. Most often encountered in statistics in the context of *maximum likelihood estimation*, where such methods are frequently needed to find the values of the parameters that maximize the *likelihood*. See also **simplex method** and **Newton–Raphson method.** [*An Introduction to Optimization Methods and Their Application in Statistics*, 1987, B. S. Everitt, Chapman and Hall/CRC Press, London.]

Option-3 scheme: A scheme of measurement used in situations investigating possible changes over time in *longitudinal data*. The scheme is designed to prevent measurement *outliers* causing an unexpected increase in falsely claiming that a change in the data has occurred. Two measures are taken initially and, if they are closer than a specified threshold, the average of the two is considered to be an estimate of the true mean; otherwise a third measurement is made, and the mean of the closest 'pair' is considered to be the estimate. [*Statistics in Medicine*, 1998, **17**, 2607–2615.]

Oracle property: A name given to methods for estimating the regression parameters in models fitted to *high-dimensional data* that have the property that they can correctly select the nonzero coefficients with probability converging to one and that the estimators of the nonzero coefficients are asymptotically normal with the same means and covariances that they would have if the zero coefficients were known in advance, i.e., the estimators are asymptotically as efficient as the ideal estimation assisted by an 'oracle' who knows which coefficients are nonzero. [*Annals of Statistics*, 2005, **32**, 928–961.]

Ordered alternative hypothesis: A hypothesis that specifies an order for a set of parameters of interest as an alternative to their equality, rather than simply that they are not all equal. For example, in an evaluation of the treatment effect of a drug at several different doses, it might be thought reasonable to postulate that the response variable shows either a monotonic increasing or monotonic decreasing effect with dose. In such a case the null hypothesis of the equality of, say, a set of m means would be tested against

$$\mathrm{H}_1 : \mu_1 \leq \mu_2 \leq \cdots \leq \mu_m,$$

using some suitable test procedure such as *Jonckheere's k-sample test*. [*Biostatistics: A Methodology for the Health Sciences*, 2nd edn, 2004, G. Van Belle, L. D. Fisher, P. J. Heagerty and T. S. Lumley, Wiley, New York.]

Ordered logistic regression: Logistic regression when the response is an *ordinal variable*. See also **proportional odds model**.

Order statistics: The ordered values of a collection of random variables, i.e. if $X_1, X_2, X_3, \ldots, X_n$ is a collection of random variables with ordered values, $X_{(1)} \leq X_{(2)} \leq \ldots \leq X_{(n)}$ then their rth order statistic is the rth smallest amongst them, $X_{(r)}$ and $X_{(1)}$ and $X_{(n)}$ are, respectively the sample minimum and maximum. The order statistics are widely used as the basis of estimators and assessment of fit; for example, two simple statistics based on them are the sample median and the *alpha (α)-trimmed mean*. [*Order Statistics*, 3rd edn, 2003, H. A. David and H. N. Nagaraja, Wiley, New York.]

Ordinal variable: A measurement that allows a sample of individuals to be ranked with respect to some characteristic but where differences at different points of the scale are not necessarily equivalent. For example, anxiety might be rated on a scale 'none', 'mild', 'moderate' and 'severe', with the values 0,1,2,3, being used to label the categories. A patient with anxiety score of one could be ranked as less anxious than one given a score of three, but patients with scores 0 and 2 do not necessarily have the same difference in anxiety as patients with scores 1 and 3. See also **categorical variable** and **continuous variable**.

Ordinary least squares (OLS): See **least squares estimation**.

Ordination: The process of reducing the dimensionality (i.e. the number of variables) of multivariate data by deriving a small number of new variables that contain much of the information in the original data. The reduced data set is often more useful for investigating possible structure in the observations. See also **curse of dimensionality, principal components analysis** and **multidimensional scaling**. [MV1 Chapter 1.]

Ornstein–Uhlenbeck process: An aspect of *Brownian motion* dealing with the velocity of the movement of a particle. [*Neural Computation*, 1999, **76**, 252–259.]

Orthant probability: The probability that n random variables $X_1, X_2, \ldots, X_n$ are all positive when the n variates have a joint *multivariate normal distribution* with all the means zero and all the variances are one. Used, for example, in *simultaneous testing.* [KA1 Chapter 15.]

Orthogonal: A term that occurs in several areas of statistics with different meanings in each case. Most commonly encountered in relation to two variables or two linear functions of a set of variables to indicate statistical independence. Literally means 'at right angles' but in applied statistics most often used as a descriptive term for the ability to disentangle individual effects.

Orthogonal contrasts: Sets of linear functions of either parameters or statistics in which the defining coefficients satisfy a particular relationship. Specifically if c_1 and c_2 are two *contrasts* of a set of m parameters such that

$$c_1 = a_{11}\beta_1 + a_{12}\beta_2 + \cdots + a_{1m}\beta_m$$
$$c_2 = a_{21}\beta_1 + a_{22}\beta_2 + \cdots + a_{2m}\beta_m$$

they are orthogonal if $\sum_{i=1}^{m} a_{1i}a_{2i} = 0$. If, in addition, $\sum_{i=1}^{m} a_{1i}^2 = 1$ and $\sum_{i=1}^{m} a_{2i}^2 = 1$. then the contrasts are said to be *orthonormal.* [*The Analysis of Variance*, 1959, H. Scheffé, Wiley, London.]

Orthogonal matrix: A square matrix that is such that multiplying the matrix by its transpose results in an identity matrix.

Orthogonal polynomials: Two polynomials $p_i(x)$ and $p_j(x)$ of degree i and j respectively are said to be orthogonal if they are uncorrelated as x varies over some distribution. An example is the series,

$$p_0(x) = 1, p_1(x) = x, p_2(x) = 2x^2 - 1, \ldots, p_{n+1}(x) = 2xp_n(x) - p_{n-1}(x), n \geq 1$$

Such polynomials are useful in many areas of statistics, for example, in *polynomial regression* where their use leads to parameter estimates that are uncorrelated which greatly simplifies estimation and interpretation. [*Linear Regression Analysis*, 1977, G. A. F. Seber, Wiley, New York.]

Orthonormal contrasts: See **orthogonal contrasts**.

Outcome variable: Synonym for **response variable**.

Outcomes research: A multidisciplinary field of study that seeks to understand and improve the end results of particular health care practices and interventions. [*Science*, 1998, **282**, 245–246.]

Outlier: An observation that appears to deviate markedly from the other members of the sample in which it occurs. In the set of systolic blood pressures, {125, 128, 130, 131, 198}, for example, 198 might be considered an outlier. More formally the term refers to an observation which appears to be inconsistent with the rest of the data, relative to an assumed model. Such extreme observations may be reflecting some abnormality in the measured characteristic of a subject, or they may result from an error in the measurement or recording. See also **log-likelihood distance, outside observation**, **five number summary**, **Wilks' multivariate outlier test, inlier** and **additive outlier**. [SMR Chapter 7.]

Outside observation: An observation falling outside the limits

$$F_L - 1.5(F_U - F_L), F_U + 1.5(F_U - F_L)$$

where F_U and F_L are the upper and lower quartiles of a sample. Such observations are usually regarded as being extreme enough to be potential *outliers*. See also **box-and-whisker plot**.

Overdispersion: The phenomenon that arises when empirical variance in the data exceeds the nominal variance under some assumed model. Most often encountered when modelling data that occurs in the form of proportions or counts, where it is often observed that there is more variation than, for example, an assumed *binomial distribution* can accomodate. There may be a variety of relatively simple causes of the increased variation, ranging from the presence of one or more *outliers*, to *mis-specification* of the model being applied to the data. If none of these explanations can explain the phenomenon then it is likely that it is due to variation between the response probabilities or correlation between the binary responses, in which case special modelling procedures may be needed. See also **clustered data** and **generalized linear model**. [*Modelling Binary Data*, 2nd edition, 2003, D. Collett, Chapman and Hall/CRC Press, London.]

Overfitted models: Models that contain more unknown parameters than can be justified by the data. See also **underfitted models**. [SMR Chapter 12.]

Overidentified model: See **identification**.

Overmatching: A term applied to studies involving matching when the matching variable is strongly related to exposure but not to disease risk. Such a situation leads to a loss of efficiency. [*Statistical Methods in Cancer Research*, Volume 1, *The Analysis of Case–Control Studies*, 1980, N. E. Breslow and N. E. Day, International Agency for Research on Cancer, Lyon.]

Overparameterized model: A model with more parameters than observations for estimation. For example, the following simple model for a *one-way design* in *analysis of variance*

$$y_{ij} = \mu + \alpha_i + e_{ij} \qquad (i = 1, 2, \ldots, g; j = 1, 2, \ldots, n_i)$$

where g is the number of groups, n_i the number of observations in group i, y_{ij} represents the jth observation in the ith group, μ is the grand mean effect and α_i the effect of group i, has $g + 1$ parameters but only g group means to be fitted. It is overparameterized unless some constraints are placed on the parameters, for example, that $\sum_{i=1}^{g} \alpha_i = 0$. See also **identification**. [ARA Chapter 8.]

Overviews: Synonym for **meta-analysis**.

Page's test: A *distribution free* procedure for comparing related samples. [*Biostatistics: A Methodology for the Health Sciences*, 2nd edn, 2004, G. Van Belle, L. D. Fisher, P. J. Heagerty and T. S. Lumley, Wiley, New York.]

Paired availability design: A design which can reduce *selection bias* in situations where it is not possible to use random allocation of subjects to treatments. The design has three fundamental characteristics:

- the intervention is the availability of treatment not its receipt;
- the population from which subjects arise is well defined with little in- or out-migration;
- the study involves many pairs of control and experimental groups.

In the experimental groups, the new treatment is made available to all subjects though some may not receive it. In the control groups, the experimental treatment is generally not available to subjects though some may receive it in special circumstances. [*Statistics in Medicine*, 1994, **13**, 2269–78.]

Paired Bernoulli data: Data arising when an investigator records whether a particular characteristic is present or absent at two sites on the same individual. [*Biometrics*, 1988, **44**, 253–7.]

Paired comparisons experiment: See **Bradley–Terry model**.

Paired samples: Two samples of observations with the characteristic feature that each observation in one sample has one and only one matching observation in the other sample. There are several ways in which such samples can arise in medical investigations. The first, *self-pairing*, occurs when each subject serves as his or her own control, as in, for example, therapeutic trials in which each subject receives both treatments, one on each of two separate occasions. Next, *natural-pairing* can arise particularly, for example, in laboratory experiments involving litter-mate controls. Lastly *artificial pairing* may be used by an investigator to match the two subjects in a pair on important characteristics likely to be related to the response variable. See also **matching**. [SMR Chapter 9.]

Paired samples *t*-test: Synonym for **matched pairs *t*-test**. [*Biometrics*, 1988, **44**, 253–7.]

Pandemic: An epidemic that spreads through populations across a large region, for example a whole continent. Examples are the HIV pandemic of the 1980s and the 2009 Swine flu pandemic. [*The AIDS Pandemic. The Collision of Epidemiology with Political Correctness*, 2007, J. Chin, Radcliffe Publishing, Oxford.]

Panel study: A study in which a group of people, the 'panel', are interviewed or surveyed with respect to some topic of interest on more than one occasion. Essentially equivalent to a *longitudinal study*. [*Analysis of Panel Data*, 2nd edition, 2003, C. Hsiao, Cambridge University Press, Cambridge.]

Papadakis analysis: A form of *analysis of covariance* used in field-plot experiments. [*Biometrika*, 1989, **76**, 253–259.]

Parallel coordinate plots: A simple but powerful technique for obtaining a graphical display of *multivariate data.* In this plot, the variable axes are arranged horizontally, each parallel to the one above it. A line is then plotted for each observation by joining the appropriate variable values on these axes. The example in Fig. 99 shows such a plot for four measurements made on plants from three species of iris. See also **Andrews' plots** and **Chernoff's faces**. [*Visual Computer*, 1985, **1**, 69–96.]

Parallel distributed processing: Information processing involving a large number of units working contemporaneously in parallel with units, like the neurons of the brain, exciting or inhibiting one another. See also **artificial neural networks**. [*Pattern Recognition and Neural Networks*, 1996, B. D. Ripley, Cambridge University Press, Cambridge.]

Parallel-dose design: See **dose-ranging trial**.

Parallel groups design: A simple experimental set-up in which two different groups of patients, for example, treated and untreated, are studied concurrently. [SMR Chapter 15.]

Parallelism in ANCOVA: One of the assumptions made in the *analysis of covariance*, namely that the slope of the regression line relating the response variable to the covariate is the same in all treatment groups: See also **Johnson-Neyman technique**.

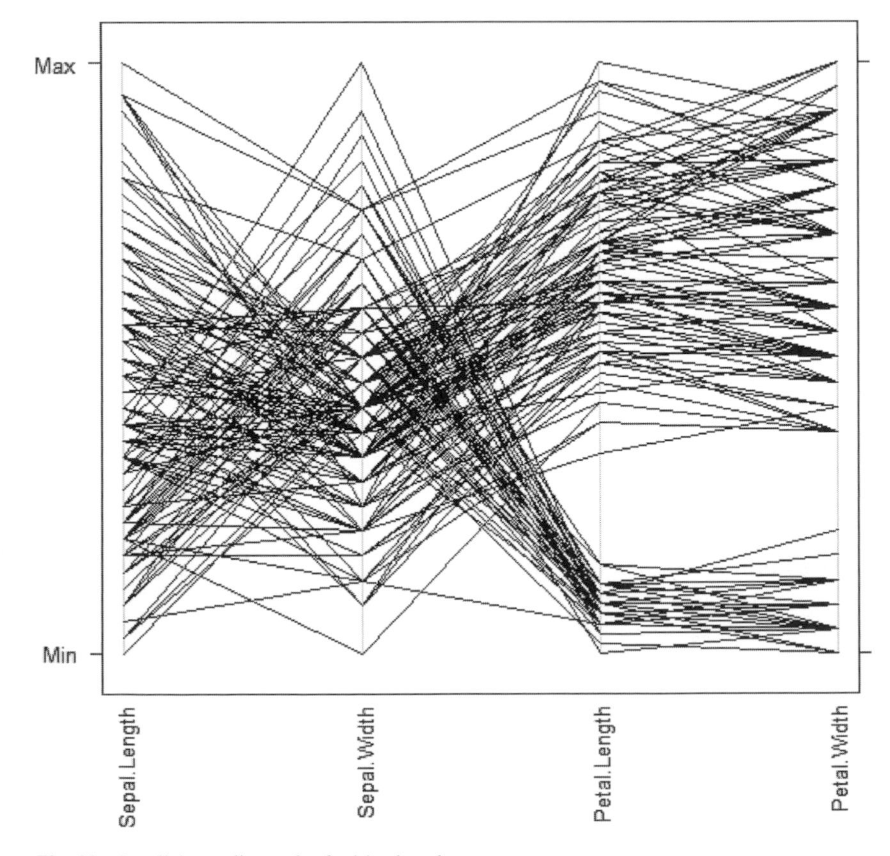

Fig. 99 Parallel coordinate plot for iris plant data.

Parallel-line bioassay: A procedure for estimating equipotent doses. The model used can be formulated by the following equations:

$$y_s = \alpha + \beta x_s$$
$$y_t = \alpha + \beta(x_t + \mu)$$

where y_s, y_t are the responses to doses x_s, x_t (usually transformed in terms of logarithms to base 10) involving a known standard preparation against a test preparation, respectively. The objective is to estimate the relative potency, ρ, of the test preparation, where $\log \rho = \mu$, i.e., the horizontal shift between the parallel lines. Note that if the test preparation is as potent as the standard preparation, then $\rho = 1$ or, equivalently, $\mu = 0$. [*Development of Biological Standards*, 1979, **44**, 129–38.]

Parallel processing: Synonym for **parallel distributed processing**.

Parameter: A numerical characteristic of a population or a model. The probability of a 'success' in a *binomial distribution*, for example. [ARA Chapter 1.]

Parameter design: See **Taguchi's parameter design**.

Parameter-driven model: See **observation-driven model**.

Parameter space: The set of allowable values for a set of parameters $\boldsymbol{\theta}' = [\theta_2, \ldots, \theta_m]$. Usually denoted by Ω or sometimes by $\Omega_{\boldsymbol{\theta}}$. In a series of *Bernoulli trials* with probability of success equal to p, for example, Ω_p is $0 \leq p \leq 1$.

Parametric hypothesis: A hypothesis concerning the parameter(s) of a distribution. For example, the hypothesis that the mean of a population equals the mean of a second population, when the populations are each assumed to have a normal distribution.

Parametric methods: Procedures for testing hypotheses about parameters in a population described by a specified distributional form, often, a normal distribution. *Student's t-test* is an example of such a method. See also **distribution free methods**.

Pareto distribution: The probability distribution, $f(x)$ given by

$$f(x) = \frac{\gamma\alpha^\gamma}{x^{\gamma+1}}, \quad \alpha \leq x < \infty, \alpha > 0, \gamma > 0$$

Examples of the distribution are given in Fig. 100.
The mean and variance of the distribution are as follows:

$$\text{mean} = \gamma\alpha/(\gamma - 1), \gamma > 1$$
$$\text{variance} = \gamma\alpha^2/[(\gamma - 1)^2(\gamma - 2)], \gamma > 2$$

Such distributions with $0 < \gamma < 2$ are known as *stable Pareto distributions*. [STD Chapter 30.]

Pareto plot: A *bar chart* with the bars ordered according to decreasing frequency enhanced by a line joining points above each bar giving the cumulative frequency. The plot is used in *quality control* applications.

Parking lot test: A test for assessing the quality of *random number generators*. [*Random Number Generation and Monte Carlo Methods*, 1998, J. E. Gentle, Springer-Verlag, New York.]

Parres plot: Acronym for **partial residual plot**.

Parsimony principle: The general principle that among competing models, all of which provide an adequate fit for a set of data, the one with the fewest parameters is to be preferred. See also **Akaike's information criterion** and **Occam's razor**.

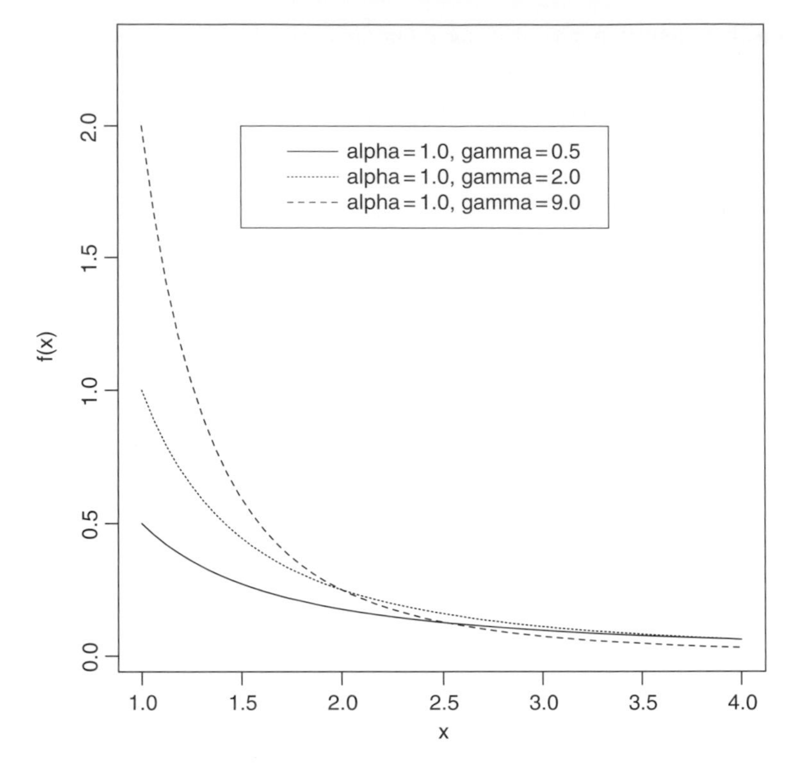

Fig. 100 Pareto distributions at three values of γ for $\alpha = 1.0$.

Partial autocorrelation: A measure of the correlation between the observations a particular number of time units apart in a *time series*, after controlling for the effects of observations at intermediate time points. [*Time Series Analysis, Forecasting and Control*, 3rd edition, 1994, G. E. P. Box, G. M. Jenkins and C. G. Reinsel, Prentice Hall, Englewood Cliffs, NJ.]

Partial common principal components: See **common principal components**.

Partial correlation: The correlation between a pair of variables after adjusting for the effect of a third. Can be calculated from the sample correlation coefficients of each of the three pairs of variables involved as

$$r_{12|3} = \frac{r_{12} - r_{13}r_{23}}{\sqrt{(1 - r_{13}^2)(1 - r_{23}^2)}}$$

[SMR Chapter 11.]

Partial least squares: An alternative to *multiple regression* which, instead of using the original q explanatory variables directly, constructs a new set of k regressor variables as linear combinations of the original variables. The linear combinations are chosen sequentially in such a way that each new regressor has maximal sample covariance with the response variable subject to being uncorrelated with all previously constructed regressors. See also **principal components regression analysis**. [*Systems under Indirect Observation*, Volumes I & II, 1982, K. G. Joreskog and H. Wold, editors, North Holland, Amsterdam.]

Partial likelihood: A product of *conditional likelihoods*, used in certain situations for estimation and hypothesis testing. The basis of estimation in *Cox's proportional hazards model.* [*The*

Statistical Analysis of Failure Time Data, 2002, J. D. Kalbfleisch and R. L. Prentice, Wiley, Chichester.]

Partial multiple correlation coefficient: An index for examining the linear relationship between a response variable, *y*, and a group of explanatory variables $x_1, x_2, \ldots, x_k$, while controlling for a further group of variables $x_{k+1}, x_{k+2}, \ldots, x_q$. Specifically given by the *multiple correlation coefficient* of the variable $y - \hat{y}$ and the variables $x_1 - \hat{x}_1, x_2 - \hat{x}_2, \ldots, x_k - \hat{x}_k$ where the 'hats' indicate the predicted value of a variable from its linear regression on $x_{k+1}, x_{k+2}, \ldots, x_q$. [*Biometrika*, 1927, **19**, 39–49.]

Partial questionnaire design: A procedure used in studies in *epidemiology* as an alternative to a lengthy questionnaire which can result in lower rates of participation by potential study subjects. Information about the exposure of interest is obtained from all subjects, but information about secondary variables is determined for only a fraction of study subjects. [*Statistics in Medicine*, 1994, **13**, 623–34.]

Partial-regression leverage plot: Synonym for **added variable plot**.

Partial residual plot: A useful diagnostic tool in *multiple regression*, particularly for assessing whether a non-linear term in one or other explanatory variable is needed in the model. Consists of a *scatterplot* of the pairs (x_{ij}, r_{ij}) where x_{ij} is the *j*th observation on the *i*th explanatory variable and r_{ij} is defined as

$$r_{ij} = \bar{y} + b_i(x_{ij} - \bar{x}_i) + r_j$$

where $\bar{y}$ is the mean of the response variable, $\bar{x}_i$ is the mean of the *i*th explanatory variable, b_i is the estimated regression coefficient for explanatory variable *i* and r_j is the *residual* for the *j*th observation. Non-linearity in the plot indicates that the regression model should contain a non-linear term in x_i. An example of such a plot is shown in Fig. 101. See also **added variable plot**. [*Regression Analysis*, Volume 2, 1993, edited by M. S. Lewis-Beck, Sage Publications, London.]

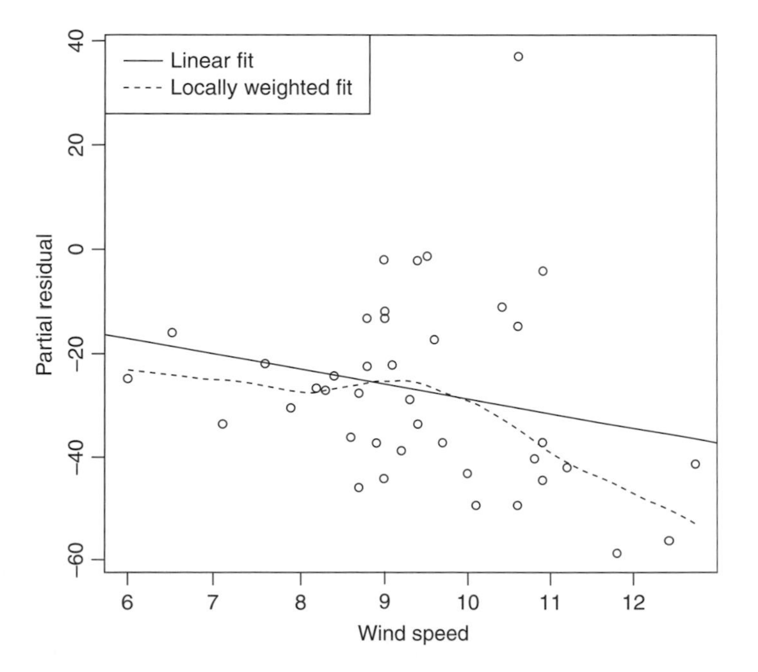

Fig. 101 An example of a partial residual plot.

Particle filters: A *simulation* technique for tracking moving target distributions and for reducing the computational burden of a dynamic Bayesian analysis. The technique uses a *Markov chain Monte Carlo method* for sampling in order to obtain an evolving distribution, i.e., to adapt estimates of *posterior distributions* as new data arrive. The technique is particularly useful in problems in which the observed data become available sequentially in time and interest centres on performing inference in an online fashion. [*Journal of the American Statistical Association*, 1998, **93**, 1032–1044.]

Partitioned algorithms: Algorithms which attempt to make the iterative estimation of the parameters in *non-linear models* more manageable by replacing the original estimation problem with a series of problems of lower dimension. Consider, for example, the model $y = \theta_1 \exp(\theta_2 x)$ to be fitted to data pairs $(x_i, y_i), i = 1, 2, \ldots, n$. Here, for any given value of θ_2 the linear least squares estimator of θ_1 is

$$\hat{\theta}_1 = \frac{\sum_{i=1}^{n} y_i \exp(\theta_2 x_i)}{\sum_{i=1}^{n} \exp(2\theta_2 x_i)}$$

This may now be minimized with respect to θ_2. [*Computers and Structures*, 2002, **80**, 1991–9.]

Partner studies: Studies involving pairs of individuals living together. Such studies are often particularly useful for estimating the *transmission probabilities* of infectious diseases. [*Biometrics*, 1993, **49**, 1110–16.]

Pascal, Blaise (1623–1662): Born in Clermont-Ferrand, France, Pascal was largely educated by his father. Introduced the concept of mathematical expectation and discovered the principle of a calculating machine when only 20. Pascal is most remembered for his *Traité du Triangle Arithmétique* discovered after his death in Paris on 19 August 1662. This deals with the equitable division of stakes problem in games of chance in association with *Pascal's triangle.*

Pascal distribution: *Negative binomial distribution* with integer *k*.

Pascal's triangle: An arrangement of numbers defined by Pascal in his *Traité du Triangle Arithmétique* published in 1665 as 'The number in each cell is equal to that in the preceding cell in the same column plus that in the preceding cell in the same row'. He placed an arbitrary number in the first cell (in the right angle of the triangle) and regarded the construction of the first row and column as special 'because their cells do not have any cells preceding them'. Assuming the number in the first cell to be unity then the arrangement produces

```
1 1 1 1 1 1
1 2 3 4 5
1 3 6 10
1 4 10
1 5
1
```

More commonly nowadays the numbers would be arranged as follows;

```
          1
        1   1
      1   2   1
    1   4   6   4   1
  1   5  10  10   5   1
```

[*Pascal's Arithmetical Triangle*, 1987, A. W. F. Edwards, Edward Arnold, London.]

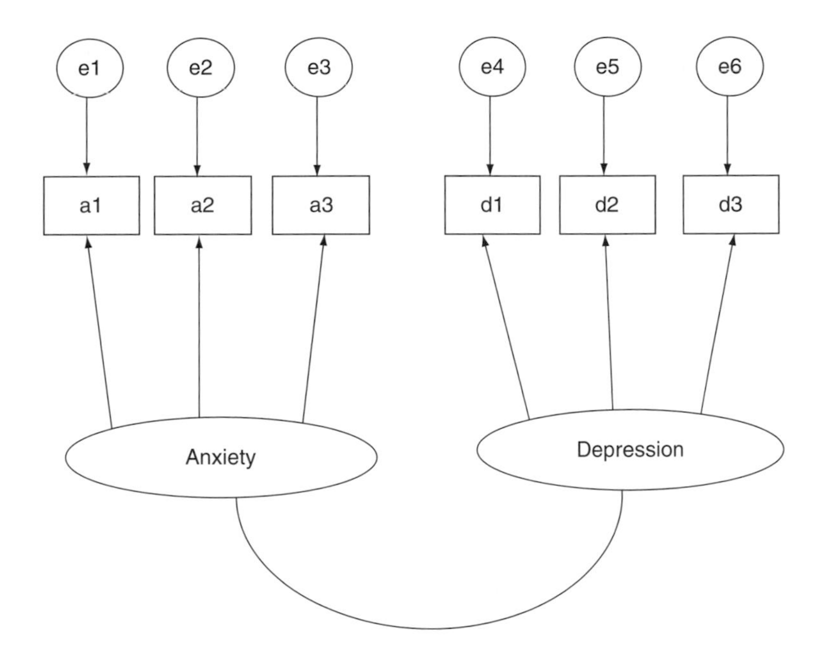

Fig. 102 A path diagram for a correlated two factor model for latent variables anxiety and depression.

Passenger variable: A term occasionally used for a variable, A, that is associated with another variable B only because of their separate relationships to a third variable C.

Pasture trials: A study in which pastures are subjected to various treatments (types of forage, agronomic treatments, animal management systems, etc.). The grazing animal is then introduced onto the pasture to serve as a forage management tool or, more generally, as a means of measuring the yield and quality of the forage. [*Agricultural Field Experiments and Analysis*, 1994, R. G. Petersen, CRC, Boca Raton.]

Path analysis: A tool for evaluating the interrelationships among variables by analysing their correlational structure. The relationships between the variables are often illustrated graphically by means of a *path diagram*, in which single headed arrows indicate the direct influence of one variable on another, and curved double headed arrows indicate correlated variables. An example of such a diagram for a correlated two factor model is shown in Fig. 102. Originally introduced for simple regression models for observed variables, the method has now become the basis for more sophisticated procedures such as *confirmatory factor analysis* and *structural equation modelling*, involving both *manifest variables* and *latent variables*. [MV2 Chapter 11.]

Path coefficient: Synonym for **standardized regression coefficient**.

Path diagram: See **path analysis**.

Path length tree: Synonym for **additive tree**.

Pattern mixture models: A class of models used to handle *missing data*. For a unit *i,* let y_i be the response variable of interest and M_i a missing indicator taking the value 1 if the response is missing and 0 if not. In a pattern mixture model, the joint distribution of y_i and M_i is decomposed as

$$f(y_i, M_i|\xi, \omega) = f(y_i|M_i, \xi)f(M_i|\omega)$$

where the first distribution (with parameters ξ) characterises the distribution of y_i in strata defined by different patterns of missing data M_i and the second distribution (with parameters

ω) models the distribution of different patterns. It has been argued that pattern mixture models are more explicit regarding unverifiable assumptions than *selection models* but more awkward to interpret. [*Statistical Analysis with Missing Data*, 2002, R. J. A. Little and D. B. Rubin, Wiley, New York.]

Pattern recognition: A term for a technology that recognizes and analyses patterns automatically by machine and which has been used successfully in many areas of application including optical character recognition. Speech recognition, remote sensing and medical imaging processing. Because 'recognition' is almost synonymous with 'classification' in this field, pattern recognition includes statistical classification techniques such as *discriminant analysis* (here known as *supervised pattern recognition* or *supervised learning*) and *cluster analysis* (known as *unsupervised pattern recognition* or *unsupervised learning*). Pattern recognition is closely related to *artificial intelligence, artificial neural networks* and *machine learning* and is one of the main techniques used in *data mining*. Perhaps the distinguishing feature of pattern recognition is that no direct analogy is made in its methodology to underlying biological processes. [*Pattern Recognition, 4th edn*, 2008, S. Theodondis and K. Koutroumbas, Academic Press.]

PDF: Abbreviation for **probability density function**.

PDP: Abbreviation for **parallel distributed processing**.

Peak value: The maximum value of (usually) a *dose–response curve*. Often used as an additional (or alternative) response measure to the *area under the curve*. [SMR Chapter 14.]

Pearson, Egon Sharpe (1896–1980): Born in Hampstead, Egon Pearson was the only son of *Karl Pearson*. Read mathematics at Trinity College, Cambridge, and finally obtained his first degree in 1919 after interruptions due to a severe bout of influenza and by war work. Entered his father's Department of Applied Statistics at University College, London in 1921, and collaborated both with *Jerzy Neyman* who was a visitor to the department in 1925–26 and with *W. S. Gosset*. The work with the former eventually resulted in the collection of principles representing a general approach to statistical and scientific problems often known as the *Neyman–Pearson theory*. Became Head of the Department of Applied Statistics on Karl Pearson's retirement in 1933. On his father's death in 1936, he became managing editor of *Biometrika*. Egon Pearson was awarded the Royal Statistical Society's Guy medal in gold in 1955 and became President of the Society from 1955 to 1957. His Presidential Address was on the use of geometry in statistics. Egon Pearson died in Sussex on 12 June 1980.

Pearson, Karl (1857–1936): Born in London, Karl Pearson was educated privately at University College School and at King's College, Cambridge, where he was Third Wrangler in the Mathematics Tripos in 1879. On leaving Cambridge, he spent part of 1879 and 1880 studying medieval and sixteenth-century German literature at Berlin and Heidelberg Universities. He then read law at Lincoln's Inn and was called to the bar by Inner Temple in 1881. He became Professor of Mathematics at King's College, London in 1881 and at University College, London in 1883. Although largely motivated by the study of evolution and heredity, his early statistical work included an assessment of the randomness of Monte Carlo roulette; he concluded that the published evidence was incompatible with a fair wheel. Of more interest scientifically was his work on skew curves, particular his investigation of mixtures of two normal curves. In the space of 15 years up to 1900, Pearson laid the foundations of modern statistics with papers on moments, correlation, maximum likelihood and the chi-squared goodness-of-fit test. Became Professor of Eugenics at University College in 1911. Founded and edited the journal *Biometrika*. Karl Pearson died in Surrey on 27 April 1936.

Pearson's chi-squared statistic: See **chi-squared statistic**.

Pearson's product moment correlation coefficient: See **correlation coefficient**.

Pearson's residual: A model diagnostic used particularly in the analysis of *contingency tables* and *logistic regression* and given by

$$r_i = \frac{O_i - E_i}{\sqrt{E_i}}$$

where O_i represents the observed value and E_i the corresponding predicted value under some model. Such residuals, if the assumed model is true, have approximately a standard normal distribution and so values outside the range –2.0 to 2.0 suggest aspects of the current model that are inadequate. [*The Analysis of Contingency Tables*, 2nd edition, 1992, B. S. Everitt, Chapman and Hall/CRC Press, London.]

Pearson's Type VI distribution: A family of probability distributions given by

$$f(x) = \frac{\Gamma(q_1)(a_2 - a_1)^{q_1 - q_2 - 1}}{\Gamma(q_1 - q_2 - 1)\Gamma(q_2 + 1)} \frac{(x - a_2)^{q_2}}{(x - a_1)^{q_1}}, \qquad q_1 > q_2 > -1, \quad x \geq a_2 > a_1$$

The *F-distribution* belongs to this family. [*International Statistical Review*, 1982, **50**, 71–101.]

Peeling: Synonym for **convex hull trimming**.

Penalized maximum likelihood estimation: An approach commonly used in curve estimation when the aim is to balance fit as measured by the *likelihood* and 'roughness' or rapid variation. For an example see the **splines** entry. [*Statistics in Medicine*, 1994, **13**, 2427–36.]

Penetrance function: The relationship between a *phenotype* and the *genotype* at a locus. For a categorically defined disease trait it specifies the probability of disease for each genotype class. [*Statistics in Human Genetics*, 1998, P. Sham, Arnold, London.]

Pentamean: A measure of location sometimes used in *exploratory data analysis*; it is given by $\frac{1}{10}$ [(sum of greatest and least observations)+median+(sum of upper and lower quartiles)]

Percentile: The set of divisions that produce exactly 100 equal parts in a series of continuous values, such as blood pressure, weight, height, etc. Thus a person with blood pressure above the 80th percentile has a greater blood pressure value than over 80% of the other recorded values.

Percentile–percentile plot: Synonym for **quantile–quantile plot**.

Perceptron: A simple classifier into two classes which computes a linear combination of the variables and returns the sign of the result; observations with positive values are classified into one group and those with negative values to the other. See also **artificial neural networks**. [MV1 Chapter 4.]

Per-comparison error rate: The significance level at which each test or comparison is carried out in an experiment. See also **per-experiment error rate**. [*Biostatistics: A Methodology for the Health Sciences*, 2nd edn, 2004, G. Van Belle, L. D. Fisher, P. J. Heagerty and T. S. Lumley, Wiley, New York.]

Per-experiment error rate: The probability of incorrectly rejecting at least one null hypothesis in an experiment involving one or more tests or comparisons, when the corresponding null hypothesis is true in each case. See also **per-comparison error rate**. [*Biostatistics, A Methodology for Health Sciences*, 2nd edition, 2004, G. Van Belle, L. D. Fisher, P. J. Heagerty and T. S. Lumley, Wiley, New York.]

Performance indicators: Properties designed to assess the impact of Government policies on public services, or to identify well performing or under performing institutions and public servants. [*Journal of the Royal Statistical Society, Series A*, 2005, **158**, 1–?27.]

Periodic survey: Synonym for **panel study**.

Periodogram: See **harmonic analysis**.

Period prevalence: See **prevalence**.

Permutation test: Synonym for **randomization test**.

Permuted block allocation See **sequential allocation procedures**.

Per protocol: A term for a subset of participants in a *clinical trial* who complied with the protocol and/or for the analysis based only on these participants. [*British Medical Journal*, 1998, **316**, 285.]

Perron–Frobenius theorem: If all the elements of a positive definite matrix **A** are positive, then all the elements of the first *eigenvector* are positive. (The first eigenvector is the one associated with the largest *eigenvalue*.) [*Mathematische Zeitschrift,* 1996, **222**, 677–98.]

Personal probabilities: A radically different approach to allocating probabilities to events than, for example, the commonly used long-term *relative frequency* approach. In this approach, probability represents a degree of belief in a proposition, based on all the information. Two people with different information and different subjective ignorance may therefore assign different probabilities to the same proposition. The only constraint is that a single person's probabilities should not be inconsistent. [KA1 Chapter 8.]

Person-time: A term used in studies in *epidemiology* for the total observation time added over subjects. [*Statistics in Medicine*, 1989, **8**, 525–38.]

Person-time incidence rate: A measure of the *incidence* of an event in some population given by

$$\frac{\text{number of events occurring during the interval}}{\text{number of person-time units at risk observed during the interval}}$$

Person-years at risk Units of measurement which combine persons and time by summing individual units of time (years and fractions of years) during which subjects in a study population have been exposed to the risk of the outcome under study. A person-year is the equivalent of the experience of one individual for one year. [*Applied Mathematical Demography*, 1977, N. Keyfitz, Wiley, New York.]

Perspective plot: See **contour plot**.

Persson–Rootzén estimator: An estimator for the parameters in a normal distribution when the sample is truncated so that all observations below some fixed value C are removed. The estimator uses both the information provided by the recorded values greater than C and by the number of observations falling below C. [*Biometrika*, 1977, **64**, 123–8.]

Perturbation methods: Methods for investigating the stability of statistical models when the observations suffer small random changes. [*Perturbation Methods*, 1992, E. J. Hinch, Cambridge University Press.]

Perturbation theory: A theory useful in assessing how well a particular algorithm or statistical model performs when the observations suffer small random changes. In very general terms the theory poses the following question; given a function $f(\mathbf{A})$ of a matrix $\mathbf{A}$ and a pertubation $\mathbf{A}+\mathbf{E}$ of $\mathbf{A}$, how does the difference $f(\mathbf{A}+\mathbf{E})-f(\mathbf{A})$ behave as a function of $\mathbf{E}$? Such an analysis gives some idea of the accuracy of a computed solution. [*Perturbation Theory for Linear Operators*, 1966, T. Kato, Springer Verlag, Berlin.]

PEST: Software for the planning and evaluation of *sequential trials*. See also **EAST**. [MPS Research Unit, University of Reading, Earley Gate, Reading RG6 6FN.]

Petersburg game: Also known as the *Petersburg paradox* or *St Petersburg paradox*, this is an illustration of a variable without an expectation and for which the *law of large numbers* is inapplicable. A single trial in the game consists in tossing an unbiased coin until it falls heads; if this occurs on the rth throw the player receives 2^r pounds (dollars etc.) The gain at each trial is a random variable with values $2^1, 2^2, 2^3, \ldots$ and corresponding probabilities $2^{-1}, 2^{-2}, 2^{-3}, \ldots$. Each term in the series for the expected gain is one, so that the gain has no finite expectation. First posed by Nicolaus Bernoulli (1695–1726) and his brother Daniel Bernoulli (1700–1782). [*An Introduction to Probability Theory and its Applications*, Volume 1, 3rd edition, 1968, W. Feller, Wiley, New York.]

Petersburg paradox: Synonym for **Petersburg game**.

Petersen estimator: See **capture–recapture sampling**.

Peto–Prentice test: See **logrank test**.

Phase I study: Initial *clinical trials* on a new compound usually conducted among healthy volunteers with a view to assessing safety. [SMR Chapter 15.]

Phase II study: Once a drug has been established as safe in a *Phase I study*, the next stage is to conduct a *clinical trial* in patients to determine the optimum dose and to assess the efficacy of the compound. [SMR Chapter 15.]

Phase III study: Large multi-centre comparative *clinical trials* to demonstrate the safety and efficacy of the new treatment with respect to the standard treatments available. These are the studies that are needed to support product licence applications. [SMR Chapter 15.]

Phase IV study: Studies conducted after a drug is marketed to provide additional details about its safety, efficacy and usage profile. [SMR Chapter 15.]

Phenotype: The observed characteristics of an individual that are influenced by *genes*, that is by the *genotype*. [*Statistics in Human Genetics*, 1998, P. Sham, Arnold, London.]

Phenotypic assortment: See **assortative mating**.

Phi-coefficient: A measure of association of the two variables forming a *two-by-two contingency table* given simply by

$$\phi = \sqrt{\frac{X^2}{N}}$$

where X^2 is the usual *chi-squared statistic* for the independence of the two variables and N is the sample size. The coefficient has a maximum value of one and the closer its value to one, the stronger the association between the two variables. See also **Cramer's V** and **contingency coefficient**. [*The Analysis of Contingency Tables*, 2nd edition, 1992, B. S. Everitt, Chapman and Hall/CRC Press, London.]

Phillips curve: A curve showing the relationship between unemployment and the rate of inflation that demonstrates that any attempt by governments to reduce unemployment is likely to lead to increased inflation. [*The Economic Record*, 1975, **51**, 303–7.].

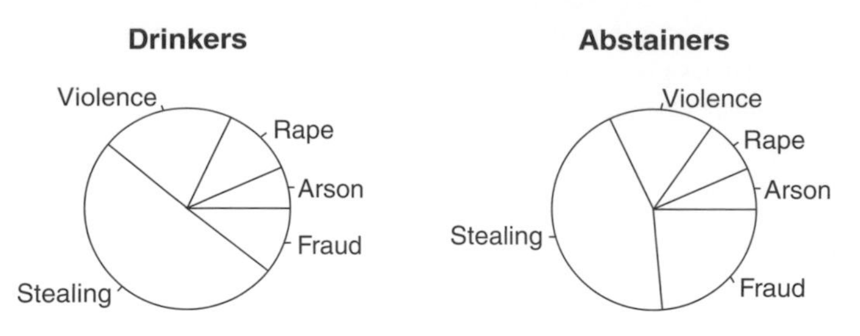

Fig. 103 Pie charts for 'drinkers' and 'abstainers' crime percentages.

Pie chart: A widely used graphical technique for presenting the relative frequencies associated with the observed values of a categorical variable. The chart consists of a circle subdivided into sectors whose sizes are proportional to the quantities (usually percentages) they represent. An example is shown in Fig. 103. Such displays are popular in the media but have little relevance for serious scientific work when other graphics are generally far more useful. See also **bar chart** and **dot plot**.

Pillai–Bartlett trace: See **multivariate analysis of variance**.

Pilot study: A small scale investigation carried out before the main study primarily to gain information and to identify problems relevant to the study proper. [SMR Chapter 15.]

Pilot survey: A small scale investigation carried out before the main survey primarily to gain information and to identify problems relevant to the survey proper.

Pitman drift: A procedure for forming asymptotic power functions by restricting attention to local alternatives in the close neighbourhood of H_0 as the sample size increases. [*Econometric Theory and Methods*, 2004, R. Davidson and J. G. MacKinnon, Oxford University Press, Oxford.]

Pitman, Edwin James George (1897–1993): Born in Melbourne, Pitman graduated from Melbourne University with a first-class honours degree in mathematics and held the Chair of Mathematics at the University of Tasmania from 1926 to 1962. Between 1936 and 1939 he introduced the concept of *permutation tests* and later the notion of *asymptotic relative efficiency*. The Statistical Society of Australia named their award for statistical achievement the Pitman medal and made Pitman the first recipient. He died on 21 July 1993 in Hobart, Tasmania.

Pitman efficiency: Synonym for **Pitman nearness criterion**.

Pitman nearness criterion: A method of comparing two competing estimators that does not introduce additional structure into the estimation problems. If the two estimators of a parameter θ, based on data $y_1, y_2, \ldots, y_n$, are $\hat{\theta}_n$ and $\tilde{\theta}_n$ the criterion is defined as

$$\Pr_\theta^{(n)}\{|\tilde{\theta}_n - \theta| < |\hat{\theta}_n - \theta|\}$$

where $\Pr_\theta^{(n)}$ represents the probability of the data under θ. [*Some Basic Theory for Statistical Inference*, 1979, E. J. G. Pitman, Chapman and Hall/CRC Press, London.]

Pitman-Yor process: A distribution over probability distributions characterized by a discount parameter $0 \leq d \leq 1$, a strength or concentration parameter $\alpha > -d$ and a base distribution H. Reduces to the *Dirichlet process* when d=0. [*Annals of Probability*, 1997, **25**, 855–900.]

Pivotal variable: A function of one or more statistics and one or more parameters that has the same probability distribution for all values of the parameters. For example, the statistic, z, given by

$$z = \frac{\bar{x} - \mu}{\sigma/\sqrt{n}}$$

has a standard normal distribution whatever the values of μ and σ. [*Biometrika*, 1980, **67**, 287–92.]

Pixel: A contraction of 'picture-element'. The smallest element of a display.

Placebo: A treatment designed to appear exactly like a comparison treatment, but which is devoid of the active component.

Placebo effect: A well-known phenomenon in medicine in which patients given only inert substances, in a *clinical trial* say, often show subsequent clinical improvement when compared with patients not so 'treated'. [SMR Chapter 15.]

Placebo reactor: A term sometimes used for those patients in a *clinical trial* who report side effects normally associated with the active treatment while receiving a placebo.

Plackett–Burman designs: A term for certain two-level *fractional factorial designs* which allow efficient estimation of the main effects of all the factors under investigation, assuming that all interactions between the factors can be ignored. See also **response surface methodology**. [*Optimum Experimental Designs*, 1993, A. C. Atkinson and A. N. Donev, Oxford Science Publications, Oxford.]

Planned comparisons: Comparisons between a set of means suggested before data are collected. Usually more powerful than a general test for mean differences.

Platykurtic curve: See **kurtosis**.

Playfair, William (1759–1823): Born at Benvie near Dundee in Scotland, Playfair, at the age of 13, was sent to serve as an apprentice to a millwright. At the age of 21 he became a draughtsman for the firm of Benilton and Walt in Birmingham. In 1787 Playfair went to Paris becoming involved in French politics. He returned to London in 1793. Widely regarded as the founder of graphic methods in statistics, between 1786 and 1821 Playfair wrote several works containing excellent charts. Most important of these was *The Commercial and Political Atlas* published in 1786, in which Playfair achieved a major conceptual breakthrough in graphical presentation by allowing spatial dimensions to represent nonspatial, quantitative, idiographic, empirical data. An example of one of Playfair's charts is shown in Fig. 104.

Play-the-winner rule: A procedure sometimes considered in *clinical trials* in which the response to treatment is either positive (a success) or negative (a failure). One of the two treatments is selected at random and used on the first patient; thereafter the same treatment is used on the next patient whenever the response of the previously treated patient is positive and the other treatment whenever the response is negative. [*Journal of the American Statistical Association*, 1969, **64**, 131–46.]

Plot dendrogram: A technique for combining a *dendrogram* from an *agglomerative hierarchical clustering method* applied to a set of data, with a scatterplot of the same data obtained perhaps from application of *multidimensional scaling*. Figure 105 shows an example. [*Data Science, Classification and Related Methods*, 1998, edited by C. Hayashi, N. Ohsumi, K. Yajima, Y. Tanaka, H. H. Bock and Y. Baba, Springer, Tokyo.]

PLP: Abbreviation for **power law process**.

Plug-in principle: A method by which some characteristic of a probability distribution f (for example, the mean, median, mode, etc.) is estimated by using the corresponding

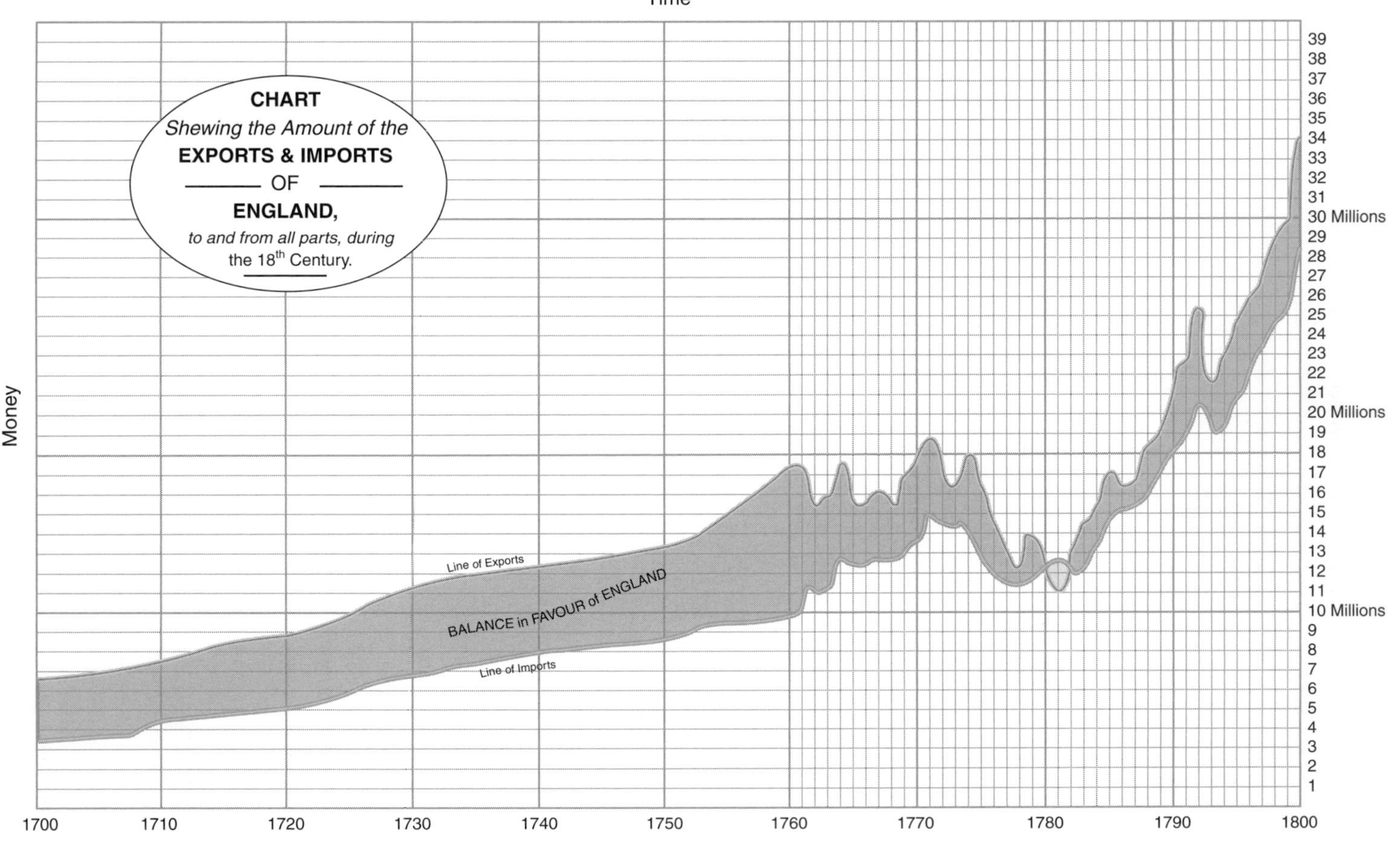

Fig. 104 An example of one of Playfair's original charts.

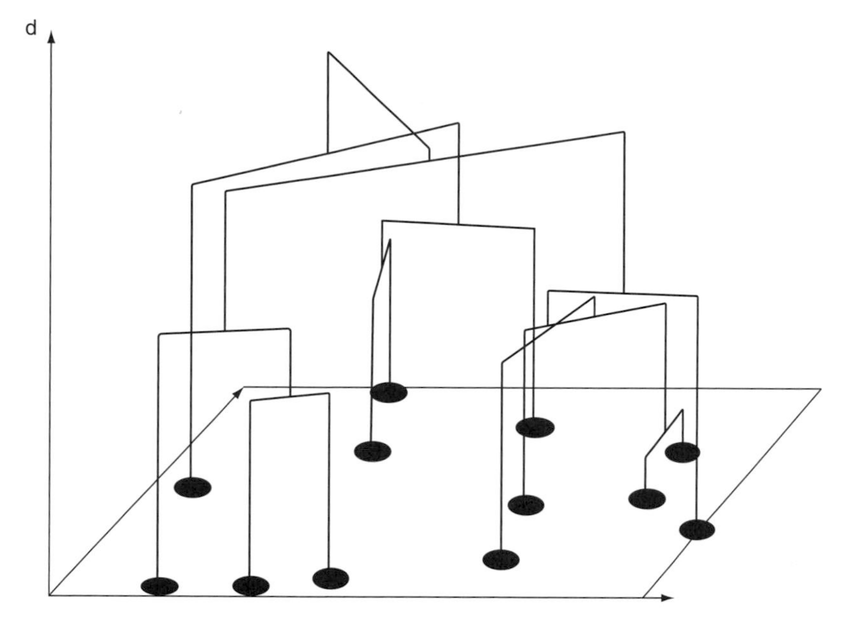

Fig. 105 Combining the output of a hierarchical cluster analysis and a two-dimensional plot. (Taken from *Data Science, Classification and Related Methods*, 1998, with permission of the publishers, Springer.)

characteristic of an empirical distribution $\hat{f}$ formed on the basis of a random sample drawn from f. The *bootstrap* is a direct application of this principle. [MV2 Chapter 9.]

PMR: Abbreviation for **proportionate mortality ratio**.

Point–biserial correlation: A special case of *Pearson's product moment correlation coefficient* used when one variable is continuous (y) and the other is a binary variable (x) representing a natural dichotomy. Given by

$$r_{pb} = \frac{\bar{y}_1 - \bar{y}_0}{s_y}\sqrt{pq}$$

where $\bar{y}_1$ is the sample mean of the y variable for those individuals with $x = 1$, $\bar{y}_0$ is the sample mean of the y variable for those individuals with $x = 0$, s_y is the standard deviation of the y values, p is the proportion of individuals with $x = 1$ and $q = 1 - p$ is the proportion of individuals with $x = 0$. See also **biserial correlation**. [KA2 Chapter 26.]

Point estimate: See **estimate**.

Point estimation: See **estimation**.

Point prevalence: See **prevalence**.

Point process: A *stochastic process* concerned with the occurrence of events at points of time or space determined by some chance mechanism. [*Point Processes*, 1994, D. R. Cox and V. Isham, Chapman and Hall/CRC Press, London.]

Point process data: A set of ordered numbers to be thought of as the times of events that occurred in some time interval, say $(0, T)$. Usual examples are the times of telephone calls and the times of particle emission by some radioactive material. [*Point Processes*, 1994, D. R. Cox and V. Isham, Chapman and Hall/CRC Press, London.]

Point scoring: A simple *distribution free method* that can be used for the prediction of a response that is a binary variable from the observations on a number of explanatory variables which are also binary. The simplest version of the procedure, often called the *Burgess method*, operates by first taking the explanatory variables one at a time and determining which level of each variable is associated with the higher proportion of the 'success' category of the binary response. The prediction score for any individual is then just the number of explanatory variables at the high level (usually only variables that are ' significant' are included in the score). The score therefore varies from 0, when all explanatory variables are at the low level, to its maximum value when all the significant variables are at the high level. The aim of the method is to divide the population into risk groups. See also **prognostic scoring system**.

Poisson distribution: The probability distribution of the number of occurrences, X, of some random event, in an interval of time or space. Given by

$$\Pr(X = x) = \frac{e^{-\lambda}\lambda^x}{x!}, \quad x = 0, 1, \ldots$$

The mean and variances of the distribution are both λ. The *skewness* of the distribution is $1/\sqrt{\lambda}$ and its *kurtosis* is $3 + 1/\lambda$. The distribution is positively skewed for all values of λ. [STD Chapter 31.]

Poisson homogeneity test: See **index of dispersion**.

Poisson process: A *point process* with independent increments at constant intensity, say λ. The count after time t has therefore a *Poisson distribution* with mean λt. [*Point Processes*, 1994, D. R. Cox and V. Isham, Chapman and Hall/CRC Press, London.]

Poisson regression: A method for the analysis of the relationship between an observed count with a *Poisson distribution* and a set of explanatory variables. [*American Statistician*, 1981, **35**, 262–3.]

Poisson, Siméon Denis (1781–1840): Born in Pithiviers, France, Poisson first studied medicine and then turned to mathematics studying under *Laplace* and Lagrange. He became professor at the École Polytechnique in 1806. Undertook research in celestial mechanics, electromagnetism as well as probability where his major work was *Recherches sur la probabilité des jugements en matiere criminelle et en matiere civile*. This book contains the germ of the two things now most commonly associated with Poisson's name, the *Poisson distribution* and a generalization of the Bernoulli *law of large numbers*. Poisson died on 25 April 1840.

Poisson-stopped-sum distributions: Probability distributions arising as the distribution of the sum of N independent and identically distributed random variables, where N is a random variable with a *Poisson distribution*. [*Univariate Discrete Distributions*, 1992, N. Johnson, S. Kotz and A. W. Kemp, Wiley, New York.]

Poker tests: A special frequency test for combinations of five or more digits in a random number. Counts of 'busts', 'pairs,' 'two pairs,' 'threes,' 'full house,' etc. are tested against expected frequencies of these occurrences. [*Handbook of Parametric and Nonparametric Statistical Procedures*, 4th edition, 2007, D.Sheskin, Chapman and Hall/CRC, Boca Raton.]

Polishing: An iterative process aimed at producing a set of residuals from a linear regression analysis that show no relationship to the explanatory variable.

Politz–Simmons technique: A method for dealing with the 'not-at-home' problem in household interview surveys. The results are weighted in accordance with the proportion of days the respondent is ordinarily at home at the time he or she was interviewed. More weight is given to respondents who are seldom at home, who represent a group with a high non-response rate. See also **probability weighting**. [*Sampling Techniques*, 3rd edition, 1977, W. G. Cochran, Wiley, New York.]

Polyá distribution: See **beta-binomial distribution**.

Polychotomous variables: Strictly variables that can take more than two possible values, but since this would include all but binary variables, the term is conventionally used for categorical variables with more than two categories. [*Categorical Data Analysis*, 1990, A. Agresti, Wiley, New York.]

Polynomial regression: A linear model in which powers and possibly cross-products of explanatory variables are included. For example

$$E(y|x) = \beta_0 + \beta_1 x + \beta_2 x^2$$

[SMR Chapter 11.]

Polynomial trend: A trend of the general form

$$y = \beta_0 + \beta_1 t + \beta_2 t^2 + \cdots + \beta_m t^m$$

often fitted to a *time series.*

Polynomial-trigonometric regression: A form of regression analysis in which the proposed model relating the response variable y to an explanatory variable x is

$$E(y|x) = \beta_0 + \sum_{j=1}^{d} \beta_j x^j + \sum_{j=1}^{\lambda} \{c_j \cos(jx) + s_j \sin(jx)\}$$

To fit a data set with such a model, values must be chosen for d and λ. Typically d is fixed, for example, $d = 2$, so that a low-order polynomial is included in the regression. The parameter λ, which defines the number of sine and cosine terms, is then manipulated to obtain a suitable amount of smoothing. The coefficients, $\beta_0, \beta_1, \ldots, \beta_d, c_1, \ldots, c_\lambda, s_1, \ldots, s_\lambda$ are estimated by least squares. See also **kernel regression smoothing** and **spline function**. [*Journal of the Royal Statistical Society, Series A*, 1977, **140**, 411–31.]

Poly-Weibull distribution: The probability distribution of the minimum of m independent random variables $X_1, \ldots, X_m$ each having a *Weibull distribution.* Given by

$$f(x) = \sum_{j-1}^{m} \frac{\gamma_j x^{\gamma_j - 1}}{\beta_j^{\gamma_j}} \exp\left\{-\sum_{k=1}^{m}\left(\frac{x}{\beta_k}\right)^{\gamma_k}\right\} \qquad x>0$$

where γ_i and β_i are parameters of the Weibull distribution associated with X_i. [*Journal of the American Statistical Association*, 1993, **88**, 1412–18.]

POPS: Abbreviation for **principal oscillation pattern analysis**.

Population: In statistics this term is used for any finite or infinite collection of 'units', which are often people but may be, for example, institutions, events, etc. See also **sample**.

Population averaged models: Models for *clustered data* in which the marginal expectation of the response variable is the primary focus of interest. An alternative approach is to use *subject-specific models* which concentrate on the modelling of changes in an individual's response. This is accomplished by introducing subject-specific random effects into the model. A *mixed effects model* or *multilevel model* is an example. There are two key points which differentiate the two types of model.

- The regression coefficients of a population averaged model describe what the average population response looks like. By contrast the regression coefficients of a subject-specific model describe what the average individual's response curve looks like. In many cases and in particular when the model is linear in the subject-specific effects, the two interpretations will coincide. In the more general non-linear setting, however, the two approaches can lead to very different conclusions.
- A further distinction lies in the specification of the underlying variance–covariance structure. In population averaged models the marginal expectations are explicitly modelled while choosing a variance–covariance structure that adequately describes the correlation pattern among the repeated measurements. In subject-specific models, however, individual heterogeneity is modelled using subject-specific effects and it is these random effects which partially determine the variance–covariance structure.

See also **clustered data, generalized estimating equations** and **mixed-effects logistic regression**. [*Analysis of Longitudinal Data*, 2nd edition, 2002, P. Diggle, P. J. Heagerty, K.-Y. Liang and S. Zeger, Oxford Science Publications, Oxford.]

Population genetics: The study of the distributions of *genes* in populations across space and time, and the factors that influence these distributions. [*Statistics in Human Genetics*, 1998, P. Sham, Arnold, London.]

Population growth models: Models intended for forecasting the growth of human populations. In the short term simple geometric or exponential growth models of the form

$$P_t = P_0(1 + r)^t \text{ or } P_t = P_0 e^{rt}$$

where P_0 is population at time 0, P_t is population at time t, r is the growth rate and t is time, can be reasonably accurate. In the long term, however, it is clear that populations cannot grow exponentially. The most common currently used methods are based on controlling explicitly for fertility, mortality and migration. The *Leslie matrix model* is the basis of most such computational procedures for making projections. [*Mathematical Models for the Growth of Human Populations*, 1973, J. H. Pollard, Cambridge University Press, Cambridge.]

Population projection matrix: See **Leslie matrix model**.

Population pyramid: A diagram designed to show the comparison of a human population by sex and age at a given time, consisting of a pair of histograms, one for each sex, laid on their sides with a common base. The diagram is intended to provide a quick overall comparison of the age and sex structure of the population. A population whose pyramid has a broad base and narrow apex has high fertility. Changing shape over time reflects the changing composition of the population associated with changes in fertility and mortality at each age. The example in Fig. 106 shows population pyramids for men and women in Australia in 2002. [*A Workbook in Demography*, 1974, R. Pressat, Methuen, London.]

Population weighting adjustments: A procedure for making estimates of the quantities of interest in a sample survey representative of a given population rather than the one actually sampled. [*Research in Higher Education*, 2008, **49**, 153–171.]

Portmanteau tests: Tests for assessing the fit of models for *time series* in the presence of *outliers*. [*Computational Statistics*, 1994, **9**, 301–10.]

Positive binomial distribution: A probability distribution that arises from the *binomial distribution* by omission of the zero value. Given explicitly as

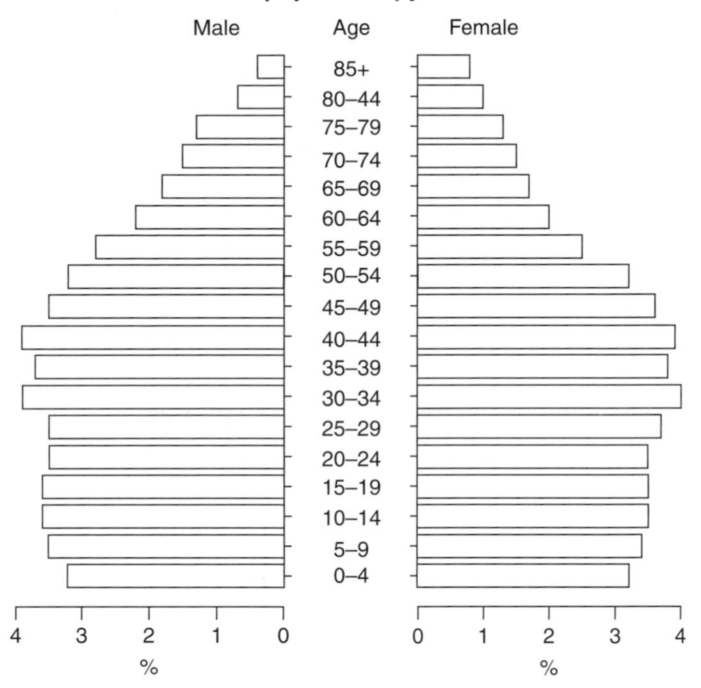

Fig. 106 Example of population pyramids; men and women in Australia.

$$\Pr(X = x) = \frac{n!}{x!(n-x)!}\frac{p^x q^{n-x}}{(1-q^n)} \quad x = 1, 2, \ldots, n$$

[*Univariate Discrete Distributions*, 2005, N. L. Johnson, A. W. Kemp and S. Kotz, Wiley, New York.]

Positive hypergeometric distribution: A probability distribution that arises from the *hypergeometric distribution* by omission of the zero value. Given explicitly by

$$\Pr(X = x) = \frac{\binom{r}{x}\binom{N-r}{n-x}}{\left[\binom{N}{n} - \binom{N-r}{n}\right]} \quad x = 1, \ldots, \min(n, r)$$

[*Univariate Discrete Distributions*, 2005, N. L. Johnson, A. W. Kemp and S. Kotz, Wiley, New York.]

Positive Poisson distribution: A probability distribution that arises from the *Poisson distribution* by omission of the zero value; for example, because the observational apparatus becomes active only when at least one event occurs. Given explicitly by

$$\Pr(X = x) = \frac{e^{-\lambda}\lambda^x}{x!(1-e^{-\lambda})} \quad x = 1, 2, \ldots$$

[*Univariate Discrete Distributions*, 2005, N. L. Johnson, A. W. Kemp and S. Kotz, Wiley, New York.]

Positive predictive value: The probability that a person having a positive result on a *diagnostic test* actually has a particular disease. See also **negative predictive value**. [SMR Chapter 4.]

Positive skewness: See **skewness**.

Positive synergism: See **synergism**.

Post-enumeration survey: A survey composed of a sample of census enumerations and a sample of the population. The two samples are based on a common area sample of census blocks and of housing units within blocks. Used in making *dual system estimates*.

Posterior distributions: Probability distributions that summarize information about a random variable or parameter after, or *a posteriori* to, having obtained new information from empirical data. Used almost entirely within the context of *Bayesian inference*. See also **prior distributions**. [KA2 Chapter 31.]

Posterior predictive checks: An approach to assessing model fit when using *Bayesian inference*. Measures of discrepancy between the estimated model and the data are constructed and their posterior predictive distribution compared to the discrepancy observed for the dataset. [*Annals of Statistics*, 1984, **12**, 1151–1172.]

Posterior probability: See **Bayes' theorem**.

Post-hoc comparisons: Analyses not explicitly planned at the start of a study but suggested by an examination of the data. Such comparisons are generally performed only after obtaining a significant overall F value. See also **multiple comparison procedures, sub-group analysis, data dredging** and **planned comparisons**.

Poststratification: The classification of a simple random sample of individuals into strata after selection. In contrast to a conventional *stratified sampling*, the stratum sizes here are random variables. [*Statistician*, 1991, **40**, 315–23.]

Poststratification adjustment: One of the most frequently used *population weighting adjustments* used in complex surveys, in which weights for elements in a class are multiplied by a factor so that the sum of the weights for the respondents in the class equals the population total for the class. The method is widely used in household surveys to control the weighted sample totals to known population totals for certain demographic subgroups. For example, in the US National Health Interview Survey, poststratification by age, sex and race is employed. [*International Statistical Review*, 1985, **53**, 221–38.]

Potthoff and Whitlinghill's test: A test for the existence of *disease clusters*. [*Biometrika*, 1966, **53**, 167–82.]

Potthoff test: A conservative test for equality of the locations of two distributions when the underlying distributions differ in shape. [*Biometrika*, 1966, **53**, 1183–90.]

Power: The probability of rejecting the null hypothesis when it is false. Power gives a method of discriminating between competing tests of the same hypothesis, the test with the higher power being preferred. It is also the basis of procedures for estimating the sample size needed to detect an effect of a particular magnitude. [SMR Chapter 8.]

Power divergence statistics: Generalized measures of the 'distance' between a vector of estimated proportions, $p' = [p_1 \ldots, p_r]$ and the vector of the corresponding 'true' values $\pi' = [\pi_1, \ldots, \pi_r]$. The general form of such measures is

$$I^{\lambda}(p, \pi) = \frac{1}{\lambda(\lambda+1)} \sum_{i=1}^{r} p_i \left[\left(\frac{p_i}{\pi_i} \right)^{\lambda} - 1 \right]$$

This is known as the directed divergence of order λ. The usual *chi-squared statistic* corresponds to $\lambda = 1$, the deviance is the limit as $\lambda \to 0$ and the *Freeman–Tukey test* to $\lambda = -0.5$. [*Journal of the Royal Statistical Society, Series B*, 1984, **46**, 440–64.]

Power exponential distribution: A probability distribution, $f(x)$ given by

$$f(x) = \frac{1}{\sigma\Gamma\left(1+\frac{1}{2\beta}\right)2^{1+\frac{1}{2\beta}}}\exp\left[-\frac{1}{2}\left|\frac{x-\mu}{\sigma}\right|^{2\beta}\right], \quad -\infty<\mu<\infty, 0<\sigma, 0<\beta\leq\infty$$

See also **multivariate power exponential distribution**. [*Communications in Statistics*, **A27**, 589–600.]

Power function distribution: A *beta distribution* with $\beta = 1$.

Power law process: A technique often used to model failure data from repairable systems, for which the expected number of failures by time t is modelled as

$$M(t) = (\alpha t)^{\gamma}$$

The expected rate of occurrence of failures, $\mathrm{d}M(t)/\mathrm{d}t$ is given by

$$\alpha\gamma(\alpha t)^{\gamma-1}$$

If $\gamma > 1$ this increases with time, as often happens with aging machinery, but if $\gamma < 1$ it decreases and so the model can be applied to reliability growth as well as systems that deteriorate with age. [*Water Resources Research*, 1997, **33**, 1567–83.]

Power spectrum: A function, $h(\omega)$, defined on $-\pi \leq \omega \leq \pi$ for a *stationary time series*, which has the following properties;

1. The function $h(\omega)$ defines the amount of 'power' or the contribution to the total variance of the time series made by the frequencies in the band $[\omega, \omega\delta\omega]$
2. Harmonic components with finite power produce spikes $h(\omega)$.
3. For real series the spectrum is symmetric, $h(\omega) = h(-\omega)$.

The function $h(\omega)$ is related to the *autocovariance function*, $\gamma(k)$, of the series by

$$h(\omega) = \frac{1}{2\pi}\sum_{k=-\infty}^{\infty}\gamma(k)\cos k\omega \quad -\pi \leq \omega \leq \pi$$

The implication is that all the information in the autocovariance is also contained in the spectrum and vice versa. As an example, consider a *white noise process* in which the autocovariances are zero apart from the first which is σ^2, so $h(\omega) = \sigma^2/2\pi$, i.e. a 'flat' spectrum. See also **harmonic analysis** and **spectral analysis**. [*Applications of Time Series Analysis in Astronomy and Meteorology*, 1997, edited by T. Subba Rao, M. B. Priestley and O. Lessi, Chapman and Hall/CRC Presss, London.]

Power transfer function: See **linear filters**.

Power transformation: A transformation of the form $y = x^m$.

P–P plot: Abbreviation for **probability–probability plot**.

Pragmatic analysis: See **explanatory analysis**.

Precision: A term applied to the likely spread of estimates of a parameter in a statistical model. Measured by the inverse of the standard deviation of the estimator; this can be decreased, and hence precision increased, by using a larger sample size. [SMR Chapter 2.]

Precision matrix: A synonym for **concentration matrix**.

Predictor variables: Synonym for **explanatory variables**.

Prentice criterion: A criterion for assessing the validity of a *surrogate endpoint* in a *clinical trial*, in the sense that the test based on the surrogate measure is a valid test of the hypothesis of interest about the true end point. Specifically the criterion is that of the *conditional independence* of the treatment and the true end point, given the surrogate end point. [*Statistics in Medicine*, 1989, **8**, 431–40.]

PRESS statistic: A measure of the generalizability of a model in a regression analysis based on the calculation of *residuals* of the form

$$\hat{e}_{(-i)} = y_i - \hat{y}_{(-i)}$$

where y_i is the observed value of the response variable for observation i and $\hat{y}_{(-i)}$ is the predicted value of the response variable for this observation found from the fitted regression equation calculated from the data after omitting the observation. From these residuals a *multiple correlation coefficient* type of statistic is obtained as

$$R^2_{\text{PRESS}} = 1 - \sum_{i=1}^{n} \hat{e}^2_{(-i)} \Big/ \sum_{i=1}^{n} (y_i - \bar{y})^2$$

This can be used to assess competing models. [ARA Chapter 7.]

Prevalence: A measure of the number of people in a population who have a particular disease at a given point in time. Can be measured in two ways, as *point prevalence* and *period prevalence*, these being defined as follows;

$$\text{point prevalence} = \frac{\text{number of cases at a particular moment}}{\text{number in population at that moment}}$$

$$\text{period prevalence} = \frac{\text{number of cases during a specified time period}}{\text{number in population at midpoint of period}}$$

Essentially measures the amount of a disease in a population. See also **incidence**. [SMR Chapter 14.]

Prevented fraction: A measure that can be used to attribute protection against disease directly to an intervention. The measure is given by the proportion of disease that would have occurred had the intervention not been present in the population, i.e.

$$\text{PF} = \frac{\text{PAI} - \text{PI}}{\text{PAI}}$$

where PAI is the risk of disease in the absence of the intervention in the population and PI is overall risk in the presence of the intervention. See also **attributable risk**. [*Statistics in Medicine*, 1995, **14**, 51–72.]

Prevention trials: *Clinical trials* designed to test treatments preventing the onset of disease in healthy subjects. An early example of such a trial was that involving various whooping-cough vaccines in the 1940s. [*Controlled Clinical Trials*, 1990, **11**, 129–46.]

Prevosti's distance: A measure of genetic distance between two populations P and Q based upon, but not restricted to, the difference between the frequencies of chromosomal arrangements in each population. It is defined as

$$D_p = \frac{1}{2r} \sum_{j=1}^{r} \sum_{i=1}^{k_j} |p_{ji} - q_{ji}|$$

where r is the number of chromosomes or loci and p_{ji} and q_{ji} are the probabilities of chromosomal arrangement or allele i in the chromosome or locus of populations P and Q respectively. The distance is estimated in the obvious way by replacing probabilities with observed sample frequencies. The distance has been used in studies of chromosome polymorphism as well as in other fields such as anthropology and sociology. [*Annual Review of Anthropology*, 1985, **14**, 343–73.]

Prewhitening: A term for transformations of *time series* intended to make their spectrum more nearly that of a *white noise process*. [TMS Chapter 7.]

Principal components analysis: A procedure for analysing *multivariate data* which transforms the original variables into new ones that are uncorrelated and account for decreasing proportions of the variance in the data. The aim of the method is to reduce the dimensionality of the data. The new variables, the principal components, are defined as linear functions of the original variables. If the first few principal components account for a large percentage of the variance of the observations (say above 70%) they can be used both to simplify subsequent analyses and to display and summarize the data in a parsimonious manner. See also **factor analysis**. [MV1 Chapter 2.]

Principal components regression analysis: A procedure often used to overcome the problem of *multicollinearity* in regression, when simply deleting a number of the explanatory variables is not considered appropriate. Essentially the response variable is regressed on a small number of principal component scores resulting from a *principal components analysis* of the explanatory variables. [ARA Chapter 9.]

Principal coordinates analysis: Synonym for **classical scaling**.

Principal curve: A smooth, one-dimensional curve that passes through the middle of a q-dimensional data set; it is nonparametric, and its shape is suggested by the data. [*Annals of Statistics*, 1996, **24**, 1511–20.]

Principal factor analysis: A method of *factor analysis* which is essentially equivalent to a *principal components analysis* performed on the *reduced covariance matrix* obtained by replacing the diagonal elements of the sample *variance–covariance matrix* with estimated communalities. Two frequently used estimates of the latter are (a) the square of the *multiple correlation coefficient* of the ith variable with all other variables, (b) the largest of the absolute values of the correlation coefficients between the ith variable and one of the other variables. See also **maximum likelihood factor analysis**. [*Applied Multivariate Data Analysis*, 2nd edition, 2001, B. S. Everitt and G. Dunn, Edward Arnold, London.]

Principal Hessian directions: A method based on the *Hessian matrix* of a regression function that can be effective for detecting and visualizing nonlinearities. [*Journal of the American Statistical Association*, 1992, **87**, 1025–39.]

Principal oscillation pattern analysis (POPS): A method for isolating spatial patterns with a strong temporal dependence, particularly in the atmospheric sciences. Based on the assumption of a first-order *Markov chain*. [*Journal of Climate*, 1995, **8**, 377–400.]

Principal points: Points $\eta_1, \eta_2, \ldots, \eta_k$ which minimize the expected squared distance of a p-variate random variable $\mathbf{x}$ from the nearest of the η_i. [*Statistics and Computing*, 1996, **6**, 187–90.]

Principal stratification: A method for adjusting for a response variable C that is intermediate on the causal pathway from a treatment to the final response y. C has potential outcomes $C(1)$ and $C(0)$, respectively, when a unit is assigned treatment 1 or 0 (see *Neyman-Rubin causal model*). A common mistake is to treat C_{OBS}, the observed value of C, as if it were a covariate and conduct analysis stratified on C_{OBS}. The correct approach is to stratify on the principal strata $(C(1),C(0))$, latent classes which are unaffected by treatment assignment and can hence be treated as a vector covariate. An important example is modelling of causal effects taking compliance into consideration, where C represents compliance status. [*Biometrics*, 2002, **58**, 21–29.]

Prior distributions: Probability distributions that summarize information about a random variable or parameter known or assumed at a given time point, *prior* to obtaining further information from empirical data. Used almost entirely within the context of *Bayesian inference*. In any particular study a variety of such distributions may be assumed. For example, *reference priors* represent minimal prior information; *clinical priors* are used to formalize opinion of well-informed specific individuals, often those taking part in the trial themselves. Finally, *sceptical priors* are used when large treatment differences are considered unlikely. See also **improper prior distribution, Jeffreys prior distribution** and **posterior distributions**. [KA2 Chapter 3.]

Probabilistic distance measures: Distance measures for two classes or groups (G_1 and G_2) based on their conditional probability density functions, $\Pr(\mathbf{x}|G_1)$ and $\Pr(\mathbf{x}|G_2)$. Such distance functions satisfy the following conditions

- $D = 0$ if $\Pr(\mathbf{x}|G_1) = \Pr(\mathbf{x}|G_2)$
- $D \geq 0$
- D attains its maximum value when the classes are disjoint, i.e., when $\Pr(\mathbf{x}|G_1) = 0$ and $\Pr(\mathbf{x}|G_2) \neq 0$. An example of such a distance measure is *Bhattacharyya's distance*. [*Statistical Pattern Recognition*, 1999, A. Webb, Arnold, London.]

Probabilistic matching: A technique developed to maximize the accuracy of linkage decisions based on the level of agreement and disagreement between the identifiers on different records in data bases. Used in *record linkage* applications when there are no unique personal identifiers present. [*Handbook of Record Linkage*, 1988, H. B. Newcombe, Oxford University Press, New York.]

Probability: A measure associated with an event A and denoted by $\Pr(A)$ which takes a value such that $0 \leq \Pr(A) \leq 1$. Essentially the quantitative expression of the chance that an event will occur. In general the higher the value of $\Pr(A)$ the more likely it is that the event will occur. If an event cannot happen $\Pr(A) = 0$; if an event is certain to happen $\Pr(A) = 1$. Numerical values can be assigned in simple cases by one of the following two methods:

(1) If the *sample space* can be divided into subsets of n ($n \geq 2$) equally likely outcomes and the event A is associated with r ($0 \leq r \leq n$) of these, then $\Pr(A) = r/n$.

(2) If an experiment can be repeated a large number of times, n, and in r cases the event A occurs, then r/n is called the *relative frequency* of A. If this leads to a limit as $n \rightarrow \infty$, this limit is $\Pr(A)$.

See also **addition rule for probabilities**, **personal probabilities**, and **law of large numbers**. [KA1 Chapter 8.]

Probability density: See **probability distribution**.

Probability distribution: For a discrete random variable, a mathematical formula that gives the probability of each value of the variable. See, for example, *binomial distribution* and *Poisson distribution*. For a continuous random variable, a curve described by a mathematical formula which specifies, by way of areas under the curve, the probability that the variable falls within a particular interval. Examples include the normal distribution and the *exponential distribution*. In both cases the term *probability density* may also be used. (A distinction is sometimes made between 'density' and 'distribution', when the latter is reserved for the probability that the random variable falls below some value. In this dictionary, however, the latter will be termed the *cumulative probability distribution* and probability distribution and probability density used synonymously.) [KA1 Chapter 8.]

Probability generating function: A function, $P(t)$, which when expanded gives the probability that a discrete random variable takes a particular value r as the coefficient of t^r. Often a useful summarization of the probability distribution of such a variable. The function can be found from:

$$P(t) = \sum_{r=0}^{\infty} \Pr(X = r)t^r$$

where $\Pr(X = r)$ is the probability that the random variable X takes the value r. For a variable with a *binomial distribution*, for example,

$$P(t) = \sum_{j=0}^{n} \binom{n}{j} p^j (1-p)^{n-j} t^j$$
$$= \{pt + (1-p)\}^n$$

[KA1 Chapter 2.]

Probability judgements: Human beings often need to assess the probability that some event will occur and the accuracy of these probability judgements often determines the success of our actions and decisions. There is strong evidence that human probability judgements are biased with overestimation of low probabilities. [*Judgment under Uncertainty: Heuristics and Biases*, 1982, D. Kahneman, P. Slovic, and A. Tversky, eds., Cambridge University Press, Cambridge.]

Probability-of-being-in-response function: A method for assessing the 'response experience' of a group of patients, by using a function of time, $P(t)$, which represents the probability of being 'in response' at time t. The purpose of such a function is to synthesize the different summary statistics commonly used to represent responses that are *binary variables*, namely the proportion who respond and the average duration of response. The aim is to have a function which will highlight the distinction between a treatment that produces a high response rate but generally short-lived responses, and another that produces a low response rate but with longer response durations. [*Biometrics*, 1982, **38**, 59–66.]

Probability paper: Graph paper structured in such a way that the values in the *cumulative frequency distribution* of a set of data from a normal distribution fall on a straight line. Can be used to assess sample data for normality. See also **probability plot**.

Probability plot: Plots for comparing two probability distributions. There are two basic types, the *probability–probability plot* and the *quantile–quantile plot*. A plot of points whose coordinates are the cumulative probabilities $\{p_x(q), p_y(q)\}$ for different values of q is a probability–probability plot, while a plot of the points whose coordinates are the quantiles $\{q_x(p), q_y(p)\}$ for different values of p is a quantile–quantile plot. The latter is the more frequently used of the two types and its use to investigate the assumption that a set of data is from a normal distribution, for example, would involve plotting the ordered sample values $y_{(1)}, \ldots, y_{(n)}$ against the quantiles of a standard normal distribution, i.e.

$$\Phi^{-1}[p_{(i)}]$$

where usually

$$p_i = \frac{i - \frac{1}{2}}{n} \quad \text{and} \quad \Phi(x) = \int_{-\infty}^{x} \frac{1}{\sqrt{2\pi}} e^{-\frac{1}{2}u^2}\, du$$

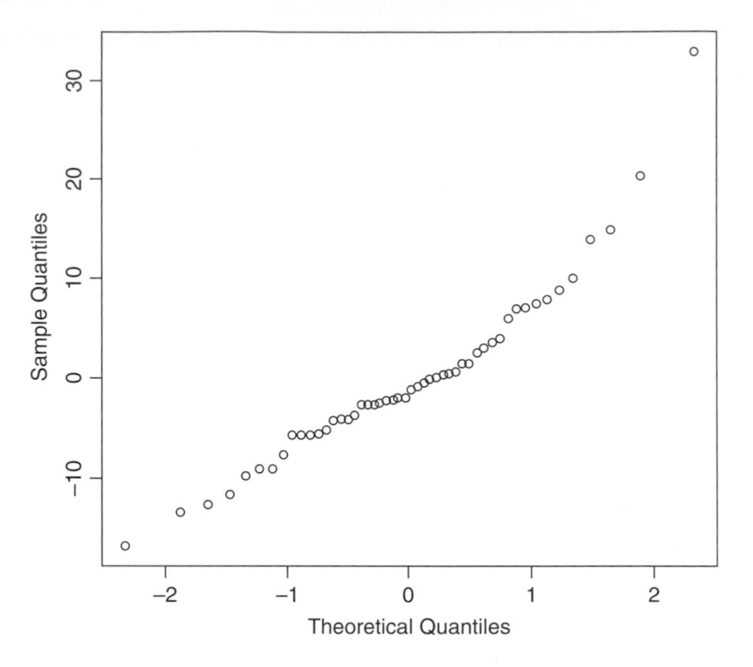

Fig. 107 An example of a normal probability plot.

(This is usually known as a *normal probability plot*, an example of which is shown in Fig. 107.) [*Methods for Statistical Analysis of Multivariate Observations*, 1977, R. Gnandesikan, Wiley, New York.]

Probability–probability (P–P) plot: See **probability plot**.

Probability sample: A sample obtained by a method in which every individual in a *finite population* has a known (but not necessarily equal) chance of being included in the sample. [*Sampling of Populations: Methods and Applications*, 3rd edition, 1999, P. S. Levy and S. Lemeshow, Wiley, New York.]

Probability weighting: The process of attaching weights equal to the inverse of the probability of being selected, to each respondent's record in a sample survey. These weights are used to compensate for the facts that sample elements may be selected at unequal sampling rates and have different probabilities of responding to the survey, and that some population elements may not be included in the list or frame used for sampling. [*Journal of Risk and Uncertainty*, 2001, **1**, 21–33.]

Probit analysis: A technique most commonly employed in bioassay, particularly toxicological experiments where sets of animals are subjected to known levels of a toxin and a model is required to relate the proportion surviving at a particular dose, to the dose. In this type of analysis the *probit transformation* of a proportion is modelled as a *linear function* of the dose or more commonly, the logarithm of the dose. Estimates of the parameters in the model are found by *maximum likelihood estimation*. [*Modelling Binary Data*, 2nd edition, 2003, D. Collett, Chapman and Hall, London.]

Probit transformation: A transformation y of a proportion p derived from the equation

$$p = 5 + \frac{1}{\sqrt{2\pi}} \int_{-\infty}^{y} \exp\left[-\frac{u^2}{2}\right] du$$

The '5' in the equation was introduced by *Fisher* to prevent negative values in the days when biologists were even more resistant to doing arithmetic than they are today! The transformation is used in the analysis of *dose-response curves*. See also **probit analysis**. [GLM Chapter 10.]

Procrustes analysis: A method of comparing alternative geometrical representations of a set of multivariate data or of a *proximity matrix*, for example, two competing *multidimensional scaling* solutions for the latter. The two solutions are compared using a residual sum of squares criterion, which is minimized by allowing the coordinates corresponding to one solution to be rotated, reflected and translated relative to the other. *Generalized Procrustes analysis* allows comparison of more than two alternative solutions by simultaneously translating, rotating and reflecting them so as to optimize a predefined goodness-of-fit measure. (The name of this type of analysis arises from Procrustes, 'The Smasher', who, according to Greek mythology, caught travellers on the borders of Athens and by stretching or pruning them using rack or sword, made them an exact fit to his lethal bed. Procrustes eventually experienced the same fate as that of his guests at the hands of Theseus.) [MV2 Chapter 14.]

Product-limit estimator: A procedure for estimating the *survival function* for a set of survival times, some of which may be *censored observations*. The idea behind the procedure involves the product of a number of conditional probabilities, so that, for example, the probability of a patient surviving two days after a liver transplant can be calculated as the probability of surviving one day, multiplied by the probability of surviving the second day given that the patient survived the first day. Specifically the estimator is given by

$$\hat{S}(t) = \prod_{j|t_{(j)} \leq t} \left(1 - \frac{d_j}{r_j}\right)$$

where $\hat{S}(t)$ is the estimated survival function at time t, $t_{(1)} \leq t_{(2)} \leq \cdots \leq t_{(n)}$ are the ordered *survival times*, r_j is the number of individuals at risk at time $t_{(j)}$ and d_j is the number of individuals who experience the event of interest at time $t_{(j)}$. (Individuals censored at time $t_{(j)}$ are included in r_j.) The resulting estimates form a step function that can be plotted to give a graphical display of survival experience. An example appears in Fig. 108. See also **Alshuler's estimator** and **Greenwood's formula**. [SMR Chapter 13.]

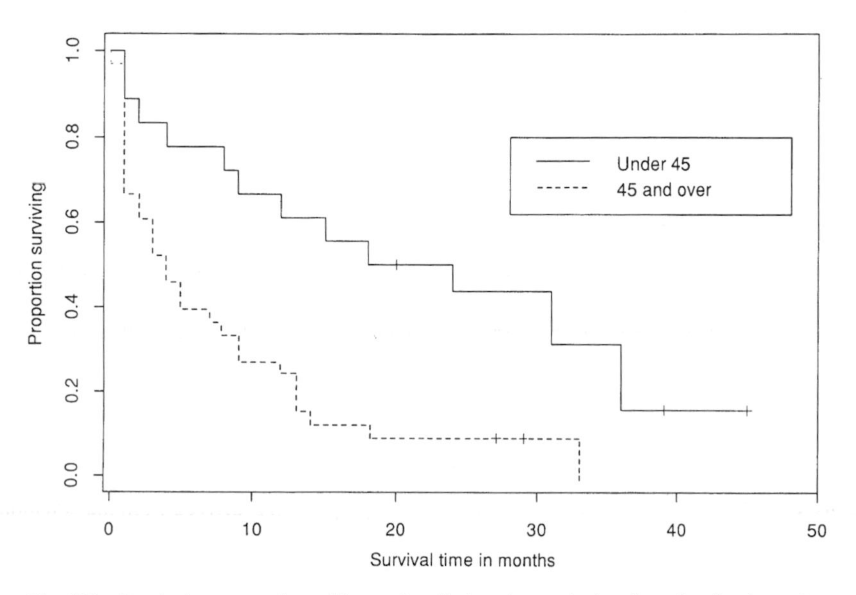

Fig. 108 Survival curves estimated by product-limit estimator: leukemia study of under and over 45s.

Profile analysis: A term sometimes used for the *analysis of variance* of *longitudinal data.*

Profile likelihood: See **likelihood**.

Profile plots: A method of representing *multivariate data* graphically. Each observation is represented by a diagram consisting of a sequence of equispaced vertical spikes, with each spike representing one of the variables. The length of a given spike is proportional to the magnitude of the variable it represents relative to the maximum magnitude of the variable across all observations. As an example, consider the data below showing the level of air pollution in four cities in the United States along with a number of other climatic and human ecologic variables. The profile plots representing these data are shown in Fig. 109(a). Chicago is identified as being very different from the other three cities. Another example of a profile plot is shown in Fig. 109(b); here the weight profiles over time for rats in three groups are all shown on the same diagram. See also **star plot**.

Air pollution data for four cities in the USA

	SO2	Temp	Manuf	Pop	Wind	Precip	Days
Atlanta	24	61.5	368	497	9.1	48.34	115
Chicago	110	50.6	3344	3369	10.4	34.44	122
Denver	17	51.9	454	515	9.0	12.95	86
San Francisco	12	56.7	453	716	8.7	20.66	67

SO2 = Sulphur dioxide content of air (microgram per cubic metre)
Temp = Average annual temperature (F)
Manuf = Number of manufacturing enterprises employing 20 or more workers
Pop = Population size
Wind = Average annual wind speed (miles per hour)
Precip = Average annual precipitation (miles per hour)
Days = Average number of days with precipitation per year

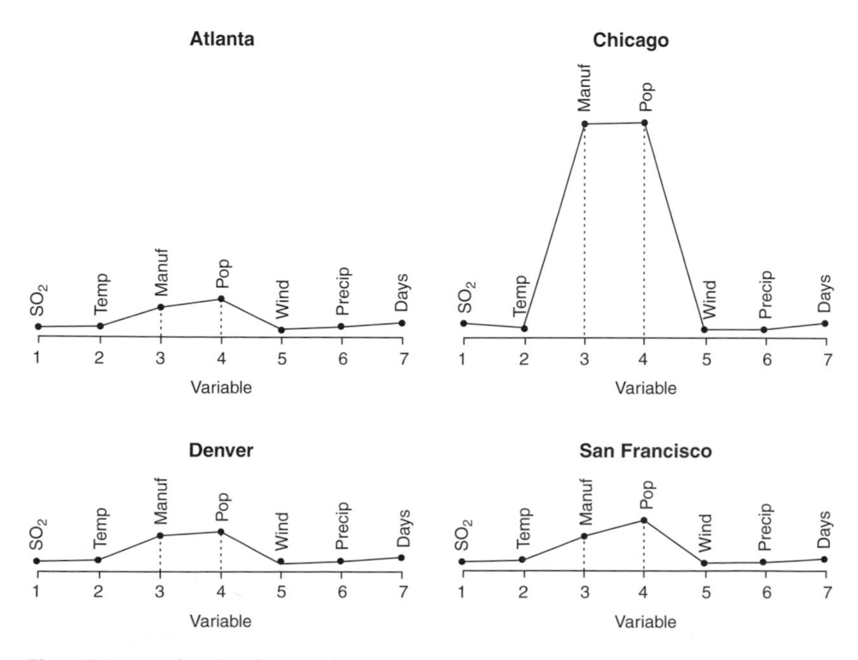

Fig. 109(a) Profile plots for air pollution data from four cities in the United States.

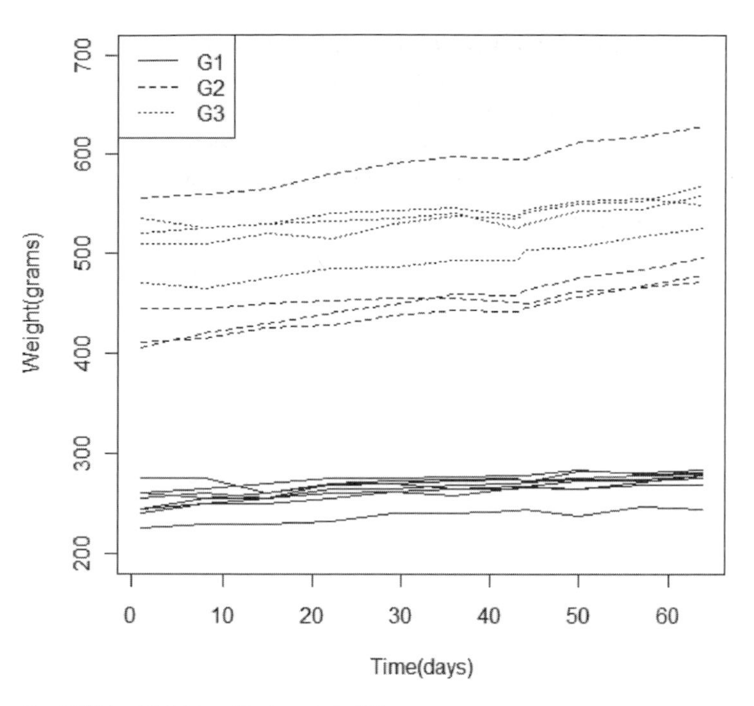

Fig. 109(b) Weight profile for rats in different groups.

Profile *t* plots: A graphical method used to construct *confidence intervals* for components of a parameter vector when fitting a *non-linear model.* [*Journal of the American Statistical Association*, 1995, **90**, 1271–6.]

Prognostic scoring system: A method of combining the prognostic information contained in a number of risk factors, in a way which best predicts each patient's risk of disease. In many cases a linear combination of scores is used with the weights being derived from, for example, a *logistic regression.* An example of such a system developed in the British Regional Heart Study for predicting men aged 40–59 years to be at risk of ischaemic heart disease over the next five years is as follows:

51 × total serum cholesterol(mmol/l)
+5 × total time man has smoked(years)
+3 × systolic blood pressure(mmHg)
+100 if man has symptoms of angina
+170 if man can recall diagnosis of IHD
+50 if either parent died of heart trouble
+95 if man is diabetic

See also **point scoring**. [SMR Chapter 12.]

Prognostic survival model: A quantification of the survival prognosis of patients based on information at the start of follow-up. [*Statistics in Medicine*, 2000, **19**, 3401–15.]

Prognostic variables: In medical investigations, an often used synonym for explanatory variables. [SMR Chapter 5.]

Programming: The act of planning and producing a set of instructions to solve a problem by computer. See also **algorithm**.

Progressively censored data: *Censored observations* that occur in *clinical trials* where the period of the study is fixed and patients enter the study at different times during that period.

Since the entry times are not simultaneous, the censored times are also different. See also **singly censored data**. [*Technometrics*, 1994, **36**, 84–91.]

Progressively type II censored sample: The sample that arises when n experimental units are placed on a life test, and at the time of the first failure, R_1 of the $n-1$ surviving units are randomly withdrawn from the experiment. At the time of the second failure, R_2 of the $n - R_1 - 2$ surviving units are randomly withdrawn. Finally at the time of the mth failure (m being set by the investigator), all of the remaining $n - m - R_1 - \cdots R_{m-1}$ surviving units are withdrawn from the experiment. [*Technometrics*, 2004, **46**, 470–81.]

Projection: The numerical outcome of a specific set of assumptions regarding future trends. See also **forecast**.

Projection plots: A general term for any technique that can be used to produce a graphical display of multivariate data by projecting the data into two dimensions enabling a scatterplot to be drawn. Examples of such techniques are *principal components analysis*, *multidimensional scaling* and *projection pursuit*.

Projection pursuit: A procedure for obtaining a low-dimensional (usually two-dimensional) representation of multivariate data, that will be particularly useful in revealing interesting structure such as the presence of distinct groups of observations. The low-dimensional representation is found by optimizing some pre-defined numerical criterion designed to reveal 'interesting' patterns. [*Pattern Recognition and Neural Networks*, 1996, B. D. Ripley, Cambridge University Press, Cambridge.]

Proof of concept trials: *Clinical trials* carried out to determine if a treatment is biologically active or inactive. [*Statistics in Medicine*, 2005, **24**, 1815–35.]

Propensity score: The conditional probability of assignment to a particular treatment given a vector of covariates. Used for causal inference in observational studies by forming the basis for matching to *balance* observed confounders. [*Biometrika*, 1983, **70**, 41–55.]

Proportional allocation: In *stratified sampling*, the allocation of portions of the total sample to the individual strata so that the sizes of these subsamples are proportional to the sizes of the corresponding strata in the total sample. [*Applied Sampling*, 1976, S. Sudman, Academic Press, New York.]

Proportional hazards model: See **Cox's proportional hazards model**.

Proportional mortality ratio: A ratio that consists of a numerator that is the number of deaths from a particular cause and a denominator of total deaths from all causes. Simply the fraction of deaths from a particular cause. [*American Journal of Individual Medicine*, 1986, **10**, 127–41.]

Proportional-odds model: A model for investigating the dependence of an ordinal variable on a set of explanatory variables. In the most commonly employed version of the model the cumulative probabilities, $\Pr(y \leq k)$, where y is the response variable with categories $1 \leq 2 \leq 3 \leq \cdots \leq c$, are modelled as linear functions of the explanatory variables via the *logistic transformation*, that is

$$\log \frac{\Pr(y \leq k)}{1 - \Pr(y \leq k)} = \tau_k + \beta_1 x_1 + \cdots + \beta_q x_q$$

where τ_k is a category specific parameter and $\beta_1, \ldots, \beta_q$ are regression coefficients. The name proportional-odds arises since the *odds ratio* of having a score of k or less for two different sets of values of the explanatory variables does not depend on k. [GLM Chapter 5.]

Proportionate mortality ratio (PMR): The number of deaths from a specific cause in a specific period of time per 100 deaths from all causes in the same time period. The ratio is used to approximate the *standardized mortality ratio* when death data are available but the population at risk is not known. See also **mortality odds ratio**. [*Journal of Chronic Diseases*, 1978, **31**, 15–22.]

Prospective study: Studies in which individuals are followed-up over a period of time. A common example of this type of investigation is where samples of individuals exposed and not exposed to a possible risk factor for a particular disease, are followed forward in time to determine what happens to them with respect to the illness under investigation. At the end of a suitable time period a comparison of the incidence of the disease amongst the exposed and non-exposed is made. A classic example of such a study is that undertaken among British doctors in the 1950s, to investigate the relationship between smoking and death from lung cancer. All *clinical trials* are prospective. See also **retrospective study, cohort study** and **longitudinal study**. [SMR Chapter 10.]

Protective efficacy of a vaccine: The proportion of cases of disease prevented by the vaccine, usually estimated as

$$\text{PE} = (\text{ARU} - \text{ARV})/\text{ARU}$$

where ARV and ARU are the *attack rates* of the disease under study among the vaccinated and unvaccinated cohorts, respectively. For example, if the rate of the disease is 100 per 10 000 in a non-vaccinated group but only 30 per 10 000 in a comparable vaccinated group, the PE is 70%. Essentially equivalent to *attributable risk.*

Protocol: A formal document outlining the proposed procedures for carrying out a *clinical trial.* The main features of the document are study objectives, patient selection criteria, treatment schedules, methods of patient evaluation, trial design, procedures for dealing with *protocol violations* and plans for statistical analysis. [*Clinical Trials in Psychiatry*, 2003, B. S. Everitt and S. Wessley, Oxford University Press, Oxford.]

Protocol violations: Patients who either deliberately or accidentally have not followed one or other aspect of the protocol for carrying out a *clinical trial.* For example, they may not have taken their prescribed medication. Such patients are said to show *non-compliance.*

Protopathic bias: A type of bias (also know as *reverse-causality*) that is a consequence of the differential misclassification of exposure related to the timing of occurrence. Occurs when a change in exposure taking place in the time period following disease occurrence is incorrectly thought to precede disease occurrence. For example, a finding that alcohol has a protective effect for clinical gallstone disease might be explained by a reduction in alcohol use because of symptoms related to gallstone disease. [*Biological Psychiatry*, 1997, **41**, 257–8.]

Proximity matrix: A general term for either a *similarity matrix* or a *dissimilarity matrix.* In general such matrices are symmetric, but *asymmetric proximity matrices* do occur in some situations. [MV1 Chapter 5.]

Pruning algorithms: Algorithms used in the design of *artificial neural networks* for selecting the right-sized network. This is important since the use of the right-sized network leads to an improvement in performance. [*IEEE Transactions in Neural Networks*, 1993, **4**, 740–7.]

Pseudo-inverse: Synonym for **Moore–Penrose inverse**.

Pseudo-likelihood: A term that has many meanings in statistics but the common theme is to signal that such a 'likelihood' differs from a standard likelihood. For instance, the term has been used in the context of inference based on complex survey data with probability weights, multi-stage estimation where estimates from previous stages are plugged in at later stages, and for approximations based on linearization for generalized linear mixed models.

Pseudorandom numbers: A sequence of numbers generated by a specific computer *algorithm* which satisfy particular statistical tests of randomness. So although not random, the numbers appear so. See also **congruential methods**. [KA1 Chapter 9.]

Pseudo R^2: An index sometimes used in assessing the fit of specific types of models particularly logistic regression and those used for modelling survival times. It is defined as

$$1 - \frac{\ln L(\text{present model})}{\ln L(\text{model without covariates})}$$

where L represents *likelihood.* Expresses the relative decrease of the *log-likelihood* of the present model as opposed to the model without covariates. [*Sociological Methods and Research*, 2002, **31**, 27–74.]

Psychometrics: The study of the measurements of psychological characteristics such as abilities, aptitudes, achievement, personality traits and knowledge.

Publication bias: The possible bias in published accounts of, for example, *clinical trials*, produced by editors of journals being more likely to accept a paper if a statistically significant effect has been demonstrated. See also **funnel plot**. [SMR Chapter 8.]

P-value: The probability of the observed data (or data showing a more extreme departure from the *null hypothesis*) when the null hypothesis is true. See also **misinterpretation of P-values, significance test** and **significance level**. [SMR Chapter 8.]

Pyke, Ronald (1932–2005): Pyke was born in Hamilton, Ontario and graduated with honours from McMaster University in 1953. In 1965 he completed a Ph.D. in mathematics at the same university, the title of his thesis being 'On one-sided distribution free statistics'. On completing his Ph.D., Pyke spent two years in Stanford and then moved to the University of Washington where he remained until he retired in 1998. Pyke made important contributions to the *Kolmogorov–Smirnov statistics and goodness of fit statistics*, and branching processes, and was editor of the *Annals of Probability* during the first four years of its existence; he was also President of the Institute of Mathematical Statistics in 1986–7. Pyke died on 22 October 2005.

Pyramids: Generalizations of hierarchical classifications in which each of the constituent classes comprises an interval of an ordering of the set of objects, and two classes may have some but not all their objects in common. [*Classification*, 2nd edition, 1999, A. D. Gordon, Chapman and Hall/CRC Press, London.]

Q

Q–Q plot: Abbreviation for **quantile–quantile plot**.

Q-technique: The class of data analysis methods that look for relationships between the individuals in a set of *multivariate data*. Includes *cluster analysis* and *multidimensional scaling*, although the term is most commonly used for a type of *factor analysis* applied to an $n \times n$ matrix of 'correlations' between individuals rather than between variables. See also **R-techniques**. [*A User's Guide to Principal Components*, 2nd edition, 2003, J. K. Jackson, Wiley, New York.]

Quadrant sampling: A sampling procedure used with *spatial data* in which sample areas (the quadrants) are taken and the number of objects or events of interest occurring in each recorded. See also **distance sampling** and **line-intersect sampling**. [*Quarterly Review of Biology*, 1992, **67**, 254.]

Quadratic discriminant function: See **discriminant analysis**.

Quadratic function: A function of the form

$$Q = \sum_{i=1}^{n} \sum_{j=1}^{n} a_{ij} x_i x_j$$

important in *multivariate analysis* and *analysis of variance*.

Quadratic loss function: See **decision theory**.

Quadrature: See **numerical integration**.

Quadrature spectrum: See **multiple time series**.

Quality-adjusted survival analysis: A methodology for evaluating the effects of treatment on survival which allows consideration of quality of life as well as quantity of life. For example, a highly toxic treatment with many side effects may delay disease recurrence and increase survival relative to a less toxic treatment. In this situation, the trade-off between the negative quality-of-life impact and the positive quantity-of-life impact of the more toxic therapy should be evaluated when determining which treatment is most likely to benefit a patient. The method proceeds by defining a *quality function* that assigns a 'score' to a patient which is a composite measure of both quality and quantity of life. In general the quality function assigns a small value to a short life with poor quality and a high value to a long life with good quality. The assigned scores are then used to calculate *quality-adjusted survival times* for analysis. [*Statistics in Medicine*, 1993, **12**, 975–88.]

Quality-adjusted survival times: The weighted sum of different time episodes making up a patient's *survival time*, with the weights reflecting the quality of life of each period. [*Statistics in Medicine*, 1993, **12**, 975–88.]

Quality assurance: Any procedure or method for collecting, processing or analysing data that is aimed at maintaining or improving the reliability or validity of the data.

Quality control angle chart: A modified version of the *cusum*, derived by representing each value taken in order by a line of some unit length in a direction proportional to its value. The line for each successive value starts at the end of the preceeding line. A change in direction on the chart indicates a change in the mean value. [*Statistics in Medicine*, 1987, **6**, 425–40.]

Quality control procedures: Statistical procedures designed to ensure that the precision and accuracy of, for example, a laboratory test, are maintained within acceptable limits. The simplest such procedure involves a chart (usually called a *control chart*) with three horizontal lines, one drawn at the target level of the relevant *control statistic*, and the others, called *action lines*, drawn at some pre-specified distance above and below the target level. The process is judged to be at an *acceptable quality level* as long as the observed control statistic lies between the two lines and to be at a *rejectable quality level* if either of these lines are crossed. [*Modern Methods of Quality Control and Improvement*, 2nd edition, 2001, H. M. Wadsworth, K. S. Stephen and A. B. Godfrey, Wiley, New York.]

Quality function: See **quality-adjusted survival analysis**.

Quangle: Synonym for **quality control angle chart**.

Quantal assay: An experiment in which groups of subjects are exposed to different doses of, usually, a drug, to which a particular number respond. Data from such assays are often analysed using the *probit transformation*, and interest generally centres on estimating the *median effective dose* or *lethal dose 50*. [*Analysis of Quantal Response Data*, 1992, B. J. T. Morgan, Chapman and Hall/CRC Press, London.]

Quantal variable: Synonym for **binary variable**.

Quantile–quantile (Q–Q) plot: See **probability plot**.

Quantile regression: An extension of classical least squares from the estimation of conditional mean models to the estimation of a variety of models for several conditional quantile functions. An example of such a model is *least absolute deviation regression*. Quantile regression is often capable of providing a more complete statistical analysis of the stochastic relationship between random variables. [*Journal of Economic Perspectives*, 2001, **15**, 143–56.]

Quantiles: Divisions of a probability distribution or frequency distribution into equal, ordered subgroups, for example, quartiles or percentiles. [SMR Chapter 7.]

Quantit model: A three-parameter non-linear *logistic regression model*. [*Biometrics*, 1977, **33**, 175–86.]

Quartiles: The values that divide a frequency distribution or probability distribution into four equal parts. [SMR Chapter 7.]

Quasi-experiment: A term used for studies that resemble experiments but are weak on some of the characteristics, particularly that allocation of subjects to groups is not under the investigator's control. For example, if interest centred on the health effects of a natural disaster, those who experience the disaster can be compared with those who do not, but subjects cannot be deliberately assigned (randomly or not) to the two groups. See also **prospective studies, experimental design** and **clinical trials**. [*Experimental and Quasi-experimental Designs for Research*, 1963, D. T. Campbell and J. C. Stanley, Houghton Mifflin, Boston.]

Quasi-independence: A form of independence for a *contingency table*, conditional on restricting attention to a particular part of the table only. For example, in the following table showing the social class of sons and their fathers, it might be of interest to assess whether, once a son has moved out of his father's class, his destination class is independent of that of his father. This would entail testing whether independence holds in the table after ignoring the entries in the main diagonal.

Father's social class		**Son's social class**	
	Upper	Middle	Lower
Upper	588	395	159
Middle	349	714	447
Lower	114	320	411

[*The Analysis of Contingency Tables*, 2nd edition, 1992, B. S. Everitt, Chapman and Hall/CRC Press, London.]

Quasi-likelihood: A function, Q, that can be used in place of a conventional *log-likelihood* when it is not possible (and/or desirable) to make a particular distributional assumption about the observations, with the consequence that there is insufficient information to construct the *likelihood* proper. The function depends on assuming some relationship between the means and variances of the random variables involved, and is given explicitly by

$$Q = \sum_{i=1}^{n} Q_i(\mu_i, y_i)$$

where $y_1, y_2, \ldots, y_n$ represent the observations which are assumed to be independent with means $\mu_1, \mu_2, \ldots, \mu_n$ and variances $\sigma^2 V_i(\mu_i)$ where $V_i(\mu_i)$ is a known function of μ_i. The components Q_i of the function Q are given by

$$Q_i = \int_{y_i}^{\mu_i} \frac{y_i - t}{\sigma^2 V_i(t)} \mathrm{d}t$$

Q behaves like a log-likelihood function for $\mu_1, \mu_2, \ldots, \mu_n$ under very mild assumptions. [GLM Chapter 8.]

Quasi-score estimating function: See **estimating functions**.

Quasi-symmetry: A term used to describe a pattern of frequencies or proportions that might be displayed by a *two-dimensional contingency table* of size 3 × 3 or larger and implies that the *odds ratios* describing the association structure are symmetric, that is for all i and j the odds ratio $(\pi_{ik}\ \pi_{kj})/((\pi_{ki}\ \pi_{jk})$ is the same for all values of k. The following is an example of a 3 × 3 table displaying quasi-symmetry

10	5	15	30
5	40	5	50
6	2	12	20
21	47	32	100

[*Biometrics*, 1979, **35**, 417–426.]

Quetelet, Adolphe (1796–1874): Born in Ghent, Belgium, Quetelet received the first doctorate of science degree awarded by the new University of Ghent in 1819 for a dissertation on the theory of conic sections. Later he studied astronomy in Paris and for much of the rest of his life his principal work was as an astronomer and a meteorologist at the Royal Observatory in Brussels. But a man of great curiosity and energy, Quetelet also built an international reputation for his work on censuses and population such as birth rates, death rates, etc. Introduced the concept of the 'average man' ('l'homme moyen') as a simple way of summarizing some characteristic of a population and devised a scheme for fitting normal distributions to grouped data that was essentially equivalent to the use of *normal probability plot.* Quetelet died on 17 February 1874 in Brussels.

Quetelet index: An index of body mass given by height divided by the square of weight. See also **body mass index**.

Queuing theory: A largely mathematical theory concerned with the study of various factors relating to queues such as the distribution of arrivals, the average time spent in the queue, etc. Used in medical investigations of waiting times for hospital beds, for example. [*Introduction to Queuing Theory*, 1990, R. B. Cooper, North Holland, New York.]

Quick and dirty methods: A term once applied to many *distribution free methods* presumably to highlight their general ease of computation and their imagined inferiority to the corresponding parametric procedure.

Quick-size: A versatile simulation-based tool for determining an exact sample size in essentially any hypothesis testing situation. It can also be modified for use in the estimation of *confidence intervals.* See also **nQuery advisor**. [*The American Statistician*, 1999, **53**, 52–5.]

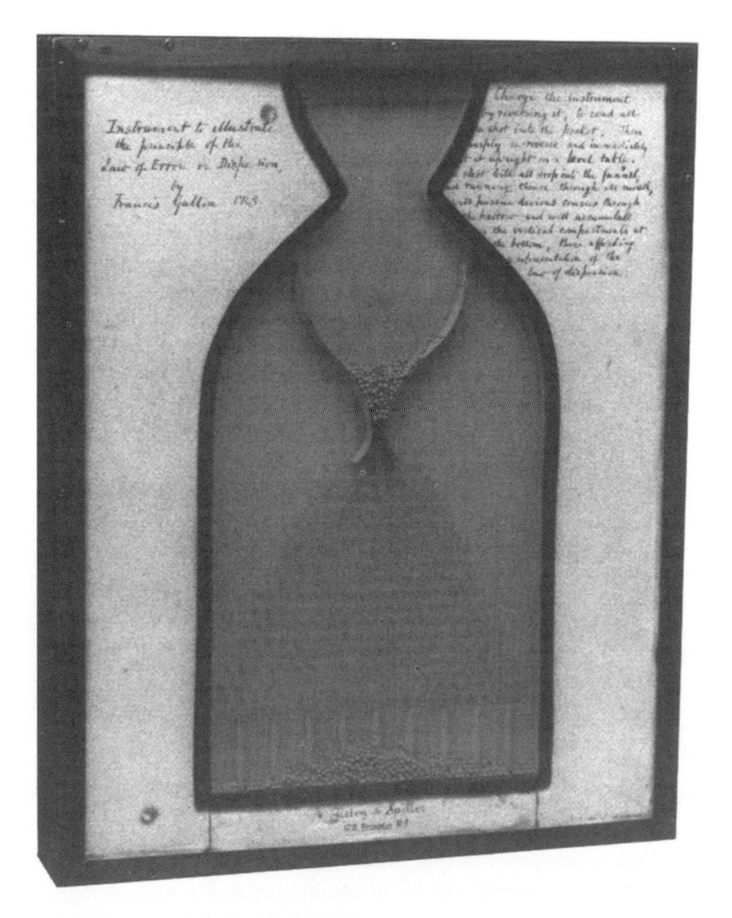

Fig. 110 The quincunx used by Galton.

Quincunx: A device used by *Galton* to illustrate his lectures, which is shown in Fig. 110. It had a glass face and a funnel at the top. Shot was passed through the funnel and cascaded through an array of pins, each shot striking one pin at each level and in principle falling left or right with equal probabilities each time a pin was hit. The shot then collected in compartments at the bottom. The resulting outline after many shot were dropped should resemble a normal curve.

Quintiles: The set of four variate values that divide a frequency distribution or a probability distribution into five equal parts.

Quitting ill effect: A problem that occurs most often in studies of smoker cessation where smokers frequently quit smoking following the onset of disease symptoms or the diagnosis of a life-threatening disease, thereby creating an anomalous rise in, for example, lung cancer risk following smoking cessation relative to continuing smokers. Such an increase has been reported in many studies. [*Tobacco Control*, 2005, **14**, 99–105.]

Quota sample: A sample in which the units are not selected at random, but in terms of a certain number of units in each of a number of categories; for example, 10 men over age 40, 25 women between ages 30 and 35, etc. Widely used in opinion polls. See also **sample survey, stratified sampling** and **random sample**. [*Elements of Sampling Theory*, 1974, V. D. Barnett, English Universities Press, London.]

R: An open-source software environment for statistical computing and graphics. The software provides a wide variety of statistical and graphical techniques and its simple and effective programming language means that it is relatively simple for users to define new functions. Full details of the software are available at http://www.r-project.org/. Many of the figures in this dictionary have been produced using **R**. [*Introductory Statistics with R*, 2002, P. Dalgaard, Springer, New York.]

Radial plot of odds ratios: A diagram used to display the *odds ratios* calculated from a number of different *clinical trials* of the same treatment(s). Often useful in a *meta-analysis* of the results. The diagram consists of a plot of $y = \hat{\Delta}/SE(\hat{\Delta})$ against $x = 1/SE(\hat{\Delta})$ where $\hat{\Delta}$ is the logarithm of the odds ratio from a particular study. An example of such a plot is shown in Fig. 111. Some features of this display are:

- Every estimate has unit standard error in the y direction. The y scale is drawn as a ± 2 error bar to indicate this explicitly.
- The numerical value of an odds ratio can be read off by extrapolating a line from (0,0) through (x, y) to the circular scale drawn, the horizontal line corresponds to an odds ratio of one. Also approximate 95% confidence limits can be read off by extrapolating lines from (0,0) through $(x, y + 2)$, $(x, y - 2)$ respectively.
- Points with large $SE(\hat{\Delta})$ fall near the origin, while points with small $SE(\hat{\Delta})$, that is the more informative estimates, fall well away from the origin and naturally look more informative.

[*Statistics in Medicine*, 1988, **7**, 889–94.]

Radical statistics group: A national network of social scientists in the United Kingdom committed to the critique of statistics as used in the policy making process. The group is committed to building the competence of critical citizens in areas such as health and education. [http://www.radstats.org.uk/]

Raking adjustments: An alternative to *poststratification adjustments* in complex surveys that ensures that the adjusted weights of the respondents conform to each of the marginal distributions of a number of auxiliary variables that are available. The procedure involves an iterative adjustment of the weights using an *iterative proportional-fitting algorithm*. The method is widely used in health and other surveys when many population control totals are available. For example, raking was used in the 1991 General Social Survey in Canada, a *random digit dialling sampling* telephone survey that concentrated on health issues. In the survey, province, age and sex control totals were used. [*International Statistical Review*, 1994, **62**, 333–47.]

Ramsey, Frank Plumpton (1903–1930): Born in Cambridge, England, Ramsey was educated at Trinity College, Cambridge. A close friend of the Austrian philosopher Ludwig Wittgenstein, whose *Tractatus* he translated into English while still an undergraduate. Ramsey worked on the foundations of probability and died at the tragically early age of 26, from the effects of jaundice.

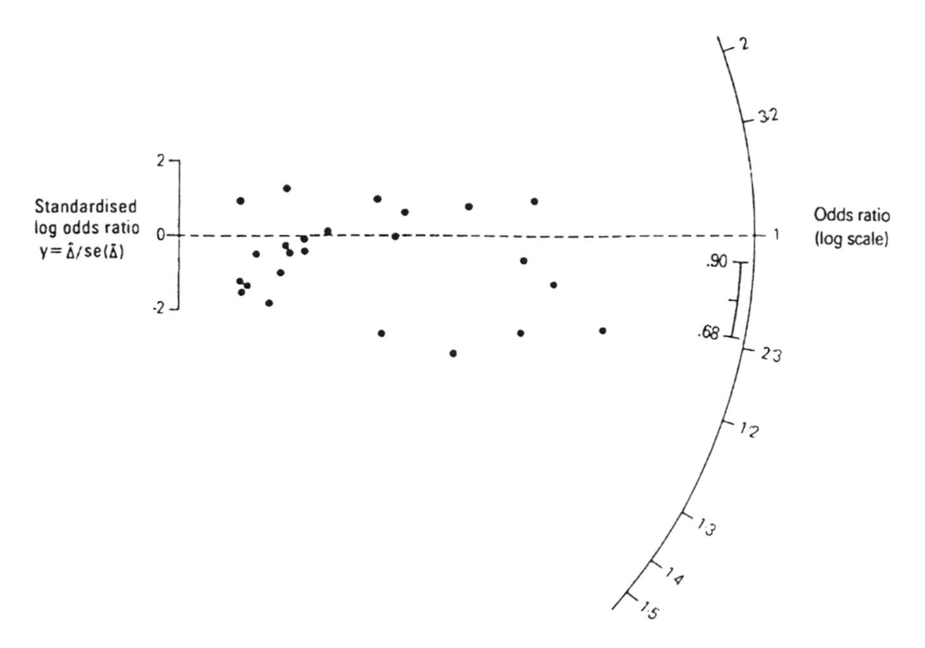

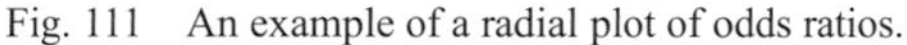
Fig. 111 An example of a radial plot of odds ratios.

Ramsey regression equation specification error test: A general specification test for linear regression models. Based on the idea that adding squared, cubic, quartic etc. components of the predicted responses as covariates should not increase the coefficient of determination if the model is correctly specified. [*Journal of the Royal Statistical Society, Series B*, 1969, **31**, 350–371.]

Rand coefficient: An index for assessing the similarity of alternative classifications of the same set of observations. Used most often when comparing the solutions from two methods of *cluster analysis* applied to the same set of data. The coefficient is given by

$$R = \left[T - 0.5P - 0.5Q + \binom{n}{2}\right] \Big/ \binom{n}{2}$$

where

$$T = \sum_{i=1}^{g} \sum_{j=1}^{g} m_{ij}^2 - n$$

$$P = \sum_{i=1}^{g} m_{i.}^2 - n$$

$$Q = \sum_{j=1}^{g} m_{.j}^2 - n$$

The quantity m_{ij} is the number of individuals in common between the ith cluster of the first solution, and the jth cluster of the second (the clusters in the two solutions may each be labelled arbitrarily from 1 to g, where g is the number of clusters). The terms $m_{i\cdot}$ and $m_{\cdot j}$ are appropriate marginal totals of the matrix of m_{ij} values and n is the number of observations. The coefficient lies in the interval (0,1) and takes its upper limit only when there is complete agreement between the two classifications. [MV2 Chapter 10.]

Random: Governed by chance; not completely determined by other factors. Non-deterministic.

Random allocation: A method for forming treatment and control groups particularly in the context of a *clinical trial*. Subjects receive the active treatment or placebo on the basis of the outcome of a chance event, for example, tossing a coin. The method provides an impartial procedure for allocation of treatments to individuals, free from personal biases, and ensures a firm footing for the application of significance tests and most of the rest of the statistical methodology likely to be used. Additionally the method distributes the effects of *concomitant variables*, both observed and unobserved, in a statistically acceptable fashion. See also **block randomization, minimization** and **biased coin method**. [SMR Chapter 5.]

Random coefficient models: See **multilevel models** and **growth models**.

Random digit dialling sampling: See **telephone interview surveys**.

Random dissimilarity matrix model: A model for lack of clustering structure in a *dissimilarity matrix*. The model assumes that the elements of the lower triangle of the dissimilarity matrix are ranked in random order, all $n(n-1)/2!$ rankings being equally likely. [*Classification*, 2nd edition, 1999, A. D. Gordon, Chapman and Hall/CRC Press, London.]

Random effects: The effects attributable to a (usually) infinite set of levels of a factor, of which only a random sample occur in the data. For example, the investigator may want to accommodate effects of subjects in a *longitudinal study* by introducing subject-specific intercepts that are viewed as a random sample from a distribution of effects. See also **fixed effects**.

Random effects model: See **multilevel models**.

Random events: Events which do not have deterministic regularity (observations of them do not necessarily yield the same outcome) but do possess some degree of statistical regularity (indicated by the statistical stability of their frequency).

Random forests: An ensemble of classification or regression trees (see *classification and regression tree technique*) that have been fitted to the same n observations, but with random weights obtained by use of the *bootstrap*. Additional randomness is supplied by selecting only a small fraction of covariates for split point determination in each inner node of these trees. Final predictions are then obtained by averaging the predictions obtained from each tree in the forest. Empirical and theoretical investigations have shown that such an aggregation over multiple tree-structured models helps to improve upon the prediction accuracy of single trees. [*Machine Learning*, 2001, **45**, 5–32.]

Random matrix theory: The study of stochastic linear algebra where the equations themselves are random. Attracting interest in *regularization* of high-dimensional problems in statistics. [*Random Matrix Theory and its Applications*, 2009, Z.Bai, Y.Chen and Y.-C. Liang, Eds., World Scientific, Singapore.]

Randomization tests: Procedures for determining statistical significance directly from data without recourse to some particular *sampling distribution*. For example, in a study involving the comparison of two groups, the data would be divided (permuted) repeatedly between groups and for each division (permutation) the relevant test statistic (for example, a t or F), is calculated to determine the proportion of the data permutations that provide as large a test statistic as that associated with the observed data. If that proportion is smaller than some significance level α, the results are significant at the α level. [*Randomization Tests*, 1986, E. S. Edington, Marcel Dekker, New York.]

Randomized block design: An experimental design in which the treatments in each block are assigned to the experimental units in random order.

Randomized clinical trial: See **clinical trials**.

Randomized consent design: A design originally introduced to overcome some of the perceived ethical problems facing clinicians entering patients in *clinical trials* involving random allocation. After the patient's eligibility is established the patient is randomized to one of two treatments A and B. Patients randomized to A are approached for patient consent. They are asked if they are willing to receive therapy A for their illness. All potential risks, benefits and treatment options are discussed. If the patient agrees, treatment A is given. If not, the patient receives treatment B or some other alternative treatment. Those patients randomly assigned to group B are similarly asked about treatment B, and transferred to an alternative treatment if consent is not given. See also **Zelen's single-consent design**. [*Contemperary Clinical Trials*, 2006, **27**, 320–332.]

Randomized encouragement trial: *Clinical trials* in which participants are encouraged to change their behaviour in a particular way (or not, if they are allocated to the control condition) but there is little expectation on the behalf of the trial investigators that participants will fully comply with the request or that there will not be a substantial minority who change their behaviour in the required way without actually being asked (by the trial investigators, at least) to do so. An example involves randomizing expectant mothers (who are also cigarette smokers) to receive advice to cut down on or completely stop smoking during pregnancy. Many of the mothers will fail to reduce their smoking even though they have received encouragement to do so. On the other hand, many will cut down whether or not they have received the advice suggesting that they should. Randomized encouragement to take part in cancer screening programmes is another example. Again, adherence to the offer of screening will be far from complete, but, on the other hand, there will always be people who will ask for the examination in any case. See also **Zelen single-consent design**. [*Biometrics*, 1989, **4**, 619–622.]

Randomized response technique: A procedure for collecting information on sensitive issues by means of a survey, in which an element of chance is introduced as to what question a respondent has to answer. In a survey about abortion, for example, a woman might be posed both the questions 'have you had an abortion' and 'have you never had an abortion', and instructed to respond to one or the other depending on the outcome of a randomizing device under her control. The response is now not revealing since no one except the respondent is aware of which question has been answered. Nevertheless the data obtained can be used to estimate quantities of interest, here, for example, the proportion of women who have had an abortion (π); if the probability of selecting the question 'have you had an abortion', P, is known and is not equal to 0.5. If y is the proportion of 'yes' answers to this question in a random sample of n respondents, one estimator of π is given by

$$\hat{\pi} = \frac{y - (1 - P)}{2P - 1}$$

This estimator is unbiased and has variance, V

$$V = \frac{\pi(1 - \pi)}{n} + \frac{P(1 - P)}{n(2P - 1)}$$

[*Randomized Response: Theory and Techniques*, 1988, A. Chaudhuri and R. Mukerjee, Marcel Dekker, New York.]

Randomly-stopped sum: The sum of a variable number of random variables.

Random model: A model containing random or probabilistic elements. See also **deterministic model**.

Random number: See **pseudorandom numbers**.

Random sample: Either a set of n independent and identically distributed random variables, or a sample of n individuals selected from a population in such a way that each sample of the same size is equally likely.

Random variable: A variable, the values of which occur according to some specified probability distribution. [KA1 Chapter 1.]

Random variation: The variation in a data set unexplained by identifiable sources.

Random walk: The motion of a 'particle' that moves in discrete jumps with certain probabilities from point to point. At its simplest, the particle would start at the point $x = k$ on the x axis at time $t = 0$ and at each subsequent time, $t = 1, 2, \ldots$, it moves one step to the right or left with probabilities p and $1 - p$, respectively. As a concrete example, the position of the particle might represent the size of a population of individuals and a step to the left corresponds to a death and to the right a birth. Here the process would stop if the particle ever reached the origin, $x = 0$, which is consequently termed an *absorbing barrier*. See also **birth–death process, Markov chain** and **Harris walk**. [TMS Chapter 1.]

Range: The difference between the largest and smallest observations in a data set. Often used as an easy-to-calculate measure of the dispersion in a set of observations but not recommended for this task because of its sensitivity to *outliers* and the fact that its value increases with sample size. [SMR Chapter 2.]

Rank adjacency statistic: A statistic used to summarize *autocorrelations* of *spatial data*. Given by

$$D = \sum\sum w_{ij}|y_i - y_j| \Big/ \sum\sum w_{ij}$$

where $y_i = \text{rank}(x_i)$, x_i is the data value for location i, and the w_{ij} are a set of weights representing some function of distance or contact between regions i and j. The simplest weighting option is to define

$$\begin{aligned} w_{ij} &= 1 \quad \text{if regions } i \text{ and } j \text{ are adjacent} \\ &= 0 \quad \text{otherwise} \end{aligned}$$

in which case D becomes the average absolute rank difference over all pairs of adjacent regions.The theoretical distribution of D is not known but spatial clustering (or positive *spatial autocorrelation*) in the data will be reflected by the tendency for adjacent data values to have similar ranks, so that the value of D will tend to be smaller than otherwise. See also **Moran's** *I*. [*Statistics in Medicine*, 1993, **12**, 1885–94.]

Rank correlation coefficients: Correlation coefficients that depend only on the ranks of the variables not on their observed values. Examples include *Kendall's tau statistics* and *Spearman's rho*. [SMR Chapter 11.]

Ranking: The process of sorting a set of variable values into either ascending or descending order. [SMR Chapter 2.]

Rank of a matrix: The number of linearly independent rows or columns of a matrix.

Rank order statistics: Statistics based only on the ranks of the sample observations, for example *Kendall's tau statistics*.

Rank regression: A method of regression analysis in which the ranks of the dependent variable are regressed on a set of explanatory variables. Since the ranks are unaffected by strictly increasing transformations, this can be formally specified by the model

$$g(y) = \mathbf{x}'\boldsymbol{\beta} + \epsilon$$

where g is any strictly increasing transformation of the dependent variable y, $\mathbf{x}$ is a vector of explanatory variables and $\boldsymbol{\beta}$ is a vector of regression coefficients. [*Journal of Machine Learning*, 2007, **8**, 2727–2754.]

Ranks: The relative positions of the members of a sample with respect to some characteristic. [SMR Chapter 2.]

Rank transform method: A method consisting of replacing the observations by their ranks in the combined sample and performing one of the standard *analysis of variance* procedures on the ranks. [*Journal of Statistical Computation and Simulation*, 1986, **23**, 231–40.]

Rao-Blackwell theorem: A theorem stating that for an unbiased estimator $f(x)$ of a parameter θ, the conditional expectation of $f(x)$, given a *sufficient statistic* $T(X)$, is typically a better estimator of θ and never worse. Transforming an estimator using the Rao-Blackwell theorem is called *Rao-Blackwellization*. [*Theory of Point Estimation*, 1998, E. L. Lehmann and G. Casella, Springer, New York.]

Rao–Hartley–Cochran method (RHC): A simple method of sampling with unequal probabilities without replacement. The method is as follows:

- Randomly partition the population of N units into n groups $\{G_g\}_{g=1}^{n}$ of sizes $\{N_g\}_{g=1}^{n}$.
- Draw one unit from each group with probability Z_j/Z_g for the gth group, where $Z_j = x_j/X$, $Z_g = \sum_{j\in G_g} Z_j$, x_j = some measure of the size of the jth unit and $X = \sum_{j=1}^{N} x_j$.

An unbiased estimator of $\bar{Y}$, the population mean, is

$$\hat{\bar{Y}} = \frac{1}{n}\sum_{y=1}^{n} w_g y_g$$

where $w_g = f/\pi_g$, $\pi_g = z_g/Z_g$, is the probability of selection of the gth sampled unit, $f = n/N$ is the probability of selection under simple random sampling without replacement, and $y_g = z_g$ denote the values for the unit selected from the gth group. Note that by definition, $\sum_{g=1}^{n} Z_g = 1$. An unbiased estimator of $\text{var}(\hat{\bar{Y}})$ is

$$\text{var}(\hat{\bar{Y}}) = \frac{\sum_{g=1}^{n} N_g^2 - N}{N^2 - \sum_{g=1}^{n} N_g^2}\sum_{g=1}^{n} Z_g\left(\frac{y_g}{N z_g} - \hat{\bar{Y}}^2\right)$$

[*Sampling Techniques*, 3rd edition, 1977, W. G. Cochran, Wiley, New York.]

Rasch, Georg (1901–1980): Born in Odense, Denmark Rasch received a degree in mathematics from the University of Copenhagen in 1925, and then became Doctor of Science in 1930. In the same year he visited University College in London to work with *Fisher*. On returning to Denmark he founded the Biostatistics Department of the State Serum Institute and was the Director from 1940 to 1956. Developed *latent trait models* including the one named after him. He died on 19 October 1980, in Byrum, Denmark.

Rasch model: A model often used in ability testing. The model proposes that the probability that an examinee i with ability quantified by a parameter, δ_i, answers item j correctly ($X_{ij} = 1$) is given by

$$P(X_{ij} = 1|\delta_i, \epsilon_j) = \delta_i\epsilon_j/(1 + \delta_i\epsilon_j)$$

where ϵ_j is a measure of the simplicity of the item. This probability is commonly rewritten in the form

$$P(X_{ij} = 1|\theta_i, \sigma_j) = \frac{\exp(\theta_i - \sigma_j)}{1 + \exp(\theta_i - \sigma_j)}$$

where $\exp(\theta_i) = \delta_i$ and $\exp(-\sigma_j) = \epsilon_j$. In this version, θ_i is still called the *ability parameter* but σ_j is now termed the *item difficulty*. [*Rasch Models: Foundations, Recent Developments and Applications*, 1995, G. H. Fischer and I. W. Molenaar, editors, Springer, New York.]

Ratchet scan statistic: A statistic used in investigations attempting to detect a relatively sharp increase in disease incidence for a season superimposed on a constant incidence over the entire year. If $n_1, n_2, \ldots, n_{12}$ are the total number of observations in January, February,. . ., December, respectively (months are assumed to be of equal lengths), then the sum of k consecutive months starting with month i can be written as

$$S_i^k = \sum_{j=1}^{(i+k-1)\bmod 12} n_j$$

The ratchet scan statistic is based on the maximum number falling within k consecutive months and is defined as

$$T^k = \max_{i=1,\ldots,12} S_i^k$$

Under the hypothesis of a *uniform distribution* of events, i.e. the expected value of n_i is $N/12$ for all $i = 1, 2, \ldots, 12$, where N=the total number of events in the year, it can be shown that

$$P(T^k \geq n) = \sum \frac{N!}{12^N \prod_{i=1}^{12} n_i!}$$

where the sum is taken over all the values of $(n_1, n_2, \ldots, n_{12})$ that yield $T^k \geq n$. Approximations for the distribution for $k = 1$ have been derived but not for $k > 1$. [*Biometrics*, 1992, **48**, 1177–85.]

Rate: A measure of the frequency per unit time of some phenomenon of interest given by

$$\text{Rate} = \frac{\text{number of events in specified period}}{\text{average population during the period}}$$

(The resulting value is often multiplied by some power of ten to convert it to a whole number.) See also **incidence**.

Ratio estimator: Suppose that the targets of inference are the population total Y and the population mean $\bar{Y}$ for a finite population with N units and assume that we have a random sample of y_i for n units. The conventional simple random sampling estimators are

$$\hat{\bar{Y}} = \frac{1}{n} \sum_{i=1}^{n} y_i$$

for the population mean and

$$\hat{Y} = \frac{N}{n} \sum_{i=1}^{n} y_i$$

for the population total. Further, assume that a variable x_i that is correlated with y_i is known for all N population units, so the population total is

$$X = \sum_{j=1}^{N} x_j$$

and the population mean is

$$\bar{X} = \frac{1}{N}\sum_{j=1}^{N} x_j$$

Using this knowledge, we can improve on the estimators $\hat{\bar{Y}}$and $\hat{Y}$ by using the ratio estimators

$$\hat{\bar{Y}}_R = \bar{X}\frac{\bar{y}}{\bar{x}}$$

for the population mean and

$$\hat{Y}_R = X\frac{\bar{y}}{\bar{x}}$$

for the population total. The rationale of the ratio estimators is to exploit correlation between x_i and y_i to increase precision compared to conventional simple random sampling estimators. [*Sampling Techniques*, 3rd edition, 1977, W. G. Cochran, Wiley, New York.]

Ratio variable: A continuous variable that has a fixed rather than an arbitrary zero point. Examples are, height, weight and temperature measured in degrees Kelvin. See also **categorical variable** and **ordinal variable**.

Rayleigh distribution: A form of *noncentral chi-squared distribution* encountered in mathematical physics. The distribution is given specifically by

$$f(x) = \left(\frac{x}{\beta}\right)e^{-x^2(2\beta^2)} \qquad 0<x<\infty \qquad \beta>0$$

The mean is $\beta(\pi/2)^{\frac{1}{2}}$ and the variance is $(2 - \pi/2)\beta^2$. [STD Chapter 34.]

Rayleigh quotient: A function that arises in the conditional maximization of a quadratic form $\mathbf{y}'\,\mathbf{A}\mathbf{y}$ subject to the constraint $\mathbf{y}'\,\mathbf{y} = c$ where c is a scalar and given by $\mathbf{x}'\,\mathbf{A}\mathbf{x}/\mathbf{x}'\mathbf{x}$ where $\mathbf{A}$ is an n x n matrix and $\mathbf{x}$ is an n x1 vector. [*Journal of Sound and Vibration*, 1997, **199**, 155–164.]

Rayleigh's test: A test of the null hypothesis that *circular random variable*, θ, has a *uniform distribution* of the form

$$f(\theta) = \frac{1}{2\pi} \qquad 0 \le \pi \le 2\pi$$

The test is based on a random sample of observed values $\theta_1, \theta_2, \ldots, \theta_n$, and the *test statistic* is

$$r^2 = \frac{1}{n}\left\{\left[\sum_{i=1}^{n}\cos\theta_i\right]^2 + \left[\sum_{i=1}^{n}\sin\theta_i\right]^2\right\}$$

Critical values of the test statistic are available. An example of the application of this test is to determine if the arrangement of bird nest entrances for a particular type of bird in a particular area differs significantly from random. [MV2 Chapter 14.]

RBD: Abbreviation for **randomized block design**.

RCT: Abbreviation for **randomized clinical trial**.

RDS: Abbreviation for **respondent-driven sampling**.

Recall bias: A possible source of bias, particularly in a *retrospective study*, caused by differential recall among cases and controls, in general by underreporting of exposure in the control group. See also **ascertainment bias**. [SMR Chapter 5.]

Receiver operating characteristic (ROC) curve: A plot of the *sensitivity* of a *diagnostic test* against one minus its *specificity*, as the cut-off criterion for indicating a positive test is varied. Often used in choosing between competing tests, although the procedure takes no account of the prevalence of the disease being tested for. As an example consider the following ratings from 1 (definitely normal) to 5 (definitely diseased) arising from 50 normal and 50 diseased subjects.

	1	2	3	4	5	**Total**
Normal	4	17	20	8	1	50
Diseased	3	3	17	19	8	50

If the rating of 5 is used as the cut-off for identifying diseased cases, then the sensitivity is estimated as 8/50=0.16, and the specificity as 49/50=0.98. Now using the rating 4 as cut-off leads to a sensitivity of 27/50=0.54 and a specificity of 41/50=0.82. The values of (sensitivity, 1− specificity) as the cut-off decreases from 5 to 1 are (0.16,0.02), (0.54,0.18), (0.88,0.58), (0.94,0.92), (1.00,1.00). These points are plotted in Fig. 112. This is the required receiver operating characteristic curve. [SMR Chapter 14.]

Reciprocal causation: See **recursive models**.

Reciprocal model: The regression model defined by $E(y|x) = (\alpha + \beta x)^{-1}$ [*Biometrics*, 1979, **35**, 41–54.]

Reciprocal transformation: A transformation of the form $y = 1/x$, which is particularly useful for certain types of variables. Resistances, for example, become conductances, and times become rates. In some cases the transformation can lead to linear relationships, for example,

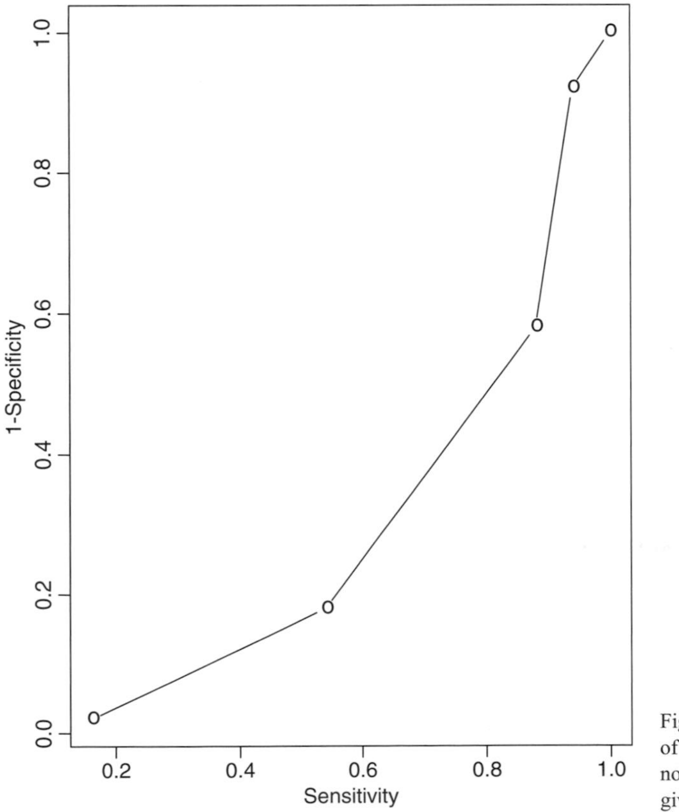

Fig. 112 An example of an ROC curve for the normal/diseased ratings given.

airways resistance against lung volume is non-linear but airways conductance against lung volume is linear. [SMR Chapter 7.]

Record linkage: The establishment of which records in a database relate to the same individual. A straightforward task if there is some form of unique identifier for each individual, but when such identifiers are not available or not feasible, probabilistic matching may be needed. [*Handbook of Record Linkage*, 1988, H. B. Newcombe, Oxford University Press, New York.]

Rectangular distribution: Synonym for **uniform distribution**.

Recurrence risk: Usually the probability that an individual experiences an event of interest given previous experience(s) of the event; for example, the probability of recurrence of breast cancer in a woman who has previously had the disease. In medical genetics, however, the term means the chance that a genetic disease present in the family will recur in that family and affect another person (or persons).

Recurrent event data: Data in which recurrent events, for example, tumours, are ordered and each event of interest is treated as being of the same nature. Other examples of such data are bugs in software and the repeated fixing of aircraft. A variety of models have been proposed for analysing such data. [*Technometrics*, 2007, **49**, 210–220.]

Recursive: A function or sequence is defined recursively if

- the value of $f(0)$ and
- the value of $f(n+1)$ given the value of $f(n)$ are both stated.

For example, the factorial function may be defined by

- $f(0) = 1$,
- $f(n+1) = nf(n)$ for $n = 0, 1, 2, \ldots$.

Recursive models: Statistical models in which causality flows in one direction, i.e. models that include only unidirectional effects. Such models do not include circular effects or *reciprocal causation*, for example, variable A influences variable B, which in turn influences variable A; nor do they permit feedback loops in which, say, variable A influences variable B, which influences variable C, which loops back to influence variable A. [*Structural Equations with Latent Variables*, 1989, K. A. Bollen, Wiley, New York.]

Recursive partitioning regression: Synonym for **classification and regression tree technique.**

Reduction of dimensionality: A generic term for the aim of methods of analysis which attempt to simplify complex *multivariate data* to aid understanding and interpretation. Examples include *principal components analysis* and *multidimensional scaling*.

Reed–Muench estimator: An estimator of the *median effective dose* in *bioassay*. See also **Dragstedt–Behrens estimator**. [*Probit Analysis*, 1971, 3rd edition, D. J. Finney, Cambridge University Press, Cambridge.]

Reference interval: A range of values for a variable that encompasses the values obtained from the majority of normal subjects. Generally calculated as the interval between two predetermined centiles (such as the fifth and the 95th) of the distribution of the variable of interest. Often used as the basis for assessing the results of *diagnostic tests* in the classification of individuals as normal or abnormal with respect to a particular variable, hence the commonly used, but potentially misleading phrase 'normal range'. [SMR Chapter 14.]

Reference population: The standard against which a population that is being studied can be compared.

Reference priors: See **prior distributions**.

Regenerative cycles: See **regenerative process.**

Regenerative process: A term for a *stochastic process* for which there exist random time points $0 = T_0 < T_1 < T_2 < \cdots$ such that at each time point, the process 'restarts' probabilistically. Formally this implies that the process $\{L_t\}$ can be split into independent identically distributed (iid) probabilistic replicas, called *regenerative cycles*, of iid random lengths $\tau_i = T_i - T_{i-1}, i = 1, 2, \ldots$. [*Regenerative Inventory Systems*, 1989, I. Sahin, Springer Verlag, New York.]

Regression analysis: Essentially a synonym for **regression modelling**.

Regression calibration: A procedure of correcting the results from a regression analysis to allow for measurement errors in the explanatory variables. See also **attenuation**. [*American Journal of Epidemiology*, 2001, **154**, 836–44.]

Regression coefficient: See **multiple regression**.

Regression diagnostics: Procedures designed to investigate the assumptions underlying particular forms of regression analysis, for example, normality, homogeneity of variance, etc., or to examine the influence of particular data points or small groups of data points on the estimated regression coefficients. See also **residuals**, **Cook's distance, COVRATIO, DFBETA, DFFIT** and **influence statistics**. [ARA Chapter 10.]

Regression dilution: The term applied when a covariate in a model cannot be measured directly and instead a related observed value must be used in the analysis. In general, if the model is correctly specified in terms of the 'true' covariate, then a similar form of the model with a simple error structure will not hold for the observed values. In such cases, ignoring the measured values will lead to biased estimates of the parameters in the model. Often also referred to as the *errors in variables problem*. See also **attenuation, latent variables** and **structural equation models**. [ARA Chapter 9.]

Regression discontinuity design: A *quasi-experimental design* in which participants in, for example, an intervention study, are assigned to treatment and control groups on the basis of a cutoff value on a pre-intervention measure that affects the outcome, rather than by randomization. If the treatment has an effect a discontinuity in the outcomes would be expected at the cutoff. A weakness is that extrapolation of counterfactual outcomes for the treated in absence of treatment is required, based on a regression model estimated for the non-treated. See Figure 113 for an illustration where treatment is given to those below a cutoff on a pretest. [*Experimental and Quasi-Experimental Designs for Generalized Causal Inference*, 2002, W. R. Shadish, T. D. Cook and D. T. Campbell, Houghton-Mifflin, Boston.]

Regression modelling: A frequently applied statistical technique that serves as a basis for studying and characterizing a system of interest, by formulating a reasonable mathematical model of the relationship between a response variable, y and a set of q explanatory variables, $x_1, x_2, \ldots, x_q$. The choice of the explicit form of the model may be based on previous knowledge of the system or on considerations such as 'smoothness' and continuity of y as a function of the x variables. In very general terms all such models can be considered to be of the form.

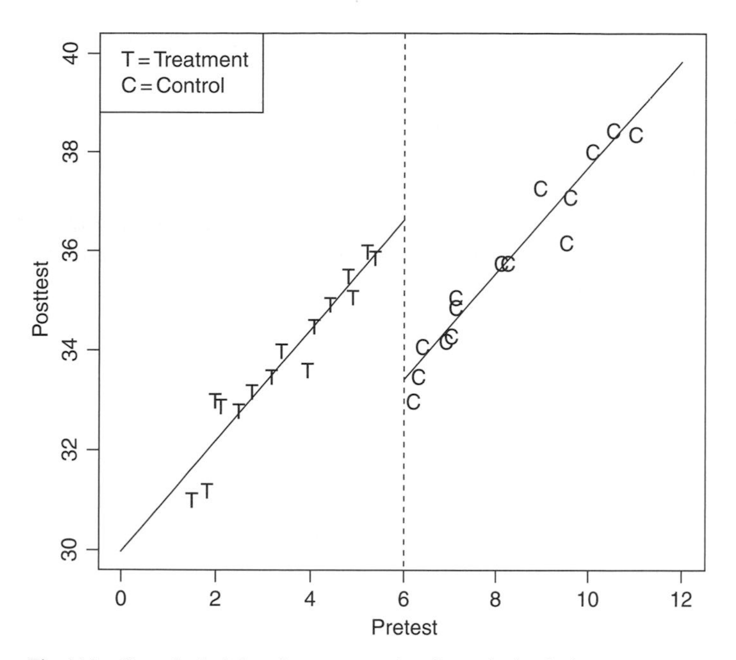

Fig. 113 Hypothetical data from a regression discontinuity design.

$$y = f(x_1, \ldots, x_q) + \epsilon$$

where the function f reflects the true but unknown relationship between y and the explanatory variables. The random additive error ϵ which is assumed to have mean zero and variance σ_ϵ^2 reflects the dependence of y on quantities other than $x_1, \ldots, x_q$. The goal is to formulate a function $\hat{f}(x_1, x_2, \ldots, x_p)$ that is a reasonable approximation of f. If the correct parametric form of f is known, then methods such as *least squares estimation* or *maximum likelihood estimation* can be used to estimate the set of unknown coefficients. If f is linear in the parameters, for example, then the model is that of *multiple regression*. If the experimenter is unwilling to assume a particular parametric form for f then *nonparametric regression modelling* can be used, for example, *kernel regression smoothing, recursive partitioning regression* or *multivariate adaptive regression splines*. [*Regression Modeling Strategies*, 2001, F. E. Harrell, Jr., Springer, New York.]

Regression through the origin: In some situations a relationship between two variables estimated by regression analysis is expected to pass through the origin because the true mean of the dependent variable is known to be zero when the value of the explanatory variable is zero. In such situations the linear regression model is forced to pass through the origin by setting the intercept parameter to zero and estimating only the slope parameter. [ARA Chapter 1.]

Regression to the mean: The process first noted by *Sir Francis Galton* that 'each peculiarity in man is shared by his kinsmen, but on the average to a less degree.' Hence the tendency, for example, for tall parents to produce tall offspring but who, on average, are shorter than their parents. The term is now generally used to label the phenomenon that a variable that is extreme on its first measurement will tend to be closer to the centre of the distribution for a later measurement. For example, in a screening programme for hypertension, only persons with high blood pressure are asked to return for a second measure. On the average, the second measure taken will be less than the first. [*Statistical Methods in Medical Research*, 1997, **6**, 103–14.]

Regression weight: Synonym for **regression coefficient**.

Regularization: Approaches designed to prevent *overfitting* of statistical models, such as *ridge regression*, the *lasso* and the L^2-norm in *support vector machines*. The approach is also used to refer to model selection procedures which penalize models according to their number of parameters, such as *Akaike's information criterion* (AIC) and the *Bayesian information criterion* (BIC). [*SIAM Review*, 1998, **40**, 636–666.]

Regularized discriminant analysis: A method of *discriminant analysis* for small, high-dimensional data sets, which is designed to overcome the degradation in performance of the *quadratic discriminant function* when there are large numbers of variables. [*Journal of the American Statistical Association*, 1989, **84**, 165–75.]

Reification: The process of naming *latent variables* and the consequent discussion of such things as quality of life and racial prejudice as though they were physical quantities in the same sense as, for example, are length and weight. [MV1 Chapter 1.]

Reinsch spline: The function, g, minimizing $\sum_i \{Y_i - g(x_i)\}^2$ over g such that

$$\int_0^1 \{g''\}^2 \mathrm{d}t \leq C$$

where C is a smoothing parameter and Y_i are observations at design points $x_i, 0 \leq x_1 < x_2 < \cdots < x_n \leq 1$ and

$$Y_i = g(x_i) + \epsilon_i$$

with the errors ϵ_i being independent and normally distributed with zero mean and variance σ^2. See also **spline function**. [*Nonparametric Regression and Generalized Linear Models*, 1994, P. J. Green and B. W. Silverman, Chapman and Hall/CRC Press, London.]

Reinterviewing: A second interview for a sample of survey respondents in which the questions of the original interview (or a subset of them) are repeated. The same methods of questioning need not be used on the second occasion. The technique can be used to estimate components in survey error models and to evaluate field work.

Rejectable quality level: See **quality control procedures**.

Relative efficiency: The ratio of the variances of two possible estimates of a parameter or the ratio of the sample sizes required by two statistical procedures to achieve the same *power*. See also **Pitman nearness criterion.**

Relative frequency: See **probability**.

Relative poverty statistics: Statistics on the properties of populations falling below given fractions of average income that play a central role in the discussion of poverty. The proportion below half national median income, for example, has been used as the basis of cross-country comparisons.

Relative risk: A measure of the association between exposure to a particular factor and risk or probability of a certain outcome, calculated as

$$\text{relative risk} = \frac{\text{risk among exposed}}{\text{risk among nonexposed}}$$

Thus a relative risk of 5, for example, means that an exposed person is 5 times as likely to have the disease than one who is not exposed. Relative risk does **not** measure the probability that someone with the factor will develop the disease. The disease may be rare among both the nonexposed and the exposed. See also **incidence**, and **attributable risk**. [SMR Chapter 10.]

Relative survival: The ratio of the observed survival of a given group of patients to the survival the group would have experienced based on the *life table* of the population from which they were diagnosed. [*Statistics is in Medicine*, 2004, **23**, 51–64.]

Reliability: The extent to which repeated measurements on units (for instance people) yield similar results. See also **intraclass correlation coefficient** and **kappa coefficient**. [*Journal of Applied Psychology*, 1993, **78**, 98–104.]

Reliability data: A term generally applied to various types of data, for example, overall equipment failure rate, common causes failure rates, fail-to-start probabilities etc., collected about some type of engineering system. [*The Reliability Data Handbook*, 2005, T. R. Moss, Wiley, New York.]

Reliability theory: A theory which attempts to determine the reliability of a complex system from knowledge of the reliabilities of its components. Interest may centre on either the lifetime or failure-free operating time of a system or piece of equipment or on broader aspects of system's performance over time. In the former situation survival analysis techniques are generally of most importance; in the latter case Markov processes are often used to model system performance. [*Statistical Theory of Reliability and Life Testing: Probability Models*, 1975, R. E. Barlow and F. Proschan, Rinehart and Winston, New York.]

Relplot: Synonymous with **bivariate boxplot.**

Remedian: A *robust estimator* of location that is computed by an iterative process. Assuming that the sample size n can be written as b^k where b and k are integers, the statistic is calculated by computing medians of groups of b observations yielding b^{k-1} estimates on which the process is iterated and so on until only a single estimate remains. [*Journal of the American Statistical Association*, 1990, **85**, 97–104.]

Remington, Richard (1931–1992): Born in Nampa, Idaho, USA, Remington received both B. A. and M. A. degrees in mathematics from the University of Montana and then in 1958 a Ph.D in biostatistics from the University of Michigan. He then joined the Department of Statistics at the University of Michigan School of Public Health becoming professor in 1965. In 1969 Remington moved to the University of Texas becoming Associate Dean for Research, Professor and Head of Biometry. In 1975 he returned to the University of Michigan; later he moved to the University of Iowa. For much of his distinguished career, Remington was involved with inquiries into the public health impact of hypertension and in 1988 was the main author of the Institute of Medicine's report, *The Future of Public Health.* He died on July 26th, 1992 in Iowa City, Iowa, USA.

REML: Acronym for **residual maximum likelihood estimation** and **restricted maximum likelihood estimation**.

Removal method: A method whereby the total number of animals in an enclosed area is estimated from the numbers in successive 'removals' (i.e. captures without replacement). The assumptions behind the removal method are as follows:

(a) There is no immigration into or emigration from the enclosed area while the removals are taking place.
(b) Each animal has an equal probability, p, of being caught.
(c) Each removal is equally efficient, i.e. the probability p is constant from one removal to the rest.

If the above assumptions are valid, then the probability of catching $c_1, \ldots, c_k$ animals in k removals, given a total of N animals, can be written as $f(c, \ldots, c_k|N, p)$ where

$$f(c,\ldots,c_k|N,p) = \frac{N!}{\prod(c_i)!(N-T)!}p^T(1-p)^{kN-Y}$$

where $T = \sum_{i=1}^{k} c_i$ and $Y = \sum_{i=1}^{k}(k-i+1)c_i$. The maximum likelihood estimator of N cannot be found exactly, but approximations can be made using *Stirling's formula*. For two removals it can be shown that

$$\hat{N} = \frac{c_1^2}{c_1 - c_2}$$

The approximate asymptotic variance of $\hat{N}$ is

$$\text{var}(\hat{N}) = \frac{\hat{N}\hat{q}^2(1+\hat{q})}{\hat{p}^3} + \frac{2\hat{q}(1-\hat{p}^2-\hat{q}^3)}{\hat{p}^5} - b^2$$

where $\hat{p} = (c_1 - c_2)/c_1, \hat{q} = 1 - \hat{p}$ and $b = \hat{q}(1+\hat{q})/\hat{p}^3$. [*Biometrics*, 1994, **50**, 501–5.]

Renewal function: See **renewal theory**.

Renewal process: See **renewal theory**.

Renewal theory: The study of *stochastic processes* counting the number of events that take place as a function of time. The events could, for example, be the arrivals of customers at a waiting line or the successive replacement of light bulbs. Formally, let $X_1, X_2, \ldots$ be a sequence of non-negative, independent random variables having a common cumulative probability distribution.

$$F(x) = \Pr(X_k \leq x_k), \quad x \geq 0, k = 1, 2, \ldots$$

Let $\mu_i = E(x_i)$ and assume $0 < \mu_i < \infty$. The random variable X_n denotes the interoccurrence time between the $(n-1)$th and nth event in some specific setting. Define

$$S_0 = 0,\ S_n = \sum_{i=1}^{n} X_i \qquad n = 1, 2, \ldots$$

then S_n is the epoch at which the nth event occurs. Define for each $t \geq 0$

$$N(t) = \text{ the largest integer } n \geq 0 \text{ for which } S_n \leq t$$

then the random variable $N(t)$ represents the number of events up to time t. The *counting process* $\{N(t), t \geq 0\}$ is called the *renewal process* generated by the interoccurrence times $x_1, x_2, \ldots$. The expected value of $N(t)$ is known as the *renewal function*. [*Biometrical Journal*, 1996, **23**, 155–64.]

Rényi, Alfréd (1921–1970): Born in Budapest, Hungary, Rényi studied mathematics and physics at the University of Budapest until 1944. In 1950 he became the Director of the Mathematical Institute of the Hungarian Academy of Sciences and also chairman of the Department of Probability. Contributed to probability theory, random graphs and *information theory* and published about 300 papers in a period of 25 years. Rényi died on 1 February 1970 in Budapest.

Repeatability: The closeness of the results obtained in the same test material by the same observer or technician using the same equipment, apparatus and/or reagents over reasonably short intervals of time.

Repeated measures data: See **longitudinal data**.

Replicate observation: An independent observation obtained under conditions as nearly identical to the original as the nature of the investigation will permit. [SMR Chapter 5.]

Reporting bias: Synonymous with **recall bias**.

Reproduced matrix: A matrix of correlations or covariances that is calculated from parameter estimates from, for example, a *structural equation model.* Often used in the assessment of the fit of a model. [*Causal Analysis: Models, Assumptions and Data*, 1982, L. R. James, S. A. Mulaik and J. M. Brett, Sage, Beverly Hills.]

Reproducibility: The closeness of results obtained on the same test material under changes of reagents, conditions, technicians, apparatus, laboratories and so on. [SMR Chapter 12.]

Reproduction rate: See **basic reproduction rate.**

Research hypothesis: Specific testable predictions made generally about the response and explanatory variables in a study. For example, if skin cancer is related to ultraviolet light, then people with a high exposure to uv light will experience a greater *incidence* of skin cancer. [*Practical Research and Design*, 7th edn, 2001, P. D. Leedy and J. E. Ormrod, Merrill Prentice Hall, New York.]

Resentful demoralization: A possible phenomenon in *clinical trials* and intervention studies in which comparison groups not obtaining a perceived desirable treatment become aware of this potential inequity and become discouraged or retaliatory, and as a result perform worse on the outcome measures. [*Quasi-Experimentation: Design and Analysis Issues for Field Setting*, 1979, T. D. Cook and D. T. Campbell, Rand-McNally, Chicago.]

RESET test: See **Ramsey regression equation specification error test.**

Residual: The difference between the observed value of a response variable (y_i) and the value predicted by some model of interest ($\hat{y}_i$). Examination of a set of residuals, usually by informal graphical techniques, allows the assumptions made in the model fitting exercise, for example, normality, homogeneity of variance, etc., to be checked. Generally, discrepant observations have large residuals, but some form of standardization may be necessary in many situations to allow identification of patterns among the residuals that may be a cause for concern. The usual *standardized residual* for observation y_i, is calculated from

$$\frac{y_i - \hat{y}_i}{s\sqrt{1 - h_i}}$$

where s^2 is the estimated residual variance after fitting the model of interest and h_i is the ith diagonal element of the *hat matrix.* An alternative definition of a standardized residual (sometimes known as the *Studentized residual*), is

$$\frac{y_i - \hat{y}_i}{s_{(-i)}\sqrt{1 - h_i}}$$

where now $s^2_{(-i)}$ is the estimated residual variance from fitting the model after the exclusion of the ith observation. See also **regression diagnostics, Cook's distance** and **influence.** [ARA Chapter 10.]

Residual maximum likelihood estimation (REML): A method of estimation in which estimators of parameters are derived by maximizing the *residual* or *restricted likelihood* rather than the *likelihood* itself. Most often used for estimating *variance components* in a *linear mixed models.* [*Analysis of Longitudinal Data*, 2nd edition, 2002, P. J. Diggle, P. J. Heagerty, K.-Y. Liang and S. L. Zeger, Oxford Science Publications, Oxford.]

Residual plots: Plots of some type of *residual* that may be helpful in assessing the assumption made by a fitted model. In regression analysis there are various ways of plotting residual values that can be helpful in assessing particular components of the regression model. The most useful plots when diagnosing linear regression models are as follows;

- A *boxplot* or *probability plot* of the residuals can be useful in checking for symmetry and specifically for normality of the error terms in the regression model.
- Plotting the residuals against the corresponding values of the explanatory variable. Any sign of curvature in the plot might suggest that say a quadratic term in the explanatory variable should be included in the model.
- Plotting the residuals against the *fitted* values of the response variable. If the variability of the residuals appears to increase with the size of the fitted values, a transformation of the response variable prior to fitting is indicated.

Figure 114 shows some idealized residual plots that indicate particular points about models.

- Figure 114(a) is what is looked for to confirm that the fitted model is appropriate,
- Figure 114(b) suggests that the assumption of constant variance is not justified so that some transformation of the response variable before fitting might be sensible,
- Figure 114(c) implies that the model needs a quadratic term in the explanantory variable.

See also **added variable plot**. [*Regression Diagnostics*, 1991, J.Fox, Sage, Newbury Park.]

Residual sum of squares: See **analysis of variance**.

Respondent-driven sampling (RDS): A form of *snowball sampling* which begins with the recruitment of a small number of people in the target population to serve as seeds. After participating the seeds are asked to recruit other people they know in the target population. The sampling continues in this way with current sample members recruiting the next wave of

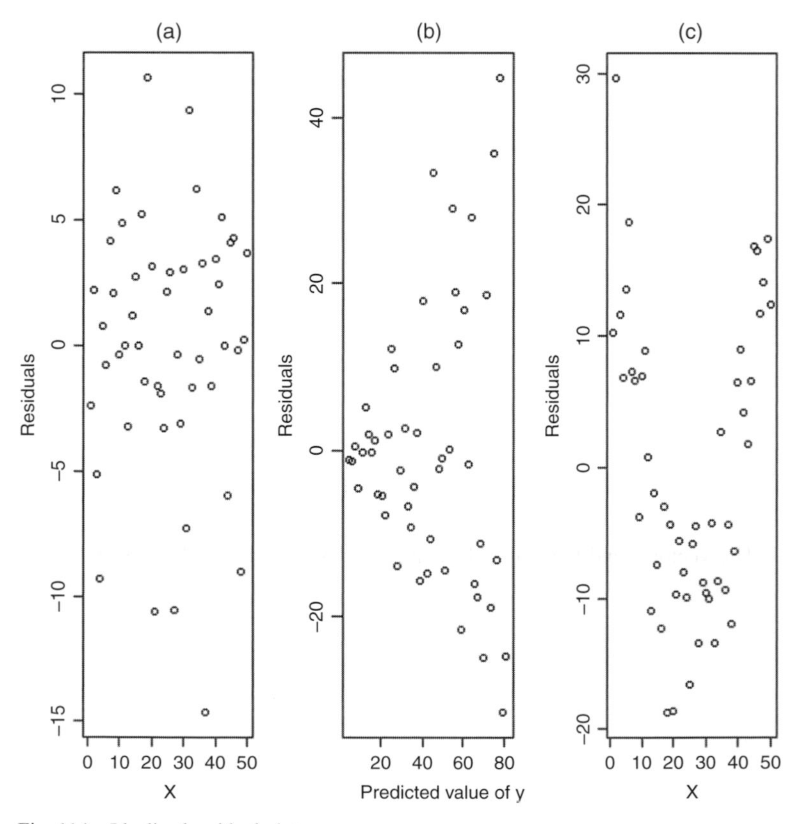

Fig. 114 Idealized residual plots.

sample members until the desired sample size is reached. By using a mathematical model that weights the sample to compensate for the fact that the sample was collected in a non-random way, the data provided by such a sampling scheme can be used to provide asymptotically unbiased estimates about the target population. An example of the use of this approach is the estimation of drug users in New York who have HIV. [*AIDS and Behavior*, 2008, **12**, 105–130.]

Responders versus non-responders analysis: A comparison of the survival experience of patients according to whether or not there is some observed response to treatment. In general such analyses are invalid because the groups are defined by a factor not known at the start of treatment. [*Psychiatric Genetics*, 2001, **11**, 41–3.]

Response bias: The systematic component of the difference between information provided by survey respondent and the 'truth'.

Response feature analysis: An approach to the analysis of *longitudinal data* involving the calculation of suitable summary measures from the set of repeated measures on each subject. For example, the mean of the subject's measurements might be calculated or the maximum value of the response variable over the repeated measurements, etc. Simple methods such as *Student's t-tests* or *Mann–Whitney tests* are then applied to these summary measures to assess differences between treatments. [*Analysis of Longitudinal Data*, 2nd edition, 2002, P. J. Diggle, P. J. Heagerty, K.-Y. Liang and S. Zeger, Oxford Science Publications, Oxford.]

Response rate: The proportion of subjects who respond to, usually, a postal questionnaire.

Response surface methodology (RSM): A collection of statistical and mathematical techniques useful for developing, improving and optimizing processes with important applications in the design, development and formulation of new products, as well as in the improvement of existing product designs. The most extensive applications of such methodology are in the industrial world particularly in situations where several input variables potentially influence some performance measure or quality characteristic of the product or process. The main purpose of this methodology is to model the response based on a group of experimental factors, and to determine the optimal settings of the experimental factors that maximize or minimize the response. Most applications involve fitting and checking the adequacy of models of the form

$$E(y) = \beta_0 + \mathbf{x}'\boldsymbol{\beta} + \mathbf{x}'\mathbf{B}\mathbf{x}$$

where $\mathbf{x} = [x_1, x_2, \ldots, x_k]'$ is a vector of k factors, $\boldsymbol{\beta} = [\beta_1, \beta_2, \ldots, \beta_k]'$ and

$$\mathbf{B} = \begin{pmatrix} \beta_{11} & \frac{1}{2}\beta_{12} & \cdots & \frac{1}{2}\beta_{1k} \\ \frac{1}{2}\beta_{12} & \beta_{22} & \cdots & \frac{1}{2}\beta_{2k} \\ \vdots & \vdots & \vdots & \vdots \\ \frac{1}{2}\beta_{1k} & \frac{1}{2}\beta_{2k} & \cdots & \beta_{kk} \end{pmatrix}$$

The vector $\boldsymbol{\beta}$ and the matrix $\mathbf{B}$ contain the parameters of the model. See also **criteria of optimality, design matrix** and **multiple regression**. [*Empirical Model-Building with Response Surfaces*, 1987, G. E. P. Box and N. D. Draper, Wiley, New York.]

Response variable: The variable of primary importance in investigations since the major objective is usually to study the effects of treatment and/or other explanatory variables on this variable and to provide suitable models for the relationship between it and the explanatory variables. [SMR Chapter 11.]

Restricted maximum likelihood estimation: Synonym for **residual maximum likelihood estimation**.

Resubstitution error rate: The estimate of the proportion of subjects misclassified by a rule derived from a *discriminant analysis*, obtained by reclassifying the *training set* using the rule. As an estimate of likely future performance of the rule, it is almost always optimistic, i.e. it will underestimate the 'true' misclassification rate. See also **jackknife** and **b632 method**. [MV2 Chapter 9.]

Resubstitution method: See **error-rate estimation**.

Retrospective cohort study: See **retrospective study**.

Retrospective study: A general term for studies in which all the events of interest occur prior to the onset of the study and findings are based on looking backward in time. Most common is the traditional *case-control study*, in which comparisons are made between individuals who have a particular disease or condition (the cases) and individuals who do not have the disease (the controls). A sample of cases is selected from the population of individuals who have the disease of interest and a sample of controls is taken from among those individuals known not to have the disease at the end of the observation period. Information about possible *risk factors* for the disease is then obtained retrospectively for each person in the study by examining past records, by interviewing each person and/or interviewing their relatives, or in some other way. In order to make the cases and controls otherwise comparable, they are frequently matched on characteristics known to be strongly related to both disease and exposure leading to a *matched case-control study*. Age, sex and socioeconomic status are examples of commonly used matching variables. Also commonly encountered is the *retrospective cohort study*, in which a past cohort of individuals are identified from previous information, for example, employment records, and their subsequent mortality or morbidity determined and compared with the corresponding experience of some suitable control group. [SMR Chapter 5.]

Reversal designs: Synonym for **switchback designs**.

Reverse-causality: Synonym for **protopathic bias**.

Reversible jump sampler: An improved *Markov chain Monte Carlo method* which allows 'jumps' between parameter subspaces of differing dimensionality. [*Biometrika*, 1995, **82**, 711–32.]

Reversibility: A term used in the context of *time series* to imply that all modelling and statistical analyses can be performed equally on the reversed time-ordered values as on the original sequence. [*International Statistical Review*, 1991, **59**, 67–79.]

RHC: Abbreviation for **Rao–Hartley–Cochran method**.

Rice distribution: Synonym for the **noncentral chi-squared distribution**.

Ridge estimator: See **ridge regression**.

Ridge regression: A method of *multiple linear regression* designed to overcome the possible problem of *multicollinearity* among the explanatory variables. Such multicollinearity often makes it difficult to estimate the separate effects of variables on the response. In such regression, the regression coefficients are obtained by solving

$$b^*(\theta) = (\mathbf{X}'\mathbf{X} + \theta\mathbf{I})^{-1}\mathbf{X}'\mathbf{y}$$

where $\theta \geq 0$ is a constant. Generally values in the interval $0 \leq \theta \leq 1$ are appropriate. The *ridge estimator* $b^*(\theta)$ is not an unbiased estimator of $\boldsymbol{\beta}$ as is the ordinary least squares estimator $\mathbf{b}$. This form of regression seeks to find a set of regression coefficients that is more 'stable' in the sense of having a small mean square error. To obtain the ridge estimator explicitly a value for θ must be specified. This is usually done by plotting the values of $b^*(\theta)$ against θ, This display is called the *ridge trace*. An example is given in Fig. 115. See also **biased estimator**. [ARA Chapter 12.]

Ridge trace: See **ridge regression**.

Ridit analysis: A method of analysis for ordinal variables that proceeds from the assumption that the ordered categorical scale is an approximation to an underlying, but not directly measurable, continuous variable. The successive categories are assumed to correspond to consecutive intervals on the underlying continuum. Numerical values called *ridits* are calculated for each category, these values being estimates of the probability that a subject's value on the underlying variable is less than or equal to the midpoint of the corresponding interval. These scores are then used in subsequent analyses involving the variable. [*Journal of Dental Research*, 1979, **58**, 2080–2084.]

Ridits: See **ridit analysis.**

Riemann integral: See **Riemann–Stieltjes integral**.

Riemann–Stieltjes integral: A more general type of integration than the usual *Cauchy integral* which involves the sum;

$$S_n = \psi(\theta_1)\{F(x_1) - F(a)\} + \psi(\theta_2)\{F(x_2) - F(x_1)\} + \cdots + \psi(\theta_{n+1})\{F(b) - F(x_n)\}$$

where $\psi(x)$ is a continuous function of x and θ_1 is in the range a to x_1, θ_2 is in the range x_1 to x_2 and so on. As the length of the intervals tend to zero uniformly S_n tends to a limit that is independent of the location of the points θ or the boundary points of the intervals. This limit is written as

$$\int_a^b \psi(x)\mathrm{d}F$$

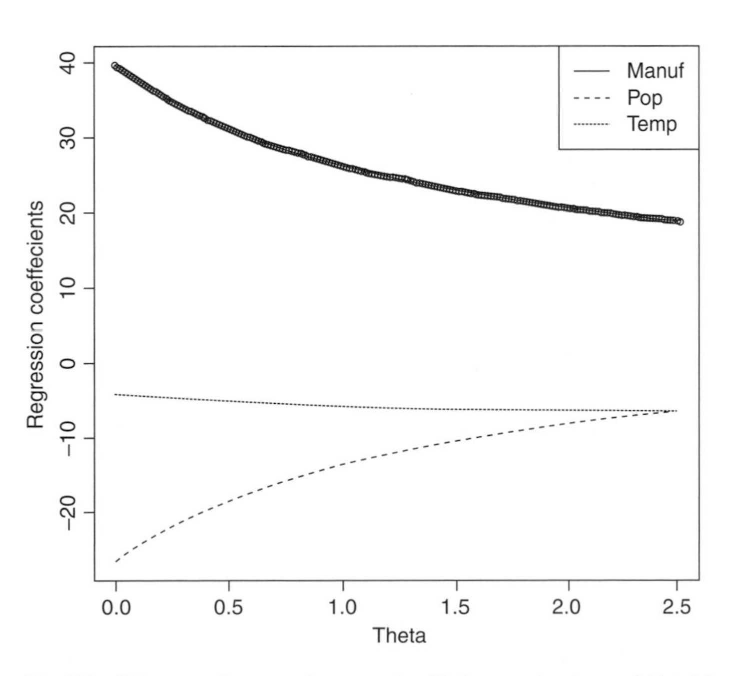

Fig. 115 Ridge trace for regression example with three explanatory variables, Manuf, Pop and Temp.

and is known as the *Stieltjes integral* of $\psi(x)$ with respect to $F(x)$. When $F(x) = x$ it becomes the *Riemann integral*. [KA1 Chapter 1.]

Right censored: See **censored observations**.

Rim estimation: Synonym for **raking adjustment**.

Risk: A term often used in medicine for the probability that an event will occur, for example, that a person will become ill, will die, etc.

Risk factor: An aspect of personal behaviour or lifestyle, an environmental exposure, or an inborn or inherited characteristic which is thought to be associated with a particular disease or condition.

Risk measure: A function that assigns real numbers to random variables representing uncertainty payoffs, for example, insurance losses. [*Insurance Mathematics and Economics*, 1999, **25**, 337–347.]

Risk set: A term used in *survival analysis* for those individuals who are alive and uncensored at a time just prior to some particular time point.

Robbins, Herbert (1915–2001): Robbins obtained a first degree from Harvard in 1935 and a Ph.D. in pure mathematics in 1938. After the war he joined the newly formed department at the University of North Carolina, moving in 1953 to Columbia University as the Higgins Professor of Mathematical Statistics. Robbins made important contributions to empirical Bayes methods and stochastic approximation, but was perhaps best known for his very popular book *What is Mathematics?* Robbins died in Princeton, NJ, on 12 February 2001.

Robinson matrix: A matrix whose elements are monotonically non-decreasing as one moves from the diagonal within each row and within each column. Such matrices are important in some areas of *cluster analysis*. [*Journal of Classification*, 1984, **1**, 75–92.]

Robust estimation: Methods of estimation that work well not only under ideal conditions, but also under conditions representing a departure from an assumed distribution or model. See also **high breakdown methods**. [*Introduction to Robust Estuiation and Hypotresis Testing*, 2005, R. R. Wilcox, Academic Press.]

Robust regression: A general class of statistical procedures designed to reduce the sensitivity of the parameter estimates to failures in the assumption of the model. For example, *least squares estimation* is known to be sensitive to *outliers*, but the impact of such observations can be reduced by basing the estimation process not on a sum-of-squares criterion, but on a sum of absolute values criterion. See also **M-estimators**. [*Robust Statistics*, 2nd edition, 2003, P. J. Huber, Wiley, New York.]

Robust statistics: Statistical procedures and tests that still work reasonably well even when the assumptions on which they are based are mildly (or perhaps moderately) violated. *Student's t-test*, for example, is robust against departures from normality. See also **high breakdown methods** and **robust estimation**. [*Robustness in Statistics*, 1979, R. Launer and G. Wilkinson, Academic Press, New York.]

ROC curve: Abbreviation for **receiver operating characteristic curve**.

Rootogram: A diagram obtained from a *histogram* in which the rectangles represent the square roots of the observed frequencies rather than the frequencies themselves. The idea behind such a diagram is to remove the tendency for the variability of a count to increase with its typical size. See also **hanging rootogram**. [*Exploratory Data Analysis*, 1977, J. W. Tukey, Addison Wesley, Reading, Massachusetts.]

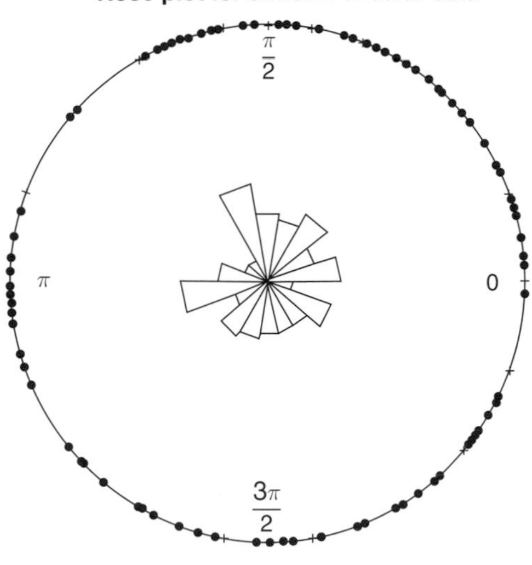

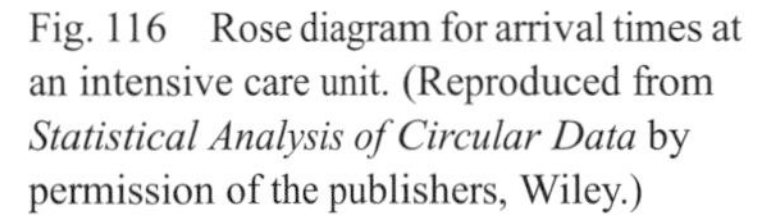
Fig. 116 Rose diagram for arrival times at an intensive care unit. (Reproduced from *Statistical Analysis of Circular Data* by permission of the publishers, Wiley.)

Rose diagram: A form of *angular histogram* in which each group is displayed as a sector. The radius of each sector is usually taken to be proportional to the square root of the frequency of the group, so that the area of the sector is proportional to the group frequency. Figure 116 shows such a diagram for arrival times on a 24 hour clock of 254 patients at an intensive care unit, over a period of 12 months. [*Statistical Analysis of Circular Data*, 1995, N. I. Fisher, Cambridge University Press, Cambridge.]

Rosenbaum's test: A *distribution free method* for testing the equality of the scale parameters of two populations known to have the same median. The test statistic is the total number of values in the sample from the first population that are either smaller than the smallest or larger than the largest values in the sample from the second population. [*Model Assisted Statistics and Applications*, 2008, **3**, 59–69.]

Rosenthal effect: The observation that investigators often find what they expect to find from a study. For example, in a reliability study in which auscultatory measurements of the foetal heart rate were compared with the electronically recorded rate, it was found that when the true rate was under 130 beats per minute the hospital staff tended to overestimate it, and when it was over 150 they tended to underestimate it. See also **Hawthorne effect**. [*Journal of Educational Psychology*, 1984, **76**, 85–97.]

Rotatable design: See **design rotatability**.

Rotation sampling: Sampling a given population at several regular intervals of time in order to estimate the manner in which the population is changing. Essentially equivalent to *panel studies* or *longitudinal studies*. [*Journal of Statistical Planning and Inference*, 2007, **137**, 549–610.]

Rothamsted Park Grass Experiment: A long term experiment started at Rothamsted Experimental Station in 1865 to study the effect of different fertilizers on the yield of hay.

Rounding: The procedure used for reporting numerical information to fewer decimal places than used during analysis. The rule generally adopted is that excess digits are simply discarded if the first of them is less than five, otherwise the last retained digit is increased by one. So rounding 127.2492341 to three decimal places gives 127.249.

Round robin study: A term sometimes used for interlaboratory comparisons in which samples of a material manufactured to well-defined specifications are sent out to the participating laboratories for analysis. The results are used to assess differences between laboratories and to identify possible sources of incompatibility or other anomalies. See also **interlaboratory studies**. [*Organohalogen Compounds*, 2004, **66**, 519–24.]

Roy's largest root criterion: See **multivariate analysis of variance**.

R-techniques: The class of data analysis methods that look for relationships between the variables in a set of *multivariate data.* Includes *principal components analysis*, and *factor analysis.* See also **Q-techniques**.

Ruben, Harold (1923–2001): Born in Rawa, Poland, Ruben's family moved to England when he was quite young, and he studied mathematics at Imperial College, London. In 1950 he was awarded a Doctor of Philosophy degree in mathematical statistics. After this Ruben worked in Aberdeen and then in the Department of Genetics at Cambridge. In 1953 he moved to the University of Manchester and then four years later to Columbia University in New York. Further moves between the USA and the UK led to posts at the University of Sheffield and the University of Minnesota. In 1969 Ruben moved to Montreal as Professor in the Department of Mathematics at McGill University where he stayed until his retirement in 1988. Ruben made important contributions in *multivariate analysis*, geometric probability and statistical physics. He died in Harlow, Essex, UK, on 30 November 2001.

Rudas–Clogg–Lindsay index of fit: A measure for evaluating goodness of fit in the analysis of *contingency tables*, which gives a directly interpretable quantitative measure of lack of fit, namely the proportion of cases incompatible with the model being tested. In this way the index allows immediate comparisons across samples or studies which is not sensitive to sample size. [*Journal of the Royal Statistical Society, Series B*, 1994, **56**, 623–39.]

Rugplot: A method of displaying graphically a sample of values on a continuous variable by indicating their positions on a horizontal line. See Fig. 117 for an example.

Ruin theory: A method for examining the solvency of an insurance company by modelling the insurer's surplus or reserve for an insurance portfolio over time, by a *stochastic process.* Of central importance are the time to ruin at which the surplus becomes negative for the first time, the surplus immediately before the time of ruin and the deficit at the time of ruin. [*Ruin Probabilities*, 2000, S. Asmussen, World Scientific, Singapore.]

Rule of three: A rule which states that if in n trials, zero events of interest are observed, an upper 95% confidence bound on the underlying rate is $3/n$. [*The American Statistician*, 1997, **51**, 137–9.]

Run-in: A period of observation prior to the formation of treatment groups in a *clinical trial* during which subjects acquire experience with the major components of a study *protocol.* Those subjects who experience difficulty complying with the protocol are excluded while the group

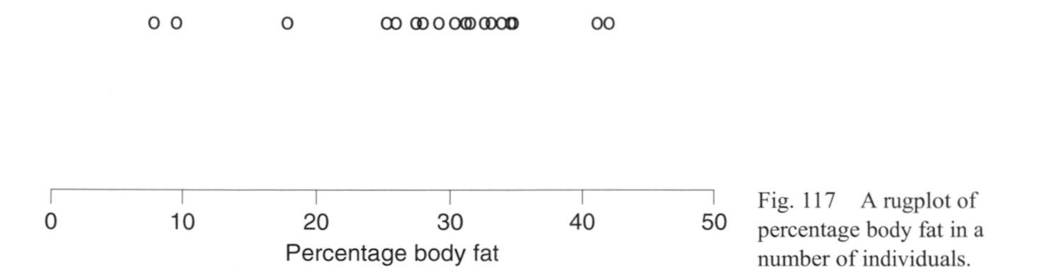

Fig. 117 A rugplot of percentage body fat in a number of individuals.

of proven compliers are randomized into the trial. The rationale behind such a procedure is that, in general, a study with higher compliance will have higher *power* because the observed effects of the difference between treatment groups will not be subjected to the diluting effects of non-compliance. [*Statistics in Medicine*, 1990, **9**, 29–34.]

Run: In a series of observations the occurrence of an uninterrupted sequence of the same value. For example, in the series

$$111224333333$$

there are four 'runs', the single value, 4, being regarded as a run of length one.

Runs test: A test frequently used to detect *serial correlations.* The test consists of counting the number of runs, or sequences of positive and negative residuals, and comparing the result to the expected value under the hypothesis of independence. If the sample observations consist of n_1 positive and n_2 negative residuals with both n_1 and n_2 greater than 10, the distribution of the length of runs under independence can be approximated by a normal distribution with mean, μ, and variance, σ^2 given by

$$\mu = \frac{2n_1 n_2}{n_1 + n_2} + 1$$
$$\sigma^2 = \frac{2n_1 n_2(2n_1 n_2 - n_1 - n_2)}{(n_1 + n_2)^2(n_1 + n_2 - 1)}$$

[ARA Chapter 10.]

RV-coefficient: A measure of the similarity of two configurations of n data points given by

$$\text{RV}(\mathbf{X}, \mathbf{Y}) = \frac{\text{tr}(\mathbf{XY}'\mathbf{YX}')}{\{\text{tr}(\mathbf{XX}')^2 \text{tr}(\mathbf{YY}')^2\}^{\frac{1}{2}}}$$

where $\mathbf{X}$ and $\mathbf{Y}$ are the matrices describing the two configurations. Several techniques of multivariate analysis can be expressed in terms of maximizing RV $(\mathbf{X}, \mathbf{Y})$ for some definition of $\mathbf{X}$ and $\mathbf{Y}$. [*Applied Statistics*, 1976, **25**, 257–65.]

Saddlepoint method: An approach to providing extremely accurate approximations to probability distributions based on the corresponding *cumulative generating function*. For a variable y with cumulative generating function $c(t)$, the first-order approximation is

$$f(y) \approx (2\pi)^{-\frac{1}{2}k} |c''(\hat{\phi})|^{-\frac{1}{2}} \exp\{c(\hat{\phi}) - \hat{\phi}'y\}$$

where $\hat{\phi}$ is determined by $c'(\hat{\phi}) = y$ and c' and c'' denote first derivative vector and second derivative matrix. [KA1 Chapter 11.]

Saint Petersburg paradox: Synonym for **Petersburg game**.

Sakoda coefficient: A measure of association, S, of the two variables forming an $r \times c$ two-dimensional contingency table, given by

$$S = \sqrt{\frac{qX^2}{(q-1)(X^2 + N)}}$$

where X^2 is the usual chi-squared statistic for testing the independence of the two variables, N is the sample size and $q = \min(r, c)$. See also **contingency coefficient, phi-coefficient** and **Tschuprov coefficient**. [*An Introduction to Categorical Data Analysis*, 1996, A. Agresti, Wiley, New York.]

Sample: A selected subset of a population chosen by some process usually with the objective of investigating particular properties of the parent population.

Sample size: The number of individuals to be included in an investigation. Usually chosen so that the study has a particular power of detecting an effect of a particular size. Software is available for calculating sample size for many types of study, for example, *SOLO* and *nQuery advisor*. See also **nomogram**. [SMR Chapter 8.]

Sample size formulae: Formulae for calculating the necessary sample size to achieve a given power of detecting an effect of some fixed magnitude. For example, the formula for determining the number of observations needed in each of two samples when using a one-tailed z-test to test for a difference in the means of two populations is

$$n = \frac{2(z_\alpha - z_\beta)^2 \sigma^2}{(\mu_1 - \mu_2)^2}$$

where σ^2 is the known variance of each population, μ_1 and μ_2 the population means, z_α the normal deviate for the chosen significance level α and z_β the normal deviate for β where $1 - \beta$ is the required power. See also **SOLO** and **nQuery advisor**. [SMR Chapter 15.]

Sample space: The set of all possible outcomes of a probabilistic experiment. For example, if two coins are tossed, the sample space is the set of possible results, HH, HT, TH and TT, where H indicates a head and T a tail.

Sample survey: A study that collects planned information from a sample of individuals about their history, habits, knowledge, attitudes or behaviour in order to estimate particular population characteristics. See also **opinion survey, random sample**, and **quota sample**. [*The American Census: a Social History*, 1988, M. Anderson, Yale University Press, New Haven.]

Sampling: The process of selecting some part of a population to observe so as to estimate something of interest about the whole population. To estimate the amount of recoverable oil in a region, for example, a few sample holes might be drilled, or to estimate the abundance of a rare and endangered bird species, the abundance of birds in the population might be estimated on the pattern of detections from a sample of sites in the study region. Some obvious questions are how to obtain the sample and make the observations and, once the sample data are to hand, how best to use them to estimate the characteristic of the whole population. See also **simple random sampling** and **cluster sampling**. [SMR Chapter 5.]

Sampling design: The procedure by which a sample of units is selected from the population. In general a particular design is determined by assigning to each possible sample S the probability $\Pr(S)$ of selecting that sample. See also **random sample**.

Sampling distribution: The probability distribution of a statistic calculated from a random sample of a particular size. For example, the sampling distribution of the arithmetic mean of samples of size n taken from a normal distribution with mean μ and standard deviation σ, is a normal distribution also with mean μ but with standard deviation $\sigma/\sqrt{n}$. [SMR Chapter 8.]

Sampling error: The difference between the sample result and the population characteristic being estimated. In practice, the sampling error can rarely be determined because the population characteristic is not usually known. With appropriate sampling procedures, however, it can be kept small and the investigator can determine its probable limits of magnitude. See also **standard error**. [*Sampling Techniques*, 1977, 3rd edition, W. G. Cochran, Wiley, New York.]

Sampling frames: The portion of the population from which the sample is selected. They are usually defined by geographic listings, maps, directories, membership lists or from telephone or other electronic formats. [*Survey Sampling*, 1995, L. Kish, Wiley, New York.]

Sampling units: The entities to be sampled by a *sampling design*. In many surveys these entities will be people, but often they will involve larger groupings of individuals. In *institutional surveys*, for example, they may be hospitals, businesses, etc. Occasionally it may not be clear what the units should be. In a survey of agricultural crops in a region, the region might be divided into a set of geographic areas, plots or segments, and a sample of units selected using a map. But such units could obviously be made alternative sizes and shapes, and such choices may affect both the cost of the survey and the precision of the estimators. [SMR Chapter 14.]

Sampling variation: The variation shown by different samples of the same size from the same population. [SMR Chapter 7.]

Sampling with and without replacement: Terms used to describe two possible methods of taking samples from a *finite population*. When each element is replaced before the next one is drawn, sampling is said to be 'with replacement'. When elements are not replaced then the sampling is referred to as 'without replacement'. See also **bootstrap**, **jackknife** and **hypergeometric distribution**. [KA1 Chapter 9.]

Sampling zeros: Zero frequencies that occur in the cells of *contingency tables* simply as a result of inadequate sample size. See also **structural zeros**. [*The Analysis of Contingency Tables*, 2nd edition, 1992, B. S. Everitt, Chapman and Hall/CRC Press, London.]

Sandwich estimators: Estimators of *variance–covariance matrices* of estimators encountered for instance in the analysis of clustered data by *generalized estimating equations* and when using the *pseudolikelihood* for *complex survey data.* The estimators are of the form

$$\mathbf{ABA}$$

where the outside pieces of the 'sandwich' (the **A**s) correspond to a model-based covariance matrix assuming a correct model and the centre term (the **B**) corresponds to an empirical covariance matrix. [*Advances in Econometrics*, 2003, **17**, 45–73.]

Sanghvi's distance: A measure of the distance between two populations described by a set of categorical variables; given by

$$G^2 = 100 \frac{\sum_{j=1}^{r} \sum_{k=1}^{s_j+1} \left[\frac{(p_{1jk} - p_{jk})^2}{p_{jk}} + \frac{(p_{2jk} - p_{jk})^2}{p_{jk}} \right]}{\sum_{j=1}^{r} s_j}$$

where p_{1jk} and p_{2jk} are the proportions in the kth class for the jth variable in populations P_1 and P_2, respectively, $P_{jk} = \frac{1}{2}(p_{1jk} + p_{2jk})$ and s_{j+1} is the number of classes for the jth character. [*Biometrics*, 1970, **26**, 525–34.]

Sargan distribution: A probability distribution, $f(x)$, given by

$$f(x) = \frac{1}{2} K \alpha e^{-\alpha|x|} \left(1 + \sum_{j=1}^{\rho} \gamma_j \alpha_i |x|^j \right)$$

where $\alpha \geq 0, \gamma_j > 0 (j = 1, 2, \ldots, \rho)$ and $K = \sum_{j=1}^{\rho} (1 + j! \gamma_j)^{-1}$. The parameter ρ is usually referred to as the order of the distribution. Order zero corresponds to a *Laplace distribution.* Used as a viable alternative to the normal distribution in econometric models. [*Journal of Econometrics*, 1987, **34**, 349–54.]

Sargan test: A test of the validity of *instrumental variables*, namely that they are uncorrelated with the appropriate set of error terms. [*Applied Economic Letters*, 2008, **15**, 349–353.]

SAS: An acronym for Statistical Analysis System, a large computer software system for data processing and data analysis. Extensive data management, file handling and graphics facilities. Can be used for most types of statistical analysis including *multiple regression*, *log-linear models*, *principal components analysis*, etc. [http://www.sas.com.]

Satterthwaite's approximation: A useful general method of approximating the probability distribution of a nonnegative variable which has a known (or well-estimated) mean and variance. [ARA Chapter 16.]

Saturated model: A model that contains all main effects and all possible interactions between factors. Since such a model contains the same number of parameters as cells (categorical data) or means (continuous data) it results in a perfect fit for a data set.

Savage, Leonard James (1917–1971): Born in Detroit, Savage studied at Wayne State and later University of Michigan, eventually gaining a Ph.D. on the application of vector methods to metric geometry. In 1944 he became a member of the Statistical Research Group at Columbia. In 1949 Savage helped to form the statistics department at Chicago and in 1954 produced *The Foundations of Statistics* a book which had great influence on later developments in the subject. He was President of the Institute of Mathematical Statistics for the period 1957–58. Savage died on 1 November 1971, in New Haven.

Savage's test: A *distribution free* procedure for the comparison of two samples of size n and m from populations with cumulative probability distributions $F_1(x)$ and $F_2(x)$. For testing the null hypothesis $F_1(x) = F_2(x)$ for all x against $F_1(x) \geq F_2(x)$ for all x and $F_1(x) > F_2(x)$ for some x. The test statistic is

$$S = \sum_{i=1}^{n} a_N(R_i)$$

where $R_1 < R_2 < \cdots < R_n$ are the ordered ranks associated with the sample of size n in the combined sample, $N = n + m$ and $a_N(k) = \sum_{j=1}^{k} (N - j + 1)^{-1}$. If m and n are large and the null hypothesis is true then S is approximately normal with mean equal to n and variance given by

$$mn(N-1)^{-1}\left[1 - N^{-1}\sum_{j=1}^{N}(1/j)\right]$$

[*Encyclopedia of Statistical Sciences*, 2006, eds. S.Kotz, C. B. Read, N.Balakrishnan and B. Vidakovic, Wiley, New York.]

Scaled deviance: A term used in *generalized linear models* for the *deviance* divided by the *scale factor.* Equals the deviance unless *overdispersion* is being modelled. [GLM Chapter 2.]

Scale factor: See **generalized linear models**.

Scale parameter: A general term for a parameter of a probability distribution that determines the scale of measurement. For example, the parameter σ^2 in a normal distribution.

Scaling: See **attitude scaling**, **multidimensional scaling** and **optimal scaling**.

Scan statistic: A statistic for evaluating whether an apparent *disease cluster* in time or in space is due to chance. The statistic employs a 'moving window' of a particular length and finds the maximum number of cases revealed through the window as it moves over the entire time period or spatial area. Approximations for an upper bound to the probability of observing a certain size cluster under the null hypothesis of a *uniform distribution* are available. The statistic has been applied to test for possible clustering of lung cancer at a chemical works, rashes following injection of a varicella virus vaccine and trisomic spontaneous abortions. See also **geographical analysis machine**. [*International Journal of Health Geographics*, 2005, **18**, 4–11.]

Scatter: Synonym for **dispersion**.

Scatter diagram: A two-dimensional plot of a sample of *bivariate observations.* The diagram is an important aid in assessing what type of relationship links the two variables. An example is shown in Fig. 118. See also **correlation coefficient** and **draughtsman's plot**. [SMR Chapter 3.]

Scattergram: Synonym for **scatter diagram**.

Scatterplot: Synonym for **scatter diagram**.

Scatterplot matrix: An arrangement of the pairwise scatter diagrams of the variables in a set of multivariate data in the form of a square grid. Such an arrangement may be extremely useful in the initial examination of the data. Each panel of the matrix is a scatter plot of one variable against another. The upper left-hand triangle of the matrix contains all pairs of scatterplots and so does the lower right-hand triangle. The reasons for including both the upper and lower triangles in the matrix, despite the seeming redundancy, is that

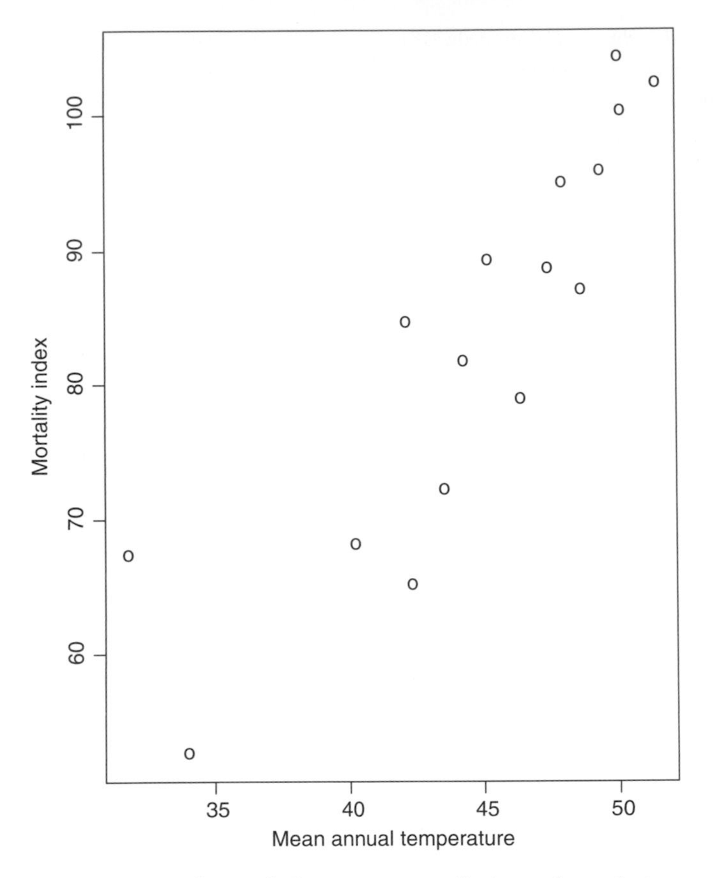

Fig. 118 Scatter diagram for breast cancer mortality in a region against mean annual temperature in the region.

it enables a row and column to be visually scanned to see one variable against all others, with the scales for the one variable lined up along the horizontal or the vertical. The plot can be made even more useful by enhancing the individual scatterplots with, for example, a fitted linear regression or a *locally weighted regression* (or both) for the corresponding pair of variables. Two examples of scatterplot matrices for data on sulphur dioxide concentration and a number of other variables for some cities in the US are shown in Figures 119 and 120. The latter shows the fitted linear regression for the two variables forming each panel in the plot. [*Visualizing Data*, 1993, W. S. Cleveland, Hobart Press, Summit, New Jersey.]

Scenario analysis: A term used to describe a range of *time series* forecasts based on different assumptions. [*Journal of Forecasting*, 1991, **10**, 549–64.]

Sceptical priors: See **prior distributions**.

Scheffé, Henry (1907–1977): Born in New York City, Scheffé studied at the University of Wisconsin, Madison receiving a Ph.D. in 1935 for his thesis entitled 'Asymptotic solutions of certain linear differential equations in which the coefficient of the parameter may have a zero'. Moved to Princeton in 1940 and then became Head of the Department of Mathematical Statistics at Columbia University in 1951. His now classic work *The Analysis of Variance* was published in 1959. Scheffé died on 5 July 1977 as a result of a tragic cycling accident.

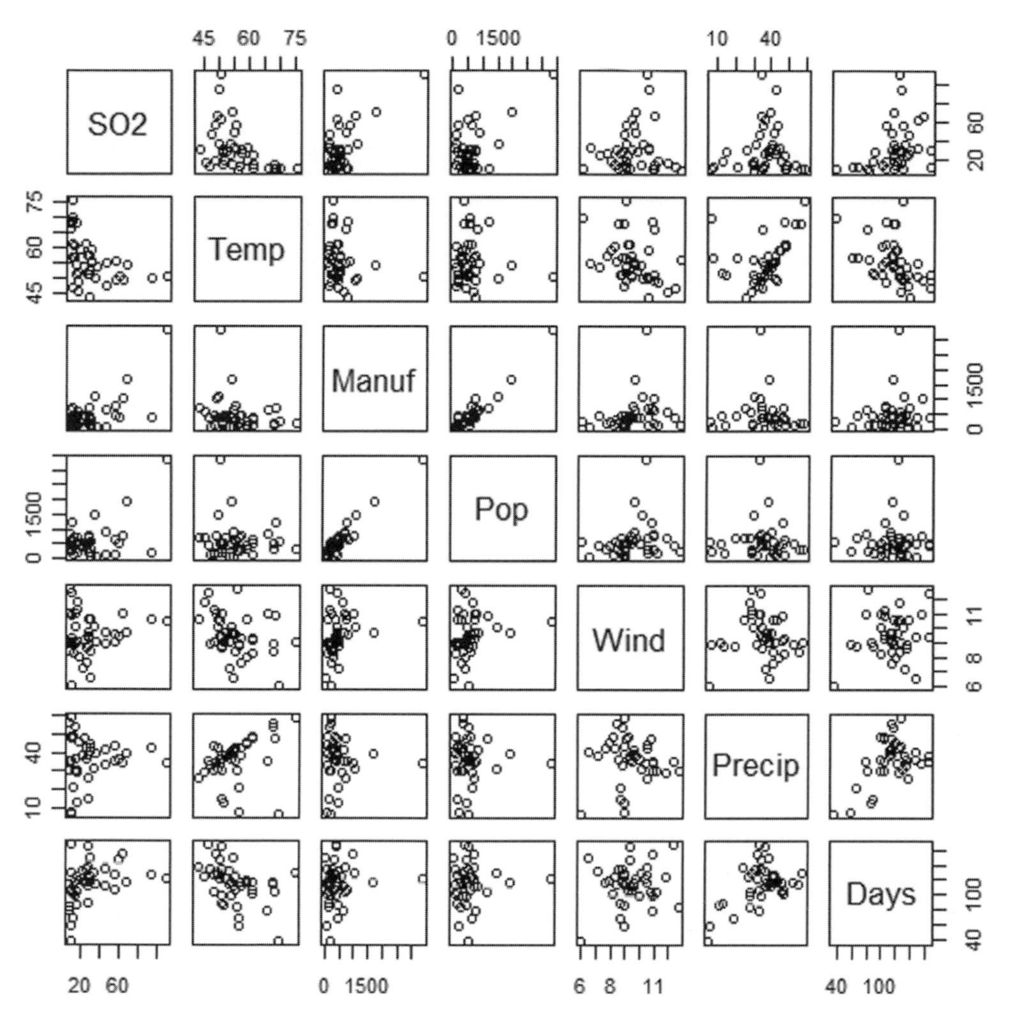

Fig. 119 Scatterplot matrix of seven variables recorded for a number of US cities.

Scheffé's test: A *multiple comparison test* which protects against a large *per-experiment error rate*. In an *analysis of variance* of a *one-way design*, for example, the method calculates the *confidence interval* for the difference between two means as

$$\bar{y}_{i.} - \bar{y}_{j.} \pm \sqrt{(g-1)F_{g-1,N-g}}\sqrt{MSE\left(\frac{1}{n_i} + \frac{1}{n_j}\right)}$$

where $\bar{y}_{i.}$ and $\bar{y}_{j.}$ are the observed means of groups i and j, g is the number of groups, MSE is the mean square error in the *analysis of variance table*, n_i and n_j are the number of observations in groups i and j, and $F_{g-1,N-g}$ is the F-value for some chosen significance level. The total sample size is represented by N. [*Biostatistics: A Methodology for the Health Sciences*, 2nd edn, 2004, G.Van Belle, L. D. Fisher, P. J. Heagerty and T. S. Lumley, Wiley, New York.]

Schemper's measures: Two measures of explained variation in *Cox's proportional hazards model*, which correspond to the proportion of explained variation in the normal case. The measures V_1 and V_2 are defined as follows

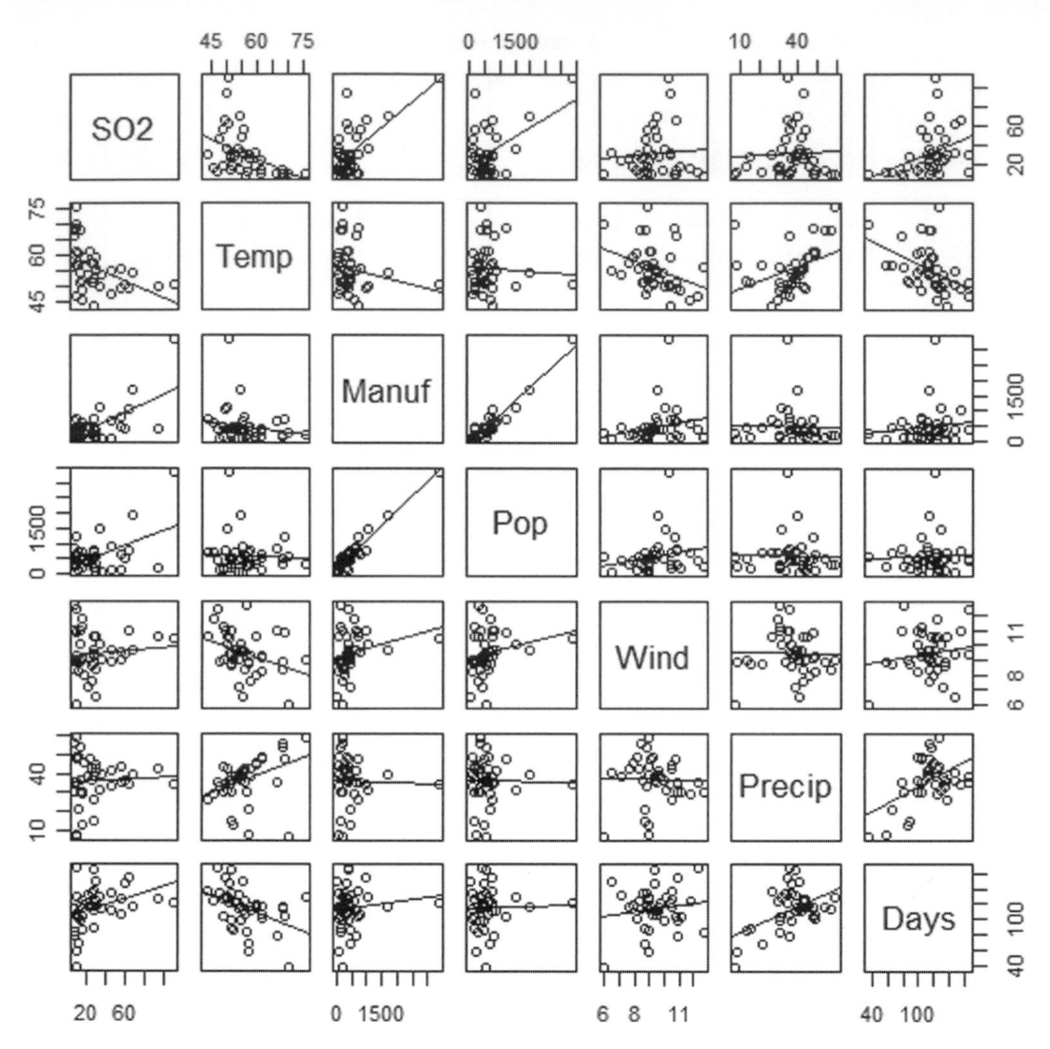

Fig. 120 Scatterplot matrix of variables recorded for a number of US cities with each panel enhanced by the linear regression fit.

$$1 - V_l = \frac{\sum k_i^{-1} \sum |S_{ij} - \bar{S}_{ij}|^l}{\sum k_i^{-1} \sum |S_{ij} - \bar{S}_j|^l} \quad l = 1, 2$$

where S_{ij}=1 for individual i at all time points t_j for which the individual is still alive, drops to 0.5 at the point at which the individual dies; and thereafter S_{ij}=0, K_i is the total number of deaths should individual i correspond to a death, otherwise it is the number of deaths before the censoring time of the individual i, $\bar{S}_j$ is the *Kaplan–Meier estimate* of survival at time t_j and $\bar{S}_{ij}$ the estimate of survival for individual i at time point t_j derived from the proportional hazards model. [*Biometrika*, 1990, **77**, 216–18.]

Schoenfield residual: A diagnostic used in applications of *Cox's proportional hazards model.* This type of *residual* is not, in fact, a single value for each subject but a set of values, one for each explanatory variable included in the fitted model. Each such residual is the difference between the *j*th explanatory variable and a weighted average of the values of the explanatory variables over individuals in the risk set at the death time of the *j*th individual. The residuals have the property that in large samples their expected values are zero and that they are uncorrelated with one another. [*Modelling Survival Data in Medical Research*, 2nd edition, 2003, D. Collett, Chapman and Hall/CRC Press, London.]

Schuster's test: A test for non-zero periodic ordinates in the *periodogram* of a *time series*, which assumes that the random component of the series is an independent zero mean normal series. Under this assumption the probability distribution of each component of the periodogram $I(\omega_p)$ is proportional to *chi-squared distribution* with two degrees of freedom. Specifically for $p \neq 0$

$$I(\omega_p)/\sigma_x^2 \sim \chi_2^2$$

where $\sigma_x^2 = \text{var}(X_t)$. Consequently the significance of a given periodogram ordinate can be tested. [*Applications of Time Series Analysis in Astronomy and Meterology*, edited by E. Subba Rao, M. B. Priestley and O. Lessi, 1997, Chapman and Hall, London.]

Schwarz's criterion: An index used as an aid in choosing between competing models. It is defined as

$$-2L_m + m \ln n$$

where n is the sample size, L_m is the maximized *log-likelihood* of the model and m is the number of parameters in the model. The index takes into account both the statistical goodness of fit and the number of parameters that have to be estimated to achieve this particular degree of fit, by imposing a penalty for increasing the number of parameters. Lower values of the index indicate the preferred model, that is, the one with the fewest parameters that still provides an adequate fit to the data. If $n \geq 8$ this criterion will tend to favour models with fewer parameters than those chosen by *Akaike's information criterion*. See also **Bayesian information criterion (BIC), parsimony principle** and **Occam's razor**. [TMS Chapter 11.]

Score residual: Synonym for **Schoenfield residual**.

Score test: Typically (but not necessarily) a test for the hypothesis that a vector of parameters, $\boldsymbol{\theta}' = [\theta_1, \theta_2, \ldots, \theta_m]$, is the null vector. The test statistic is

$$s = \mathbf{S}'\mathbf{V}\mathbf{S}$$

where $\mathbf{S}$ is the vector with elements $\partial L/\partial\theta_i$ and L is the *log-likelihood*. $\mathbf{V}$ is the asymptotic *variance–covariance matrix* of the parameters. In contrast to the *likelihood ratio test*, the score test only requires estimation under the null hypothesis. See also **likelihood ratio test** and **Wald's test**. [GLM Chapter 5.]

Scott, Elizabeth Leonard (1917–1988): Born in Fort Sills, Oklahoma, USA, Scott studied astronomy at the University of California, Berkeley. During World War II she worked with *Jerzy Neyman* in the Statistical Laboratory at Berkeley on improving the precision of air bombing. In 1962 Scott became Professor of Statistics in Berkeley. In her career she worked on a variety of statistical problems including models for carcinogenesis and cloud seeding experiments. She was President of the Institute of Mathematical Statistics in 1977. Scott died on 20 December 1988 in Berkeley.

Scram data: Data produced by the nuclear power industry relating to unplanned shutting down of nuclear plants by the reactor protection system following some transient event, such as the loss of offsite power.

Scree diagram: A plot of the ordered eigenvalues of a *correlation matrix*, used to indicate the appropriate number of factors in a *factor analysis* or *principal components analysis*. The critical feature sought in the plot is an 'elbow', the number of factors then being taken as

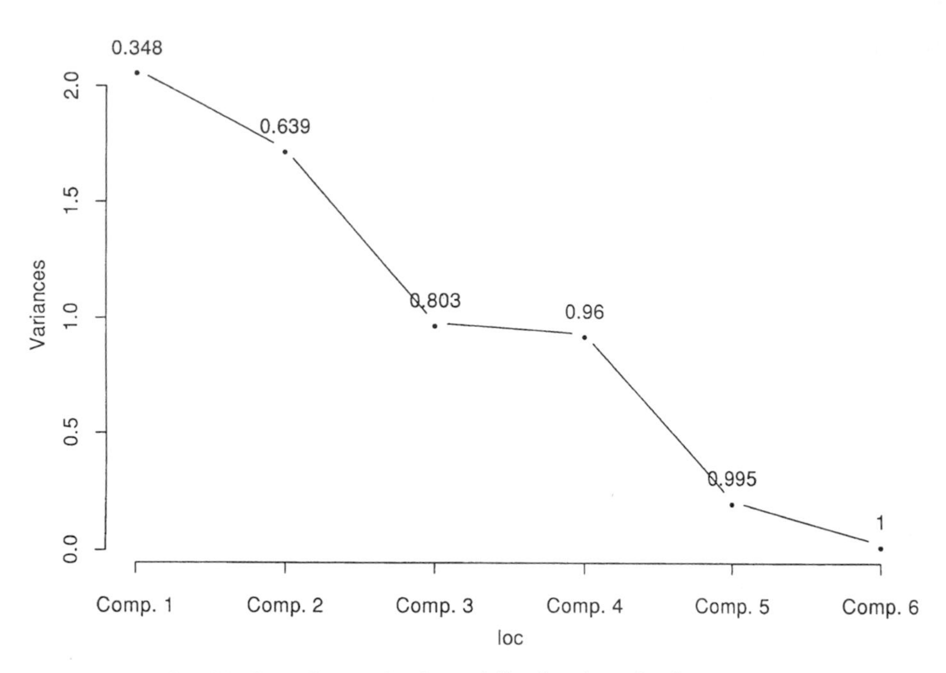

Fig. 121 Scree diagram showing an 'elbow' at eigenvalue three.

the number of eigenvalues up to this point. An example of such a diagram is given in Fig. 121. See also **Kaiser's rule**. [MV2 Chapter 12.]

Screened-to-eligible ratio: The number of subjects that have to be examined in a *clinical trial* to identify one protocol-eligible subject. [*Critical Care Medicine*, 2000, **28**, 867–71.]

Screening experiments: Experiments in which many factors are considered with the purpose of identifying those factors (if any) that have large effects. The factors that are identified as important are then investigated more throughly in subsequent experiments. Such experiments frequently employ *fractional factorial designs.* See also **response surface methodology**. [*American Journal of Public Health*, 2008, **98**, 1354–1359.]

Screening studies: Studies in which *diagnostic tests* are applied to a symptomless population in order to diagnose disease at an early stage. Such studies are designed both to estimate disease prevalence and to identify for treatment patients who have particular diseases. The procedure is usually concerned with chronic illness and aims to detect disease not yet under medical care. Such studies need to be carefully designed and analysed in order to avoid possible problems arising because of *lead time bias* and *length biased sampling.* Most suitable designs are based on *random allocation.* For example, in the *continuous screen design* subjects are randomized either to a group that are given periodic screening throughout the study, or to a group that do not get such screening but simply follow the usual medical care practices. One drawback of this type of design is that the cost involved in screening all the patients in the 'intervention' arm of the trial for the duration of the trial may be prohibitive; if this is so an alternative approach can be used, namely the *stop screen design*, in which screening is offered only for a limited time in the intervention group. [SMR Chapter 14.]

SD: Abbreviation for **standard deviation**.

S_B-distribution: See **Johnson's system of distributions**.

S_L-distribution: See **Johnson's system of distributions**.

S_U-distribution: See **Johnson's system of distributions**.

SE: Abbreviation for **standard error**.

Seasonally adjusted: A term applied to *time series* from which periodic oscillations with a particular period, for example, one year have been removed. [SMR Chapter 4.]

Seasonal variation: Although strictly used to indicate the cycles in a *time series* that occur yearly, also often used to indicate other periodic movements. [SMR Chapter 7.]

Seber's model: A model for a ring recovery experiment in which R_i birds are ringed in the *i*th year of the study $1 \leq i \leq k$ and m_{ij} of these are found and reported dead in the *j*th year $i \leq j \leq k$. Letting ϕ_j be the probability that a bird survives its *i*th year of life conditional on it having survived *i*-1 years, and λ be the probability that a bird, having died, is found and reported, then given that a bird is ringed in year *i* the probability that it is recovered in year $i + j$ is $\lambda\phi_1 \ldots \phi_j(1 - \phi_{j+1})$ and the probability that it is never recovered is $1 - \lambda(1 - \phi_1 - \cdots - \phi_{k+1-i}),\ 1 \leq i \leq k,\ 0 \leq j \leq k - i.$ [*Statistical Inference from Capture Recapture Experiments*, 1990, K. H. Pollock, J. D. Nicholls, C. Brownie and J. E. Hines, The Wildlife Society.]

Secondary attack rate: The degree to which members of some collective or isolated unit, such as a household, litter or colony, become infected with a disease as a result of coming into contact with another member of the collective unit who became infected. [*Statistics in Medicine*, 1996, **15**, 2393–404.]

Second-level significance testing: See **higher criticism**.

Second-order stationarity: See **stationarity**.

Secretary problem: A problem in which *n* individuals applying for a job (for example, a secretarial position) arrive sequentially in random order. Upon arrival each individual is measured on a characteristic desirable for the position (for example, typing skill). The goal is to select the applicant with the highest value for the characteristic assessed, but the individual must be told immediately after the interview whether she or he has been hired. The optimal strategy is to reject the first *s(n)* candidates and to choose the first, if any, among applicants $s(n) + 1, \ldots, n$ who is the best seen so far, where $s(n)/n \approx e^{-1}$. [*Chance Rules*, 2nd edn, 2008, B. S. Everitt, Springer, New York.]

Secular trend: The underlying smooth movement of a *time series* over a fairly long period of time. [*Proceedings of the Nutrition Society*, 2000, **59**, 317–324.]

Seemingly unrelated regression (SUR): A set of regression models that appear unrelated because they have different response and explanatory variables but which are estimated from the same set of data and so may have correlated error terms. *Ordinary least squares* will yield unbiased and consistent estimates for the parameters of each separate model but because this approach ignores the correlation of the error terms the estimates will not be efficient. [*Seemingly Unrelated Regression Equations Models*, 1987, V. K. Srivastava and D. E. A. Giles, CRC, Boca Raton.]

Segmentation: The division of an image into regions or objects. Often a necessary step before any quantitative analysis can be carried out. Also used for finding market segments in market research usually by some form of *cluster analysis*. [*Statistics for Spatial Data*, 1991, N. A. C. Cressie, Wiley, New York.]

Segregation analysis: A method of collecting evidence for the presence of discrete phenotypic classes that are probabilistically determined by the underlying *genotype* at a locus by examining the joint distribution of *phenotype* in families. [*Statistics in Human Genetics*, 1998, P. Sham, Arnold, London.]

Selection bias: The bias that may be introduced into all types of scientific investigations whenever a treatment is chosen by the individual involved or is subject to constraints that go unobserved by the researcher. If there are unobserved factors influencing health outcomes and the type of treatment chosen, any direct links between treatment and outcome are confounded with unmeasured variables in the data. A classic example of this problem occurred in the Lanarkshire milk experiment of the 1920s. In this trial 10 000 children were given free milk supplementation and a similar number received no supplementation. The groups were formed by random allocation. Unfortunately, however, well-intentioned teachers decided that the poorest children should be given priority for free milk rather than sticking strictly to the original groups. The consequence was that the effects of milk supplementation were indistinguishable from the effects of poverty. The term may also be used in the context of multiple regression when subsets of variables chosen by the same procedure may be optimal for the original data but may perform poorly in future data and in general when individuals included in a study are not representative of the *target population* for the study. [*Statistics in Medicine*, 1988, **7**, 417–22.]

Selection methods in regression: In regression analysis an *underfitted model* can lead to severely biased estimation and prediction. In contrast, an *overfitted model* can seriously degrade the efficiency of of the resulting parameter estimates and predictions. Consequently a variety of techniques all with the aim of selecting the most important explanatory variables for predicting the response variable and thereby obtaining a parsimonious and effectively predictive model have been developed. In *multiple linear regression*, for example, three commonly used methods are *forward selection, backward elimination* and a combination of both of these known as *stepwise regression*. The criterion used for assessing whether or not a variable should be added to an existing model in forward selection or removed from an existing model in backward elimination is, essentially the change in the residual sum-of-squares produced by the inclusion or exclusion of the variable. Specifically in forward selection an 'F-statistic' known as the *F-to-enter* is calculated as

$$F = \frac{RSS_m - RSS_{m+1}}{RSS_{m+1}/(n - m - 2)}$$

where RSS_m and RSS_{m+1} are the residual sums of squares when models with m and $m+1$ explanatory variables have been fitted. The F-to-enter is then compared with a preset term; calculated Fs greater than the preset value lead to the variable under consideration being added to the model. In backward selection a calculated F less that a corresponding F-to-remove leads to a variable being removed from the current model. In the stepwise procedure, variables are entered as with forward selection, but after each addition of a new variable, those variables currently in the model are considered for removal by the backward elimination process. In this way it is possible that variables included at some earlier stage might later be removed, because the presence of new variables has made their contribution to the regression model no longer important. It should be stressed none of these automatic procedures for selecting variables is foolproof and they must be used with care.

Other methods that can be used for selecting variables are *all subsets regression* and *Mallow's C_p* statistic. Akaike's information criterion might also be useful in comparing competing regression models. But all these methods suffer from a major drawback because parameter estimation and model selection are considered as two different processes and this can result in instability and complicated stochastic properties.. And when applying regression analysis to *high-dimensional data* special selection techniques that are computationally efficient and statistically accurate are needed for retaining relevant and deleting perhaps thousands of irrelevant variables. Examples of such methods are the *Dantzing selector*, the *lasso, sure screening* and *smoothly clipped absolute deviation.* [ARA Chapter 6.]

Selection models: A class of models used to handle *missing data.* For a unit *i,* let y_i be the response variable of interest and M_i a missing indicator taking the value 1 if the response is missing and 0 if not. Let θ be the parameters of the model for y_i, which are of main interest, and ψ the parameters of the missingness mechanism. In a selection model, the joint distribution of y_i and M_i is decomposed as

$$f(y_i, M_i|\theta, \psi) = f(y_i|\theta)f(M_i|y_i, \psi)$$

where the first distribution characterises the distribution of y_i in the population and the second distribution represents the probability of missing data as a function of y_i, and the parameter vectors θ and ψ are distinct. The most famous model of this type is the *Heckman selection model.* Selection models are more natural and easier to interpret than pattern mixture models but have been criticised for being sensitive to distributional assumptions. See also **Pattern mixture model**. [*Statistical Analysis with Missing Data*, 2002, R. J. A. Little and D. B. Rubin, Wiley, New York.]

Self-modeling regression (SEMOR): A method developed to model groups of one-dimensional response curves when the curves demonstrate a similar shape with individual differences in scale and location. [*Statistica, Sinica*, 2004, **14**, 695–711.]

Self-pairing: See **paired samples**.

Self-similarity: A term that applies to geometric shapes if the same geometric structures are observed independently of the distance from which one looks at the shape. Also used to describe certain *stochastic processes,* $\{X_t\}$ that are such that for every sequence of time points $t_1, t_2, \ldots, t_k$ and for any positive constant *c*, then $c^{-H}(X_{ct_1}, X_{ct_2}, \ldots, X_{ct_k})$ has the same distribution as $X_{t_1}, X_{t_2}, \ldots, X_{t_k}$, where H is known as the *self-similarity parameter.* [*The Fractal Nature of Geometry*, 1982, B. B. Mandelbrot, Freeman, San Francisco.]

Self-similarity parameter: See **self-similarity**.

Semi-interquartile range: Half the difference between the upper and lower quartiles.

Semi-continuous variables: Variables that take on a single discrete value (often zero) with positive probability but are otherwise continuously distributed. Such variable occur in a wide variety of contexts; for example, variables in economic surveys relating to income or expenditure where many individuals will have no income or expenditure of a certain type in a given year. A commonly used approach to modelling such variables is one that involves two stages: (a) a *logistic regression model* for predicting whether the response is zero or non-zero and (b) a continuous, often *linear regression model,* for predicting the level of response among those individuals whose response are non-zero. [*Analysis of Incomplete Data*, 1997, J. Schafer, Wiley, New York.]

Semi-parametric efficiency bounds: Extensions of efficiency bounds such as the *Cramér-Rao lower bound* and the *Gauss-Markov theorem* to statistical models with a nonparametric component. [*Journal of Applied Econometrics*, 1990, **5**, 99–135.]

Semi-parametric regression: Regression models that are a compromise between parametric and nonparametric models, which aim to offer the flexibility of the latter whilst retaining a certain amount of the parsimony and structure of the former. The most widely known semiparametric regression model is *Cox's proportional hazards model.* [*Econometrica*, 1996, **64**, 103–37.]

Semi-variogram: Synonym for **variogram**.

SEMOR: Acronym for **self-modeling regression**.

Sensitivity: An index of the performance of a *diagnostic test*, calculated as the percentage of individuals with a disease who are correctly classified as having the disease, i.e. the *conditional probability* of having a positive test result given having the disease. A test is sensitive to the disease if it is positive for most individuals having the disease. See also **specificity, ROC curve** and **Bayes' theorem**. [SMR Chapter 14.]

Sensitivity analysis: See **uncertainty analysis**.

Sequence models: A class of statistical models for the simultaneous analysis of multiple ordered events so as to identify their sequential patterns. An extension of *log-linear models* in which a set of parameters is used to characterize marginal *odds* and *odds ratios* of frequencies summed across sequence patterns for each combination of the occurrence/non-occurrence of events. These parameters are used for the analysis of the occurrence and association of events. Another set of parameters characterizes conditional odds and odds ratios among sequence patterns within each combination of the occurrence/non-occurrence of events. The parameters are used for the analysis of sequencing of events.

Sequential allocation procedures: Procedures for allocating patients to treatments in a prospective clinical trial in which they enter the study sequentially. At the time of entry values of prognostic factors that might influence the outcome of the trial are often known and procedures for allocation that utilize this information have received much attention. One of the most widely used of these procedures is the *permuted block allocation* in which strata are defined in terms of patients at allocation having the same values of all prognostic factors. In its simplest form this method will randomly allocate a treatment to an incoming patient when balance exists among the treatments within the stratum to which the new patient belongs. If balance does not exist the treatment that will achieve balance will be allocated. A problem is that, in principle, an investigator with access to all previous allocations can calculate, for a known set of prognostic factors, the treatment allocation for the next patient and consequently makes possible conscious selection bias. [*Statistics in Medicine*, 1986, **5**, 211–29.]

Sequential analysis: A procedure in which a statistical test of significance is conducted repeatedly over time as the data are collected. After each observation, the cumulative data are analysed and one of the following three decisions taken:

- stop the data collection, reject the null hypothesis and claim statistical significance;
- stop the data collection, do not reject the null hypothesis and state that the results are not statistically significant;
- continue the data collection, since as yet the cumulated data are inadequate to draw a conclusion.

In some cases, *open sequential designs*, no provision is made to terminate the trial with the conclusion that there is no difference between the treatments, in others, *closed sequential designs*, such a conclusion can be reached. In *group sequential designs*, *interim analyses* are undertaken after each accumulation of a particular number of subjects into the two groups. Suitable values for the number of subjects can be found from the overall significance level, the true treatment difference and the required power. [SMR Chapter 15.]

Sequential importance sampling (SIS): A method for approximating the *posterior distribution* of the state vector for a possible nonlinear dynamic system. Used for prediction and smoothing of nonlinear and non-Gaussian state space models. [*Journal of the American Statistical Association*, 1998, **93**, 1032–44.]

Sequential sums of squares: A term encountered primarily in regression analysis for the contributions of variables as they are added to the model in a particular sequence. Essentially the difference in the residual sum of squares before and after adding a variable. [ARA Chapter 4.]

Serial correlation: The correlation often observed between pairs of measurements on the same subject in a *longitudinal study*. The magnitude of such correlation usually depends on the time separation of the measurements–typically the correlation becomes weaker as the time separation increases. Needs to be properly accounted for in the analysis of such data if appropriate inferences are to be made. [ARA Chapter 10.]

Serial interval: The period from the observation of symptoms in one case to the observation of symptoms in a second case directly infected from the first.

Serial measurements: Observations on the same subject collected over time. See also **longitudinal data**. [SMR Chapter 14.]

Seriation: The problem of ordering a set of objects chronologically on the basis of dissimilarities or similarities between them. An example might be a *similarity matrix* for a set of archaeological graves, the elements of which are simply the number of varieties of artefacts shared by each pair of graves. The number of varieties in common is likely to decrease as the time gap between pairs of graves increases and a solution produced by say a *multidimensional scaling* of the data might be expected to show a linear ordering of graves according to age. In practice, however, the solution is more likely to show the *horseshoe shape* because for time intervals above some threshold pairs of graves will have no artefacts in common. Figure 122 shows such a scaling solution. [MV1 Chapter 5.]

Shaded distance matrix: A rough but simple way of graphically displaying a solution obtained from a *cluster analysis* of a set of observations, so that the effectiveness of the solution can be assessed. The individuals are rearranged so that those in the same cluster are adjacent to one another in the distance matrix. Distances within a cluster will be small for tight and well-separated clusters, while the distances between individuals in different clusters will be large. Coding increasing distance by decreasing gray levels, should result in a series of dark triangles under each tight well-separated cluster, while clusters that are simply artifacts of the clustering procedure, will not exhibit such behaviour. See Fig. 123 for an example. [*International Journal of Systematic Evolutionary Microbiology*, 2004, **54**, 7–13.]

Shannon, Claude Elwood (1916–2001): Born in Petoskey, Michigan, Shannon obtained a first degree in mathematics and electrical engineering from the University of Michigan in 1936, and Ph.D. in mathematics from MIT in 1940. After graduating from MIT he took a job at AT&T's Bell Laboratories in New Jersey. It was in 1948 that Shannon published one of the most significant scientific contributions of the twentieth century, 'A Mathematical Theory of Communication', which considered how to transmit messages while keeping them from becoming garbled by noise. Shannon proposed that the information content of a message had nothing to do with its contents but simply with the number of 1's and 0's that it takes to transmit it. In 1958 he moved to MIT and his ideas spread beyond communications engineering and computer science to the mathematics of probability, DNA replication and cryptography. Shannon died in Medford, Massachusetts on 24 February 2001.

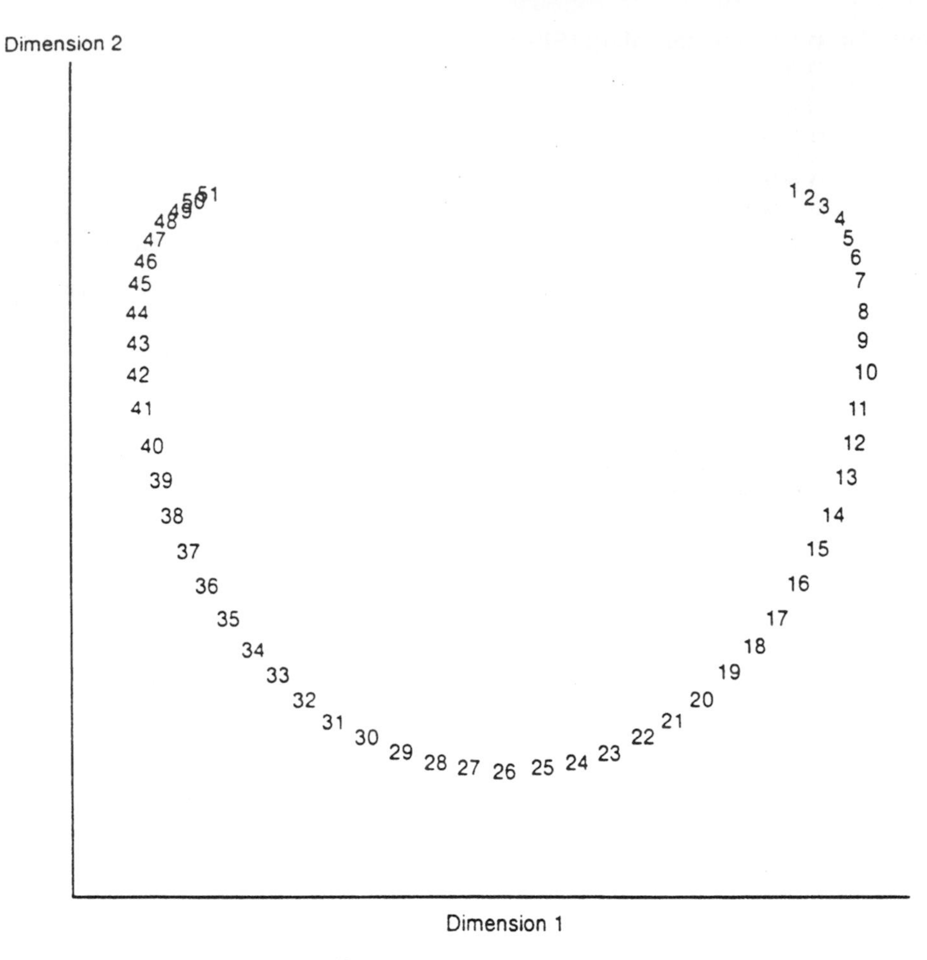

Fig. 122 Horseshoe effect.

Shannon's information measure: A measure of the average information in an event with probability *P*, given by

$$(-\log P)$$

The measure is intuitively reasonable in the sense that the more unlikely the event, the more information is provided by the knowledge that the event has occurred. The presence of a logarithm ensures that the information is additive. The logarithmic base is arbitrary and determines the unit of information. Usually base 2 is used so that information is measured in *bits*. See also **information theory** and **entropy index**. [*A First Course in Information Theory*, 2006, R. W. Yeung, Springer New York.]

Shape parameter: A general term for a parameter of a probability distribution function that determines the 'shape' (in a sense distinct from location and scale) of the distribution within a family of shapes associated with a specified type of variable. The parameter, γ, of a *gamma distribution*, for example.

Shapiro–Wilks W test: Tests that a set of random variables arise from a specified probability distribution. Most commonly used to test for depatures from the normal distribution and the *exponential distribution*. For the latter, the *test statistic* is

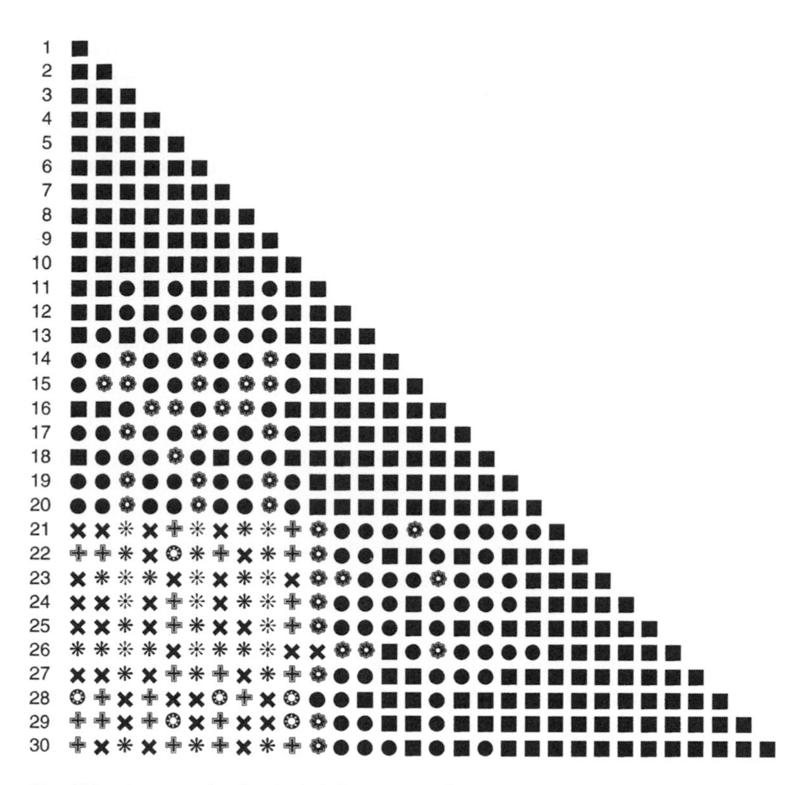

Fig. 123 An example of a shaded distance matrix.

$$W = \frac{n}{n-1} \frac{(\bar{x} - x_{(1)})^2}{\sum_{i=1}^{n}(x_{(i)} - \bar{x})^2}$$

where $x_{(1)} \leq x_{(2)} \leq \cdots \leq x_{(n)}$ are the ordered sample values and $\bar{x}$ is their mean. Critical values of W based on *simulation* studies are available in many statistical tables. [ARA Chapter 10.]

Shattuck, Lemuel (1793–1859): Born in Ashby, Massachusetts, Shattuck studied in Detroit, Michigan. At the age of 30 he returned to Concord, Massachusetts to operate a store, but eventually moved to Boston where he became a bookseller and publisher. Shattuck became interested in statistics while preparing a book, *A History of the Town of Concord*, one of the chapters of which deals with statistical history. He became very influential in publicizing the importance of high standard population statistics, and also founded the American Statistical Association.

Shepard diagram: A type of plot used in *multidimensional scaling* in which observed *dissimilarity coefficients* are plotted against the distances derived from the scaling solution. By joining together consecutive points in the diagram insight can be gained into the transformation needed to convert the observed dissimilarities into distances. [*Analysis of Proximity Data*, 1997, B. S. Everitt and S. Rabe-Hesketh, Edward Arnold, London.]

Sheppard's corrections: Corrections to the sample *moments* when those are calculated from the values of a grouped frequency distribution as if they were concentrated at the mid-points of the chosen class intervals. In particular the variance calculated from such data should be corrected by subtracting from it $h^{2/12}$ where h is the length of the interval. [KA1 Chapter 2.]

Shewhart chart: A control chart designed to identify the time at which a significant deviation in a process occurs. See also **cusum** and **exponentially weighted moving average control chart**. [*Clinical Chemistry*, 1981, **27**, 493–501.]

Shewhart, Walter A (1891-1967): Born in New Canton, Illinois, Shewhart received his Ph.D in physics from the University of California, Berkeley in 1917. He began his professional life as an engineer with the Western Electric Company and joined Bell Laboratories in 1925. Most remembered for his invention of the statistical control of quality, Shewhart died on 13 March 1967.

Shifted hats procedure: A hybrid approach to classification that uses both *kernel estimation* techniques and *finite mixture models*, essentially alternating between the two. [*Pattern Recognition*, 1993, **26**, 771–85.]

Shift outlier: An observation than needs a 'shift' of a particular magnitude for it to be consistent with the rest of the sample. [*Annals of the Institute of Statistical Mathematics*, 1994, **46**, 267–78.]

Shift tables: Tables used to summarize and interpret laboratory data usually showing, by treatment, the number of patients who have 'shifted' from having 'normal' to having 'abnormal' values during the trial and vice versa. [*Analysis of Clinical Trials using SAS, A Practical Guide*, 2005, A.Dmitrienko, G. Molenberghs, C. Chuang-Stein and W. Offen, SAS Publishing.]

Shock models: Probabilistic models of importance in the analysis of different risks in which the effects of instantaneous harmful random events that represent danger to human beings, the environment or to economic value are assessed. [*Statistics and Probability Models*, 2005, **74**, 187–204.]

Shrinkage: The phenomenon that generally occurs when an equation derived from, say, a *multiple regression*, is applied to a new data set, in which the model predicts much less well than in the original sample. In particular the value of the *multiple correlation coefficient* becomes less, i.e. it 'shrinks'. See also **shrinkage formulae**. [*Biometrics*, 1976, **32**, 1–49.]

Shrinkage estimators: Estimators obtained by some common method of estimation such as *maximum likelihood estimation* or *least squares estimation*, modified in order to minimize (maximize) some desirable criterion function such as *mean square error*. See also **ridge regression** and **James–Stein estimators**. [*Biometrics*, 1976, **32**, 1–49.]

Shrinkage formulae: Usually used for formulae which attempt to estimate the amount of 'shrinkage' in the *multiple correlation coefficient* when a regression equation derived on one set of data is used for prediction on another sample. Examples are *Wherry's formula*,

$$\hat{R}_s^2 = 1 - (n-1)/(n-k-1)(1-R^2)$$

and Herzberg's formula,

$$\hat{R}_s^2 = 1 - \left(\frac{n-1}{n-k-1}\right)\left(\frac{n-2}{n-k-2}\right)\left(\frac{n+1}{n}\right)(1-R^2)$$

where R^2 is the multiple correlation coefficient in the original sample, n is the sample size, k is the number of variables in the equation and $\hat{R}_s^2$ is the estimated value of the multiple correlation coefficient when the regression equation is used for prediction in the new sample. [*Statistical Methods in Medical Research*, 1997, **6**, 167–83.]

Sibling estimators: Estimators based on sibling data that are designed to control for common genetic and environmental family variables using sibling-specific effects. [*International Economic Review*, 1975, **16**, 422–449.]

Siegel–Tukey test: A *distribution free* test for the equality of variance of two populations having the same median. See also **Ansari–Bradley test, Conover test** and **Klotz test**. [*Handbook of Parametic and Nonparametric Statistical Procedures*, 3rd edn. D. J. Sheskin, Chapman and Hall/CRC Press, Boca Raton.]

Sigmoid: A description of a curve having an elongated 'S'-shape. [*Modelling Binary Data*, 2nd edition, 2003, D. Collett, Chapman and Hall/CRC Press, London.]

Signed rank test: See **Wilcoxon's signed rank test.**

Signed root transformation: A useful procedure for constructing *confidence intervals* when the observed *likelihood function* is noticeably non-normal. The transformation is

$$z_n(\theta) = \sqrt{[2\{l_n(\hat{\theta}) - l_n(\theta)\}]}\,\mathrm{sign}(\theta - \hat{\theta})$$

where $l_n(\theta)$ is the *log-likelihood* for n observations and $\hat{\theta}$ is the maximum likelihood estimator of θ. The quantity z_n is approximately standard normal for n large for each fixed θ. [*Biometrika*, 1973, **60**, 457–65.]

Significance level: The level of probability at which it is agreed that the null hypothesis will be rejected. Conventionally set at 0.05. [SMR Chapter 8.]

Significance test: A statistical procedure that when applied to a set of observations results in a *p-value* relative to some hypothesis. Examples include *Student's t-test, z-test* and *Wilcoxon's signed rank test.* [SMR Chapter 8.]

Sign test: A test of the null hypothesis that positive and negative values among a series of observations are equally likely. The observations are often differences between a response variable observed under two conditions on a set of subjects. [SMR Chapter 9.]

Silhouette plot: A graphical method of assessing the relative compactness and isolation of groups arising from a *cluster analysis*. For each object i an index $s(i)\ \epsilon[-1, 1]$ is defined measuring the (standardized) difference between the average dissimilarity (distance) of object i to all other objects in its own cluster and the average dissimilarity (distance) of object i to all objects in the nearest cluster to the one it is in. When $s(i)$ is close to the value 1, object i is nearer its own cluster than a neighbouring cluster and so is 'well classified'. When the opposite is the case and $s(i)$ is close to –1 the object i is taken to be 'misclassified'. In the silhouette plot the $s(i)$ are dispalyed as horizontal bars, ranked in decreasing order for each cluster. An example is shown in Fig. 124. See also **icicle plot** [*Computational Statistics and Data Analysis*, 2006, **51**, 526–544.]

Simes modified Bonferroni procedure: An improved version of the *Bonferroni correction* for conducting multiple tests of significance. If $H = \{H_1, H_2, \ldots, H_n\}$ is a set of null hypotheses with corresponding test statistics $T_1, T_2, \ldots, T_n$, P-values, $\mathrm{P}_1, \ldots, \mathrm{P}_n$ and H_0 is the hypothesis that all H_i, $i = 1, 2, \ldots, n$ are true, the suggested procedure rejects H_0 if and only if there exists some value of j $(1 \leq j \leq n)$ such that $\mathrm{P}_{(j)} \leq j\alpha/n$ where $\mathrm{P}_{(1)} \leq \cdots \leq \mathrm{P}_{(n)}$ are the ordered values of $\mathrm{P}_1, \ldots, \mathrm{P}_n$. [*Biometrika*, 1996, **83**, 928–33.]

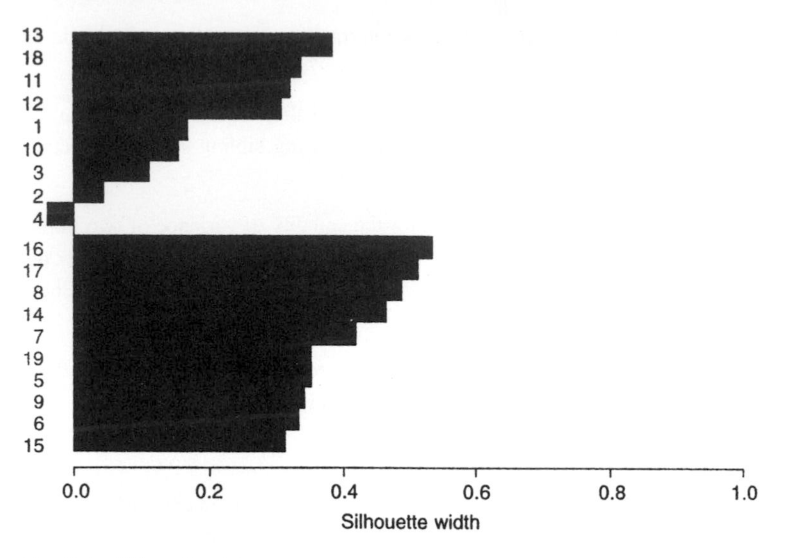

Fig. 124 Silhouette plot.

SIMEX: Abbreviation for **simulation and extrapolation procedure**.

Similarity coefficient: Coefficients ranging usually from zero to one used to measure the similarity of the variable values of two observations from a set of multivariate data. Most commonly used on binary variables. Example of such coefficients are the *matching coefficient* and *Jaccard's coefficient.* [MV1 Chapter 5.]

Similarity matrix: A *symmetric matrix* in which values on the main diagonal are one and off-diagonal elements are the values of some *similarity coefficient* for the corresponding pair of individuals. [MV1 Chapter 5.]

Simple random sampling: A form of *sampling design* in which n distinct units are selected from the N units in a finite population in such a way that every possible combination of n units is equally likely to be the sample selected. With this type of sampling design the probability that the ith population unit is included in the sample is $\pi_i = n/N$, so that the *inclusion probability* is the same for each unit. Designs other than this one may also give each unit equal probability of being included, but only here does each possible sample of n units have the same probability. [SMR Chapter 8.]

Simple structure: See **factor analysis**.

Simplex algorithm: A procedure for maximizing or minimizing a function of several variables. The basic idea behind the algorithm is to compare the values of the function being minimized at the vertices of a simplex in the parameter space and to move this simplex gradually towards the minimum during the iterative process by a combination of reflection, contraction and expansion. See also **Newton–Raphson method**. [MV2 Chapter 12.]

Simplex models: Models for the analysis of relationships among variables that can be arranged according to a logical ordering. [*Psychometrika*, 1962, **27**, 155–62.]

Simpson's paradox: The observation that a measure of association between two variables (for example, type of treatment and outcome) may be identical within the levels of a third variable (for example, sex), but can take on an entirely different value when the third variable is disregarded, and the association measure calculated from the pooled data. Such a situation can only occur if the third variable is associated with both of the other two variables. As an example consider the following pair of *two-by-two contingency tables* giving information about amount of pre-natal care and survival in two clinics.

Clinic A

		Infant's survival		
		Died	Survived	Total
	Less	3	176	179
Amount of Care				
	More	4	293	297
	Total	7	469	476

Clinic B

		Infant's survival		
		Died	Survived	Total
	Less	17	197	214
Amount of Care				
	More	2	23	25
	Total	19	220	239

In both clinics A and B, the *chi-squared statistic* for assessing the hypothesis of independence of survival and amount of care leads to acceptance of the hypothesis. (In both cases the statistic is almost zero.) If, however, the data are collapsed over clinics the resulting chi-squared statistic takes the value 5.26, and the conclusion would now be that amount of care and survival are related. See also **collapsing categories** and **log-linear models**. [*Applied Categorical Data Analysis*, 1987, D. H. Freeman, Jr., Marcel Dekker, New York.]

Simulated annealing: An *optimization* technique, which is based on an analogy with the physical process of annealing, the process by which a material undergoes extended heating and is slowly cooled. Can be helpful in overcoming the local minimum problem by allowing some probability of change in parameter values that lead to a local increase. [MV1 Chapter 4.]

Simulation: The artificial generation of random processes (usually by means of *pseudorandom numbers* and/or computers) to imitate the behaviour of particular statistical models. See also **Monte Carlo methods**. [SMR Chapter 7.]

Simulation and extrapolation (SIMEX) procedure: A procedure for calibration that is computationally intensive but is applicable to highly *non-linear models*. The method can be most clearly illustrated in simple *linear regression* when the explanatory variable is subject to measurement error. If the regression model is

$$E(Y|X) = \alpha + \beta X$$

but with $W = X + \sigma U$ observed rather than X, where U has mean zero and variance 1 and the measurement error variance σ^2 is known. For any fixed $\lambda > 0$ suppose that one repeatedly 'adds on' via simulation, additional error with mean zero and variance $\sigma^2\lambda$ to W, computes the ordinary least squares slope each time and then takes the average. The simulation estimator consistently estimates

$$g(\lambda) = \frac{\sigma_x^2}{\sigma_x^2 + \sigma^2(1+\lambda)}\beta$$

where σ_x^2 is the variance of X. Plotting $g(\lambda)$ against $\lambda \geq 0$, fitting a suitable model and then extrapolating back to $\lambda = -1$ will yield a consistent estimate of β. [*Measurement Error in Nonlinear Models,: A Modern Perspective*, 2nd edn, 2006, R. J. Carroll, D. Ruppert, L. A. Stefanski, and C. Crainiceau, Chapman and Hall/CRC, Boca Raton,]

Simulation envelope: A *confidence interval* about the expected pattern of *residuals* if the fitted model is true, constructed by simulating the distribution of the residuals taking estimated model parameters as true values. [*Statistical Inference*, 2nd edn, 2002, P. H. Garthwaite, I. T. Joliffe and B. Jones, Oxford. University Press, Oxford.]

Simultaneous confidence interval: A *confidence interval* (perhaps more correctly a region) for several parameters being estimated simultaneously. [KA2 Chapter 20.]

Simultaneous inference: Inference on several parameters simultaneously, when the parameters are considered to constitute a family in some sense. See also **multiple comparisons**. [*Biometrical Journal*, 2008, **50**, 346–363.]

Sinclair, John (1754–1835): Born in Thurso Castle, Caithness, Scotland, Sinclair was educated at the high school of Edinburgh and at Edinburgh, Glasgow and Oxford universities where he read law. His lifelong enthusiasm for collecting 'useful information' led to his work *The Statistical Account of Scotland* which was published between 1791 and 1799 in 21 volumes each of 600–700 pages. One of the first data collectors, Sinclair first introduced the words 'statistics' and 'statistical' as now understood, into the English language. He died on 21 December 1835 in Edinburgh.

Single-blind: See **blinding**.

Single-case study: Synonym for **N of 1 clinical trial**.

Single index model: A generalization of the linear regression model and given by $E(y|x) = g(x)$ where g is an unknown function. Often used in the finance industry to measure risk and return of a stock. [*Biometrika*, 2007, **94**, 217–229.]

Single-linkage clustering: A method of *cluster analysis* in which the distance between two clusters is defined as the least distance between a pair of individuals, one member of the pair being in each group. [*Cluster Analysis*, 4th edition, 2001, B. S. Everitt, S. Landau and M. Leese, Arnold, London.]

Single-masked: Synonym for **single-blind**.

Single sample *t*-test: See **Student's *t*-tests**.

Singly censored data: *Censored observations* that occur in *clinical trials* where all the patients enter the study at the same time point, and where the study is terminated after a fixed time period. See also **progressively censored data**. [*Applied Life Data Analysis*, 2004, W. Nelson, Wiley, New York.]

Singular matrix: A square matrix whose determinant is equal to zero; a matrix whose inverse is not defined. See also **Moore–Penrose inverse**.

Singular value decomposition: The decomposition of an $r \times c$ matrix, $\mathbf{A}$ into the form

$$\mathbf{A} = \mathbf{USV}'$$

where $\mathbf{U}$ and $\mathbf{V}'$ are orthogonal matrices and $\mathbf{S}$ is a diagonal matrix. The basis of several techniques of multivariate analysis including *correspondence analysis*. [MV1 Chapter 4.]

Sinh-normal distribution: The probability distribution, $f(x)$, given by

$$f(x) = [2/(\alpha\sigma\sqrt{2\pi})]\cosh[(x-\gamma)/\sigma]\exp\{(-2/\alpha^2)\sinh^2[(x-\gamma)/\sigma]\}$$

The distribution is symmetric about the location parameter γ, is strongly unimodal for $\alpha \leq 2$ and bimodal for $\alpha > 2$. The distribution of the logarithm of a random variable having a

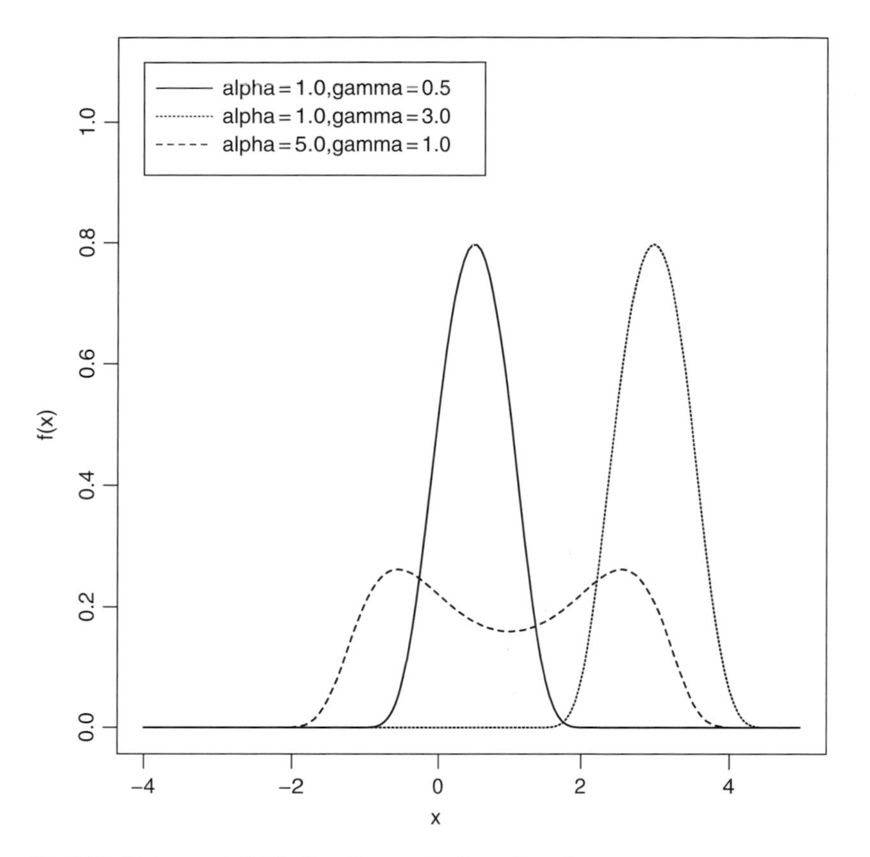

Fig. 125 Sinh-normal distributions for several values of α and γ.

Birnbaum–Saunders distribution. Some examples are given in Fig. 125. [*International Maths Forum*, 2006, **35**, 1709–1727.]

Six sigma initiative: A programme aimed at the near elimination of defects from every product, process and transaction within a company by adopting a highly disciplined and statistically based approach. [*Quality Progress*, 1988, May, 60–4.]

SiZer map: A graphical device for use in association with smoothing methods in data analysis, that helps to answer which observed features are 'really there' as opposed to being spurious sampling artifacts. [*Journal of the American Statistical Association*, 1999, **94**, 807–23.]

Skewness: The lack of symmetry in a probability distribution. Usually quantified by the index, s, given by

$$s = \frac{\mu_3}{\mu_2^{3/2}}$$

where μ_2 and μ_3 are the second and third *moments* about the mean. The index takes the value zero for a *symmetrical distribution*. A distribution is said to have *positive skewness* when it has a long thin tail to the right, and to have *negative skewness* when it has a long thin tail to the left. Examples appear in Fig. 126. See also **kurtosis**. [KA1 Chapter 3.]

Skew-normal distribution: A probability distribution, $f(x)$, given by

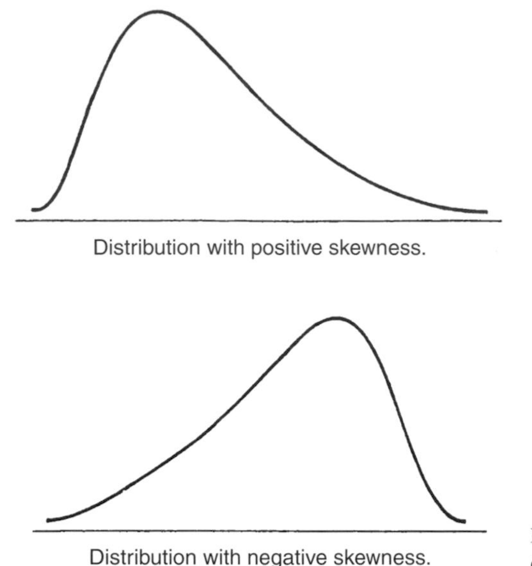

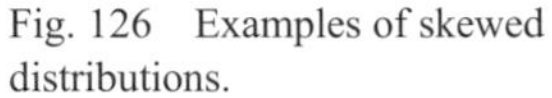
Fig. 126 Examples of skewed distributions.

$$f(x) = 2\frac{1}{\sqrt{2\pi}}e^{-\frac{1}{2}x^2}\int_{-\infty}^{\lambda x}\frac{1}{\sqrt{2\pi}}e^{-\frac{1}{2}u^2}\mathrm{d}u, \quad -\infty < x < \infty$$

For $\lambda = 0$ this reduces to the standard normal distribution. [*Biometrika*, 1996, **83**, 715–26.]

Skew-symmetric matrix: A matrix in which the elements a_{ij} satisfy

$$a_{ii} = 0; \quad a_{ij} = -a_{ji} \qquad i \neq j$$

An example of such a matrix is **A** given by

$$\mathbf{A} = \begin{pmatrix} 0 & 1 & -3 \\ -1 & 0 & 2 \\ 3 & -2 & 0 \end{pmatrix}$$

Sliced inverse regression: A data-analytic tool for reducing the number of explanatory variables in a regression modelling situation without going through any parametric or nonparametric model-fitting process. The method is based on the idea of regressing each explanatory variable on the response variable, thus reducing the problem to a series of one-dimensional regressions. [*Journal of the American Statistical Association*, 1994, **89**, 141–8.]

Sliding square plot: A graphical display of *paired samples* data. A scatterplot of the n pairs of observations (x_i, y_i) forms the basis of the plot, and this is enhanced by three *box-and-whisker plots*, one for the first observation in each pair (i.e. the control subject or the measurement taken on the first occasion), one for the remaining observation and one for the differences between the pairs, i.e. $x_i - y_i$. See Fig. 127 for an example. [*The American Statistician*, 1994, **48**, 249–253.]

Slime plot: A method of plotting *circular data* recorded over time which is useful in indicating changes of direction. [*Statistical Analysis of Circular Data*, 1995, N. I. Fisher, Cambridge University Press, Cambridge.]

Slope ratio assay: A general class of biological assay where the *dose–response* lines for the standard test stimuli are not in the form of two parallel regression lines but of two different lines with

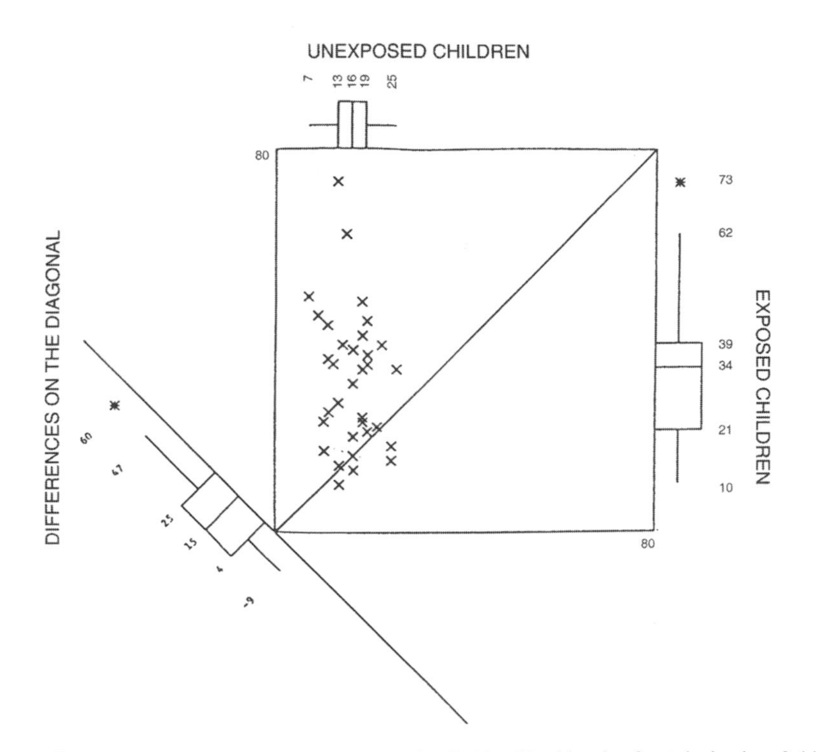

Fig. 127 An example of a sliding square plot for blood lead levels of matched pairs of children.

different slopes intersecting the ordinate corresponding to zero dose of the stimuli. The relative potency of these stimuli is obtained by taking the ratio of the estimated slopes of the two lines. [*Bioassay*, 2nd edition, 1984, J. J. Hubert, Kendall-Hunt, Dubuque.]

Slutsky, Eugen (1880–1948): Born in Yaroslaval province, Slutsky entered the University of Kiev as a student of mathematics in 1899 but was expelled three years later for revolutionary activities. After studying law he became interested in political economy and in 1918 received an economic degree and became a professor at the Kiev Institute of Commerce. In 1934 he obtained an honorary degree in mathematics from Moscow State University, and took up an appointment at the Mathematical Institue of the Academy of Sciences of the Soviet Union, an appointment he held until his death. Slutsky was one of the originators of the theory of *stochastic processes* and in the last years of his life studied the problem of compiling tables for functions of several variables.

Slutsky's theorem: If $X_1, X_2 \ldots, X_n$ are a sequence of random variables such that $\lim_{n\to\infty} P(X_n \leq x) = P(X \leq x)$ for some random variable x for which $P(X \leq x)$ is continuous everywhere, then for any continuous function g: $\lim_{n\to\infty} P[g(X_n) \leq y] = P[g(X) \leq y]$

Slutsky–Yule effect: The introduction of correlations into a *time series* by some form of smoothing. If, for example, $\{x_t\}$ is a *white noise sequence* in which the observations are completely independent, then the series $\{y_t\}$ obtained as a result of applying a *moving average* of order 3, i.e.

$$y_t = (x_{t-1} + x_t + x_{t+1})/3$$

consists of correlated observations. The same is true if $\{x_t\}$ is operated on by any *linear filter*. [TMS Chapter 2.]

Small area estimation: The application of model-based or indirect estimators to link survey outcome variables such as disease or substance use available for a national or regional study, for example, a *census*, to local area predictors such as countydemographic and socioeconomic variables, to estimate local area disease or substance use prevalence rates. The 'areas' in small area estimation may be defined by geographical domains such as a state or county and by socio-demographic characteristics such as income, race, age, or gender subgroups. Such an approach can be applied to cases where the number of area-specific sample observations is not large enough to produce reliable direct estimates. [*Small Area Estimation*, 2003, J. N. K. Rao, Wiley, New York.]

Small expected frequencies: A term that is found in discussions of the analysis of *contingency tables*. It arises because the derivation of the *chi-squared distribution* as an approximation for the distribution of the *chi-squared statistic* when the hypothesis of independence is true, is made under the assumption that the expected frequencies are not too small. Typically this rather vague phrase has been interpreted as meaning that a satisfactory approximation is achieved only when expected frequencies are five or more. Despite the widespread acceptance of this 'rule', it is nowadays thought to be largely irrelevant since there is a great deal of evidence that the usual chi-squared statistic can be used safely when expected frequencies are far smaller. See also **STATXACT**. [*The Analysis of Contingency Tables*, 2nd edition, 1992, B. S. Everitt, Chapman and Hall/CRC Press, London.]

Smear-and-sweep: A method of adjusting death rates for the effects of confounding variables. The procedure is iterative, each iteration consisting of two steps. The first entails 'smearing' the data into a two-way classification based on two of the confounding variables, and the second consists of 'sweeping' the resulting cells into categories according to their ordering on the death rate of interest. [*Encyclopedia of Statistical Sciences*, 2006, eds. S.Kotz, C. B. Read, N.Balakrishnan and B.Vidakovic, Wiley, New York.]

Smirnov, Nikolai Vasil'yevich (1900–1966): Born in Moscow, Russia, Smirnov graduated from the University of Moscow in 1926 and then taught at Moscow University, Timoryazev Agricultural Academy and Moscow City Pedagogical Institute. In 1938 he obtained his doctorate with his dissertation, 'On approximation of the distribution of random variables'. From 1938 until his death Smirnov worked at the Steklov Mathematical Institute of the USSR Academy of Sciences in Moscow making significant contributions to the distributions of statistics used in *nonparametric tests* and the limiting distributions of *order statistics*. He died on June 2nd, 1966 in Moscow.

Smith, Cedric Austen Bardell (1917–2002): Born in Leicester, UK, Smith won a scholarship to Trinity College, Cambridge, in 1935 from where he graduated in mathematics with first-class honours in 1938. He then began research in statistics under *Bartlett, J. Wishart* and *Irwin*, taking his doctorate in 1942. After World War II Smith became Assistant Lecturer at the Galton Laboratory, eventually becoming Weldon Professor in 1964. It was during this period that he worked on linkage analysis, introducing 'lods' (log-odds) to linkage studies and showing how to compute them. Later he introduced a Bayesian approach to such studies. Smith died on 16 January 2002.

Smoothing methods: A term that could be applied to almost all techniques in statistics that involve fitting some model to a set of observations, but which is generally used for those methods which use computing power to highlight unusual structure very effectively, by taking advantage of people's ability to draw conclusions from well-designed graphics. Examples of such techniques include **kernel methods, spline functions, nonparametric regression** and **locally weighted regression**. [TMS Chapter 2.]

Smoothly clipped absolute deviation: A method for estimating the parameters in a regression model and simultaneously selecting important variables consistently, whilst producing parameter estimates that are as efficient as if the true model was known. The method is particularly important for *high-dimensional data*. See also **lasso** and **sure screening methods**. [*Biometrika*, 2007, **94**, 553–568.]

SMR: Acronym for **standardized mortality rate**.

***S–N* curve:** A curve relating the effect of a constant stress (S) on the test item to the number of cycles to failure (N). [*Statistical Research on Fatigue and Fracture*, 1987, edited by T. Tanaka, S. Nishijima and M. Ichikawa, Elsevier, London.]

Snedecor, George Waddel (1881–1974): Born in Memphis, Tennessee, USA, Snedecor studied mathematics and physics at the Universities of Alabama and Michigan. In 1913 he became Assistant Professor of Mathematics at Iowa State University and began teaching the first formal statistics course in 1915. In 1927 Snedecor became Director of a newly created Mathematical Statistical Service in the Department of Mathematics with the remit to provide a campus-wide statistical consultancy and computation service. He contributed to design of experiments, sampling and *analysis of variance* and in 1937 produced a best selling book *Statistical Methods*, which went through seven editions up to 1980. Snedecor died on 15 February 1974 in Amherst, Massachusetts.

Snedecor's F-distribution: Synonym for **F-distribution**.

Snow, John (1813–1858): Born in York, England, Snow was, at the age of 14, apprenticed to a surgeon in Newcastle and later worked as a colliery surgeon. Snow was a pioneer in epidemiology and his investigation of a cholera outbreak in Soho, a district of London, in 1854 in which he plotted the number of deaths in different streets and identified the popular water pump in Broad Street as the source of the infection, probably saved many lives; Snow's findings also demonstrated that the 'miasma' (bad air) theory of how cholera was spread to be wrong. Snow died in London on the 16th June, 1858.

Snowball sampling: A method of sampling that uses sample members to provide names of other potential sample members. For example, in sampling heroin addicts, an addict may be asked for the names of other addicts that he or she knows. Little is known about the statistical properties of such samples. See also **respondent driven sampling (RDS)**. [*International Journal of Epidemology*, 1996, **25**, 1267–70.]

Snowflakes: Synonymous for **star plots**.

Sobel and Weiss stopping rule: A procedure for selecting the 'better' (higher probability of success, p) of two independent *binomial distributions*. The observations are obtained by a *play-the-winner rule* in which trials are made one at a time on either population and a success dictates that the next observation be drawn from the same population while a failure causes a switch to the other population. The proposed stopping rule specifies that play-the-winner sampling continues until r successes are obtained from one of the populations. At that time sampling is terminated and the population with r successes is declared the better. The constant r is chosen so that the procedure satisfies the requirement that the probability of correct selection is at least some pre-specified value whenever the difference in the P-values is at least some other specified value. See also **play-the-winner rule**. [*Biometrika*, 1970, **57**, 357–65.]

Soft methods: Statistical modelling often needs to make qualitative and subjective judgements that cannot be easily translated into precise probability values. Such judgements give rise to a number of different types of uncertainty which classical statistics may not be equipped to deal with. Soft methods are a range of powerful techniques developed in the area of *artificial intelligence* that attempt to address these problems when the encoding of subjective information is unavoidable. See also **belief functions** and **imprecise probabilities**. [*Soft Methods in Probability, Statistics and Data Analysis*, 2002, P. Grzegorzewski, O. Hryniewicz and M. A. Gil, Springer, New York.]

Sojourn time: The total time spent in a condition or state. Often used in medicine for the interval during which a particular condition is potentially detectable but not yet diagnosed. [*Statistics in Medicine*, 1989, **8**, 743–56.]

SOLAS: Software for *multiple imputation*. [Statistical Solutions, 8 South Bank, Crosse's Green, Cork, Ireland.]

SOLO: A computer package for calculating sample sizes to achieve a particular power for a variety of different research designs. See also **nQuery advisor**. [Statistical Solutions, 8 South Bank, Crosse's Green, Cork, Ireland.]

Somer's d: A measure of association for a *contingency table* with ordered row and column categories that is suitable for the asymmetric case in which one variable is considered the response and one explanatory. See also **Kendall's tau statistics**. [*The Analysis of Contingency Tables*, 2nd edition, 1992, B. S. Everitt, Chapman and Hall/CRC Press, London.]

Sorted binary plot: A graphical method for identifying and displaying patterns in *multivariate data* sets. [*Technometrics*, 1989, **31**, 61–7.]

Sources of data: Usually refers to reports and government publications giving, for example, statistics on cancer registrations, number of abortions carried out in particular time periods, number of deaths from AIDS, etc. Examples of such reports are those provided by the World Health Organization, such as the *World Health Statistics Annual* which details the seasonal distribution of new cases for about 40 different infectious diseases, and the *World Health Quarterly* which includes statistics on mortality and morbidity.

Space–time clustering: An approach to the analysis of epidemics that takes account of three components:

- the time distribution of cases;
- the space distribution;
- a measure of the space–time interaction.

The analysis uses the simultaneous measurement and classification of time and distance intervals between all possible pairs of cases. [*Statistics in Medicine*, 1995, **14**, 2383–92.]

Spaghetti plot: A name occasionally used for a plot of individual subjects' profiles of response values in a longitudinal study. An example for data consisting of plasma inorganic phosphate measurements obtained from 13 control and 20 obese patients 0.5, 1,1.5, 2 and 3 hours after an oral glucose challenge is given in Figure 128 Such plots quickly become 'messy' as the number of subjects increases.

Sparsity-of-effect principle: The belief that in many industrial experiments involving several factors the system or process is likely to be driven primarily by some of the main effects and low-order interactions. See also **response surface methodology**.

Spatial autocorrelation: See **autocorrelation**.

Spatial automodels: *Spatial processes* whose probability structure is dependent only upon contributions from neighbouring observations and where the conditional probability

Spectral density matrix: See **multiple time series**.

Spectral radius: A term sometimes used for the size of the largest eigenvalue of a *variance–covariance matrix*.

Spectrum: See **spectral analysis**.

Spherical variable: An angular measure confined to be on the unit sphere. Figure 129 shows a representation of such a variable. [*Multivariate Analysis*, 1979, K. V. Marda, J. T. Kent and J. B. Bibby, Academic Press, London.]

Sphericity: See **Mauchly test**.

Sphericity test: Synonym for **Mauchly test**.

Spiegelman, Mortimer (1901–1969): Born in Brooklyn, New York, Spiegelman received a masters of engineering degree from the Polytechnic Institute of Brooklyn in 1923 and a masters of business administration degree from Harvard University in 1925. He worked for 40 years for the Metropolitan Life Insurance Office and made important contributions to biostatistics particularly in the areas of *demography* and public health. Spiegelman died on 25 March 1969 in New York.

Spineplot: An alternative to the *bar chart* for graphically representing categorical data; in this type of plot the category count is represented by the width of the bar rather than by the height. An example showing the number of failures and non-failures of the O-rings in space shuttle flights against temperature is shown in Figure 130. [*Computational Statistics*, 1996, **11**, 23–33.]

Splicing: A refined method of smoothing out local peaks and troughs, while retaining the broad ones, in data sequences contaminated with noise.

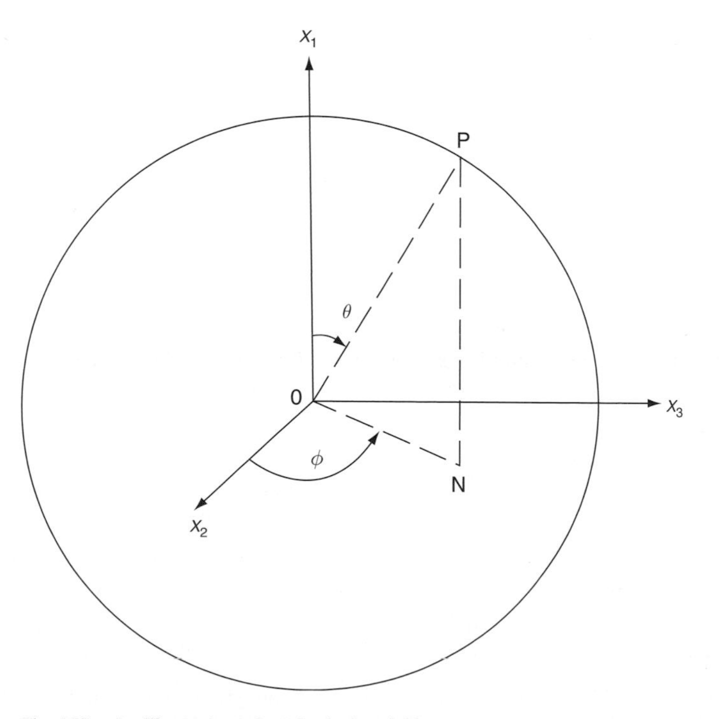

Fig. 129 An illustration of a spherical variable.

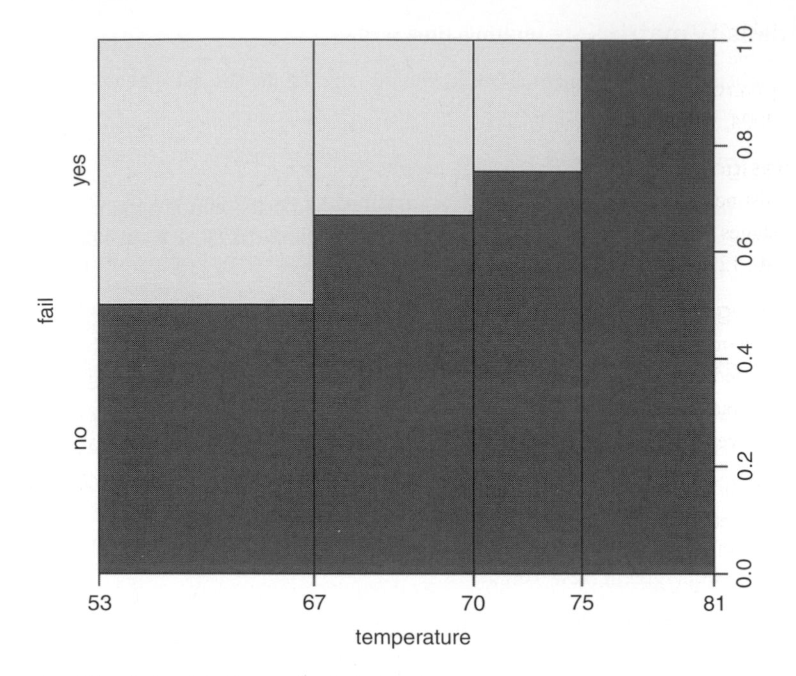

Fig. 130 Spineplot of space shuttle O-ring data.

Spline function: A smoothly joined piecewise polynomial of degree n. For example, if $t_1, t_2, \ldots, t_n$ are a set of n values in the interval a,b, such that $a < t_1 \leq t_2 \leq \cdots \leq t_n \leq b$, then a *cubic spline* is a function g such that on each of the intervals $(a, t_1), (t_1, t_2), \ldots, (t_n, b)$, g is a cubic polynomial, and secondly the polynomial pieces fit together at the points t_i in such a way that g itself and its first and second derivatives are continuous at each t_i and hence on the whole of a, b. The points t_i are called *knots*. A commonly used example is a *cubic spline* for the smoothed estimation of the function g in the following model for the dependence of a response variable y on an explanatory variable x

$$y = f(x) + \epsilon$$

where ϵ represents an error variable with expected value zero. The starting point for the construction of the required estimator is the following minimization problem; find f to minimize

$$\sum_{i=1}^{n}(y_i - f(x_i))^2 + \lambda \int_{-\infty}^{\infty} [f''(u)]^2 \mathrm{d}u$$

where primes represent differentiation. The first term is the residual sum of squares which is used as a distance function between data and estimator. The second term penalizes roughness of the function. The parameter $\lambda \geq 0$ is a smoothing parameter that controls the trade-off between the smoothness of the curve and the bias of the estimator. The solution to the minimization problem is a cubic polynomial between successive x-values with continuous first and second derivatives at the observation points. Such curves are widely used for interpolation for smoothing and in some forms of regression analysis. See also **Reinsch spline**. [*Journal of the Royal Statistical Society, Series B*, 1985, **47**, 1–52.]

Split-half method: A procedure used primarily in psychology to estimate the reliability of a test. Two scores are obtained from the same test, either from alternative items, the so-called odd–even technique, or from parallel sections of items. The correlation of these scores, or some transformation of them gives the required reliability. See also **Cronbach's alpha**. [*Statistical Evaluation of Measurement Errors: Design and Analysis of Reliability Studies*, 2004, G. Dunn, Arnold, London.]

Split-lot design: A design useful in experiments where a product is formed from a number of distinct processing stages. Each factor is applied to one and only one of the processing stages with at each of these stages a split-plot structure being used. [*Technometrics*, 1998, **40**, 127–40.]

Split-plot design: A term originating in agricultural field experiments where the division of a testing area or 'plot' into a number of parts permitted the inclusion of an extra factor into the study. In medicine similar designs occur when the same patient or subject is observed at each level of a factor, or at all combinations of levels of a number of factors. See also **longitudinal data** and **repeated measures data**. [MV2, Chapter 13.]

S-PLUS: A high level programming language with extensive graphical and statistical features that can be used to undertake both standard and non-standard analyses relatively simply. [Insightful, 5th Floor, Network House, Basing View, Basingstoke, Hampshire, RG21 44G, UK; Insightful Corporation, 1700 Westlake Avenue North, Suite 500, Seattle, Washington, WA 98109-3044, USA; www.insightful.com]

Spread: Synonym for **dispersion**.

Spreadsheet: In computer technology, a two-way table, with entries which may be numbers or text. Facilities include operations on rows or columns. Entries may also give references to other entries, making possible more complex operations. The name is derived from the sheet of paper employed by an accountant to set out financial calculations, often so large that it had to be spread out on a table. [*Journal of Medical Systems*, 1990, **14**, 107–17.]

SPSS: A statistical software package, an acronym for Statistical Package for the Social Sciences. A comprehensive range of statistical procedures is available and, in addition, extensive facilities for file manipulation and re-coding or transforming data. [SPSS UK, St Andrew's House, West St., Woking, Surrey, GU21 6EB, UK; SPSS Inc., 233 S.Wacker Drive, Chicago, Illinois 60606-6307, USA; www.spss.com.]

Spurious correlation: Commonly used for a correlation between two variables that disappears when it is 'controlled' for a third variable but is also used for the introduction of correlation due to computing rates using the same denominator. Specifically if two variables X and Y are not related, then the two ratios X/Z and Y/Z will be related, where Z is a further random variable. [*Causality, Models, Reasoning and Inference*, 2000, J. Pearl, Cambridge University Press, Cambridge.]

Spurious precision: The tendency to report results to too many significant figures, largely due to copying figures directly from computer output without applying some sensible *rounding*. See also **digit preference**.

SQC: Abbreviation for **statistical quality control**.

Square contingency table: A *contingency table* with the same number of rows as columns.

Square matrix: A matrix with the same number of rows as columns. *Variance–covariance matrices* and *correlation matrices* are statistical examples.

Square root rule: A rule sometimes considered in the allocation of patients in a *clinical trial* which states that if it costs r times as much to study a subject on treatment A than B, then one should allocate $\sqrt{r}$ times as many patients to B than A. Such a procedure minimizes the cost of a trial while preserving power. [*Randomization in Clinical Trials: Theory and Practice*, 2002, W. F. Rosenberger and J. M. Lachin, Wiley, New York.]

Square root transformation: A transformation of the form $y = \sqrt{x}$ often used to make random variables suspected to have a *Poisson distribution* more suitable for techniques such as *analysis of variance* by making their variances independent of their means. See also **variance stabilizing transformations**. [SMR Chapter 3.]

Stable Pareto distribution: See **Pareto distribution**.

Stage line diagrams: A type of reference diagram useful for tracking developmental processes over time, for example, in oncology. In the diagram transition probabilities between successive stages are modelled as smoothly varying functions of age. [*Statistics in Medicine*, 2009, **28**, 1569–1579.]

Staggered entry: A term used when subjects are entered into a study at times which are related to their own disease history (e.g, immediately following diagnosis) but which are unpredictable from the point-of-view of the study. [*Annals of Statistics*, 1997, **25**, 662–682.]

Stahel–Donoho robust multivariate estimator: An estimator of multivariate location and scatter obtained as a weighted mean and a weighted *variance–covariance matrix* with weights of the form *W(r)*, where *W* is a weight function and *r* quantifies the extent to which an observation may be regarded as an *outlier*. The estimator has high breakdown point. See also **minimum volume ellipsoid estimator**. [*Journal of the American Statistical Association*, 1995, **90**, 330–41.]

Staircase method: Synonym for **up-and-down method**.

Stalactite plot: A plot useful in the detection of multiple *outliers* in *multivariate data*, that is based on *Mahalanobis distances* calculated from means and covariances estimated from increasing sized subsets of the data. The aim is to reduce the masking effect that can arise due to the influence of outliers on the estimates of means and covariances obtained from all the data. The central idea is that, given distances using m observations for estimation of means and covariances, the $m+1$ observations to be used for this estimation in the next stage are chosen to be those with the $m+1$ smallest distances. Thus an observation can be included in the subset used for estimation for some value of m, but can later be excluded as m increases. The plot graphically illustrates the evolution of the set of outliers as the size of the fitted subset m increases. Initially m is usually chosen as $q+1$ where q is the number of variables, since this is the smallest number allowing the calculation of the required Mahalanobis distances. The cut-off point generally employed to define an outlier is the maximum expected value from a sample of n (the sample size) random variables having a *chi-squared distribution* on q degrees of freedom given approximately by $\chi_p^2\{(n-0.5)/n\}$. by $\chi_p^2\{(n-0.5)/n\}$. The example shown in Fig. 131 arises for a data set of seven variables for climate and ecology measured for 41 cities in the USA. Initially most cities are indicated as outliers (a 'one' in the plot) but as the number of observations on which the Mahalanobis distances are calculated is increased the number of outliers indicated by the plot decreases. Only two cities are indicated to be outliers for all stages of the plot; these are Chicago and Phoenix which both have large values on one or other of the variables. [*Journal of the American Statistical Association*, 1994, **89**, 1329–39.]

STAMP: Structural Time series Analyser, Modeller and Predictor, software for constructing a wide range of *structural time series models*. [*STAMP 5.0; Structural Time Series Analyser,*

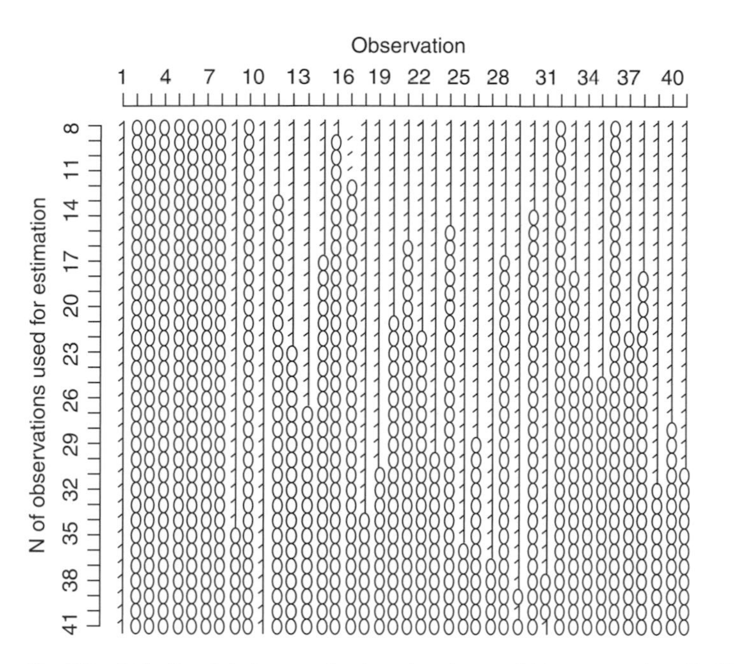

Fig. 131 Stalactite plot of seven climate and ecology variables measured on 41 cities in the USA.

Modeller and Predictor, 1995, S. J. Koopman, A. C. Harvey, J. A. Doornik and N. Shephard, Chapman and Hall/CRC Press, London.]

Standard curve: The curve which relates the responses in an assay given by a range of standard solutions to their known concentrations. It permits the analytic concentration of an unknown solution to be inferred from its assay response by interpolation.

Standard design: Synonym for **Fibonacci dose escalation scheme**.

Standard deviation: The most commonly used measure of the spread of a set of observations. Equal to the square root of the variance. [SMR Chapter 3.]

Standard error: The standard deviation of the *sampling distribution* of a statistic. For example, the standard error of the sample mean of n observations is $\sigma/\sqrt{n}$, where σ^2 is the variance of the original observations. [SMR Chapter 8.]

Standard gamble: An alternative name for the **von Neumann–Morgensten standard gamble**.

Standard gamma distribution: See **gamma distribution**.

Standardization: A term used in a variety of ways in medical research. The most common usage is in the context of transforming a variable by dividing by its standard deviation to give a new variable with standard deviation 1. Also often used for the process of producing an index of mortality, which is adjusted for the age distribution in a particular group being examined. See also **standardized mortality rate, indirect standardization** and **direct standardization**. [SMR Chapter 2.]

Standardized mortality rate (SMR): The number of deaths, either total or cause- specific, in a given population, expressed as a percentage of the deaths that would have been expected if the age and sex-specific rates in a 'standard' population had applied. [SMR Chapter 2.]

Standardized range procedure: See **many outlier detection procedures**.

Standardized regression coefficient: See **beta coefficient**.

Standardized residual: See **residual**.

Standard logistic distribution: See **logistic distribution**.

Standard normal variable: A variable having a normal distribution with mean zero and variance one. [SMR Chapter 4.]

Standard scores: Variable values transformed to zero mean and unit variance.

Star plot: A method of representing *multivariate data* graphically. Each observation is represented by a 'star' consisting of a sequence of equiangular spokes called radii, with each spoke representing one of the variables. The length of a spoke is proportional to the value of the variable it represents relative to the maximum value of the variable across all the observations in the sample. The star plots for each of the 41 cities in the USA constructed from seven climate and ecology variables measured on each city are shown in Fig. 132. Chicago is clearly identified as being very different from the other four cities.

STATA: A comprehensive software package for many forms of statistical analysis; particularly useful for epidemiological and *longitudinal data*. [STATA Corporation, 4905 Lakeway Drive, College Station, TX 77845, USA; Timberlake Consultants, Unit B3, Broomsley Business Park, Worsley Bridge Rd., London SE26 5BN, UK; www.stata.com.]

Fig. 132 Star plots for 41 cities in the United States.

State space: See **stochastic process**.

State-space representation of time series: A compact way of describing a *time series* based on the result that any finite-order linear difference equation can be rewritten as a first-order vector difference equation. For example, consider the following autoregressive model

$$X_t + a_1 X_{t-1} + a_2 X_{t-2} = \epsilon_t$$

and write $X_t^{(2)} = X_t$, $X_t^{(1)} = -a_2 X_{t-1} (= -a_2 X_{t-1}^{(2)})$ then the model may be rewritten as

$$\begin{pmatrix} X_t^{(1)} \\ X_t^{(2)} \end{pmatrix} = \begin{pmatrix} 0 & -a_2 \\ 1 & -a_1 \end{pmatrix} \begin{pmatrix} X_{t-1}^{(1)} \\ X_{t-1}^{(2)} \end{pmatrix} + \begin{pmatrix} 0 \\ 1 \end{pmatrix} \epsilon_t$$

To recover X_t from the vector $[X_t^{(1)}, X_t^{(2)}]'$ we use

$$X_t = [0, 1] \begin{pmatrix} X_t^{(1)} \\ X_t^{(2)} \end{pmatrix}$$

The original model involves a two-stage dependence but the rewritten version involves only a (vector) one-stage dependence. [*Applications of Time Series Analysis in Astronomy and Meterology*, 1997, edited by T. Subba Rao, M. B. Priestley and O. Lessi, Chapman and Hall/CRC Press, London.]

Stationarity: A term applied to *time series* or *spatial data* to describe their equilibrium behaviour. For such a series represented by the random variables, $X_{t_1}, X_{t_2}, \ldots, X_{t_n}$, the key aspect of the term is the invariance of their joint distribution to a common translation in time. So the requirement of *strict stationarity* is that the joint distribution of $\{X_{t_1}, X_{t_2}, \ldots, X_{t_n}\}$ should be identical to that of $\{X_{t_1+h}, X_{t_2+h}, \ldots, X_{t_n+h}\}$ for all integers n and all allowable h, $-\infty < h < \infty$. This form of stationarity is often unnecessarily rigorous. Simpler forms are used in practice, for example, stationarity in mean which requires that $E\{X_t\}$ does not depend on t. The most used form of stationarity, *second-order stationarity*, requires that the moments up to the second order, $E\{X_t\}$, $\text{var}\{X_t\}$ and $\text{cov}\{X_{t_i+h}, X_{t_j+h}\}$, $1 \le i, j \le n$ do not depend on translation time. [TM2 Chapter 2.]

Stationary distribution: See **Markov chain**.

Stationary point process: A *stochastic process* defined by the following requirements:

(a) The distribution of the number of events in a fixed interval (t_1, t_2) is invariant under translation, i.e. is the same for $(t_1 + h, t_2 + h]$ for all h.

(b) The joint distribution of the number of events in fixed intervals $(t_1, t_2], (t_3, t_4]$ is invariant under translation, i.e. is the same for all pairs of intervals $(t_1 + h, t_2 + h], (t_3 + h, t_4 + h]$ for all h.

Consequences of these requirements are that the distribution of the number of events in an interval depends only on the length of the interval and that the expected number of events in an interval is proportional to the length of the interval. [*Spatial Statistics*, 2nd edition, 2004, B. D. Ripley, Wiley, New York.]

Statistic: A numerical characteristic of a sample. For example, the sample mean and sample variance. See also **parameter**.

Statistical disclosure limitation: Procedures whose purpose is to ensure that the risk of disclosing confidential information about identifiable persons, businesses etc., will be very small. The goal of disclosure limitation is to achieve an acceptable balance between data utility and disclosure risk. Data utility is a measure of the usefulness of a dataset for making accurate inferences and disclosure risk measures the degree to which a dataset and its realised statistics reveal sensitive information. A variety of methodologies have been developed for this data protection task, for example, adding random noise to the original data (*data perturbation*) and choosing random pairs of respondents and exchanging a fraction of their data (*data swapping*). [*Statistical Science*, 2006, **21**, 143–154.]

Statistical expert system: A computer program that leads a user through a valid statistical analysis, choosing suitable tools by examining the data and interrogating the user, and explaining its actions, decisions, and conclusions on request.

Statistical graphics: Graphics that display measured quantities by means of the combined use of points, lines, a coordinate system, numbers, symbols, words, shading and colour. Graphical displays are very popular; it has been estimated that between 900 billion (9×10^{11}) and 2 trillion (2×10^{12}) images of statistical graphics are printed each year. Perhaps one of the main reasons for such popularity is that graphical presentation of data often provides the vehicle for discovering the unexpected; the human visual system is very powerful in detecting patterns, although the following caveat from the late Carl Sagan should be kept in mind namely, 'Humans are good at discerning subtle patterns that are really there, but equally so at imagining them when they are altogether absent.' The prime objective of a graphical display is to communicate to ourselves and others. Graphic design must do everything it can to help people understand. In some cases a graphic is required to give an overview of the data and perhaps to tell a story about the data. In other cases a researcher may want a graphical display to suggest possible hypotheses for testing on new data and after some model has been fitted to the data a graphic that criticizes the model may be what is needed (for example, a plot of *residuals*.) Examples of statistical graphics are given in the following entries; *histogram, bar chart, pie chart, dot plot, scatterplot, scatterplot matrix*, and *coplot*. See also **graphical deception**. [*Visual Revelations*, 1997, H. Wainer, Springer, New York.]

Statistical journals: A list of journals which publish articles in statistical science is given on the web site, www.statsci.org/jourlist.html. Amongst these are journals that publish primarily theoretical papers, for example, *Biometrika*, *Journal of the Royal Statistical Society, Series B* and *Annals of Statistics*, journals that also publish more applied papers like *Journal of the Royal Statistical Society, Series C* and *Journal of the American Statistical Association* and journals that publish papers in one particular area, for example medical statistics with journals like *Statistics in Medicine* and *Statistical Methods in Medical Research.*

Statistical quality control (SPC): The inspection of samples of units for purposes relating to quality evaluation and control of production operations, in particular to:

(1) determine if the output from the process has undergone a change from one point in time to another;
(2) make a determination concerning a finite population of units concerning the overall quality;
(3) screen defective items from a sequence or group of production units to improve the resulting quality of the population of interest.

[*Statistical Methods for Quality Improvement*, 1989, T. P. Ryan, Wiley, New York.]

Statistical quotations: These range from the well known, for example, 'a single death is a tragedy, a million deaths is a statistic' (Joseph Stalin) to the more obscure 'facts speak louder than statistics' (Mr Justice Streatfield). Other old favourites are 'I am one of the unpraised, unrewarded millions without whom statistics would be a bankrupt science. It is we who are born, marry and who die in constant ratios.' (Logan Pearsall Smith) and 'thou shalt not sit with statisticians nor commit a Social Science' (W. H. Auden).

Statistical societies: A list of professional societies with statistical interests and links to their respective web sites is given on www.statsci.org/soc.html.

Statistical software: Computer programs that implement a wide range of statistical techniques and which are reasonably easy to use even for non-statisticians. The use of statistical computer packages by statisticians and other started in the 1950s and today there is a plethora of packages, all of which can be used on a researchers PC. Perhaps the most widely used are **SAS**, **STATA, SPSS** and **STATISTICA**, although **R** is favoured by many statisticians. Such software enables even naive researchers to apply sophisticated statistical methodology to their data which is, of course, not without its dangers. In all current statistical software there is a heavy emphasis on graphical methods. See also **BMDP, GLIM, GENSTAT, MINITAB, S-PLUS, EGRET, BUGS, STATXACT** and **LOGXACT** . [*American Statistician*, 1994, **48**, 254–255.]

Statistical surveillance: The continual observation of a *time series* with the goal of detecting an important change in the underlying process as soon as possible after it has occurred. An example of where such a procedure is of considerable importance is in monitoring foetal heart rate during labour. [*Journal of the Royal Statistical Society, Series A*, 1996, **159**, 547–63.]

Statistics: Either the plural of statistic or the name of a discipline that many have tried to define; some examples are

- Statistics may be regarded as (i) the study of populations, (ii) as the study of variation, (iii) as the study of methods for the reduction of data.
- Statistics is concerned with the inferential process, in particular with the planning and analysis of experiments or surveys, with the nature of observational errors and sources of variability that obscure underlying patterns, and with the efficient summarizing of sets of data.
- The technology of the scientific method
- Statistics is a general intellectual method that applies wherever data, variation, and chance appear. It is a fundamental method because data, variation and chance are omnipresent in modern life. It is an independent discipline with its own core ideas, rather than, for example, a branch of mathematics... Statistics offers general, fundamental and independent ways of thinking.

There is clearly no consensus but certain elements appear in most definitions namely, variation, uncertainty, and inference. One thing that statistics is *not* is simply a branch of mathematics.

STATISTICA: A comprehensive package for many forms of statistical analysis. [StatSoft, Inc., 2300 East 14th Street, Tulsa OK 74104, USA; www.statsoft.com.]

Statistics for the terrified: A computer-aided learning package for statistics. See also **Activ Stats**. [*Statistics for the Terrified, version 3.0*, 1998, Radcliffe Medical Press, Oxford.]

STAT/TRANSFER: Software for moving data from one proprietary format to another. [Circle System, 1001 Fourth Ave., Suite 3200, Seattle, WA 98154.]

STATXACT: A specialized statistical package for analysing data from *contingency tables* that provides exact p-values, which, in the case of sparse tables may differ considerable from the values given by *asymptotic statistics* such as the *chi-squared statistic.* [www.Cytel.com/products/statxact/]

Steepest descent: A procedure for finding the maximum or minimum value of several variables by searching in the direction of the positive (negative) gradient of the function with respect to the parameters. See also **simplex method** and **Newton–Raphson method**. [MV1 Chapter 4.]

```
14 : 2
14 : 555
14 : 67777
14 : 889
15 : 000000111111
15 : 22222222222233333333333333333
15 : 44444444444555555555555555555555
15 : 6666666666666666666777777777777777777777
15 : 888888888888888888888888888888899999999999999999
16 : 00000000000000000000011111111111111111111
16 : 222222222222222222333333333333333333333333333.333
16 : 44444444444444444555555555555555555
16 : 666666666667777777
16 : 88888899999999
17 : 00000000000111
17 : 333
17 : 4
17 : 67
17 : 88
```

Fig. 133 A stem-and-leaf plot for the heights of 351 elderly women.

Stem-and-leaf plot: A method of displaying data in which each observation is split into two parts labelled the 'stem' and the 'leaf'. A tally of the leaves corresponding to each stem has the shape of a *histogram* but also retains the actual observation values. See Fig. 133 for an example. See also **back-to-back stem-and-leaf plot**. [SMR Chapter 3.]

Stepwise regression: See **selection methods in regression**.

Stereology: The science of inference about three-dimensional structure based on two-dimensional or one-dimensional probes. Has important applications in mineralogy and metallurgy. [*Quantitative Stereology*, 1970, E. E. Underwood, Addison-Wesley, Reading.]

Stieltjes integral: See **Riemann–Stieltjes integral**.

Stirling's formula: The formula

$$n! \approx (2\pi)^{\frac{1}{2}} n^{n+\frac{1}{2}} e^{-n}$$

The approximation is remarkably accurate even for small n. For example 5! is approximated as 118.019. For 100! the error is only 0.08%. [*Handbook of Mathematical and Computational Science*, 1998, J. W. Harris and H. Stocker, Springer, New York.]

Stochastic approximation: A procedure for finding roots of equations when these roots are observable in the presence of statistical variation. [*Stochastic Approximation*, 2004, M. T. Wasson, Cambridge University Press, Cambridge.]

Stochastic frontier models: Models that postulate a function $h(\cdot)$ relating a vector of explanatory variables, $\mathbf{x}$, to an output, y

$$y = h(\mathbf{x})$$

where the function $h(\cdot)$ is interpreted as reflecting best practice, with individuals typically falling short of this benchmark. For an individual i, who has a measure of this shortfall, τ_i with $0 < \tau_i < 1$

$$y_i = h(\mathbf{x})\tau_{\mathbf{i}}$$

The model is completed by adding measurement error (usually assumed to be normal) choosing a particular functional form for $h(\cdot)$ and a distribution for τ_i. [*Journal of Econometrics*, 1977, **6**, 21–37.]

Stochastic matrix: See **Markov chain**.

Stochastic ordering: See **univariate directional ordering**.

Stochastic process: A series of random variables, $\{X_t\}$, where t assumes values in a certain range T. In most cases x_t is an observation at time t and T is a time range. If $T = \{0, 1, 2, \ldots\}$ the process is a *discrete time stochastic process* and if T is a subset of the nonnegative real numbers it is a *continuous time stochastic process*. The set of possible values for the process, T, is known as its *state space*. See also **Brownian motion, Markov chain** and **random walk**. [*Theory of Stochastic Processes*, 1977, D. R. Cox and H. D. Miller, Chapman and Hall/CRC Press, London.]

Stopping rules: Procedures that allow *interim analyses* in *clinical trials* at predefined times, while preserving the type I error at some pre-specified level. See also **sequential analysis**.

Stop screen design: See **screening studies**.

Strata: See **stratification**.

Stratification: The division of a population into parts known as strata, particularly for the purpose of drawing a sample. In addition in *epidemiology* for example, stratification may be used to asses associations between exposure and disease in strata of a third variable or possibly strata defined by several variables in an investigation of confounding and effect modification.

Stratified Cox models: An extension of *Cox's proportional hazards model* which allows for multiple strata which divide the units into distinct groups, each of which has a distinct baseline hazard function but common values for the coefficient vector $\boldsymbol{\beta}$. [*Modelling Survival Data*, 2000, T. M. Therneau and P. M. Grambsch, Springer, New York.]

Stratified logrank test: A method for comparing the survival experience of two groups of subjects given different treatments, when the groups are stratified by age or some other prognostic variable. [*Modelling Survival Data in Medical Research*, 2nd edition, 2003, D. Collett, Chapman and Hall/CRC Press, London.]

Stratified randomization: A procedure designed to allocate patients to treatments in *clinical trials* to achieve approximate balance of important characteristics without sacrificing the advantages of random allocation. See also **minimization**. [*Journal of Clinical Epidemology*, 1999, **52**, 19–26.]

Stratified random sampling: Random sampling from each strata of a population after *stratification*. [*Survey Sampling Principles*, 1991, E. K. Foreman, Marcel Dekker, New York.]

Streaky hypothesis: An alternative to the hypothesis of independent *Bernoulli trials* with a constant probability of success for the performance of athletes in baseball, basketball and other sports. In this alternative hypothesis, there is either nonstationarity where the probability of success does not stay constant over the trials or autocorrelation where the probability of success on a given trial depends on the player's success in recent trials. [*The American Statistician*, 2001, **55**, 41–50.]

Stress: A term used for a particular measure of goodness-of-fit in *multidimensional scaling*. [*Analysis of Proximity Data*, 1997, B. S. Everitt and S. Rabe-Hesketh, Arnold, London.]

Strict stationarity: See **stationarity**.

Stripes plots: A simple but effective plot for visualizing the distance of each object for its closest and second-closest cluster centroids after applying a *k-means cluster analysis*. For each cluster, $k = 1, \ldots, K$ there is a rectangular area which is vertically divided into K smaller rectangles and horizontal lines are plotted for the distances of each object to its own cluster and then for the distances to the nearest other cluster. An example where the clusters are well separated is show in Figure 134 and one where the clusters are not so well separated in Figure 135 [*Computational Statistics and Data Analysis*, 2006, **51**, 526–544.]

Strip-plot designs: A design sometimes used in agricultural field experiments in which the levels of one factor are assigned to strips of plots running through the block in one direction. A separate randomization is used in each block. The levels of the second factor are then applied to strips of plots that are oriented perpendicularly to the strips for the first factor. [*The Design of Experiments*, 1988, R. Mead, Cambridge University Press, Cambridge.]

Structural equation model: A statistical model where exogenous variables (explanatory variables) can potentially affect endogenous variables (response variables) both directly and indirectly via intervening variables. There could also be feedback effects although this is uncommon. A structural equation model for a vector of observed endogenous variables **y** and a vector of observed exogenous variables **x** is usually specified as

$$\mathbf{y} = \mathbf{B}\mathbf{y} + \Gamma\mathbf{x} + \zeta$$

Here **B**, which governs the relations among the endogenous variables, has zeros on the diagonal and is of full rank. The regression parameter matrix Γ governs the regressions of endogenous on exogenous variables and ζ is a vector of random disturbances. The "reduced form" of a structural equation model is obtained by solving for endogenous variables in terms of exogenous variables to give

$$\mathbf{y} = (\mathbf{I} - \mathbf{B})^{-1}\Gamma\mathbf{x} + (\mathbf{I} - \mathbf{B})^{-1}\zeta$$

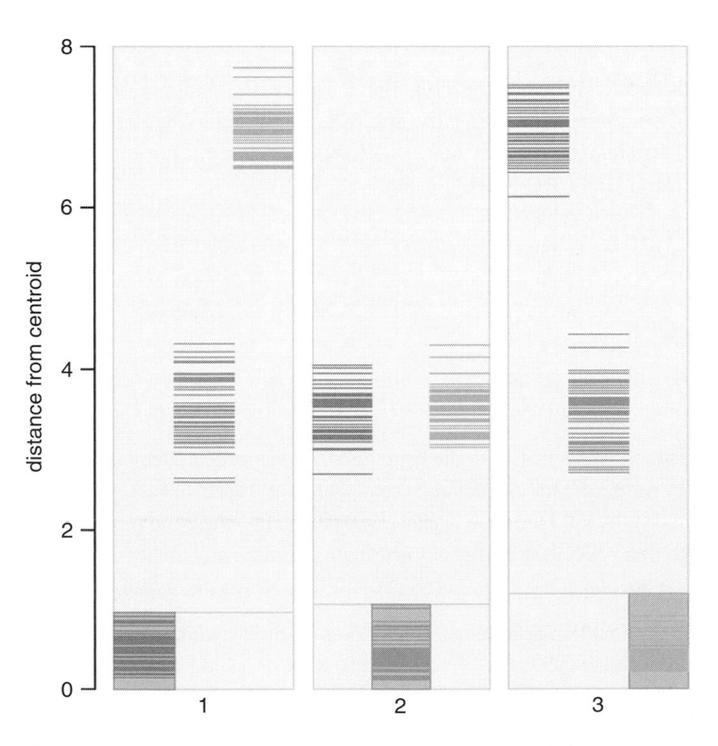

Fig. 134 Stripes plot for data containing three well separated clusters.

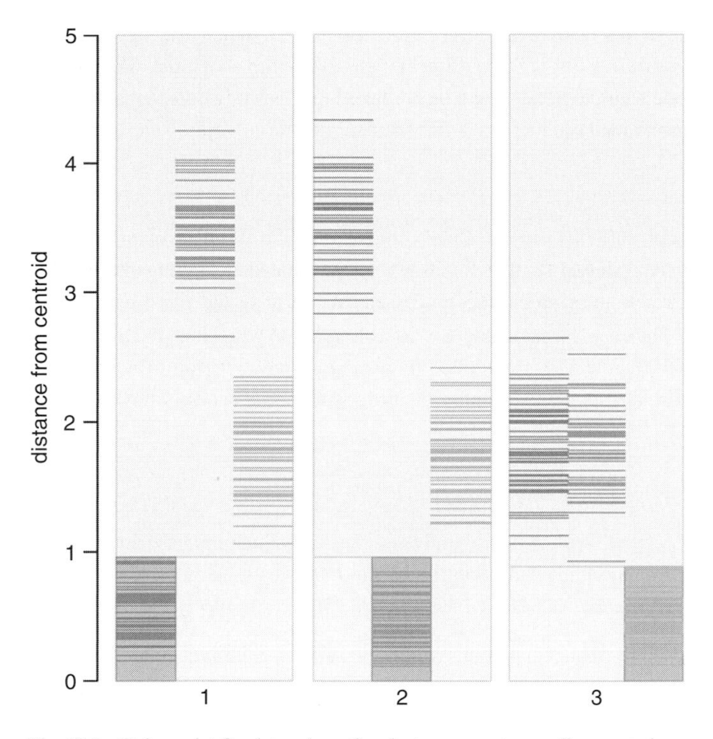

Fig. 135 Stripes plot for data where the clusters are not so well separated.

where $\mathbf{\Pi} = (\mathbf{I} - \mathbf{B})^{-1}\Gamma$ gives the reduced form regression parameters.

Structural equation models with latent rather than observed variables are also of great importance. In this case the model is composed of two submodels:

(1) a "structural model" specifying how different latent variables affect one another; explicitly the submodel is

$$\boldsymbol{\eta} = \mathbf{B}\boldsymbol{\eta} + \Gamma\xi + \zeta$$

where now $\boldsymbol{\eta}$ is a vector of latent endogenous variables, ξ is a vector of latent exogenous variables and ζ is again a vector of random disturbances.

(2) a "measurement model" specifying how the latent variables are measured by observed variables or indicators $\mathbf{x}$ and $\mathbf{y}$; explicitly this second submodel is specified as

$$\mathbf{x} = \Lambda_{\mathbf{x}}\boldsymbol{\xi} + \boldsymbol{\delta}$$
$$\mathbf{y} = \Lambda_{\mathbf{y}}\boldsymbol{\eta} + \epsilon$$

i.e., two confirmatory *factor models* in which $\Lambda_{\mathbf{x}}$ and $\Lambda_{\mathbf{y}}$ are factor loading matrices and $\boldsymbol{\delta}$ and ϵ vectors of unique factors. [*Structural Equations with Latent Variables*, 1989, K. A. Bollen, Wiley, New York]

Structural nested model: A model for estimation of the causal effect of a time-dependent exposure in the presence of *time-dependent covariates* that may simultaneously be confounders and intermediate variables. Causal effects are estimated using the *G-estimator*. More complex but also more generally applicable approach than *marginal structural model*. [*Epidemiology*, 1992, **3**, 319–336.]

Structural time series models: Regression models in which the explanatory variables are functions of time, but with coefficients which change over time. Thus within a regression framework a simple trend would be modelled in terms of a constant and time with a random disturbance added on, i.e.

$$x_t = \alpha + \beta t + \epsilon_t, \quad t = 1, \ldots, T$$

This model suffers from the disadvantage that the trend is deterministic, which is too restrictive in general so that flexibility is introduced by letting the coefficients α and β evolve over time as *stochastic processes*. In this way the trend can adapt to underlying changes. The simplest such model is for a situation in which the underlying level of the series changes over time and is modelled by a *random walk*, on top of which is superimposed a *white noise* disturbance. Formally the proposed model can be written as

$$x_t = \mu_t + \epsilon_t$$
$$\mu_t = \mu_{t-1} + \zeta_t$$

for $t = 1, \ldots, T$, $\epsilon_t \sim N(0, \sigma_\epsilon^2)$ and $\zeta_t \sim N(0, \sigma_\zeta^2)$. Such models can be used for *forecasting* and also for providing a description of the main features of the series. See also **STAMP**. [*Statistical Methods in Medical Research*, 1996, **5**, 23–49.]

Structural zeros: Zero frequencies occurring in the cells of *contingency tables* which arise because it is theoretically impossible for an observation to fall in the cell. For example, if male and female students are asked about health problems that cause them concern, then the cell corresponding to say menstrual problems for men will have a zero entry. See also **sampling zeros**. [*The Analysis of Contingency Tables*, 2nd edition, 1992, B. S. Everitt, Chapman and Hall/CRC Press, London.]

Stuart, Alan (1922–1998): After graduating from the London School of Economics (LSE), Stuart began working there as a junior research officer in 1949. He spent most of his academic career at the LSE, working in particular on nonparametric tests and sample survey theory. Stuart is probably best remembered for his collaboration with *Maurice Kendall* on the *Advanced Theory of Statistics*.

Stuart–Maxwell test: A test of *marginal homogeneity* in a *square contingency table*. The test statistic is given by

$$X^2 = \mathbf{d}'\mathbf{V}^{-1}\mathbf{d}$$

where **d** is a column vector of any r–1 differences of corresponding row and column marginal totals with r being the number of rows and columns in the table. The $(r–1) \times (r–1)$ matrix **V** contains variances and covariances of these differences, i.e.

$$v_{ii} = n_{i.} + n_{.j} - 2n_{ij}$$
$$v_{ij} = -(n_{ij} + n_{ji})$$

where n_{ij} are the observed frequencies in the table and $n_{i.}$ and $n_{.j}$ are marginal totals. If the hypothesis of marginal homogeneity is true then X^2 has a *chi-squared distribution* with r-1 degrees of freedom. [SMR Chapter 10.]

Studentization: The removal of a *nuisance parameter* by constructing a statistic whose *sampling distribution* does not depend on that parameter.

Studentized range statistic: A statistic that occurs most often in *multiple comparison tests*. It is defined as

$$q = \frac{\bar{x}_{\text{largest}} - \bar{x}_{\text{smallest}}}{\sqrt{\text{MSE}/n}}$$

where $\bar{x}_{\text{largest}}$ and $\bar{x}_{\text{smallest}}$ are the largest and smallest means among the means of k groups and MSE is the *error mean square* from an *analysis of variance* of the groups. [*Biostatistics*, 2nd edition, 2004, G.van Belle, L. D. Fisher, P. J. Heagerty and T. S. Lumley, Wiley, New York.]

Studentized residual: See **residual**.

Student's *t*-distribution: The distribution of the variable

$$t = \frac{\bar{x} - \mu}{s/\sqrt{n}}$$

where $\bar{x}$ is the arithmetic mean of n observations from a normal distribution with mean μ and s is the sample standard deviation. Given explicitly by

$$f(t) = \frac{\Gamma\{\frac{1}{2}(\nu+1)\}}{(\nu\pi)^{\frac{1}{2}}\Gamma(\frac{1}{2}\nu)}(1+\frac{t^2}{\nu})^{-\frac{1}{2}(\nu+1)}$$

where $\nu = n - 1$. The shape of the distribution varies with ν and as ν gets larger the shape of the t-distribution approaches that of the standard normal distribution. Some examples of such distributions are shown in Fig. 136. A mutivariate version of this distribution arises from considering a q-dimensional vector $\mathbf{x}' = [x_1, x_2, \ldots, x_q]$ having a *multivariate normal distribution* with mean vector $\boldsymbol{\mu}$ and *variance–covariance matrix* $\boldsymbol{\Sigma}$ and defining the elements u_i of a vector $\mathbf{u}$ as $u_i = \mu_i + x_i/y^{1/2}, i = 1, 2, \ldots, q$ where $\nu y \sim \chi^2_\nu$. Then $\mathbf{u}$ has a *multivariate Student's t-distribution* given by

$$g(\mathbf{u}) = \frac{\Gamma\{\frac{1}{2}(\nu+q)\}}{(\nu\pi)^{q/2}\Gamma(\frac{1}{2}\nu)}|\boldsymbol{\Sigma}|^{-\frac{1}{2}}[1+\frac{1}{\nu}(\mathbf{u}-\boldsymbol{\mu})'\boldsymbol{\Sigma}^{-1}(\mathbf{u}-\boldsymbol{\mu})]^{-\frac{1}{2}(\nu+q)}.$$

This distribution is often used in *Bayesian inference* because it provides a multivariate distribution with 'thicker' tails than the multivariate normal. For $\nu = 1$ the distribution is known as the *multivariate Cauchy distribution*. [STD Chapter 37.]

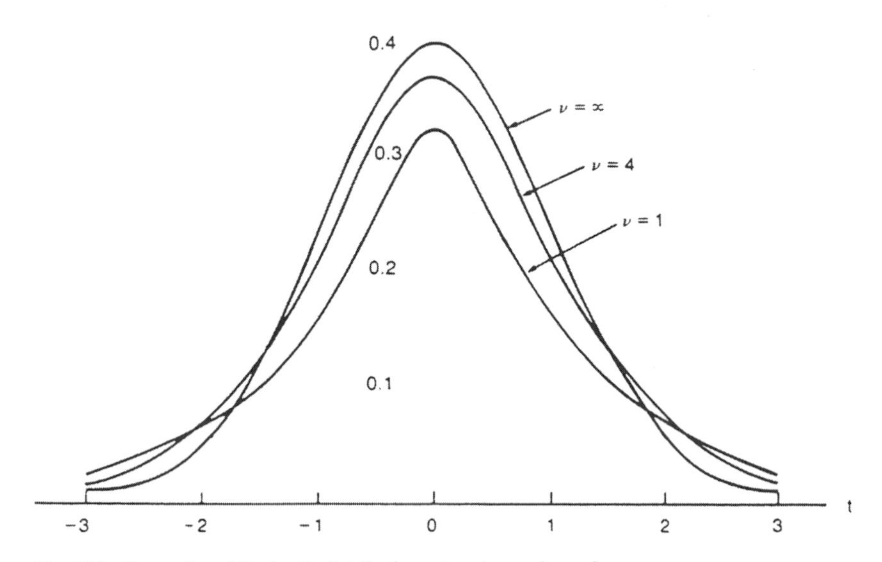

Fig. 136 Examples of Student's distributions at various values of ν.

Student's *t*-tests: Significance tests for assessing hypotheses about population means. One version is used in situations where it is required to test whether the mean of a population takes a particular value. This is generally known as a *single sample t-test*. Another version is designed to test the equality of the means of two populations. When independent samples are available from each population the procedure is often known as the *independent samples t-test* and the test statistic is

$$t = \frac{\bar{x}_1 - \bar{x}_2}{s\sqrt{\frac{1}{n_1} + \frac{1}{n_2}}}$$

where $\bar{x}_1$ and $\bar{x}_2$ are the means of samples of size n_1 and n_2 taken from each population, and s^2 is an estimate of the assumed common variance given by

$$s^2 = \frac{(n_1 - 1)s_1^2 + (n_2 - 1)s_2^2}{n_1 + n_2 - 2}$$

If the null hypothesis of the equality of the two population means is true t has a *Student's t-distribution* with $n_1 + n_2 - 2$ degrees of freedom allowing P-values to be calculated. The test assumes that each population has a normal distribution but is known to be relatively insensitive to departures from this assumption. See also **Behrens–Fisher problem** and matched **pairs t-test**. [SMR Chapter 9.]

Sturdy statistics: Synonym for **robust statistics**.

Sturges' rule: A rule for calculating the number of classes to use when constructing a *histogram* and given by

$$\text{no. of classes} = \log_2 n + 1$$

where n is the sample size. See also **Doane's rule**. [*Encyclopedia of Statistical Sciences*, 2006, eds. S.Kotz, C. B.Read, N.Balakrishnan and B.Vidakovic, Wiley, New York.]

Subgroup analysis: The analysis of particular subgroups of patients in a *clinical trial* to assess possible treatment–subset interactions. An investigator may, for example, want to understand whether a drug affects older patients differently from those who are younger. Analysing many subgroupings for treatment effects can greatly increase overall type I error rates. See also **fishing expedition** and **data dredging**. [SMR Chapter 15.]

Subjective endpoints: Endpoints in *clinical trials* that can only be measured by subjective clinical rating scales. [*Nature Medicine*, 2004, **10**, 909–15.]

Subjective probability: Synonym for **personal probability**.

Subject-specific models: See **population averaged models**.

SUDAAN: A software package consisting of a family of procedures used to analyse data from *complex survey data* and other observational and experimental studies including *clustered data*. [www.rti.org/SUDAAN/]

Sufficiency principle: If two different observations x and y have the same values $T(x)=T(y)$ of a *sufficient statistic* for a family of distributions $f(.|\theta)$, then the inference about θ based on x and y should be the same. [*Journal of the American Statistical Association*, 1962, **57**, 269–306.]

Sufficient statistic: A statistic that, in a certain sense, summarizes all the information contained in a sample of observations about a particular parameter. In more formal terms this can be expressed using the conditional distribution of the sample given the statistic and the parameter $f(y|s,\theta)$, in the sense that s is sufficient for θ if this conditional distribution does not depend on θ. As an example consider a random variable x having the following gamma distribution;

$$f(x) = x^{\gamma-1} \exp(-x)/\Gamma(\gamma)$$

The *likelihood* for γ given a sample $x_1, x_2, \ldots, x_n$ is given by

$$\exp\left(-\sum_{i=1}^{n} x_i\right)\left[\prod_{i=1}^{n} x_i\right]^{\gamma-1} / [\Gamma(\gamma)]^n$$

Thus in this case the geometric mean of the observations is sufficient for the parameter. Such a statistic, which is a function of all other such statistics, is referred to as a *minimal sufficient statistic*. [KA2 Chapter 17.]

Sukhatme, Panurang Vasudeo (1911–1997): Born in Budh, India, Sukhatme graduated in mathematics in 1932 from Fergusson College in Pine. He received a Ph.D. degree from London University in 1936 and D.Sc. in statistics from the same university in 1939. Sukhatme started his career as statistical advisor to the Indian Council of Agricultural Research, and then became Director of the Statistics Division of the Food and Agriculture Organisation in Rome. He played a leading role in developing statistical techniques that were relevant to Indian conditions in the fields of agriculture, animal husbandry and fishery. In 1961 Sukhatme was awarded the Guy medal in silver by the Royal Statistical Society. He died in Pun on 28 January 1997.

Summary measure analysis: Synonym for **response feature analysis**.

Sunflower plot: A modification of the usual *scatter diagram* designed to reduce the problem of overlap caused by multiple points at one position (particularly if the data have been rounded to integers). The scatterplot is first partitioned with a grid and the number of points in each cell of the grid is counted. If there is only a single point in a cell, a dot is plotted at the centre of the cell. If there is more than one observation in a cell, a 'sunflower' icon is drawn on which the number of 'petals' is equal to the number of points falling in that cell. An example of such a plot is given in Fig. 137. [*Journal of the American Statistical Association*, 1984, **79**, 807–822.]

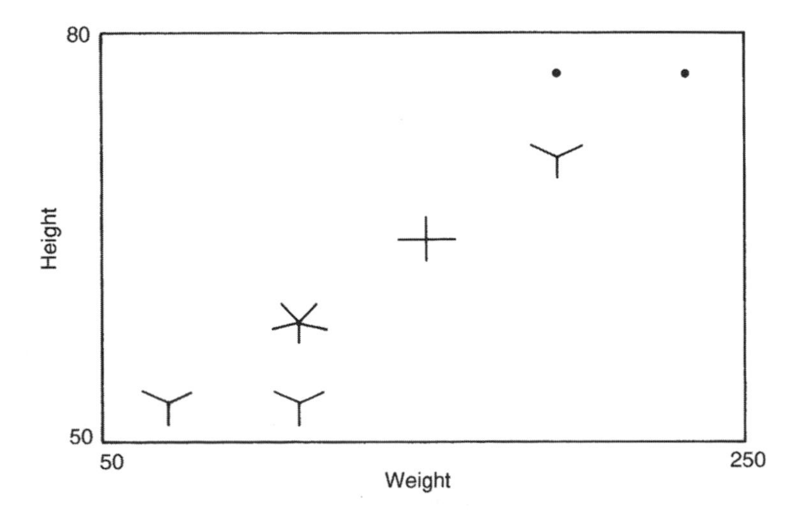

Fig. 137 A sunflower plot for height and weight.

Supercomputers: Various definitions have been proposed, some based on price, 'any computer costing more than 10 million dollars', others on performance, 'any computer where performance is limited by input/output rather than by CPU'. In essence supercomputing relates to mass computing at ultra high speeds.

Super-Duper: A random number generator arising from combining a number of simpler techniques. [*Computer Science and Statistics*, edited by W. J. Kennedy, 1973, Statistical Laboratory, Iowa State University.]

Superefficient: A term applied to an estimator which for all parameter values is asymptotically normal around the true value with a variance never exceeding and sometimes less than the *Cramér–Rao lower bound.*

Superiority trial: A *clinical trial* conducted to establish that an experimental treatment is superior to a control treatment by some measure of treatment effect or efficacy. An example of a superiority trial is the Salk vaccine trial of 1954. A *noninferiority trial* is typically conducted to establish that the treatment effect of an experimental treatment is not much worse than that of an accepted standard treatment. [*Federal Register*, 1998, **63**, 49583–49598.]

Superleverage: A term applied to observations in a *non-linear model* which have *leverage* exceeding one.

Supernormality: A term sometimes used in the context of *normal probability plots* of *residuals* from, for example, a regression analysis. Because such residuals are linear functions of random variables they will tend to be more normal than the underlying error distribution if this is not normal. Thus a straight plot does not necessarily mean that the error distribution is normal. Consequently the main use of probability plots of this kind should be for the detection of unduly influential or outlying observations.

Superpopulation: A hypothetical infinite population from which a finite population is a sample. A superpopulation model provides an alternative framework for inference in sampling by focusing not on the target finite-population parameters but on the superpopulation parameters associated with a stochastic mechanism hypothesized to generate the observations in the population, parameters that are usually more relevant for scientific questions. [*Statistical Science*, 2002, **17**, 73–96.]

Supersaturated design: A *factorial design* with n observations and k factors with $k > n - 1$. If a main-effects model is assumed and if the number of significant factors is expected to be small, such a design can save considerable cost. [*Technometrics*, 1995, **37**, 213–25.]

Supervised learning: See **pattern recognition**.

Supervised pattern recognition: See **pattern recognition**.

Support: A term sometimes used for the *likelihood* to stress its role as a measure of the evidence produced by a set of observations for a particular value(s) of the parameter(s) of a model.

Support function: Synoymous with **likelihood**.

Support vector machines (SVM): A group of supervised learning methods that can be applied to classification or regression. For classification, for example, the SVM constructs a hyperplane that optimally separates the data into two categories. Closely related to *neural networks.* [*Support Vector Machines*, 2008, I. Steinwart and A. Christmann, Springer, New York.]

Suppressor variables: A variable in a regression analysis that is not correlated with the dependent variable, but that is still useful for increasing the size of the *multiple correlation coefficient* by virtue of its correlations with other explanatory variables. The variable 'suppresses' variance that is irrelevant to prediction of the dependent variable. [*Encyclopedia of Statistical Sciences*, 2006, eds. S.Kotz, C.B. Read, N.Balakrishnan and B.Vidakovic, Wiley, New York.]

SUR: Abbreviation for **seemingly unrelated regression**.

Sure screening methods: Methods for *selection of variables* in regression that have the property that all the important variables survive with probability tending to one after the method has been applied. Such methods are important when fitting regression models to *high-dimensional data.* See also **lasso**, **oracle property** and **smoothly clipped absolute deviation**. [*Journal of the Royal Statistical Society, Series B*, 2008, **70**, 849–911.]

Surface models: A term used for those models for *screening studies* that consider only those events that can be directly observed such as disease incidence, prevelance and mortality. See also **deep models**.

Surrogate endpoint: A term often encountered in discussions of *clinical trials* to refer to an outcome measure that an investigator considers is highly correlated with an endpoint of interest but that can be measured at lower expense or at an earlier time. In some cases ethical issues may suggest the use of a surrogate. Examples include measurement of blood pressure as a surrogate for cardiovascular mortality, lipid levels as a surrogate for arteriosclerosis, and, in cancer studies, time to relapse as a surrogate for total survival time. Considerable controversy in interpretation can be generated when doubts arise about the correlation of the surrogate endpoint with the endpoint of interest or over whether or not the surrogate endpoint should be considered as an endpoint of primary interest in its own right. [*The Evaluation of Surrogate Endpoints*, 2005, eds. T.Burzykowski, G.Molenberghs and M.Buyse, Springer, New York.]

Survey: Synonym for **sample survey**.

Survival curve: See **survival function**.

Survival function: The probability that the *survival time* of an individual is longer than some particular value. A plot of this probability against time is called a *survival curve* and is a useful component in the analysis of such data. See also **product-limit estimator** and **hazard function**. [SMR Chapter 13.]

Survival time: Observations of the time until the occurrence of a particular event, for example, recovery, improvement or death. [SMR Chapter 13.]

Survivor function: Synonym for **survival function**.

Suspended rootogram: Synonym for **hanging rootogram**.

Sverdrup, Erling (1917–1994): Born in Bergen, Norway, Sverdrup's studies in actuarial mathematics were interrupted by the outbreak of World War II and were only completed in 1945. He was recruited to work in the econometrics group of the University of Oslo and in 1948 he was appointed research assistant at the Seminar of Insurance Mathematics also at the University of Oslo. In 1952 Sverdrup obtained his doctoral degree with a thesis on minimax methods and as Professor of Insurance Mathematics and Mathematical Statistics at the University of Oslo he built a modern educational program in mathematical statistics. He died on March 15th, 1994 in Oslo.

Switch-back designs: *Repeated measurement designs* appropriate for experiments in which the responses of experimental units vary with time according to different rates. The design adjusts for the different rates by switching the treatments given to units in a balanced way.

Switching effect: The effect on estimates of treatment differences of patients' changing treatments in a *clinical trial.* Such changes of treatment are often allowed in, for example, cancer trials, due to lack of efficacy and/or disease progression. [*Statistics in Medicine*, 2005, **24**, 1783–90.]

Switching regression model: A model linking a response variable y and q explanatory variables $x_1, \ldots, x_q$ and given by

$$y = \beta_{10} + \beta_{11}x_1 + \beta_{12}x_2 + \cdots + \beta_{1q}x_q + \epsilon_1 \text{ with probability } \lambda$$
$$y = \beta_{20} + \beta_{21}x_1 + \beta_{22}x_2 + \cdots + \beta_{2q}x_q + \epsilon_2 \text{ with probability } 1 - \lambda$$

where $\epsilon_1 \sim N(0, \sigma_1^2)$, $\epsilon_2 \sim N(0, \sigma_2^2)$ and $\lambda, \sigma_1^2, \sigma_2^2$ and the regression coefficients are unknown. See also **change point problems**. [*Sociological Methods & Research*, 1993, **22**, 248–273.]

Sylvan's estimator: An estimator of the *multiple correlation coefficient* when some observations of one variable are missing. [*Continuous Univariate Distributions*, Volume 2, 2nd edition, 1995, N. L. Johnson, S. Kotz and N. Balakrishnan, Wiley, New York.]

Symmetrical distribution: A probability distribution or frequency distribution that is symmetrical about some central value. [KA1 Chapter 1.]

Symmetric matrix: A square matrix that is symmetrical about its leading diagonal, i.e. a matrix with elements a_{ij} such that $a_{ij} = a_{ji}$. In statistics, *correlation matrices* and *variance–covariance matrices* are of this form.

Symmetry in square contingency tables: See **Bowker's test for symmetry**.

Synchronicity: A term used by some people (including Carl Jung and Arthur Koestler) for what they see as an acausal connecting principle needed to explain *coincidences*, arguing that such events occur far more frequently than can be explained by chance. But Jung and Koestler get little support from Fisher who suggested that the one chance in a million will undoubtedly occur with no more and no less that its appropriate frequency, however surprised we may be that it should occur to us. [*Chance Rules, 2nd edn*, 2008, B. S. Everitt, Springer, New York.]

Synergism: A term used when the joint effect of two treatments is greater than the sum of their effects when administered separately (*positive synergism*), or when the sum of their effects is less than when administered separately (*negative synergism*).

Synthesis analysis: A method that can be used to estimate the multivariate relations between multiple predictors (Xs) and an outcome variable (Y) from the univariate relation of each X with Y and the two-way correlations between each pair of Xs. [*Statistics in Medicine*, 2009, **28**, 1620–1635.]

Synthetic estimator: A *small-area estimator* is called synthetic if a reliable direct estimator for a large area, covering several small areas, is used to derive an indirect estimator for a small area under the assumption that the small areas have the same characteristics as the large area. [*Small Area Estimation*, 2003, J. N. K. Rao, Wiley, New York.]

Synthetic risk maps: Plots of the scores derived from a *principal components analysis* of the correlations among cancer mortality rates at different body sites. [*Statistics in Medicine*, 1993, **12**, 1931–42.]

SYSTAT: A comprehensive statistical software package with particularly good graphical facilities. [Systat Software Inc, 501 Canal Blvd Suite E, Point Richmond, CA 94804-2028, USA; Systat Software UK Ltd., 24 Vista Centre, 50 Salsbury Rd., Hounslow, London TW4 6JQ, UK.]

Systematic allocation: Procedures for allocating treatments to patients in a *clinical trial* that attempt to emulate random allocation by using some systematic scheme such as, for example, giving treatment A to those people with even birth dates, and treatment B to those with odd dates. While in principle unbiased, problems arise because of the openness of the allocation system, and the consequent possibility of abuse. [SMR Chapter 15.]

Systematic error: Errors that produce results that differ from the true value of some quantity by a fixed amount. Such errors can occur because of biases in the measuring instrument or by the incorrect use of the instrument by and investigator or both.

Systematic review: Synonym for **meta-analysis**.

T

Taboo probabilities: Probabilities $P^{(n)}_{ij,H}$, associated with *discrete time Markov chains*, that give the conditional probability that the chain is in state j at time n and that, in moving from initial state i to j, it has not visited the set of states H. [*Journal of Applied Probability*, 1994, **31A**, 251–67.]

Taguchi's idle column method: A procedure that enables a two-level orthogonal array to be used as the bases of a mixed-level fractional factorial experiment. The resulting designs are not orthogonal but can in some circumstances be described as nearly orthogonal. [*Systems of Experimental Design* (Volume 1), 1987, G. Taguchi, Kraus International, New York.]

Taguchi's minute accumulating analysis method: A method for the analysis of *interval-censored observations.* [*Systems of Design*, 1987, G. Taguchi, Krause International, New York.]

Taguchi's parameter design: An approach to reducing variation in products and processes that emphasizes proper choice of levels of controllable factors in a process for the manufacture of a product. The principle of choice of levels focuses to a great extent on variability around a prechosen target for the process response. [*Introduction to Off-line Quality Control*, 1980, G. Taguchi and Y. I. Wu, Central Japan Quality Control Association, Nagnja, Japan.]

Takeuchi's information criterion (TIC): An index that can be used as an aid in choosing between competing models. It is defined as

$$\text{TIC} = -2L_m + 2\text{trace}[J(\theta)I(\theta)^{-1}]$$

where L_m is the maximized *log-likelihood* and m is the number of parameters in the model. The matrices $J(\theta)$ and $I(\theta)$ contain the first and second-order mixed partial derivatives of the log-likelihood function. Thought to be useful in cases where the candidate models are not particularly close to the 'true' model. See also **Akaike's information criterion** and **Schwartz's criterion**. [*Model Based inference in the Life Sciences: A Primer on Evidence*, 2007, D. R. Anderson, Springer, New York.]

Tango's index: An index for summarizing the occurrences of cases of a disease in a stable geographical unit when these occurrences are grouped into discrete intervals. The index is given by

$$C = \mathbf{r}'\mathbf{A}\mathbf{r}$$

where $\mathbf{r}' = [r_1, \ldots, r_m]$ is the vector of relative frequencies of cases in successive periods, and $\mathbf{A}$ is an $m \times m$ matrix, the elements of which represent a measure of the closeness of interval i and j. Can be used to detect *disease clusters* occurring over time. See also **scan statistic**. [*Statistics in Medicine*, 1993, **12**, 1813–28.]

Tapering: A method used in the analysis of *time series* whereby the series is transformed in some way prior to *spectral analysis*, in an effort to reduce bias; the price of less bias is generally an increase in variance. [*Practical Time Series*, 2001, G. Janacek, Arnold, London.]

TAR: Acronym for **threshold autoregression**

Target population: The collection of individuals, items, measurements, etc., about which it is required to make inferences. Often the population actually sampled differs from the target population and this may result in misleading conclusions being made. The target population requires a clear precise definition, and that should include the geographical area (country, region, town, etc.) if relevant, the age group and gender. [*Sampling Techniques*, 3rd edition, 1977, W. G. Cochran, Wiley, New York.]

Tarone-Ware test: See **logrank test**.

Taylor's expansion: The expression of a function, $f(x)$, as the sum of a polynomial and a remainder. Specifically given by

$$f(x) = f(a) + f'(a)(x-a) + f''(x-a)^2/2! + f'''(x-a)^3/3! + \cdots$$

where primes on f denote differentiation. Used in the *delta method* for obtaining variances of functions of random variables. [*Introduction to Optimization Methods and Their Application in Statistics*, 1987, B. S. Everitt, Chapman and Hall/CRC Press, London.]

Taylor's power law: A convenient method for finding an appropriate transformation of grouped data to make them satisfy the homogeneity of variance assumption of techniques such as the *analysis of variance.* The method involves calculating the slope of the regression line of the logarithm of the group variances against the logarithm of the group means, i.e. b in the equation

$$\log_{10} s_i^2 = a + b \log_{10} \bar{x}_i$$

The value of $1 - b/2$ indicates the transformation needed, with non-zero values corresponding to a particular *power transformation*, and zero corresponding to a *log transformation.* Most often applied in ecology studies in investigations of the relationship bertween the abundance of a species and its variability. [*Nature*, 2003, **422**, 65–8.]

TD50: Abbreviation for **tumorigenic dose 50**.

Telephone interview surveys: Surveys which generally involve some form of random digit dialling. For example, in the *Mitofsky–Waksberg scheme*, telephone exchanges are first sampled. Then telephone numbers within sampled exchanges are sampled and called. If there is success in locating a residential telephone number on the first call for an exchange, additional numbers from the exchange are sampled and called—otherwise the next sample exchange is tried. Since some exchanges tend to contain all non-residential numbers, there are efficiency gains in using this scheme over selecting phone numbers completely at random (*random digit dialling*). [*Survey Response by Telephone*, 1983, J. H. Frey, Sage Publications, Beverley Hills, California.]

Telescoping: A term used in the collection of *event history data* for an apparent shifting of events from the near distant past towards the present or more precisely towards the time at which the recollection was made.

TES: Abbreviation for **transform-expand sample**.

Test statistic: A statistic used to assess a particular hypothesis in relation to some population. The essential requirement of such a statistic is a known distribution when the null hypothesis is true. [SMR Chapter 8.]

Tetrachoric correlation: The correlation between two variables that originally arise from a bivariate normal distribution but are only observed as variables that have been dichotomized at some threshold value, leading to a data set that is simply a 2×2 table of counts. From the four frequencies in this table the tetrachoric correlation can be estimated by *maximum likelihood*, the likelihood function involving bivariate-normal probability integrals. Such correlations are widely used in behavioural genetics research and as the basis of the factor analysis of multivariate binary data. [MV2 Chapter 12.]

Theil-Sen estimator: An estimator for the slope of a linear regression equation linking two variables which is based on the median of the slopes of all pairs of points. [*Fundamentals of Modern Statistical Methods*, 2001, R. R. Wilcox, Springer, New York.]

Theil's test: A *distribution free* test that the slope parameter in a simple *linear regression* model $y_i = \beta_0 + \beta_1 x_i + \varepsilon_i$ takes a particular value $\beta_1^{(0)}$ where it is assumed only that the error terms follow a distribution with zero median. The test statistic T is defined as

$$T = \sum_{i<j} c(D_j - D_i)$$

where

$$c(x) = 1 \text{ if } x>0$$
$$= 0 \text{ if } x = 0$$
$$= -1 \text{ if } x<0$$

and $D_i = y_i - \beta_1^{(0)}$. The approximate variance of T under the null hypothesis is

$$\text{var}_0(T) = \frac{n(n-1)(2n+5)}{18}$$

The statistic $T/\sqrt{\text{var}_0(T)}$ can be compared to a standard normal variable to assess the null hypothesis. A point estimate of β_1 is given by the median of $n(n-1)/2$ estimates of the slope

$$\frac{y_j - y_i}{x_j - x_i} \qquad i<j$$

[*Biometrika*, 1984, **71**, 51–6.]

Therapeutic trial: Synonym for **clinical trial**.

Thiel, Thorvald Nicolai (1838–1910): Born in Copenhagen, Denmark, Thiel studies Astronomy at the University of Copenhagen and became Professor of Astronomy and Director of the Copenhagen Observatory in 1875. He founded the Danish Society of Actuaries in 1901 and served as its president. Thiel's main contributions to statistics were in the areas of *skew distributions* and *cumulants*. He died on September 26th, 1910 in Copenhagen.

Thomas distribution: A probability distribution used in constructing models for the distribution of number of plants of a given species in randomly placed quadrants. [*Biometrika* 1949, **36**, 18–25.]

Thomson, Sir Godfrey Hilton (1881–1955): Born in Carlisle, England on March 27, 1881 Thomson studied for a MSc in mathematics and physics at Durham University and completed a PhD at Strasbourg University in Germany in 1906. His interests later turned from science to psychology and he returned to obtain a DSc in Psychology in 1913 at Durham University where he became a professor of education in 1920. From 1925 until his retirement in 1951 Thomson served as professor of education in Edinburgh. He played an important role in the development of *factor analysis* and the theory of intelligence and apparently formalized the idea of *maximum likelihood estimation* earlier than Fisher. Thomson was knighted in 1949 and died in Edinburgh on 9th of February 1955.

Three-group resistant line: A method of *linear regression* that is resistant to *outliers* and observations with large influence. Basically the method involves dividing the data into three groups and finding the median of each group. A straight line is then fitted through these medians.

Three-mode analysis: A term for techniques of *multivariate analysis* that can be used to analyse data that can be classified in three ways, for example, by subjects, variables and time. The general aim of such techniques is to fit a model to the data with a low-dimensional representation so that the basic underlying structure can be more readily discerned and interpreted. *Individual differences scaling* is an example of such a technique. [*Analysis of Proximity Data*, 1997, B. S. Everitt and S. Rabe-Hesketh, Edward Arnold, London.]

Three-period crossover design: A design in which two treatments, A and B, are given to subjects in the order A:B:B or B:A:A. Two sequence groups are formed by random allocation. The additional third observation period alleviates many of the problems associated with the analysis of the *two-period crossover design.* In particular, an appropriate three-period crossover design allows for use of all the data to estimate and test direct treatment effects even when *carry-over effects* are present. [*Statistics in Medicine*, 1996, **15**, 127–44.]

Three-stage sampling: A process that involves sampling the subunits from a two-stage sampling instead of enumerating them completely. [*Sampling Techniques*, 3rd edition, 1977, W. G. Cochran, Wiley, New York.]

Threshold autoregressive model: A non-linear model for *time series* capable of capturing asymmetric cyclical behaviour, for example, an unemployment rate which exhibits a tendency to increase at a faster rate than it decreases. A simple example of such a model is

$$\begin{aligned} x_t &= -1.5x_{t-1} + \varepsilon_t \text{ if } x_{t-1} \leq 0 \\ &= 0.6x_{t-1} + \varepsilon_t \text{ if } x_{t-1} > 0 \end{aligned}$$

[*Threshold Models in Nonlinear Time Series Analysis*, 1983, H. Tong, Springer-Verlag, New York.]

Threshold-crossing data: Measurements of the time when some variable of interest crosses a threshold value. Because observations on subjects occur only periodically, the exact time of crossing the threshold is often unknown. In such cases it is only known that the time falls within a specified interval, so the observation is *interval censored.* [*Statistics in Medicine*, 1993, **12**, 1589–1603.]

Threshold-crossing model: A model for a response that is a binary variable, z, which assumes that the value of the variable is determined by an observed random variable, $\mathbf{x}$, and an unobservable random variable u, such that

$$\begin{aligned} z &= 1 \quad \text{if} \quad \mathbf{x}'\boldsymbol{\beta} + u \geq 0 \\ &= 0 \quad \text{otherwise} \end{aligned}$$

where $\boldsymbol{\beta}$ is a vector of parameters. In medical research z might be an observable binary indicator of health status, and $\mathbf{x}'\boldsymbol{\beta} + u$ a latent continuous variable determining health status. [*Network Computation in Neural System*, 1996, **7**, 325–33.]

Threshold limit value: The maximum permissible concentration of a chemical compound present in the air within a working area (as a gas, vapour or particulate matter) which, according to current knowledge, generally does not impair the health of the employee or cause undue annoyance. [*Occupational Medicine*, 1989, **31**, 910.]

Threshold model: A model that postulates that an effect occurs only above some threshold value. For example, a model that assumes that the effect of a drug is zero below some critical dose level.

Thurstone, Louis Leon (1887–1955): Thurstone (an anglization of the Swedish name Torsten) was born in 1887 in Chicago. He received a masters degree in mechanical engineering from Cornell in 1912 and was offered a brief assistantship in the laboratory of Thomas Edison. After two years as an instructor of geometry and drafting at Minnesota he moved to the University of Chicago. Thurstone completed a PhD in Psychology in 1917 and worked at Chicago between 1924 and 1952. A major contributor to the development of *factor analysis*, psychophysics and the theory of intelligence, he was one of the founders of the journal *Psychometrika* and the Psychometric Society, becoming the society's first President in 1936. Thurstone died in 1955.

TIC: Abbreviation for **Takeuchi's information criterion**.

Tied observations: A term usually applied to ordinal variables to indicate observations that take the same value on a variable. Also often used for identical *survival times*.

Time-by-time ANOVA: A flawed approach to the analysis of *longitudinal data* in which groups are compared by *analysis of variance* at each time point. Ignores correlations between the repeated measurements and gives no overall answer to the question of whether or not there is a group difference. [*Encyclopedia of Statistical Sciences*, 2006, eds S.Kotz, C. B. Read, N.Balakrishnan and B.Vidakovic, Wiley, New York.]

Time-dependent covariates: Covariates whose values change over time as opposed to covariates whose values remain constant over time (*time-independent covariates*). A pretreatment measure of some characteristic is an example of the latter, age and weight, examples of the former. [*Journal of the Royal Statistical Society, Series B*, 1972, **34**, 187–220.]

Time-frequency distribution: A description of the energy or power of a signal as a two-dimensional function of both time and frequency. Such distributions are used to study *nonstationary time series*. See also **Chow-Williams distribution** and **Wigner-Ville distribution**. [*Proceedings of the IEEE*, 1989, **77**, 941–981.]

Time-homogeneous Markov chain: See **Markov chain**.

Time-independent covariates: See **time-dependent covariates**.

Time series: Values of a variable recorded, usually at a regular interval, over a long period of time. The observed movement and fluctuations of many such series are composed of four different components, secular trend, seasonal variation, cyclical variation, and irregular variation. An example from medicine is the incidence of a disease recorded yearly over several decades (see Fig. 138). Such data usually require special methods for their analysis because of the presence of *serial correlation* between the separate observations. Most often time series are analysed by linear models such the classic family of *autoregressive moving average models*. But there are many observable phenomena that cannot be accounted for adequately by linear models and which give rise to *nonlinear time series*, for which special models have been

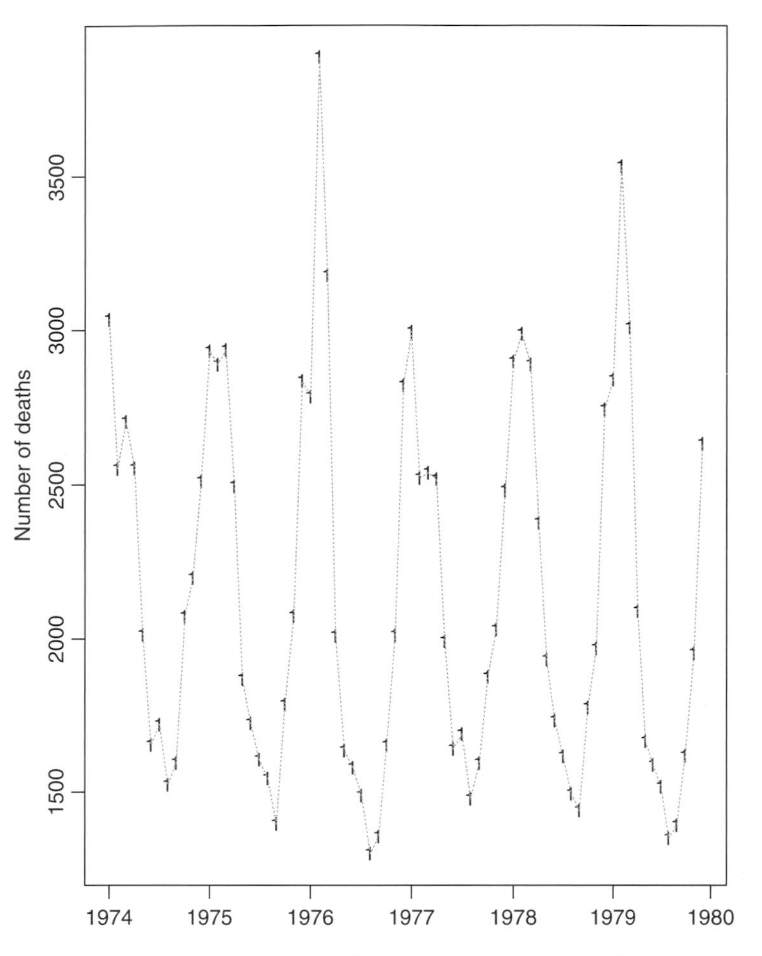

Fig. 138 A time series of monthly deaths from lung cancer in the UK 1974–1979.

developed, for example, **autoregressive conditional heteroscedastic models**. See also **autocorrelation, harmonic analysis, spectral analysis** and **fast Fourier transformation**. [TMS.]

Time-to-event data: Synonym for **survival data**.

Time trade-off technique: See **Von Neumann–Morgenstern standard gamble**.

Time-varying covariates: Synonym for **time-dependent covariates**.

T_{max}: A measure traditionally used to compare treatments in *bioequivalence trials*. The measure is simply the time at which a patient's highest recorded value occurs. See also **C_{max}**.

Tobit model: Synonym for **censored regression model**.

Toeplitz matrix: A matrix in which the element in the ith row and jth column depends only on $|i-j|$. The *variance–covariance matrices* of *stationary series* have this property. [*SIAM Journal on Matrix Analysis and Applications*, 1996, **16**, 40–57.]

Tolerance: A term used in *stepwise regression* for the proportion of the sum-of-squares about the mean of an explanatory variable not accounted for by other variables already included in the regression equation. Small values indicate possible *multicollinearity* problems.

Tolerance interval: Statistical intervals that contain at least a specified proportion of a population either on average, or else with a stated confidence value. Used to summarize uncertainty about values of a random variable, usually a future observation. [*Statistical Tolerance Regions*, 1970, I. Guttman, Hafner, Darien.]

Total matrix of sums of squares and cross-products: See **multivariate analysis of variance**.

Total sum of squares: The sum of the squared deviations of all the observations from their mean.

Tour: A motion graphic designed to study the joint distribution of *multivariate data* in search of relationships that may involve several variables. A tour is created by generating a sequence of low-dimensional projections of *high-dimensional data*. See also **brushing scatterplots**, **dynamic graphics** and **projection pursuit**. [*Interactive and Dynamic Graphics for Data Analysis*, 2007, D. Cook and D. F. Swayne, Springer, New York.],

Townsend index: An index designed to measure material deprivation of an area. The index is calculated as an unweighted sum of four standardized census variables corresponding to area-level proportions of unemployment, car ownership, crowding and home ownership. [*International Journal of Health Services*, 1985, **15**, 637–63.]

Trace of a matrix: The sum of the elements on the main diagonal of a *square matrix*; usually denoted as tr(**A**). So, for example,

$$\text{if } \mathbf{A} = \begin{pmatrix} 3 & 2 \\ 4 & 1 \end{pmatrix} \text{ then } \mathrm{tr}(\mathbf{A}) = 4$$

Tracking: A term sometimes used in discussions of data from a *longitudinal study*, to describe the ability to predict subsequent observations from earlier values. Informally this implies that subjects that have, for example, the largest values of the response variable at the start of the study tend to continue to have the larger values. More formally a population is said to track with respect to a particular observable characteristic if, for each individual, the expected value of the relevant deviation from the population mean remains unchanged over time. [*Statistics in Medicine*, 1992, **11**, 2027.]

Training set: See **discriminant analysis**.

Transfer function model: A model relating two *time series*, $\{x_t\}$ and $\{y_t\}$ and given by

$$y_t = a_0 x_t + a_1 x_{t-1} + \cdots + \varepsilon_t$$

where $a_0, a_1, \ldots$ are a set of parameters and $\{\varepsilon_t\}$ is a *white noise sequence*. In general $\{x_t\}$ is referred to as the 'input' series and $\{y_t\}$ as the 'output' series. An example might be monthly employment statistics related to inflation rate. Such models have proved very useful for prediction and forecasting. [*Time Series Analysis: Forecasting and Control*, 1994, G. E. Box, G. M. Jenkins and G. Reinsel, Prentice Hall, San Francisco.]

Transformation: A change in the scale of measurement for some variable(s). Examples are the *square root transformation* and *logarithmic transformation*. [SMR Chapter 3.]

Transform-expand sample (TES): A versatile methodology specifically designed to fit models to autocorrelated empirical observations with general marginal distribution. With this approach the aim is to fit both the empirical marginal distribution as well as the empirical *autocorrelation function*. [*Simulation*, 1995, **64**, 353–70.]

Transition matrix: See **Markov chain**.

Transition models: Models used in the analysis of *longitudinal data* in which the distribution of the observations at each time point is considered conditional on previous outcomes. Unlike *marginal models*, these models have no representation in terms of cross-sectional data. The models are comparatively straightforward to use in practice and applications with binary data, for which the models represent *Markov chains*, are widespread. See also **ante-dependence model**. [*Analysis of Longitudinal Data*, 2nd edition, 2002, P. J. Diggle, P. J. Heagerty, K.-Y. Liang and S. Zeger, Oxford Science Publications, Oxford.]

Transition probability: See **Markov chain**.

Transmission probability: A term used primarily in investigations of the spread of AIDS for the probability of contracting infection from an HIV-infected partner in one intercourse. [*Journal of the Royal Statistical Society, Series C*, 2001, **50**, 1–14.]

Transmission rate: The rate at which an infectious agent is spread through the environment or to another person.

Trapezium rule: A simple rule for approximating the integral of a function, $f(x)$, between two limits, using the formula

$$\int_a^{a+h} f(x)\mathrm{d}x = \frac{1}{2}h[f(a) + f(a+h)]$$

See also **Gaussian quadrature**. [SMR Chapter 14.]

Trapezoidal rule: Synonym for **trapezium rule**.

Trapping web design: A technique for obtaining density estimates for small mammal and arthropod populations which consists of a number of lines of traps emanating at equal angular increments from a centre point, like the spokes of a wheel. Such a design provides the link between capture data and distance sampling theory. The estimator of density is $D = M_{t+1}f(0)$, where M_{t+1} is the number of individuals captured and $f(0)$ is calculated from the M_{t+1} distances from the web centre to the traps in which those individuals were first captured. [*Biometrics*, 1994, **50**, 733–45.]

Treatment allocation ratio: The ratio of the number of subjects allocated to the two treatments in a *clinical trial.* Equal allocation is most common in practice, but it may be advisable to allocate patients randomly in other ratios when comparing a new treatment with an old one, or when one treatment is much more difficult or expensive to administer. The chance of detecting a real difference between the two treatments is not reduced much as long as the ratio is not more extreme than 2:1. [SMR Chapter 15.]

Treatment cross-contamination: Any instance in which a patient assigned to receive a particular treatment in a *clinical trial* is exposed to one of the other treatments during the course of the trial. [*Genetic Molecular Research*, 2004, **3**, 456–62.]

Treatment received analysis: Analysing the results of a *clinical trial* by the treatment received by a patient rather than by the treatment allocated at randomization as in *intention-to-treat analysis.* Not to be recommended because patient *compliance* is very likely to be related to outcome. [SMR Chapter 15.]

Treatment trial: Synonym for **clinical trial**.

Tree: A term from a branch of mathematics known as *graph theory*, used to describe any set of straight-line segments joining pairs of points in some possibly multidimensional space, such that

- every point is connected to every other point by a set of lines,
- each point is visited by at least one line,
- no closed circuits appear in the structure.

If the length of any segment is given by the distance between the two points it connects, the length of a tree is defined to be the sum of the lengths of its segments. Particularly important in some aspects of *multivariate analysis* is the *minimum spanning tree* of a set of n points which is simply the tree of shortest length among all trees that could be constructed from these points. See also **dendrogram**. [*Tree Models of Similarity and Association*, 1996, J. E. Corter, Sage University Papers 112, Sage Publications, London.]

Tree-top alternative: A term used in multivariate 'one-sided' testing problems in which the null hypothesis is that the means of k independent normal populations are equal, and the alternative hypothesis is that all the first $(k - 1)$ means are greater than or equal to the mean of the last population, with strict inequality for at least one.

Trellis graphics: Graphical displays in which scatterplots, contour plots, perspective plots etc., are constructed for a series of ranges of a separate variable; the component plots are essentially 'conditioned' on the values of the latter. An example is shown in Fig. 139. See also **coplots** and **S-PLUS**.

Trend: Movement in one direction of the values of a variable over a period of time. See also **linear trend**. [SMR Chapter 9.]

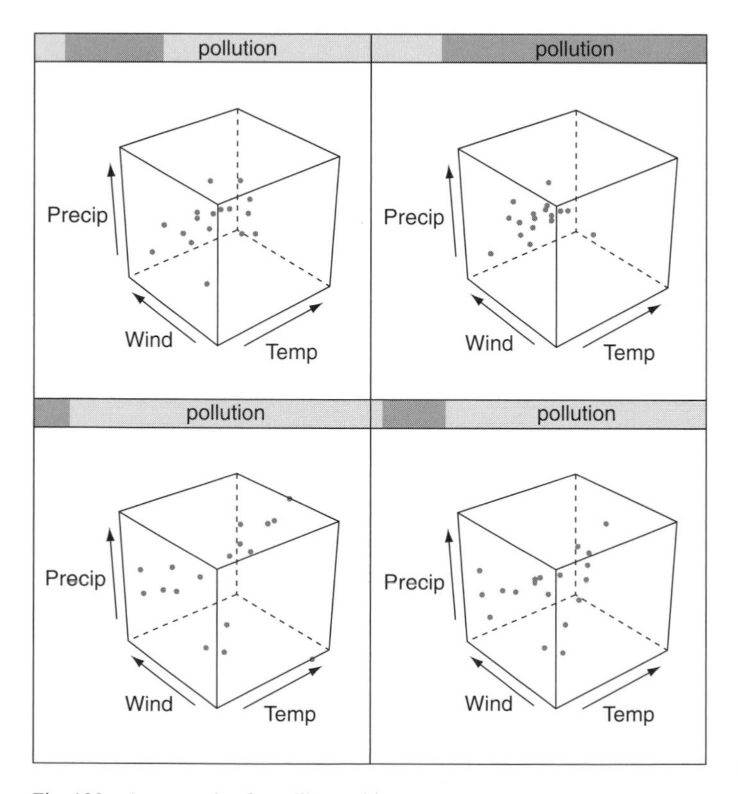

Fig. 139 An example of a trellis graphic.

Triangular contingency table: A *contingency table* in which there are *structural zeros* in a particular portion of the table. The table below, for example, gives stroke patient's initial and final disability states graded on a five point scale, A–E, of increasing severity. If a hospital never discharges a patient whose condition is worse than at admission, then no discharged patient's final state can be worse than their initial state. Such tables are generally investigated for quasi-independence. [*Biometrics*, 1975, **31**, 233–8.]

Initial rating	**Number with following final rating**					**Totals**
	A	**B**	**C**	**D**	**E**	
E	11	23	12	15	8	69
D	9	10	4	1	–	24
C	6	4	4	–	–	14
B	4	5	–	–	–	9
A	5	–	–	–	–	5
Totals	35	42	20	16	8	121

Triangular distribution: A probability distribution of the form

$$f(x) = 2(x-a)/[(\beta-\alpha)(\gamma-\alpha)] \quad \alpha \leq x \leq \gamma$$
$$= 2(\beta-x)/[\beta-\alpha)(\beta-\gamma)] \quad \gamma \leq x \leq \beta, \quad \alpha \leq x \leq \beta$$

The mean of the distribution is $(\alpha+\beta+\gamma)/3$ and its variance is $(\alpha^2+\beta^2+\gamma^2-\alpha\beta-\alpha\gamma-\beta\gamma)/18$. [STD Chapter 40.]

Triangular test: A term used for a particular type of *closed sequential design* in which the boundaries that control the procedure have the shape of a triangle as shown in Fig. 140. [*Statistics in Medicine*, 1994, **13**, 1357–68.]

Trimmed mean: See **alpha-trimmed mean**.

Triogram method: A method for function estimation using piecewise linear, bivariate *splines* based on an adaptively constructed triangulation. [*Journal of the American Statistical Association*, 2004, **93**, 101–19.]

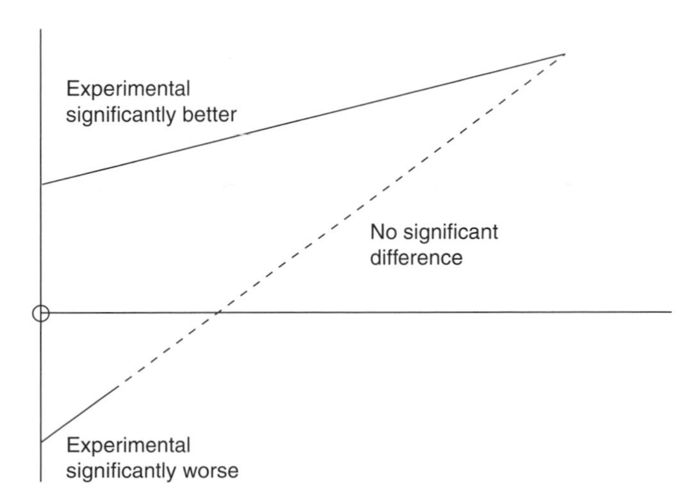

Fig. 140 The triangular region of a triangular test.

Triograms: A family of continuous piecewise linear functions defined over adaptively selected triangulations of the plane and used in a general approach to the statistical modelling of bivariate densities, to regression and to the estimation of *hazard functions*. [*Journal of the Royal Statistical Society, Series B*, **66**, 145–163.]

Triple scatterplot : Synonym for **bubble plot**.

Triples test: An asymptotically *distribution free method* for testing whether a univariate population is symmetric about some unknown value versus a broad class of asymmetric distribution alternatives. [*Statistical Methodology*, 2007, **4**, 423–433.]

Trohoc study: A term occasionally used for *retrospective study*, derived from spelling cohort backwards. To be avoided at all costs!

Trojan squares: Row-and-columns designs that arrange n replicates of nk treatments in n complete rows and n complete columns. They are constructed from k mutually orthogonal $n \times n$ *Latin squares*. Often used in crop research. [*Agricultural Science Cambridge*, 1998, **131**, 135–42.]

Trough-to-peak ratio: A measure most often used in *clinical trials* of anti-hypertensive drugs and their effect on blood pressure. This usually achieves a maximum (the peak effect P) and then decreases to a minimum (the trough effect T). The FDA recommends that the ratio should be at least 0.5. Statistical properties of the ratio are complicated by the known correlation of trough and peak effects. [*Hypertension*, 1995, **26**, 942–9.]

TRUE EPISTAT: Software for *epidemiology* and *clinical trials*. [Epistat Services, 2011 Cap Rock Circle, Richardson, TX 75080-3417, USA.]

Truncated binary data: Data arising when a group of individuals, who each have a binary response, are observed only if one or more of the individuals has a positive response. [*Biometrics*, 2000, **56**, 443–50.]

Truncated data: Data for which sample values larger (truncated on the right) or smaller (truncated on the left) than a fixed value are either not recorded or not observed. See also **censored observations** and **truncated binary data**.

Truncated Poisson distribution: Synonymous with **positive Poisson distribution**.

Tschuprov coefficient: A measure of association, T, of the two-variables forming an $r \times c$ *contingency table*, given by

$$T = \sqrt{\frac{X^2}{N\sqrt{(r-1)(c-1)}}}$$

where X^2 is the usual *chi-squared statistic* for testing the independence of the two variables and N is the sample size. See also **contingency coefficient, phi-coefficient** and **Sakoda coefficient**.

T-square test: A test for randomness of a *spatial process* which is based on comparing the distance of the nearest event of the process from a randomly chosen point with the nearest neighbour distance between randomly chosen events.

Tukey, John (1915–2000): Born in New Bedford, Massachusetts, Tukey was educated largely at home before entering Brown University to study chemistry. On obtaining his degree he moved to Princeton to take up mathematics and was awarded a doctorate in 1939. He spent his entire academic career at Princeton becoming the founding Chair of the Statistics Department. Tukey's contributions to statistics were immense and he carried out fundamental work in *time series*, *robustness*, *analysis of variance*, *experimental design* and *multiple comparisons*. He invented *exploratory data analysis* and was forever devising new terms for his new ideas. Tukey is credited with inventing the word 'software'. His collected works fill eight volumes and in 1960 he was made President of the Institute for Mathematical Statistics. Tukey died on 26 July 2000.

Tukey's lambda distributions: A family of probability distributions defined by the transformation

$$Z = [U^{\lambda} - (1-U)^{\lambda}]/\lambda$$

where U has a *uniform distribution* in (0,1). [*Technometrics*, 1985, **27**, 81–4.]

Tukey's quick test: A two-sample *distribution free method* test of location. The test statistic T is derived from merging the two samples, A and B, and arranging the values in ascending order; if there is an A observation at either end of the combined sample and a B at the other, then the number of A's at the first-mentioned end that have values more extreme than all the B's is added to the number of B's at the other end that are more extreme than all the A's to give the value of T. If, on the other hand, there are A's at both ends or B's at both ends, we put $T=0$. The test rejects the null hypothesis of no difference in location for large values of T. [*Technometrics*, 1959, **1**, 31–48.]

Tukey's test for nonadditivity: A method of testing for interaction in a two-way design with only one observation per treatment combination. Based on the sum of squares S_{AB} given by

$$S_{AB} = \frac{[\sum_{i=1}^{r}\sum_{j=1}^{c}\{y_{ij}(\bar{y}_{i.} - \bar{y}_{...})(\bar{y}_{.j} - \bar{y}_{...})\}]^2}{S_A S_B}$$

where r is the number of rows, c is the number of columns, y_{ij} is the observation in the ijth cell, $\bar{y}_{i.}$ is the mean of the ith row, $\bar{y}_{.j}$ is the mean of the jth column and $\bar{y}_{..}$ is the mean of all observations. S_A and S_B are the main effects sums of squares. [*Encyclopedia of Statistical Sciences*, 2006, eds. S. Kotz, C. B. Read, N. Balakrishnan and B. Vidakovic, Wiley, New York.]

Tumorigenic dose 50 (TD50): The daily dose of a chemical of interest that gives 50% of the test animals tumours by some fixed age. See also **lethal dose 50**. [*Biometrics*, 1984, **40**, 27–40.]

Turnbull estimator: A method for estimating the *survival function* for a set of *survival times* when the data contain *interval censored observations*. See also **product limit estimator**. [*Journal of the Royal Statistical Society, Series B*, 1976, **38**, 290–5.]

Tweedie, Richard (1947–2001): Born in Leeton, New South Wales, Australia, Tweedie won a scholarship to attend the Australian National University (ANU), obtaining a first class honours degree in statistics in 1969. From Australia he travelled to Cambridge to undertake graduate work with David Kendall. Returning to Australia in 1972 Tweedie became Post Doctoral Fellow in Statistics at the Institute of Advanced Studies at the ANU. In 1978 he took up the post of Director of Victorian Regional Office of the Commonwealth Scientific and Research Organization. Three years later he became head of the software consultancy company, *Siromath*. Despite the demands of this post, Tweedie managed to continue his fundamental research in Markov chains, for which he was awarded a Doctor of Science degree from the ANU in 1986. The final stages of his career were spent in the USA, first in the Department of Statistics at Colorado State University and then at the University of Minnesota, continuing his work on the convergence theory of *Markov chains*. Tweedie died on 7 June 2001.

Twin analysis: The analysis of data on identical and fraternal twins to make inferences about the extent and overlaps of genetic involvement in the determinants of one or more traits. It usually makes the assumption that the shared environment experiences relevant to the traits in question are equally important for identical and fraternal twins. [*Statistics in Human Genetics*, 1998, P. Sham, Arnold, London.]

Two-armed bandit allocation: A procedure for forming treatment groups in a *clinical trial*, in which the probability of assigning a patient to a particular treatment is a function of the observed difference in outcomes of patients already enrolled in the trial. The motivation behind the procedure is to minimize the number of patients assigned to the inferior treatment. See also **minimization**and **play-the-winner rule**. [*Journal of the Royal Statistical Society, Series B*, 1979, **41**, 148–77.]

Two-bend transformation: Transformation which straightens out more than a single bend in the relationship between two variables. Most common of such transformations are the *logistic transformation, arcsin transformation* and the *probit transformation*. See also **one-bend transformation**.

Two-by-two (2 × 2) contingency table: A *contingency table* with two rows and two columns formed from cross-classifying two binary variables. The general form of such a table is

		Variable 1	
		0	1
Variable 2	0	a	b
	1	c	d

[SMR Chapter 10.]

Two-by-two (2 × 2) crossover design: See **crossover design**.

Two-dimensional tables: See **contingency tables**.

Two-parameter exponential distribution: A probability distribution, $f(t)$, given by

$$f(t) = \lambda e^{-\lambda(t-G)}, \quad t \geq G$$

The term G is known as the *guarantee time* and corresponds to the time during which no events can occur. In the analysis of survival times, for example, it would represent the minimum survival time. [*Journal of Statistical Planning and Inference*, 1997, **59**, 279–89.]

Two-phase sampling: A sampling scheme involving two distinct phases, in the first of which information about particular variables of interest is collected on all members of the sample, and in the second, information about other variables is collected on a sub-sample of the individuals in the original sample. An example of where this type of sampling procedure might be useful is when estimating prevalance on the basis of results provided by a fallible, but inexpensive and easy to use, indicator of the true disease state of the sampled individuals. The diagnosis of a subsample of the individuals might then be validated through the use of an accurate diagnostic test. This type of sampling procedure is often wrongly referred to as *two-stage sampling* which in fact involves a completely different design. [*Statistical Methods in Medical Research*, 1995, **4**, 73–89.]

Two-sided test: A test where the alternative hypothesis is not directional, for example, that one population mean is either above or below the other. See also **one-sided test**. [SMR Chapter 8.]

Two-stage least squares: A way of implementing the *instrumental variable estimator*. In the first stage, each endogenous covariate is regressed on all exogenous variables in the model, including both exogenous covariates in the equation of interest and excluded instruments. The predicted values from these regressions are then obtained. In the second stage, the regression model of interest is estimated as usual, except that each endogenous covariate is replaced with the predicted values from the first stage. [*Microeconometrics*, 2005, A. C. Cameron and P. K. Trivedi, Cambridge University Press, Cambridge.]

Two-stage sampling: Sampling of units from a population where the first stage consists of sampling clusters of units from the population and the second stage consists of sampling units from each cluster. [*Practical Methods for Design and Analysis of Complex Surveys*, 2nd edition, 2004, R. Lehtonen and E. Pahkinen, Wiley, New York.]

Two-stage stopping rule: A procedure sometimes used in *clinical trials* in which results are first examined after only a fraction of the planned number of subjects in each group have completed the trial. The relevant test statistic is calculated and the trial stopped if the difference between the treatments is significant at Stage-1 level α_1. Otherwise, additional subjects in each treatment group are recruited, the test statistic calculated once again and the groups compared at Stage-2 level α_2, where α_1 and α_2 are chosen to give an overall level of α. [*Biometrics*, 1984, **40**, 791–5.]

Two-way classification: The classification of a set of observations according to two criteria, as, for example, in a *contingency table* constructed from two variables.

Type H matrix: A *variance–covariance matrix* $\mathbf{\Sigma}$ having the form

$$\mathbf{\Sigma} = \begin{pmatrix} \lambda + 2\alpha_1 & \alpha_1 + \alpha_2 & \cdots & \alpha_1 + \alpha_q \\ \alpha_2 + \alpha_1 & \lambda + 2\alpha_2 & \cdots & \alpha_2 + \alpha_q \\ \vdots & \vdots & \vdots & \vdots \\ \alpha_q + \alpha_1 & \alpha_q + \alpha_2 & \cdots & \lambda + 2\alpha_q \end{pmatrix}$$

where $\alpha_1, \alpha_2, \ldots, \alpha_q$ and λ are unknown parameters. Such a structure is a necessary and sufficient condition for the F-tests used in a univariate analysis of variance of repeated measures designs to be valid. *Compound symmetry* is a special case of this type with $\alpha_1 = \alpha_2 = \cdots = \alpha_q$. See also **sphericity**.

Type I error: The error that results when the null hypothesis is falsely rejected. [SMR Chapter 8.]

Type II error: The error that results when the null hypothesis is falsely accepted. [SMR Chapter 8.]

Type III error: It has been suggested by a number of authors that this term be used for identifying the poorer of two treatments as the better. [*American Journal of Public Health*, 1999, **89**, 1175–80.]

U

UK retail prices index: The average measure of change in the prices of goods and services bought for the purpose of consumption by the vast majority of households in the UK. [*The Retail Prices Index Technical Manual*, 1998, M. Baxter, The Stationery Office, London.]

Ultrametric inequality: See **ultrametric tree**.

Ultrametric tree: An *additive tree* in which each terminal node is equidistant from some specific node called the root of the tree. Each level of such a tree defines a partition of the objects and in terms of a series of partitions, the distance between two terminal nodes i and j can be defined as the height corresponding to the smallest subset containing both i and j. These distances, d_{ij}, satisfy the *ultrametric inequality*

$$d_{ij} \leq \max[d_{ik}, d_{jk}]$$

An example is shown in Fig. 141. [*Tree Models of Similarity and Association*, 1996, J. E. Corter, Sage University Paper 112, Sage Publications, London.]

Umbrella ordering: A commonly observed response pattern in a one-factor design with ordered treatment levels, in which the response variable increases with an increase in treatment level up to a point, then decreases with further increase in the treatment level. [*Journal of the American Statistical Association*, 1981, **76**, 175–81.]

Umbrella problem: A problem, often used to demonstrate the principles of decision making under uncertainty, which refers to the situation in which an individual must decide whether to take an umbrella in the face of uncertainty concerning whether it will rain today. [*The American Statistician*, 1987, **41**, 187–9.]

Unanimity rule: A requirement that all of a number of *diagnostic tests* yield positive results before declaring that a patient has a particular complaint. See also **majority rule**. [*Statistics in Medicine*, 1988, **7**, 549–58.]

Unbalanced design: Synonym for **nonorthogonal design** and also used for **longitudinal data** where the number of measurement occasions differs between individuals.

Unbiased: See **bias**.

Uncertainty analysis: A method for assessing the variability in an outcome variable that is due to the uncertainty in estimating the values of the input parameters. A *sensitivity analysis* can extend an uncertainty analysis by identifying which input parameters are important in contributing to the prediction imprecision of the outcome variable. Consequently a sensitivity analysis quantifies how changes in the values of the input parameters alter the value of the outcome variable. [SMR Chapter 14.]

Underfitted models: Models where some parameters or terms that would appear in a correctly specified model are missing either by mistake or by design. See also **overfitted models**.

Undirected graph: See **graph theory**.

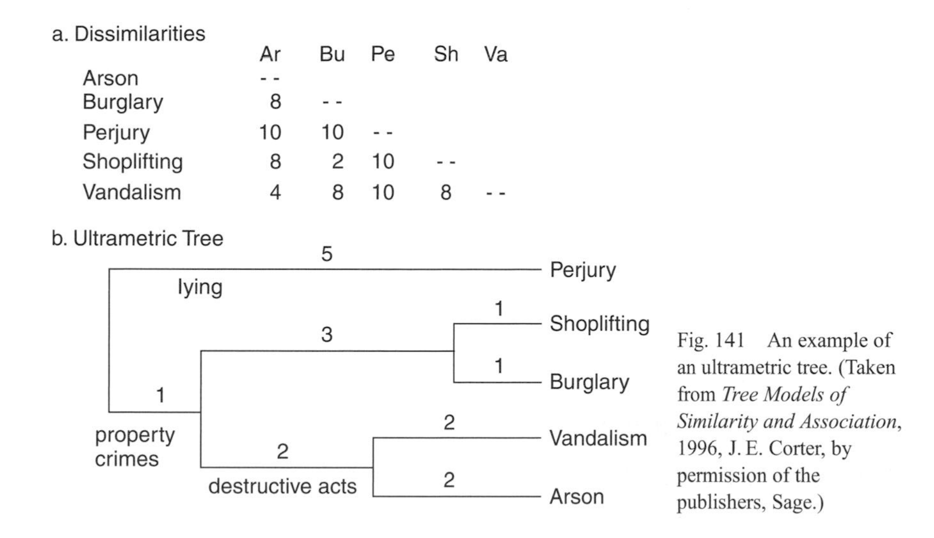

Fig. 141 An example of an ultrametric tree. (Taken from *Tree Models of Similarity and Association*, 1996, J. E. Corter, by permission of the publishers, Sage.)

Unequal probability sampling: A *sampling design* in which different *sampling units* in the population have different probabilities of being included in the sample. The differing *inclusion probabilities* may result from some inherent feature of the sampling process, or they may be deliberately imposed in order to obtain better estimates by including 'more important' units with higher probability. The unequal inclusion probabilities need to be accounted for in order to come up with reasonable estimates of population quantities. An example of such a design is *line-intercept sampling* of vegetation cover, in which the size of a patch of vegetation is measured whenever a randomly selected line intersects it. The larger the patch, the higher the probabilities of inclusion in the sample. This type of sampling may be carried out by assigning to each unit an interval whose length is equal to the desired probability and selecting random numbers from the *uniform distribution*: a unit is included if the random number is in its interval. [*Biometrika*, 1998, **85**, 89–101.]

Unfolding: An approach to simultaneous scaling of subjects and stimuli based on subjects' rankings of the stimuli. The subjects' ideal points and the points representing the stimuli are thought of as arranged along an unobserved scale where distances between ideal points and stimuli would reproduce the observed rankings. Called unfolding because the analyst must "unfold" the rankings to obtain the scale. See also **multidimensional unfolding**. [*Applied Psychological Measurement*, 1989, **13**, 193–296.]

Unidentified model: See **identification**.

Uniform distribution: The probability distribution, $f(x)$, of a random variable having constant probability over an interval. Specifically given by

$$f(x) = \frac{1}{\beta - \alpha} \qquad \alpha < x < \beta$$

The mean of the distribution is $(\alpha + \beta)/2$ and the variance is $(\beta - \alpha)^2/12$. The most commonly encountered such distribution is one in which the parameters α and β take the values 0 and 1, respectively. [STD Chapter 35.]

Uniformity trial: A method often used in agronomy to determine the 'best' size of a plot, i.e. a size such that the efficiency of a field experiment is optimized. Basically the aim is to obtain the maximal information for the lowest cost. [*Agriculture Research*, 1997, **28**, 621–7.]

Uniformly most accurate interval: A *confidence interval* that has the smallest probability of covering every possible value of a parameter except the true value. [*Journal of Econometrics*, 2001, **103**, 155–181.]

Uniformly most powerful test: A test of a given hypothesis that is least as powerful as another for all values of the parameter under consideration, and more powerful for at least one value of the parameter. [KA2 Chapter 21.]

Unimodal: A probability distribution or frequency distribution having only a single mode. A normal distribution is an example. [KA1 Chapter 1.]

Unimodal ordering: Synonym for **umbrella ordering**.

Union-intersection principle: A procedure for constructing statistical tests for *multivariate data*. [MV1 Chapter 6.]

Unit normal variate: Synonym for **standard normal variate**.

Unit root test: A test applied to *time series* to determine whether a stochastic or a deterministic trend is present in the series. The test is based on the asymptotic distribution of the ϕ obtained by *least squares estimation* in the model

$$y_i = \mu + \phi y_{t-1} + \epsilon_i$$

where ϵ_i is a stationary process. The null hypothesis is that $\phi = 1$. [*Journal of the American Statistical Association*, 1995, **90**, 268–81.]

Univariate data: Data involving a single measurement on each subject or patient.

Univariate directional orderings: A term applied to the question of whether one distribution is in some sense to the right of another. In this connection many different ordering concepts have been proposed, for example, *stochastic ordering* where the random variable Y is said to be stochastically larger then X if their *cumulative probability distributions* F and G satisfy

$$F(x) \geq G(x) \text{ for all } x$$

[*Annals of Statistics*, 1992, **20**, 2100–2110.]

Universe: A little used alternative term for population.

Unreplicated factorial: See **factorial designs**.

Unsupervised learning: See **pattern recognition**.

Unsupervised pattern recognition: See **pattern recognition**.

Unweighted means analysis: An approach to the analysis of two-way and higher-order *factorial designs* when there are an unequal number of observations in each cell. The analysis is based on cell means, using the *harmonic mean* of all cell frequencies as the sample size for all cells.

Up-and-down method: A method sometimes use for estimating the *lethal dose 50*. The method consists of the following steps; after a series of equally spaced dosage levels is chosen, the first trial is performed at some dosage level and then trials take place sequentially. Each subsequent trial is performed at the next lower or the next higher dosage level according as the immediately preceding trial did or did not evoke a positive response. [*Clinical Chemistry*, 1995, **41**, 198.]

U-shaped distribution: A probability distribution or frequency distribution shaped more or less like a letter U, though not necessarily symmetrical. Such a distribution has its greatest

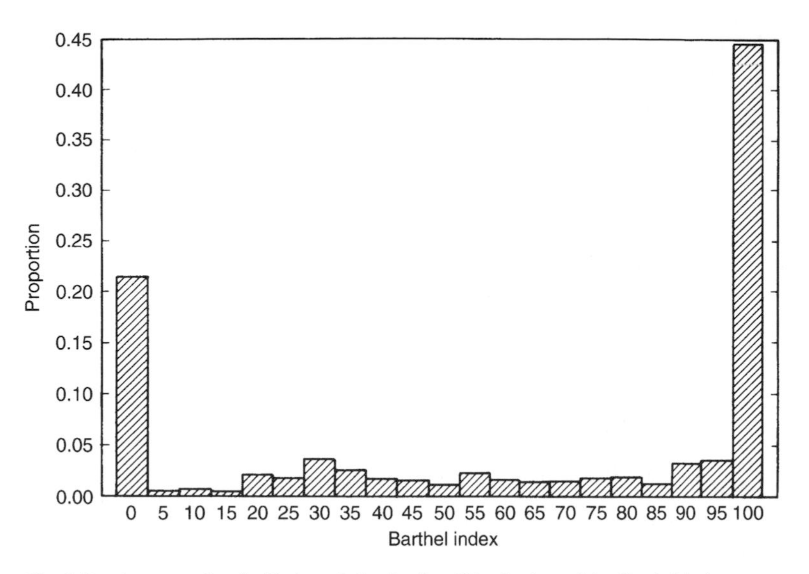

Fig. 142 An example of a U-shaped distribution. Distribution of the Barthel index.

frequencies at the two extremes of the range of the variable. See Fig. 142 for an example. [KA1 Chapter 1.]

Utility analysis: A method for decision-making under uncertainty based on a set of axioms of rational behaviour. [*Utility Theory for Decision Making*, 1970, P.C. Fishburn, Wiley, New York.]

V

Vague prior: A term used for the *prior distribution* in *Bayesian inference* in the situation when there is complete ignorance about the value of a parameter. See also **non-informative prior**. [*Subjective and Objective Bayesian Statistics*, 2002, S. J. Press, Wiley, New York.]

Validity: The extent to which a measuring instrument is measuring what was intended.

Validity checks: A part of *data editing* in which a check is made that only allowable values or codes are given for the answers to questions asked of subjects. A negative height, for example, would clearly not be an allowable value.

VALT: Abbreviation for **variable-stress accelerated life testing trials**.

Van Dantzig, David (1900–1959): One of Holland's foremost mathematical statisticians who was head of the Department of Mathematical Statistics at the University of Amsterdam. A severe critic of *Bayesian inference* his major contribution was an extension of the theory of *characteristic functions.*

Van der Waerden's rank statistics: A *distribution free method* for assessing whether two populations have the same location. The test statistic, c, is given by

$$c = \sum_{j=1}^{n} \Phi^{-1}\left(\frac{S_j}{N+1}\right)$$

where $S_1, S_2, \ldots, S_n$ are the ranks of the observations in the smaller of the two samples having n observations, and N is the total number of observations in the two samples. $\Phi^{-1}(t)$ is the tth percentile of the standard normal distribution. Tables of critical values of c are available. An alternative to *Wilcoxon's rank sum test.* [*Nonparametric Statistical Inference*, 2003, J. Dickinson and S. Chakraborti, CRC, Boca Raton.]

Variable: Some characteristic that differs from subject to subject or from time to time. [SMR Chapter 3.]

Variable sampling interval cusum charts: A *cusum* in which the time between samples is varied as a function of the value of the cusum statistic. If there is an indication that the process mean may have shifted the chart uses short sampling intervals, if there is no indication of such a shift longer sampling intervals are used. In many circumstances this type of chart is more suitable than the standard version. [*Journal of Quality Technology*, 2001, **33**, 66–81.]

Variable selection: The problem of selecting subsets of variables, in regression analysis, that contain most of the relevant information in the full data set. See also **adequate subset, all subsets regression** and **selection methods in regression**. [*Subset Selection in Regression*, 1990, A. Miller, Chapman and Hall/CRC Press, London.]

Variable-stress accelerated life testing trials (VALT): Experiments in which each of the units in a random sample of units of a product is run under increasingly severe conditions to get information quickly about the distribution of lifetimes of the units. Each unit in the experiment is subjected to a certain pattern of specified stresses, each for a specified time interval, until the unit fails. See also **fatigue models**. [*TVP*, 1996, **37**, 139–42.]

Variance: In a population, the second *moment* about the mean. An unbiased estimator of the population value is provided by s^2 given by

$$s^2 = \frac{1}{n-1}\sum_{i=1}^{n}(x_i - \bar{x})^2$$

where $x_1, x_2, \ldots, x_n$ are the n sample observations and $\bar{x}$ is the sample mean. [SMR Chapter 3.]

Variance-balanced designs: Designs for *crossover trials* with more than two treatments which are such that all pairwise difference between the treatment are estimated with the same precision. [*Design and Analysis of Cross-Over Trials*, 1989, B. Jones and M. Kenward, Chapman and Hall/CRC Press, London.]

Variance components: Variances of *random effect* terms in *multilevel models.* For example, in a simple *random intercept model* for *longitudinal data,* both subject effects and error terms are random and estimation of their variances is of some importance. In the case of a *balanced design,* estimation of these variances is usually achieved directly from the appropriate *analysis of variance table* by equating mean squares to their expected values. When the data are unbalanced a variety of estimation methods might be used, although *maximum likelihood estimation* and *restricted maximum likelihood estimation* are most often used. [*Variance Components*, 1992, S. R. Searle, G. Casella and C. E. McCulloch, Wiley, New York.]

Variance–covariance matrix: A *symmetric matrix* in which the off-diagonal elements are the covariances (sample or population) of pairs of variables and the elements on the main diagonal are the variances (sample or population) of the variables. [MV1 Chapter 2.]

Variance inflation factor: An indicator of the effect the other explanatory variables have on the variance of a regression coefficient of a particular variable, given by the reciprocal of the square of the *multiple correlation coefficient* of the variable with the remaining variables. [*Regression Analysis*, Volume 2, 1993, edited by M. S. Lewis-Beck, Sage Publications, London.]

Variance ratio distribution: Synonym for ***F*-distribution**.

Variance ratio test: Synonym for ***F*-test**.

Variance stabilizing transformations: Transformations designed to give approximate independence between mean and variance as a preliminary to, for example, *analysis of variance.* The *arc sine transformation* is an example. [*Bioinformatics*, 2004, **20**, 660–7.]

Variety trials: A term used for agricultural field experiments carried out as part of a plant breeding programme. They range from the initial studies involving small plots (single plants or single rows) to replicated yield trials involving fairly large plots. [*Progress Report 407*, Kentucky Hybrid Corn Performance Tests, College of Agriculture, University of Kentucky, USA.]

Varimax rotation: A method for *factor rotation* that by maximizing a particular function of the initial *factor loadings,* attempts to find a set of orthogonal factors satisfying, approximately at least, *simple structure.* [MV2 Chapter 12.]

Variogram: A function, $V(u)$, of a *time series* $\{x_t\}$ given by

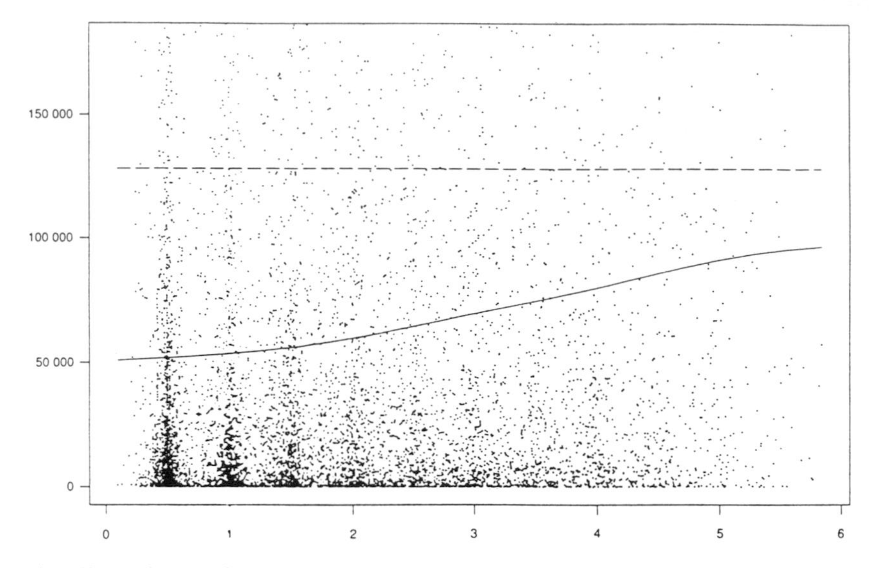

Fig. 143 Variogram plots.

$$V(u) = E\left[\frac{1}{2}\{x_t - x_{t-u}\}^2\right]$$

Gives an alternative description to the *autocovariance function* of the series that may have advantages particularly when the series is observed at irregular time-points. The *empirical variogram* is the set of points $(u_{ijk}, v_{ijk}) : k > j : i = 1, \ldots, m$ where $u_{ijk} = |t_{ij} - t_{ik}|$ and $v_{ijk} = \frac{1}{2}(y_{ij} - y_{ik})^2; t_{ij}, j = 1, \ldots, h_i$ are the n_i times at which observations are made on subject i, with the corresponding observations being y_{ij}. A scatterplot of the points (u_{ijk}, v_{ijk}) is called the *variogram cloud*, which can be used to suggest a parametric model for $V(u)$. Examples of an empirical variogram and a variogram cloud are shown in Fig. 143. [TMS Chapter 2.]

Variogram cloud: See **variogram**

Varying-coefficient models: Models that are linear in the explanatory variables but whose coefficients are allowed to change smoothly with the value of other variables. For example, if the relationship between maternal weight change and birth weight is linear but with differing shapes and intercepts that correspond to differing levels of maternal pre-pregnancy weight, then a suitable model might be:

$$\mathrm{BW} = \beta_0(\mathrm{MPPW}) + \mathrm{MWC}\beta_1(\mathrm{MPPW}) + \epsilon$$

Thus birth weights are modelled linearly in maternal weight changes but with coefficients that vary as functions of maternal pre-pregnancy weight. Often these functions are fitted non-parametrically by, for example, *cubic splines*. [*Statistics in Medicine*, 1997, **16**, 1603–16.]

Vec operator: An operator that creates a column vector from a matrix $\mathbf{A}$ by simply stacking the column vectors of $\mathbf{A}$ below one another. Hence for the $m \times n$ matrix $\mathbf{A} = [\mathbf{a}_1 \ldots \mathbf{a}_n]$ with column vectors $\mathbf{a}_1, \ldots, \mathbf{a}_n$

$$\mathrm{vec}(\mathbf{A}) = \begin{pmatrix} \mathbf{a}_1 \\ \mathbf{a}_2 \\ \vdots \\ \mathbf{a}_n \end{pmatrix}$$

See also **vech operator**.

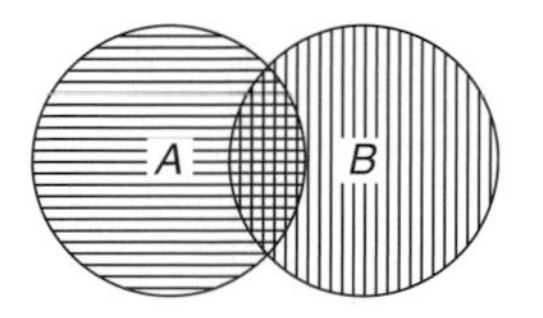

Fig. 144 The intersection of the two sets A and B appears as the cross-shaded area.

Vech operator: An operator that creates a column vector from a symmetric matrix by stacking the lower triangular elements below one another. For example,

$$\text{vech}\begin{pmatrix} a_{11} & a_{12} & a_{13} \\ a_{21} & a_{22} & a_{23} \\ a_{31} & a_{32} & a_{33} \end{pmatrix} = \begin{pmatrix} a_{11} \\ a_{21} \\ a_{31} \\ a_{22} \\ a_{32} \\ a_{33} \end{pmatrix}$$

See also **vec operator**.

Vector time series: Synonym for **multiple time series**.

Venn diagram: A graphical representation of the extent to which two or more quantities or concepts are mutually inclusive and mutually exclusive. See Fig. 144 for an example. Named after John Venn (1834–1923) who employed the device in 1876 in a paper on Boolean logic.

Vertical transmission: Transmission of a disease from mother to unborn child. [*Pediatrics and Child Health*, 1997, **2**, 227–31.]

Violin plots: Plots that add information to the simple structure of a *box plot*. The extra information consists of a density estimate of the distribution of the data. An example is shown in Fig. 145. [*The American Statistician*, 1998, **52**, 181–4.]

Virtual population analysis (VPA): A technique most often used in the analysis of exploited fish populations to estimate the past numbers and exploitation rates of distinct cohorts of a population when the history of catch, in terms of individuals, is available from these cohorts. [*Fishery Bulletin*, 1997, **95**, 280–92.]

Vital index: Synonym for **birth–death ratio**.

Vital statistics: A select group of statistical data concerned with events related to the life and death of human beings, for example, mortality rates, death rates, divorce rates etc. Central to *demography*.

Volunteer bias: A possible source of bias in *clinical trials* involving volunteers, but not involving random allocation, because of the known propensity of volunteers to respond better to treatment than other patients. [*An Introduction to Medical Statistics*, 3rd edition, 2000, M. Bland, Oxford University Press, Oxford.]

Von Mises distribution: A *circular distribution* of the form

$$f(x) = \frac{1}{2\pi I_0(\kappa)} e^{\kappa \cos(x-\mu_0)} \qquad 0 < x \leq 2\pi \qquad \kappa > 0 \qquad 0 < \mu_0 < 2\pi$$

where $I_0(z)$ is

$$I_0(z) = \sum_{r=0}^{\infty} \frac{(z/2)^{2r}}{r!\Gamma(r+1)}$$

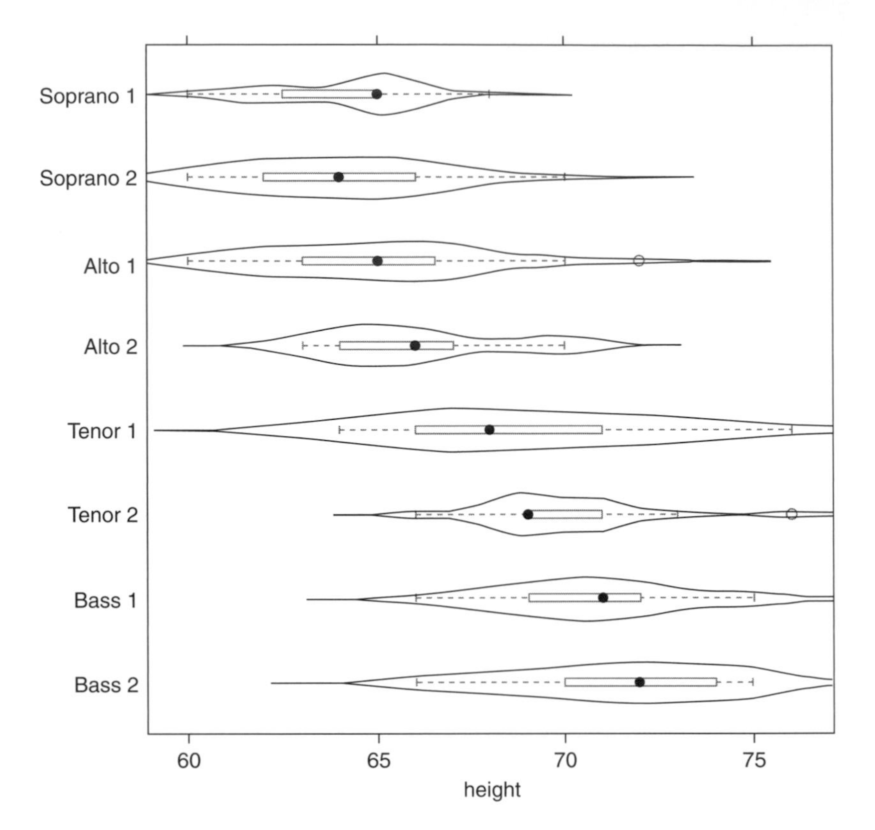

Fig. 145 Violin plots for the heights of singers with different types of voice.

The parameter μ_0 is known as the mean direction and κ is called the concentration parameter. The circular analogue to the normal distribution. Some examples of the distribution appear in Fig. 146. [STD Chapter 40.]

Von Mises, Richard Martin Edler (1883–1953): Born in Lemberg, Russia, Von Mises studied mathematics, physics and mechanical engineering at the Vienna Technical University from 1901 to 1906. At the beginning of World War I he joined the Flying Corps of the Austro-Hungarian Army and in 1919 became professor at the Technical University in Dresden and later Professor of Applied Mathematics and the first Director of the newly founded Institute of Applied Mathematics in Berlin. In 1933 Von Mises left Germany and after 6 years at the University of Istanbul he became a lecturer in the School of Engineering at Harvard University and then Gordon McKay Professor of Aerodynamics and Applied Mathematics. He made important contributions to the mathematical theory of fluids, hydrodynamics and boundary value problems. His greatest contribution was a new approach to the foundations of the theory of probability. Von Mises also worked on the laws of large numbers. He died on 14 July 1953 in Boston.

Von Neumann–Morgenstern standard gamble: A suggested procedure for assessing the risk that seriously ill patients will take when offered treatment that offers potential benefit in quality-of-life, but with the trade-off that there is a finite possibility that the patient may not survive the treatment. The patient is asked to consider the following situation:

You have been suffering from angina for several years. As a result of your illness you experience severe chest pain after even minor physical exertion such as climbing the

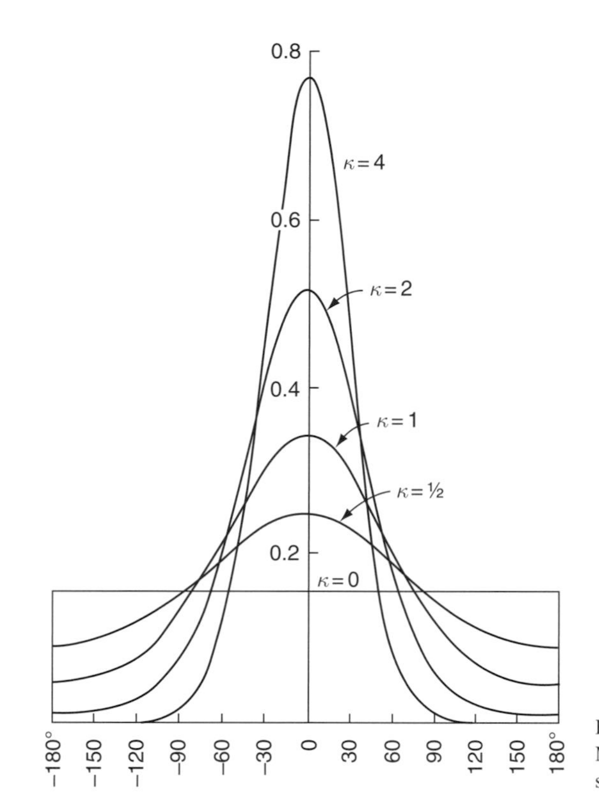

Fig. 146 Some examples of Von Mises distributions for $\mu_0 = 0$ and several values for κ.

> stairs, or walking one block in cold weather. You have been forced to quit your job and spend most days at home watching TV. Imagine that you are offered a possibility of an operation that will result in complete recovery from your illness. However the operation carries some risk. Specifically there is a probability P that you will die during the course of the operation. How large must P be before you will decline the operation and choose to remain in your present state?

Because few patients are accustomed to dealing in probabilities, an alternative procedure called the *time trade-off technique* is often suggested, which begins by estimating the likely remaining years of life for a healthy subject, using actuarial tables. The previous question is rephrased as follows:

> Imagine living the remainder of your natural span (an estimated number of years would be given) in your present state. Contrast this with the alternative that you remain in perfect health for fewer years. How many years would you sacrifice if you could have perfect health?

[*Psychiatric Services*, 2000, **51**, 1171–6.]

Von Neumann ratio: The ratio of the mean sequence successive difference to the sample variance for successive observations of a *time series*, $x_1, x_2, \ldots, x_n$. Suggested as a test statistic for the independence of the observations. [*Annals of Mathematical Statistics*, 1941, **12**, 367–95.]

Voronoi diagram: A diagram defined for two-dimensional point patterns in which each point is associated with the region of space that is closer to it than any other point. See also **Delauney triangulation**. [*Pattern Recognition*, 1980, **12**, 261–8.]

VPA: Abbreviation for **virtual population analysis**.

Vuong statistic: A statistic, V, for deciding between competing models for *count data*. The statistic is defined as

$$V = \frac{\sqrt{n}\bar{m}}{s_m}$$

where $\bar{m}$ is the mean and s_m the standard deviation of the terms, $m_j, j = 1, \ldots, n$ defined as follows

$$m_j = 1n\left[\frac{\hat{P}r_1(y_j|\mathbf{x}_j)}{\hat{P}r_2(y_j|\mathbf{x}_j)}\right]$$

where $\hat{P}r_k(y_j|\mathbf{x}_j)$ is the predicted probability of observing y_j based on model k, and $\mathbf{x}_j$ is the vector of covariate values for the jth observation. V is asymptotically distributed as a standard normal variable. Values greater than the chosen critical value lead to favoring the first model and values less than the value of minus the critical value lead to favoring the second model. Otherwise neither model is preferred. [*Econometrica*, 1989, **57**, 307–33.]

Wagner's hypothesis: A hypothesis advanced in the late 19th century by the German economist Adolph Wagner that asserts that as a country's level of development increases so does the relative size of its public sector. [*Toxicological Sciences*, 2000, **58**, 182–94.]

Wald, Abraham (1902–1950): Born in Cluj, Hungary (now part of Romania) Wald studied mathematics at the University of Vienna and in 1931 received his Ph.D. degree. Emigrated to the USA in 1938 where he studied statistics at Columbia University with *Harold Hotelling*. Wald made important contributions to *decision theory* and in particular to *sequential analysis*. He died on 13 December 1950 in Travancore, India as a result of a plane crash.

Wald–Blackwell theorem: Suppose $X_1, X_2, \ldots$ is a sequence of random variables from a distribution with mean $E(x)$ and let S be a sequential test for which $E(N)$ is finite. If $X_1 + X_2 + \cdots + X_N$ denotes the sum of the Xs drawn until S terminates then

$$E(X_1 + \cdots + X_N) = E(X)E(N)$$

[KA2 Chapter 24.]

Wald distribution: Synonym for **inverse normal distribution**.

Wald estimator: Consider a regression model with a single explanatory variable x which is correlated with the error term. When a binary *instrumental variable* z exists, an *instrumental variable estimator* for the regression parameter β is given by

$$\beta_{\text{Wald}} = \frac{\bar{y}_1 - \bar{y}_0}{\bar{x}_1 - \bar{x}_0}$$

where $\bar{y}_1$ and $\bar{x}_1$ are the subsample averages of the y and x variables when z=1, and $\bar{y}_0$ and $\bar{x}_0$ are the corresponding values when z=0. [*Microeconometrics*, 2005, A. C. Cameron and P. K. Trivedi, *Cambridge University Press*, Cambridge.]

Wald's fundamental identity: A result of key importance in *sequential analysis*, which begins with a series of mutually independent and identically distributed random variables, $Z_1, Z_2, \ldots$ such that $\Pr(|Z_i| > 0) > 0$ $(i = 1, 2, \ldots)$. If $S_n = Z_1 + Z_2 + \cdots + Z_n$, and N is the first value of n such that the inequalities $b < S_n < a$ are violated then

$$E[e^{tS_n}\{M(t)\}^{-N}] = 1$$

where $M(t)$ is the common *moment generating function* of the Zs. [KA2 Chapter 24.]

Wald's test: Typically (but not necessarily) a test for the hypothesis that a vector of parameters, $\boldsymbol{\theta}' = [\theta_1, \theta_2, \ldots, \theta_m]$, is the null vector. The test statistic is

$$W = \hat{\boldsymbol{\theta}}'\mathbf{V}^{-1}\hat{\boldsymbol{\theta}}$$

where $\hat{\boldsymbol{\theta}}'$ contains the estimated parameter values and $\mathbf{V}$ is the asymptotic variance–*covariance matrix* of $\hat{\boldsymbol{\theta}}$. Under the hypothesis, W has an asymptotic *chi-squared distribution* with degrees of freedom equal to the number of parameters. In contrast to the *likelihood ratio test*, for which Wald's test is a quadratic approximation, the wald test only requires estimation under the alternative hypothesis. See also **likelihood ratio test** and **score test**. [KA2 Chapter 25.]

Wald–Wolfowitz test: A *distribution free method* for testing the null hypothesis that two samples come from identical populations. The test is based on a count of number of *runs*. [KA2 Chapter 30.]

Walker's test: A test for non-zero periodic ordinates in a *periodogram*. The test statistic is

$$\gamma = \max_{1 \leq p \leq [N/2]} I(\omega_p)/\sigma_x^2$$

where N is the length of the series, σ_x^2 is the variance of the series and $I(\omega)$ represents the periodogram. Under the hypothesis that the ordinate is zero, γ is the maximum of $[n/2]$ independent exponential variables. [*Applications of Time Series Analysis in Astronomy and Meteorology*, 1997, edited by T. Subba Rao, M. B. Priestley and O. Lessi, Chapman and Hall/CRC Press, London.]

Walsh averages: For a random variable $[x_1, x_2, \ldots, x_n]$ from a symmetric distribution the $n(n+1)/2$ pairwise averages of the form $(x_i + x_j)/2$ computed for all i and j from 1 to n. The basis for the *Hodges–Lehmann estimator* and also used in the construction of a *confidence interval* for the median. [*Encyclopedia of Statistical Sciences*, 2006, eds. S. Kotz, C. B. Read, N. Balakrishnan and B. Vidakovic, Wiley, New York.]

Ward's method: An *agglomerative hierarchical clustering method* in which a sum-of-squares criterion is used to decide on which individuals or which clusters should be fused at each stage in the procedure. See also **single linkage, average linkage, complete linkage** and **K-means cluster analysis**. [MV2 Chapter 10.]

Waring distribution: A distribution for $x = 0, 1, 2 \ldots$ given as

$$p(x; c, a) = \frac{(c-a)(a+x-1)!c!}{c(a-1)!(c+x)}$$

where $c > 0$ and $a > 0$ are shape parameters with $c > a$. The *Yule distribution* is the special case when $a = 1$. [*Journal of the Royal Statistical Society, Series A*, 1963, **126**, 1–44.]

Waring's theorem: A theorem giving the probability of occurence of events $A_1, A_2, \ldots, A_n$ that are not *mutually exclusive*. The theorem states that

$$\Pr(A_1 \cup A_2 \cup \cdots \cup A_n) = \sum_{i=1}^{n} \Pr(A_i) - \sum_{i_1}^{n-1} \sum_{<i_2}^{n} \Pr(A_{i_1} \cap A_{i_2}) + \sum_{i_1}^{n-2} \sum_{<i_2}^{n-1} \sum_{<i_3}^{n} \Pr(A_{i_1} \cap A_{i_2} \cap A_{i_3}) - \cdots + (-1)^{n+1} \Pr(\cap_{i=1}^{n} A_i)$$

For example, for two events A_1 and A_2

$$\Pr(A_1 \cup A_2) = \Pr(A_1) + \Pr(A_2) - \Pr(A_1 \cap A_2)$$

See also **addition rule**. [KA1 Chapter 8.]

Warning lines: Lines on a *quality control chart* indicating a mild degree of departure from a desired level of control.

Warping functions: See **curve registration**.

Wash-out model: see **compartment models**.

Wash-out period: An interval introduced between the treatment periods in a *crossover design* in an effort to eliminate possible *carryover effects*. [SMR Chapter 15.]

Wasserstein metrics: Measures of the discrepancy between two cumulative distribution functions given by

$$\left[\int_0^1 |F_1^{-1}(u) - F_2^{-1}(u)|^p du\right]^{\frac{1}{p}}$$

where $F_i^{-1}(u) = \min\{t : F_j(t) \geq u\}$, $j = 1, 2$. Most often used to compare two *survival functions*. When $p = 1$ the distance measure between two estimated survival curves is the shaded area in Fig. 147. [*Recursive Partitioning in the Health Sciences*, 1999, H. Zhang and B. Singer, Springer, New York.]

Watson, Geoffrey Stuart (1921–1998): Watson was born in Bendigo, Australia. He was educated at Bendigo High School and Scotch College, Melbourne and later read mathematics at the University of Melbourne, graduating in 1942. After two years at the Institute of Statistics in North Carolina, he returned to Australia and became Senior Fellow at the Australian National University. In 1962 he became professor of statistics at the Johns Hopkins University and in 1970 Chairman of Statistics at Princeton University from where he retired in 1991. Watson made wide range of contributions to statistics including to *times series, directional data analysis* and *least squares estimation*. He died on 3 January 1998.

Watson's A-test: A test of the uniformity of a *circular distribution*. [*Biometrika*, 1967, **54**, 675–7.]

Watson's test: A test that a *circular random variable* θ has an *angular uniform distribution*. Given a set of observed values $\theta_1, \theta_2, \ldots, \theta_n$ the test statistic is

$$U^2 = \sum_{j=1}^{n}[x_j - (2j-1)/(2n)]^2 - n\left(\bar{x} - \frac{1}{2}\right) + 1/(12n)$$

where $x_j = \theta_j/2\pi$. This statistic has an approximate standard normal distribution under the hypothesis of an angular uniform distribution. See also **Kuiper's test**. [MV2 Chapter 14.]

Wavelet analysis: An approach to representing signals that is analogous to the *Fourier series* approximation but which uses a linear combination of *wavelet functions* rather than trignometric sine and cosine functions. The strength of such methods lie in their ability to

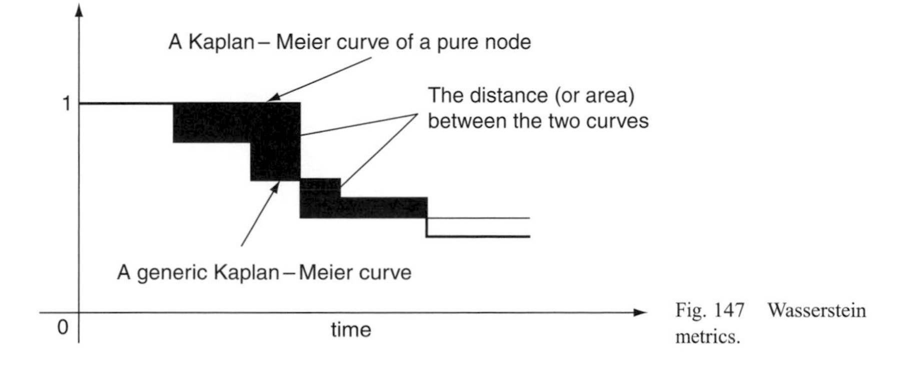

Fig. 147 Wasserstein metrics.

describe local phenomena more accurately than can a traditional expansion in sines and cosines. Thus wavelets are ideal in many fields where an approach to transient behaviour is needed, for example, in considering acoustic or seismic signals, or in image processing.

As with a sine or cosine wave, wavelet functions oscillate about zero, but the oscillations damp down to zero so that the function is localized in time or space. Such functions can be classified into two types, usually referred to as *mother wavelets* (ψ) and *father wavelets* (ϕ). The former integrates to 0 and the latter to 1. Roughly speaking, the father wavelets are good at representing the smooth and low frequency parts of a signal and the mother wavelets are good at representing the detail and high frequency parts of a signal.

The orthogonal *wavelet series approximation* to a continuous time signal $f(t)$ is given by

$$f(t) \approx \sum_k s_{J,k}\phi_{J,k}(t) + \sum_k d_{J,k}\psi_{J,k}(t) + \sum_k d_{J-1,k}\psi_{J-1,k}(t) + \cdots + \sum_k d_{1,k}\psi_{1,k}(t)$$

where J is the number of multiresolution components and k ranges from 1 to the number of coefficients in the specified component. The first term provides a smooth approximation of $f(t)$ at scale J by the so-called scaling function, $\phi_{J,k}(t)$. The remaining terms represent the wavelet decomposition proper and $\psi_{j,k}(t),\ j = 1, \ldots, J$ are the approximating wavelet functions; these are obtained through translation, k, and dilation, j, of a prototype wavelet $\psi(t)$ (an example is the *Haar wavelet* shown in Fig. 148) as follows

$$\psi_{j,k}(t) = 2^{-j/2}\psi(2^{-j}t - k)$$

The *wavelet coefficients* $s_{J,k}, d_{J,k}, \ldots, d_{1,k}$ are given approximately by the integrals

$$s_{J,k} \approx \int \phi_{J,k} f(t)\mathrm{d}t$$

$$d_{J,k} \approx \int \psi_{j,k}(t) f(t)\mathrm{d}t \qquad j = 1, 2, \ldots, J$$

Their magnitude gives a measure of the contribution of the corresponding wavelet function to the approximating sum. Such a representation is particularly useful for signals with features that change over time and signals that have jumps or other non-smooth features for which traditional Fourier series approximations are not well suited. See also **discrete wavelet transform**. [*Applications of Time Series Analysis in Astronomy and Meteorology*, 1997, edited by T. Subba Rao, M. B. Priestley and O. Lessi, Chapman and Hall/CRC Press, London.]

Wavelet functions: See **wavelet analysis**.

Wavelet series approximation: See **wavelet analysis**.

Wavelet transform coefficients: See **wavelet analysis**.

WaveShrink: An approach to function estimation and *nonparametric regression* which is based on the principle of shrinking *wavelet transform coefficients* toward zero to remove noise. The

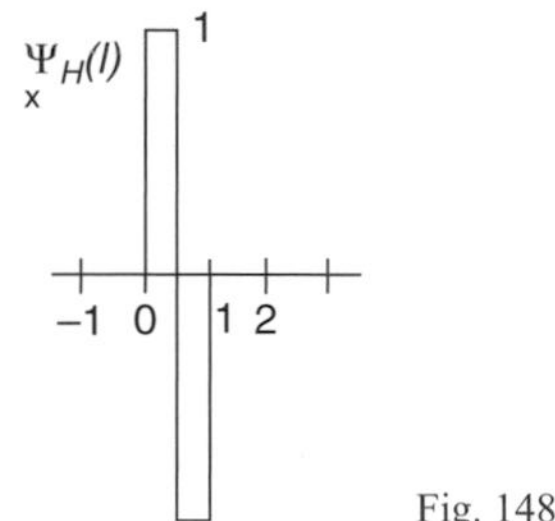

Fig. 148 The Haar wavelet.

procedure has very broad asymptotic near-optimality properties. [*Biometrika*, 1996, **83**, 727–45.]

Weakest-link model: A model for the strength of brittle material that assumes that this will be determined by the weakest element of the material, all elements acting independently and all equally likely to be the cause of failure under a specified load. This enables the strength of different lengths of material to be predicted, provided that the strength of one length is known. Explicitly the model is given by

$$S_l(x) = \{S_1(x)\}^l$$

where $S_l(x)$ is the probability distribution that a fibre of length l survives stress x, with $S_1(x)$ being a *Weibull distribution.* [*Scandinavian Journal of Metallurgy*, 1994, **23**, 42–6.]

Weathervane plot: A graphical display of multivariate data based on the *bubble plot.* The latter is enhanced by the addiction of lines whose lengths and directions code the values of additional variables. Figure 149 shows an example of such a plot on some air pollution data. [*Methods for the Statistical Analysis of Multivariate Observations*, 2nd edition, 1997, R. Gnanadesikan, Wiley, New York.]

Wedderburn, Robert William Maclagan (1947–1975): Born in Edinburgh, Wedderburn attended Fettes College and then studied mathematics and statistics at Cambridge. He then joined the Statistics Department at Rothamsted Experimental Station. During his tragically curtailed career Wedderburn made a major contribution to work on *generalized linear models*, developing the notion of *quasi-likelihood.* He died in June 1975.

Weibull distribution: The probability distribution, $f(x)$, given by

$$f(x) = \frac{\gamma x^{\gamma-1}}{\beta^{\gamma}} \exp\left[-\left(\frac{x}{\beta}\right)^{\gamma}\right] \qquad 0 \le x < \infty \qquad \beta > 0 \qquad \gamma > 0$$

Examples of the distribution are given in Fig. 150. The mean and variance of the distribution are as follows

$$\text{mean} = \beta\Gamma[(\gamma+1)/\gamma]$$
$$\text{variance} = \beta^2(\Gamma[(\gamma+2)/\gamma] - \{\Gamma[(\gamma+1)/\gamma]\}^2)$$

The distribution occurs in the analysis of *survival data* and has the important property that the corresponding *hazard function* can be made to increase with time, decrease with time, or remain constant, by a suitable choice of parameter values. When $\gamma = 1$ the distribution reduces to the *exponential distribution.* [STD Chapter 41.]

Weighted average: An average of quantities to which have been attached a series of weights in order to make proper allowance for their relative importance. For example a weighted arithmetic mean of a set of observations, $x_1, x_2, \ldots, x_n$, with weights $w_1, w_2, \ldots, w_n$, is given by

$$\frac{\sum_{i=1}^{n} w_i x_i}{\sum_{i=1}^{n} w_i}$$

Weighted binomial distribution: A probability distribution of the form

$$f_{w(x)}(x) = \frac{w(x)B_n(x;p)}{\sum_{x=0}^{n} w(x)B_n(x;p)}$$

where $w(x) > 0 (x = 1, 2, \ldots, n)$ is a positive weight function and $B_n(x;p)$ is the *binomial? distribution.* Such a distribution has been used in a variety of situations including describing

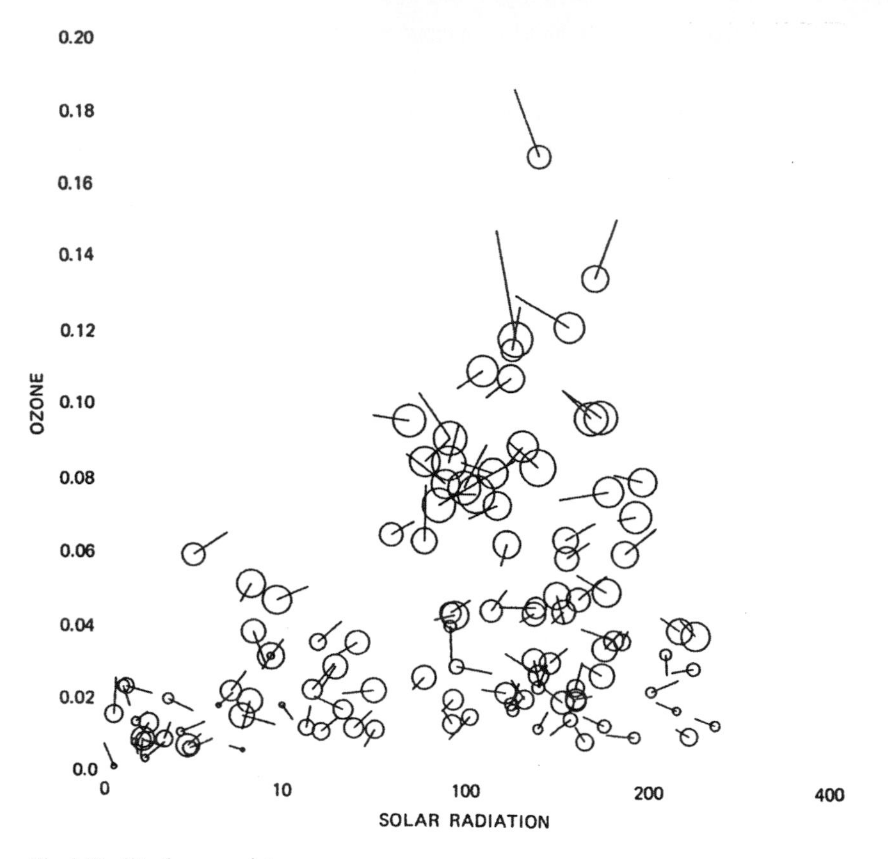

Fig. 149 Weathervane plot.

the distribution of the number of albino children in a family of size n. [*Biometrics*, 1990, **46**, 645–56.]

Weighted generalized estimating equations (WGEE): See **generalized estimating equations**.

Weighted kappa: A version of the *kappa coefficient* that permits disagreements between raters to be differentially weighted to allow for differences in how serious such disagreements are judged to be. [*Statistical Evaluation of Measurement Errors*, 2004, G. Dunn, Arnold, London.]

Weighted least squares: A method of estimation in which estimates arise from minimizing a weighted sum of squares of the differences between the response variable and its predicted value in terms of the model of interest. Often used when the variance of the response variable is thought to change over the range of values of the explanatory variable(s), in which case the weights are generally taken as the reciprocals of the variance. See also **heteroscedasticity, least squares estimation** and **iteratively weighted least squares**. [ARA Chapter 11.]

Weight of evidence (WE): A term used in the context of the use of nuclear deoxyribonucleic (DNA) for identification purposes. The joint probability of observing the genotypes that constitute evidence is quantified in turn by conditioning on pairs of hypotheses which are of forensic interest. The ratio of the two probabilities is the WE. [*Journal of the Royal Statistical Society, Series A*, 2003, **166**, 425–440.]

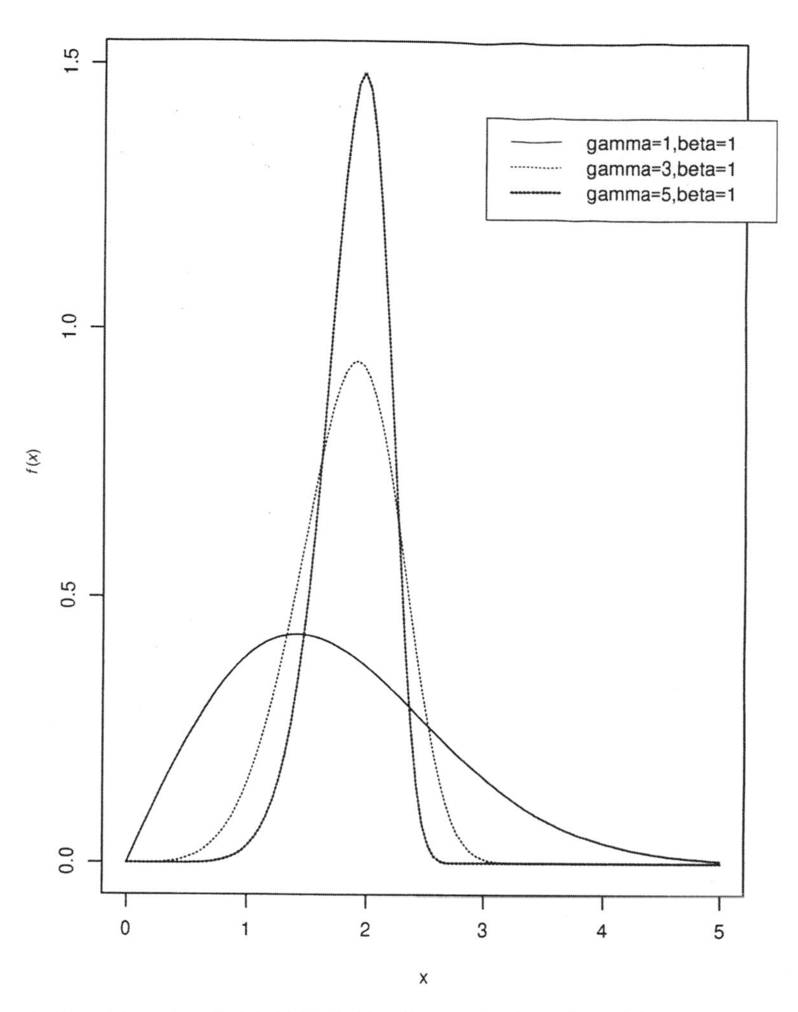

Fig. 150 Examples of Weibull distributions for several values of γ at β=2.

Wei–Lachin test: A *distribution free method* for the equality of two multivariate distributions. Most often used in the analysis of *longitudinal data* with missing observations. [*Journal of the American Statistical Association*, 1984, **79**, 653–61.]

Welch's statistic: A test statistic for use in testing the equality of a set of means in a *one-way design* where it cannot be assumed that the population variances are equal. The statistic is defined as

$$W = \frac{\sum_{i=1}^{g} w_i[(\bar{x}_i - \tilde{x})^2/(g-1)]}{1 + \frac{2(g-2)}{g^2-1}\sum_{i=1}^{g}[(1 - w_i/u)^2(n_i - 1)]}$$

where g is the number of groups, $\bar{x}_i, i = 1, 2, \ldots, g$ are the group means, $w_i = n_i/s_i^2$ with n_i being the number of observations in the ith group and s_i^2 being the variance of the ith group, $u = \sum_{i=1}^{g} w_i$ and $\tilde{x} = \sum_{i=1}^{g} w_i\bar{x}_i/u$. When all the population means are equal (even if the variances are unequal), W has, approximately, an *F-distribution* with $g - 1$ and f degrees of freedom, where f is defined by

$$\frac{1}{f} = \frac{3}{g^2 - 1}\sum_{i=1}^{g}[(1 - w_i/u)^2/(n_1 - 1)]$$

When there are only two groups this approach reduces to the test discussed under the entry for *Behrens–Fisher problem*. [*Biometrika*, 1951, **38**, 330–6.]

Westergaard, Harald (1853–1936): After training as a mathematician, Westergaard went on to study political economy and statistics. In 1883 he joined the University of Copenhagen as a lecturer in political science and the theory of statistics, the first to teach the latter subject at the university. Through his textbooks Westergaard exerted a strong influence on Danish statistics and social research for many years. After his retirement in 1924 he published *Contributions to the History of Statistics* in 1932 showing how much statistical knowledge has increased, from its small beginnings in the 17th century to its considerable scope at the end of the 19th century.

WE-test: A test of whether a set of *survival times* $t_1, t_2, \ldots, t_n$ are from an *exponential distribution*. The test statistic is

$$WE = \frac{\sum_{i=1}^{n}(t_i - \bar{t})^2}{(\sum_{i=1}^{n} t_i)^2}$$

where $\bar{t}$ is the sample mean. *Critical values* of the test statistic have been tabulated.

Wherry's formula: See **shrinkage formulae**.

Whipple index: An index used to investigate the possibility of *age heaping* in the reporting of ages in surveys. The index is obtained by summing the age returns between 23 and 62 years inclusive and finding what percentage is borne by the sum of the returns ending with 5 and 0 to one-fifth of the total sum. The results will vary between a minimum of 100, consistent with no age heaping, and a maximum of 500, if no returns were recorded with any digits other than 0 or 5. [*Demography*, 1985, W. P. Mostert, B. E. Hofmeyer, J. S. Oostenhuizen, J. A. van Zyl, Human Sciences Research Council, Pretoria.]

White noise: Term often used in *time series* to refer to an error term that has expectation zero and constant variance at all time-points and is uncorrelated over time.

White's information matrix test: A specification test for parametric models which are estimated by maximum *likelihood*. Based on the fact that the Hessian and outer-product of the gradients forms of the *information matrix* are equal if the model is correctly specified. Under the null hypothesis of correct specification the test statistic has an asymptotic χ^2 distribution. However, simulations have shown that the null distribution can be very different from χ^2 in small samples. [*Econometrica*, 1982, **50**, 1–26.]

White's homoscedasticity test: A test that assesses whether the error term in a linear regression model has a constant variance. [*Econometrica*, 1980, **48**, 817–838.]

Whittle likelihood: An approximate *likelihood function* used to estimate the *spectral density* and certain parameters of a variety of time series models. [*Communication in Statistics: Simulation and Computation*, 2006, **35**, 857–875.]

WGEE: Abbreviation for **weighted generalized estimating equations**.

Wichman/Hill generator: A *random number generator* with good randomness properties. [*Applied Statistics*, 1982, **31**, 188–90.]

Wiener, Norbert (1894–1964): Wiener was a child prodigy who entered Tufts College at the age of 11, graduating three years later. He began his graduate studies at Harvard aged 14, and received a Ph.D. in mathematical logic at the age of 18. From Harvard Wiener travelled first to Cambridge, UK, to study under Russell and Hardy and then on to Gottingen to work on differential equations under Hilbert. Returning to the USA Wiener took up a mathematics post at MIT, becoming professor in 1932, a post he held until 1960. Wiener's mathematical work included relativity and *Brownian motion*, and it was during World War II that he produced his most famous book, *Cybernetics*, first published in 1947. *Cybernetics* was the forerunner of *artificial intelligence*. Wiener died on 18 March 1964 in Stockholm, Sweden.

Wigner-Ville distribution: A *time-frequency distribution* developed for the analysis of time-varying spectra. See also **Choi-Williams distribution**. [*Proceedings of the IEEE*, 1989, **77**, 941–981.]

Wilcoxon, Frank (1892–1965): Born in County Cork, Ireland, Wilcoxon received part of his education in England, but spent his early years in Hudson River Valley at Catskill, New York. He studied at the Pennsylvania Military College in 1917 and in 1924 received a Doctor of Philosophy degree in inorganic chemistry from Cornell University. Much of his early work was on research related to chemistry, but his interest in statistics developed from reading *Fisher's* book *Statistical Methods for Research Workers*. In the 1940s Wilcoxon made major contributions to the development of *distribution free methods* introducing his famous two-sample rank tests and the signed-rank test for the paired-samples problem. Wilcoxon died on 18 November 1965 in Tallahassee, Florida.

Wilcoxon's rank sum test: A *distribution free method* used as an alternative to the *Student's t-test* for assessing whether two populations have the same location. Given a sample of observations from each population, all the observations are ranked as if they were from a single sample, and the test statistic W is the sum of the ranks in the smaller group. Equivalent to the *Mann–Whitney test*. Tables giving critical values of the test statistic are available, and for moderate and large sample sizes, a normal approximation can be used. [SMR Chapter 9.]

Wilcoxon's signed rank test: A *distribution free method* for testing the difference between two populations using matched samples. The test is based on the absolute differences of the pairs of observations in the two samples, ranked according to size, with each rank being given the sign of the original difference. The test statistic is the sum of the positive ranks. [SMR Chapter 9.]

Wilks' lambda: See **multivariate analysis of variance**.

Wilks' multivariate outlier test: A test for detecting *outliers* in *multivariate data* that assumes that the data arise from a *multivariate normal distribution*. The test statistic is

$$W_j = |\mathbf{A}^{(j)}|/|\mathbf{A}|$$

where

$$\mathbf{A} = \sum_{i=1}^{n} (\mathbf{x}_i - \bar{\mathbf{x}})(\mathbf{x}_i - \bar{\mathbf{x}})'$$

and $\mathbf{A}^{(j)}$ is the corresponding matrix with $\mathbf{x}_j$ removed from the sample. $\mathbf{x}_1, \ldots, \mathbf{x}_n$ are the n sample observations with mean vector $\bar{\mathbf{x}}$. The potential outlier is that point whose removal leads to the greatest reduction in $|\mathbf{A}|$. Tables of critical values are available. [*Sankhyā A*, 1963, **25**, 407–26.]

Wilks, Samuel Stanley (1907–1964): Born in Little Elms, Texas, USA, Wilks received a bachelor's degree in industrial arts from North Texas State Teachers College in 1926. He began his career as a teacher but studied mathematics at the University of Texas in parallel, receiving a master's degree in 1928. Wilks then moved to the University of Iowa to work for a doctorate, which he was awarded in 1931. In 1932 he was appointed a National Research Council International Research Fellow and travelled to the UK to work at both the University of London and Cambridge University, where he met both *Karl Pearson* and *John Wishart*. Returning to the USA in 1934 Wilks joined the Department of Mathematics at Princeton University from where he made many important contributions to mathematical statistics particularly in multivariate analysis. He also made major contributions to educational statistics during his time working with the Educational Testing Service. Wilks was President of the American Statistical Association in 1950 and President of the Institute of Mathematical Statistics in 1940. He died on March 7th, 1964, in Princeton, USA.

Wilks' theorem: A theorem stating that, under certain regularity conditions, the asymptotic null distribution of a *likelihood-ratio* test statistic comparing two nested models, one of which has m more parameters than the other, is a χ^2 distribution with m degrees of freedom. [*Essentials of Statistical Inference*, 2005, G. A. Young and R. L. Smith, Cambridge University Press, Cambridge]

Willcox, Walter Francis (1861–1964): Born in Reading, Massachusetts, Willcox spent most of his working life at Cornell University. He was President of the American Statistical Association in 1912 and of the International Statistical Institute in 1947. Willcox worked primarily in the area of demographic statistics. He died on 30 October 1964 in Ithaca, New York.

Williams' agreement measure: An index of *agreement* that is useful for measuring the reliability of individual raters compared with a group. The index is the ratio of the proportion of agreement (across subjects) between the individual rater and the rest of the group to the average proportion of agreement between all pairs of raters in the rest of the group. Specifically the index is calculated as

$$I_n = P_0/P_n$$

where

$$P_0 = \frac{1}{n}\sum_{j=1}^{n} P_{0,j}$$

and

$$P_n = \frac{2}{n(n-1)}\sum_{j<j'} P_{j,j'}$$

and $P_{j,j'}$ represents the proportion of observed agreements (over all subjects) between the raters j and j', with $j = 0$ indicating the individual rater of particular interest. The number of raters is n. [*British Journal of Psychology*, 1976, **2**, 89–108.]

Williams' designs: A type of *latin square* which requires fewer observations to achieve balance when there are more than three treatments. [*Australian Journal of Scientific Research*, 1949, **2**, 149–68.]

Williams' test: A test used for answering questions about the toxicity of substances and at what dose level any toxicity occurs. The test assumes that the mean response of the variate is a monotonic function of dose. To describe the details of the test assume that k dose levels are to be compared with a control group and an upward trend in means is suspected. *Maximum likelihood estimation* is used to provide estimates of the means, $M_i, i = 1, \ldots, k$ for each dose group, which are found under the constraint $M_1 \leq M_2 \leq \cdots \leq M_k$ by

$$\hat{M}_i = \max_{1 \le u \le i} \min_{i \le v \le k} \sum_{j=u}^{v} r_j X_j \Big/ \sum_{j=u}^{v} r_j$$

where X_i and r_i are the sample mean and sample size respectively for dose group i. The estimated within group variance, s^2, is obtained in the usual way from an *analysis of variance* of drug group. The test statistic is given by

$$t_k = (\hat{M}_k - X_0)(s^2/r_k + s^2/c)^{-\frac{1}{2}}$$

where X_0 and c are the control group sample mean and sample size. Critical values of t_k are available. The statistic t_k is tested first and if it is significant, t_{k-1} is calculated in the same way and the process continued until a non-significant t_i is obtained for some dose i. The conclusion is then that there is a significant effect for dose levels i+1 and above and no significant evidence of a dose effect for levels i and below. [*The Basis of Toxicity Testing*, 1997, D. J. Ecobichon, CRC, Boca Raton.]

Wilson-Hilferty method: A method for finding an approximation to the percentage points of the *chi-squared distribution*. [*Applied Statistics*, 1978, **27**, 280–290.]

Window estimates: A term that occurs in the context of both frequency domain and time domain estimation for *time series*. In the former it generally applies to the weights often applied to improve the accuracy of the *periodogram* for estimating the *spectral density*. In the latter it refers to statistics calculated from small subsets of the observations after the data has been divided up into segments. [*Density Estimation in Statistics and Data Analysis*, 1986, B. W. Silverman, Chapman and Hall/CRC Press, London.]

Window variables: Variables measured during a constrained interval of an observation period which are accepted as proxies for information over the entire period. For example, many statistical studies of the determinants of children's attainments measure the circumstances or events occurring over the childhood period by observations of these variables for a single year or a short period during childhood. Parameter estimates based on such variables will often be biased and inconsistent estimates of true underlying relationships, relative to variables that reflect cicumstances or events over the entire period. [*Journal of the American Statistical Association*, 1996, **91**, 970–82.]

Window width: See **kernel methods**.

Winsorising: See **M-estimators**.

Wishart distribution: The joint distribution of variances and covariances in samples of size n from a *multivariate normal distribution* for q variables. Given by

$$f(\mathbf{S}) = \frac{(\frac{1}{2}n)^{\frac{1}{2}q(n-1)} |\mathbf{\Sigma}^{-1}|^{\frac{1}{2}(n-1)} |\mathbf{S}|^{\frac{1}{2}(n-q-2)}}{\pi^{\frac{1}{4}q(q-1)} \prod_{j=1}^{q} \Gamma\{\frac{1}{2}(n-j)\}} \exp(-\tfrac{1}{2}n\ \mathrm{tr}(\mathbf{\Sigma}^{-1}\mathbf{S}))$$

where $\mathbf{S}$ is the sample *variance–covariance matrix* (with n rather than n-1 in the denominator) with elements s_{jk}, and $\mathbf{\Sigma}$ the population variance–covariance matrix. [MV1 Chapter 2.]

Wishart, David (1928–1998): Born in Stockton-on-Tees, England, Wishart began his university studies in chemistry at St. Andrew's, Scotland, but soon changed to mathematics. From St. Andrews he moved to Princeton to study with David Kendall, and completed a doctoral thesis on application of probability theory to queuing. Wishart was an editor for the *Journal of the Royal Statistical Society* and (in Russian) for *Mathematical Reviews*. He was awarded the Chambers Medal of the Royal Statistical Society for his contributions and work for the society.

Wishart, John (1898–1956): Born in Montrose, Wishart studied mathematics and natural philosophy at Edinburgh University obtaining a first class honours degree in 1922. He began his career as a mathematics master at the West Leeds High School (1922–1924) and then joined University College, London as Research Assistant to *Karl Pearson*. In 1928 he joined *Fisher* at Rothamsted and worked on a number of topics crucial to the development of multivariate analysis, for example, the derivation of the generalized product moment distribution and the properties of the distribution of the *multiple correlation coefficient*. Wishart obtained a readership at Cambridge in 1931 and worked on building up the Cambridge Statistical Laboratory. He died on 14 July 1956 in Acapulco, Mexico.

Within groups matrix of sums of squares and cross-products: See **multivariate analysis of variance**.

Within groups mean square: See **mean squares**.

Within groups sum of squares: See **analysis of variance**.

Wöhler curve: Synonymous with ***S–N* curve**.

Wold, Herman Ole Andreas (1908–1992): Born at Skien, Norway on Christmas Day 1908. Wold's family moved to Sweden in 1912 and he began his studies at Stockholm University in 1927. His doctoral thesis entitled 'A study in the analysis of stationary time series' was completed in 1938, and contained what became known as Wold's decomposition, the decomposition of a time series into the sum of a purely non-deterministic component and a deterministic component. In 1942 Wold became Professor of Statistics at the University of Uppsala where he remained until 1970. During this time he developed a mechanism which explained the *Pareto distribution* of wealth and, somewhat disguised, introduced the concept of latent variable models. Wold died on 16 February 1992.

Wold's decomposition theorem: A theorem that states that any *stationary stochastic process* can be decomposed into a deterministic part and a non-deterministic part, where the two components are uncorrelated and the non-deterministic part can be represented by a moving average process. [*Time Series Analysis*, 1994, J. D. Hamilton, Princeton University Press, Princeton, NJ.]

Wolfowitz, Jacob (1910–1981): Born in Warsaw, Poland, Wolfowitz came to the United States with his family in 1920. He graduated from the College of the City of New York in 1931 and obtained a Ph.D. in mathematics from New York University in 1942. In 1945 Wolfowitz became an associate professor at the University of North Carolina at Chapel Hill, and in 1951 joined the Department of Mathematics at Cornell University. Wolfowitz made many important contributions in mathematical statistics, particularly in the area of asymptotics, but also contributed to probability theory and coding theory. He died in Tampa, Florida on 16 July 1981.

Woolf's estimator: An estimator of the common *odds ratio* in a series of *two-by-two contingency tables*. Not often used because it cannot be calculated if any cell in any of the tables is zero. [*Biometrical Journal*, 2007, **29**, 369–374.]

Worcester, Jane (1910–1989): Worcester gained a first degree in 1931 and joined the Department of Biostatistics at Harvard School of Public Health as a mathematical computing assistant. She remained in the department until her retirement in 1977, becoming during this time the first female Chair of Biostatistics. During her 46 years in the department she devoted her time to research, teaching and consultancy. Worcester died on 8 October 1989 in Falmouth, Massachussetts.

Working correlation matrix: See **generalized estimating equations**.

World Health Quarterly: See **sources of data**.

World Health Statistics Annual: See **sources of data**.

Worm plot: A plot for visualizing differences between two distributions, conditional on the values of a covariate. The plot can be used as a general diagnostic tool for the analysis of residuals from the fit of a statistical model. [*Statistics in Medicine*, 2001, **20**, 1259–77.]

Wrapped Cauchy distribution: A probability distribution, $f(\theta)$, for a *circular random variable*, θ, given by

$$f(\theta) = \frac{1}{2\pi} \frac{1 - \rho^2}{1 + \rho^2 - 2\rho\cos(\theta - \mu)} \qquad 0 \leq \theta \leq 2\pi \qquad 0 \leq \rho \leq 1$$

[*Statistical Analysis of Circular Data*, 1995, N. I. Fisher, Cambridge University Press, Cambridge.]

Wrapped distribution: A probability distribution on the real line that has been wrapped around the circle of unit radius. If X is the original random variable, the random variable after 'wrapping' is $X(\text{mod})2\pi$. See also **wrapped normal distribution**. [*Statistical Analysis of Circular Data*, 1995, N. I. Fisher, Cambridge University Press, Cambridge.]

Wrapped normal distribution: A probability distribution resulting from 'wrapping' the normal distribution around the unit circle. The distribution is given by

$$f(x) = \frac{1}{\sigma\sqrt{2\pi}} \sum_{k=-\infty}^{\infty} \exp\left\{ -\frac{1}{2} \frac{(x + 2\pi k)^2}{\sigma^2} \right\} \qquad 0 < x \leq 2\pi$$

[*Statistics of Directional Data*, 1972, K. V. Mardia, Academic Press, New York.]

X

X^2-statistic: Most commonly used for the *test statistic* used for assessing independence in a *contingency table*. For a two-dimensional table it is given by

$$X^2 = \sum \frac{(O - E)^2}{E}$$

where O represents an observed count and E an *expected frequency*. [*The Analysis of Contingency Tables*, 2nd edition, 1992, B. S. Everitt, Chapman and Hall/CRC Press, London.]

X-11: A computer program for seasonal adjustment of quarterly or monthly economic *time series* used by the US Census Bureau, other government agencies and private businesses particularly in the United States. [*Seasonal Analysis of Economic Time Series*, 1976, A. Zellner, US Department of Commerce, Bureau of the Census, Washington, DC.]

X-Gvis: An interactive visualization system for *multidimensional scaling*. [*Journal of Computational and Graphical Statistics*, 1996, **5**, 78–99.]

Y

Yates' continuity correction: When testing for independence in a *contingency table*, a continuous probability distribution, namely the *chi-squared distribution*, is used as an approximation to the discrete probability of observed frequencies, namely the *multinomial distribution*. To improve this approximation Yates suggested a correction that involves subtracting 0.5 from the positive discrepancies (observed – expected) and adding 0.5 to the negative discrepancies before these values are squared in the calculation of the usual *chi-square statistic*. If the sample size is large the correction will have little effect on the value of the test statistic. [SMR Chapter 10.]

Yates, Frank (1902–1994): Born in Manchester, Yates read mathematics at St. John's College Cambridge, receiving a first-class honours degree in 1924. After a period working in the Gold Coast (Ghana) on a geodetic survey, he became *Fisher's* assistant at Rothamsted in 1931. Two years later he became Head of the Department of Statistics, a position he held until his retirement in 1968. Yates made major contributions to a number of areas of statistics particularly the design and analysis of experiments and the planning and analysis of sample surveys. He was also quick to realize the possibilities provided by the development of electronic computers in the 1950s, and in 1954 the first British computer equipped with effective magnetic storage, the Elliot 401, was installed at Rothamsted. Using only machine code Yates and other members of the statistics department produced programs both for the *analysis of variance* and to analyse data from surveys. He was made a Fellow of the Royal Society in 1948 and in 1960 was awarded the Royal Statistical Society's Guy medal in gold. In 1963 he was awarded the CBE. Yates died on 17 June 1994.

Yates–Grundy variance estimator: An estimator of the variance of the *Horvitz–Thompson estimator.* [*Survey Methodology*, 2007, **33**, 87–94.]

Yea-saying: Synonym for **acquiescence bias**.

Youden, William John (1900–1972): Born in Townsville, Australia, Youden's family emigrated to America in 1907. He obtained a first degree in chemical engineering from the University of Rochester in 1921 and in 1924 a Ph.D. in chemistry from Columbia University. Youden's transformation from chemist to statistician came early in his career and was motivated by reading *Fisher's Statistical Methods for Research Workers* soon after its publication in 1925. From 1937 to 1938 he worked under Fisher's direction at the Galton Laboratory, University College, London. Most remembered for his development of *Youden squares*, Youden was awarded the Wilks medal of the American Statistical Association in 1969.

Youden's index: An index derived from the four counts, a, b, c, d in a *two-by-two contingency table*, particularly one arising from the application of a *diagnostic test*. The index is given by

$$\frac{a}{a+c} + \frac{d}{b+d} - 1$$

Essentially the index seeks to combine information about the *sensitivity* and *specificity* of a test into a single number, a procedure not generally to be recommended. [*An Introduction to Epidemiology*, 1983, M. Alderson, Macmillan, London.]

Youden square: A row and column design where each treatment occurs exactly once in each row and where the assignment of treatments to columns is that of a symmetric *balanced incomplete-block design*. The simplest type of such design is obtained by deleting just a single row from a *Latin square*. [*Statistics in Medicine*, 1994, **13**, 11–22.]

Yule–Furry process: A *stochastic process* which increases in unit steps. Provides a useful description of many phenomena, for example, neutron chain reactions and cosmic ray showers. [*Stochastic Modelling of Scientific Data*, 1995, P. Guttorp, Chapman and Hall/CRC Press, London.]

Yule, George Udny (1871–1951): Born near Haddington, Scotland, Yule was educated at Winchester College and then studied engineering at University College, London followed by physics in Bonn, before becoming a demonstrator under *Karl Pearson* in 1893. His duties included assisting Pearson in a drawing class and preparing specimen diagrams for papers. Contributed to Pearson's papers on skew curves and later worked on the theory of correlation and regression. Yule died on 26 June 1951.

Yule distribution: The probability distribution, $P(x)$, given by

$$P(x) = \frac{\rho\Gamma(\rho+1)\Gamma(x+1)}{\Gamma(x+\rho+2)}, x = 0, 1, 2 \ldots, \text{and } \rho > 0$$

The distribution has increasingly long tails as ρ goes to zero. The distribution is a realization of *Zipf's law* that has been used to model the relative frequency of the kth most frequent word in a large collection of text. See also **Waring distribution**. [*Mathematical Medicine and Biology*, 1989, **6**, 133–136.]

Yule's characteristic K: A measure of the richness of vocabulary based on the assumption that the occurrence of a given word is based on chance and can be regarded as a *Poisson distribution*. Given by

$$K = 10^4\left(\sum_i i^2 V_i - N\right)/N^2, \qquad i = 1, 2, \ldots$$

where N is the number of words in the text, V_1 is the number of words used only once in the text, V_2 the number of words used twice, etc. [*Re-Counting Plato: A Computer Analysis of Plato's Style*, 1989, G. R. Ledger, Clarendon Press, Oxford.]

Yule–Walker equations: A set of linear equations relating the parameters $\phi_1, \phi_2, \ldots, \phi_p$ of an *autoregressive model* of order p to the *autocorrelation function*, ρ_k and given by

$$\begin{aligned}
\rho_1 &= \phi_1 + \phi_2\rho_1 + \cdots + \phi_p\rho_{p-1} \\
\rho_2 &= \phi_1\rho_1 + \phi_2 + \cdots + \phi_p\rho_{p-2} \\
&\vdots \\
\rho_p &= \phi_1\rho_{p-1} + \phi_2\rho_{p-2} + \cdots + \phi_p
\end{aligned}$$

[TMS Chapter 3.]

Z

Zelen's exact test: A test that the *odds ratios* in a series of *two-by-two contingency tables* all take the value one. [*Biometrika*, 1971, **58**, 129–37.]

Zelen's single-consent design: An alternative to simple random allocation for forming treatment groups in a *clinical trial.* Begins with the set of N eligible patients. All N of these patients are then randomly subdivided into two groups say G_1 and G_2 of sizes n_1 and n_2. The standard therapy is applied to all the patients assigned to G_1. The new therapy is assigned *only* to those patients in G_2 who consent to its use. The remaining patients who refuse the new treatment are treated with the standard. [SMR Chapter 15.]

Zero-inflated binomial regression: An adaptation of *zero-inflated Poisson regression* applicable when there is an upper bound count situation. It assumes that with probability p an observation is zero and with probability $1-p$ a random variable with a *binomial distribution* with parameters n and π. [*Biometrics*, 2000, **56**, 1030–9.]

Zero-inflated Poisson regression: A model for count data with excess zeros. It assumes that with probability p the only possible observation is 0 and with probability $1-p$ a random variable with a *Poisson distribution* is observed. For example, when manufacturing equipment is properly aligned, defects may be nearly impossible. But when it is misaligned, defects may occur according to a Poisson distribution. Both the probability p of the perfect zero defect state and the mean number of defects λ in the imperfect state may depend on covariates. The parameters in such models can be estimated using *maximum likelihood estimation.* [*Technometrics*, 1992, **34**, 1–14.]

Zero sum game: A game played by a number of persons in which the winner takes all the stakes provided by the losers so that the algebraic sum of gains at any stage is zero. Many decision problems can be viewed as zero sum games between two persons. [*The Mathematics of Games of Strategy*, 1981, M. Dresher, Dover, New York.]

Zero-truncated Poisson distribution: See **Zero-truncated probability distributions**.

Zero-truncated probability distributions: Adaptations of probability distributions for count data, for example the *Poisson distribution*, used in situations where the random variable cannot take the value zero. Example of such a count variable are days spent in hospital for people admitted to hospital for some reason or other and number of authors listed on a scientific paper. An example of such a distribution is the *zero-truncated Poisson distribution* which has the form, $(e^{-\lambda}\lambda^x/x!)/(1-e^{-\lambda}), x = 1, 2, \ldots$ [*Journal of Statistical Computation and Simulation*, 2007, **77**, 585–591.]

ZIP: Abbreviation for **zero-inflated Poisson regression**.

Zipf distribution: See **Zipf's law**.

Zipf–Estoup law: Synonymous with **Zipf's law**.

Zipf's law: A term applied to any system of classification of units such that the proportion of classes with exactly s units is approximately proportional to $s^{-(1+\alpha)}$ for some constant $\alpha > 0$. It is familiar in a variety of empirical areas, including linguistics, personal income distributions and the distribution of biological genera and species. A probability distribution appropriate to such situations can be constructed by taking

$$Pr(X = s) = cs^{-(1+\alpha)} \quad s = 1, 2, \ldots$$

where

$$c = \left[\sum_{s=1}^{\infty} s^{-(1+\alpha)}\right]^{-1}$$

This is known as the *Zipf distribution*. [*Human Behaviour and the Principle of Least Effort*, 1949, G. K. Zipf, Addison-Wesley, Reading, MA.]

Z-scores: Synonym for **standard scores**.

z-test: A test for assessing hypotheses about population means when their variances are known. For example, for testing that the means of two normally distributed populations are equal, i.e. $H_0 : \mu_1 = \mu_2$, when the variance of each population is known to be σ^2, the test statistic is

$$z = \frac{\bar{x}_1 - \bar{x}_2}{\sigma\sqrt{\frac{1}{n_1} + \frac{1}{n_2}}}$$

where $\bar{x}_1$ and $\bar{x}_2$ are the means of samples of size n_1 and n_2 from the two populations. If H_0 is true, z, has a standard normal distribution. See also **Student's *t*-tests**.

z transformation: See **Fisher's *z* transformation**.

Zygosity determination: The determination of whether a pair of twins is identical or fraternal. It can be achieved to a high (over 95%) level of certainty by questionnaire, but any desired level of accuracy can be achieved by typing a number of genetic markers to see if the genetic sharing is 100% or 50%. Important in *twin analysis*. [*Statistics in Human Genetics*, 1998, P. Sham, Arnold, London.]